Maurizio Benaglia (Ed.)
Organocatalysis

Also of interest

Organoselenium Chemistry
Ranu, Banerjee (Eds.), 2020
ISBN 978-3-11-062224-9, e-ISBN 978-3-11-062511-0

Chemical Technologies and Processes
Staszak, Wieszczycka, Tylkowski (Eds.), 2020
ISBN 978-3-11-065627-5, e-ISBN 978-3-11-065636-7

Polymer Synthesis.
Modern Methods and Technologies
Wang, Junjie, 2021
ISBN 978-3-11-059634-2, e-ISBN 978-3-11-059709-7

Chemistry and Biochemistry of Food
Perez-Castineira, 2020
ISBN 978-3-11-059547-5, e-ISBN 978-3-11-059548-2

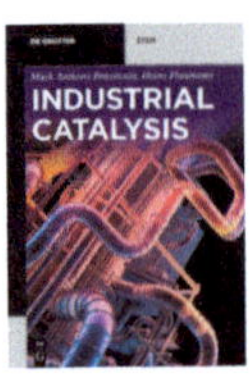

Industrial Catalysis
Benvenuto, Plaumann, 2021
ISBN 978-3-11-054284-4, e-ISBN 978-3-11-054286-8

Organocatalysis

Stereoselective Reactions and Applications
in Organic Synthesis

Edited by
Maurizio Benaglia

DE GRUYTER

Editor
Prof. Maurizio Benaglia
Università Degli Studi Di Milano
Dipartimento di Chimica
Via Golgi 19 / I-20133
20133 Milan
Italy

ISBN 978-3-11-058803-3
e-ISBN (PDF) 978-3-11-059005-0
e-ISBN (EPUB) 978-3-11-058872-9

Library of Congress Control Number: 2021934374

Bibliographic information published by the Deutsche Nationalbibliothek
The Deutsche Nationalbibliothek lists this publication in the Deutsche Nationalbibliografie; detailed bibliographic data are available on the Internet at http://dnb.dnb.de.

Cover image: Cosminxp Cosmin/iStock/Getty Images Plus
Typesetting: Integra Software Services Pvt. Ltd.
Printing and binding: CPI books GmbH, Leck

www.degruyter.com

Preface

Organocatalysis is considered today one of the three "pillars" in asymmetric catalysis, along with biocatalysis and organometallic catalysis. In the last fifteen years, an impressive number of new reactions, catalysts and activation strategies have been reported in, arguably, the most intensively developed field in organic chemistry.

Although numerous examples of what we call today "organocatalytic reactions" have been reported already in XX century, it was only in the last 20 years that Organocatalysis has been recognized as a well defined research area. The progress in the field was, and still it is incredibly fast, further accelerated by the possibility to combine organocatalysis with radical chemistry, photocatalysis and enabling technologies. New avenues in organic synthesis were opened, unprecedented catalytic activation methods developed, and some longstanding synthetic challenges have been solved.

Now, the field faces new challenges: more detailed and "*ad hoc*" designed mechanistic studies are needed for a deeper understanding of the action modes of the catalysts; that is prodromic to an optimization of old and new organocatalysts, mainly with the goal to lower the catalyst loading and realize a true process intensification; application of organocatalytic methodologies in total synthesis, in the preparation of complex and highly functionalized molecules will allow to further expand and establish the synthetic applicability of metal-free catalytic methodologies. Finally, the use of organocatalysis in industrial research and in process chemistry would represent a crucial step to move forward and bring the area to a higher level, where organocatalysis could be truly applied to the synthesis of industrially relevant molecules.

What can be expected from this volume? The book aims to cover the very recent developments in asymmetric organocatalysis, with a special focus on the works published after 2015. However, in order to guide the newcomers of the field into the fascinating but multifaceted topic of organocatalysis, in the single chapters an introductory part offers to the reader the fundamentals of that specific topic.

In the first part of the book, the most popular and successful organocatalysts are presented, to introduce the non-expert reader to the most important mechanism, modes of action and class of catalysts in aminocatalysis: proline and prolinol derivatives, ammino Cinchona-derived catalysts, chiral imidazolidinones are presented and their mechanisms and applications discussed. Other two chapters introduce the reader into the non-covalent bond activation mode: hydrogen bonding directed catalysis, chiral Brönsted acids and phase transfer catalysis, as example of ion pair catalysis, are discussed in details, with a special attention to the mechanistic aspects of the reactions. Then, another chapter presents the most recent and significant progresses in a fundamental organic transformation such as catalytic, asymmetric oxidation reactions.

https://doi.org/10.1515/9783110590050-202

The chapters of the second part of the volume look at the future, at some of the frontier topics: the use of organocatalysts in multistep, cascade reactions, able to assemble highly functionalized and diversified molecules, or the combination of organocatalysis with light, in organo-photoredox transformations, and the use of organocatalysts in packed bed reactors and in continuous flow reactors, to combine metal-free catalysis with some of the enabling technologies. The last chapter relates to the application of metal free chiral catalysts in industrial synthesis; some selected R&D process are presented to highlight how organocatalysis may indeed play a role in the prospective manufacture of commercial Active Pharmaceutical Ingredients.

I am really grateful to the authors which have agreed to contribute to the volume and I am honored to act as editor of a book whose chapters were delivered all by worldwide recognized experts in the area. The book is intended to reach Industrial and academic synthetic organic chemists, researchers (chemists and chemical engineers) interested and active in catalysis, especially on asymmetric reactions, PhD students, scientists from fine chemicals industries and companies engaged in custom synthesis. For the neophyte in the field the book wants to be an easy to consult guide, that introduces the reader in the very wide and rich area of organocatalysis. For the experts in the field, the book aims to be the occasion to establish some general, important milestones, as starting points from which the future investigations will move towards new challenges and objectives.

I do hope that reading of the volume will inspire the development of new catalysts, the design of unforeseen transformations and will stimulate many more applications in organic synthesis. I wish the reader will enjoy reading this book, that presents a stimulating chemistry developed with fantasy, creativity and the curious spirit of an explorer looking always to go beyond the borders.

Maurizio Benaglia

Contents

Contributors List

Florenci V. González Adelantado
Departament de Química Inorgànica i Orgànica
Universitat Jaume I
Avda. Sos Baynat, s/n
Castelló 12071, Spain,
E-mail: fgonzale@uji.es

Anthony Burke
Chemistry Department
University of Evora
Rua Romão Ramalho 59
7000-671 Evora, Portugal
E-mail: ajb@uevora.pt

Armando Carlone
Department of Physical and Chemical Sciences
Università degli Studi dell'Aquila, Coppito, via Vetoio
L'Aquila 67100, Italy
E-mail: armando.carlone@univaq.it

Pier Giorgio Cozzi
Università degli Studi di Bologna
Dipartimento di Chimica "G. Ciamician"
via Selmi 2
40126 Bologna, Italy
E-mail: piergiorgio.cozzi@unibo.it

Yujiro Hayashi
Department of Chemistry
Graduate School of Science
Tohoku University
Aoba-ku, Sendai 980–8578, Japan
E-mail: yhayashi@m.tohoku.ac.jp

Christoforos Kokotos
Laboratory of Organic Chemistry
Department of Chemistry
National and Kapodistrian University of Athens
Panepistimiopolis,
Athens 15771, Greece
E-mail: ckokotos@chem.uoa.gr

Alessandra Lattanzi
Dipartimento di Chimica e Biologia "A. Zambelli"
Università di Salerno
via Giovanni Paolo II 132
Fisciano 84084, Italy
Email: lattanzi@unisa.it

Elisabetta Massolo
Dipartimento di Chimica
Università degli Studi di Milano
Via Golgi, 19
Milano 20133, Italy
E-mail: elisabetta.massolo@unimi.it

Manuel Orlandi
Dipartimento di Scienze Chimiche
Università degli Studi di Padova
via Marzolo 1, Padova 35131, Italy
E-mail: manuel.orlandi@unipd.it

Alessandra Puglisi
Dipartimento di Chimica
Università degli Studi di Milano
via Golgi 19
Milano, 20133 Italy
E-mail: alessandra.puglisi@unimi.it

Laura Raimondi
Dipartimento di Chimica
Università degli Studi di Milano
via Golgi, 19
Milano, Lombardia 20133, Italy
Email: lauramaria.raimondi@unimi.it

https://doi.org/10.1515/9783110590050-204

Yujiro Hayashi

1 Domino and one-pot syntheses of biologically active compounds using diphenylprolinol silyl ether

Abstract: The successful application of diphenylprolinol silyl ether, which is one of the widely used organocatalysts, to the synthesis of natural products and drugs, is described mostly focusing on the author's results. The molecules that are explained in this paper are baclofen, telcagepant, oseltamivir, ABT-341, prostaglandins, estradiol, horsfiline and coerulescine.

Keywords: diphenylprolinol silyl ether, enamine, iminium ion

1.1 Introduction

In recent years, asymmetric reactions using organocatalysts have been extensively explored because employing organocatalysts have several synthetic advantages [1]. In general, organocatalysts are inexpensive, stable in water and oxygen, and possess low toxicity. Therefore, organocatalysts are becoming widely used as attractive catalysts in the total synthesis of optically active drugs and biologically active substances.

In total synthesis, multistep reactions are usually employed for the construction of molecular skeleton, bond formation, functional group transformation, and in the introduction and removal of protecting groups. Therefore, it is important to shorten the number of steps (step economy) in the total synthesis [2] and to synthesize the desired compounds efficiently [3].

Organocatalysts are generally weak acids or bases and do not usually interfere with successive reactions. Thus, they are suitable catalysts for domino and one-pot reactions. A domino reaction was defined by Tietze [4] as a process involving two or more bond-forming transformations that take place under the same reaction conditions without adding additional reagents and catalysts, and in which the subsequent reactions result as a consequence of the functionality formed in the previous step. However, the one-pot reaction is a strategy to improve the efficiency of a chemical reaction whereby a reactant is subjected to successive chemical reactions in just one reactor. One of the early epoch-making examples of an organocatalyst-

This article has previously been published in the journal Physical Sciences Reviews. Please cite as: Hayashi, Y. Domino and one-pot syntheses of biologically active compounds using diphenylprolinol silyl ether *Physical Sciences Reviews* [Online] 2020, 6. DOI: 10.1515/psr-2018-0088

https://doi.org/10.1515/9783110590050-001

mediated domino reaction was reported by Enders et al. [5]. They developed a domino reaction of Michael/Michael reactions of aldehyde, nitroalkene, and α, β-unsaturated aldehyde to afford optically active tetrasubstituted cyclohexene derivatives with excellent enantioselectivity using diphenylprolinol silyl ether **1** as a catalyst (eq. 1).

Organocatalyst is effective not only in domino reactions [6] but also in one-pot reactions. One successful examples of an organocatalyst-mediated one-pot reaction is the three-pot synthesis of Tamiflu, which was developed by our research group in 2009 [7]. Thus, one-pot reactions catalyzed by organocatalysts have received much attention in recent years [8]. In this paper, the total synthesis of valuable drugs and natural products [9], in which asymmetric reactions catalyzed by organocatalyst such as **1** and **2** (Figure 1.1) with a combination of domino or one-pot reactions are successfully employed, will be described with examples of the author's original work.

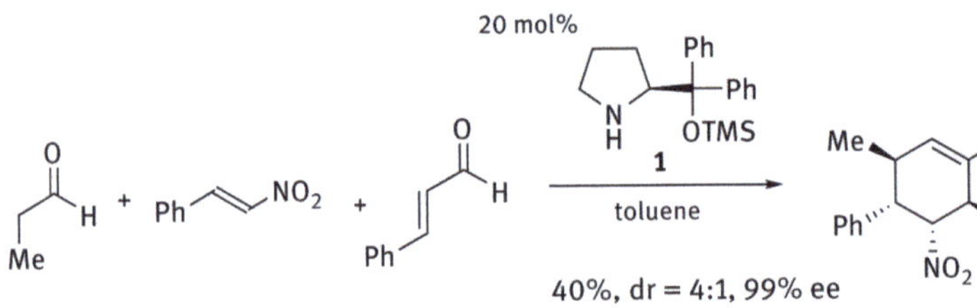

(1.1)

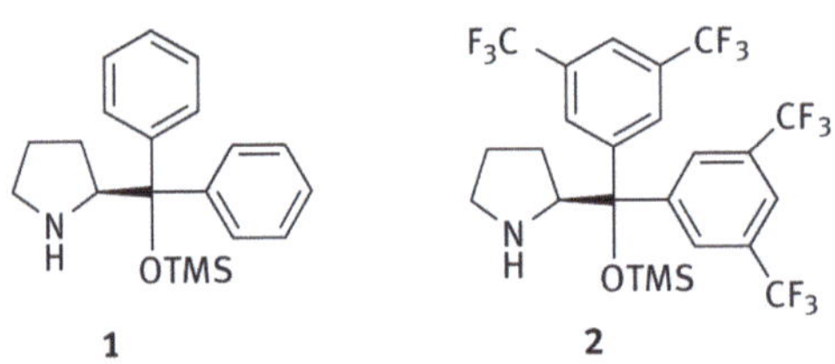

Figure 1.1: Diarylprolinol silyl ether catalysts.

1.2 Diphenylprolinol silyl ether catalyst

1.2.1 The general reaction model

Secondary or primary amines are known to react with aldehyde to generate enamine, which reacts with an electrophile to afford an α-substituted aldehyde, while secondary or primary amine reacts with α, β-unsaturated aldehyde to generate iminium ion, which reacts with a nucleophile to provide a β-substituted aldehyde (Figure 1.2) [1]. These reactions proceed using a catalytic amount of amine. If the effective chiral environment can be generated using chiral amine, a highly enantioselective reaction can be possible. Several chiral primary and secondary amine catalysts have already been developed. In 2005, our group [10] and Jørgensens' group [11] independently developed diphenylprolinol silyl ether **1** (Figure 1.1), which is now widely used as an effective organocatalyst in the reactions of enamine and iminium ions as reactive intermediates [12].

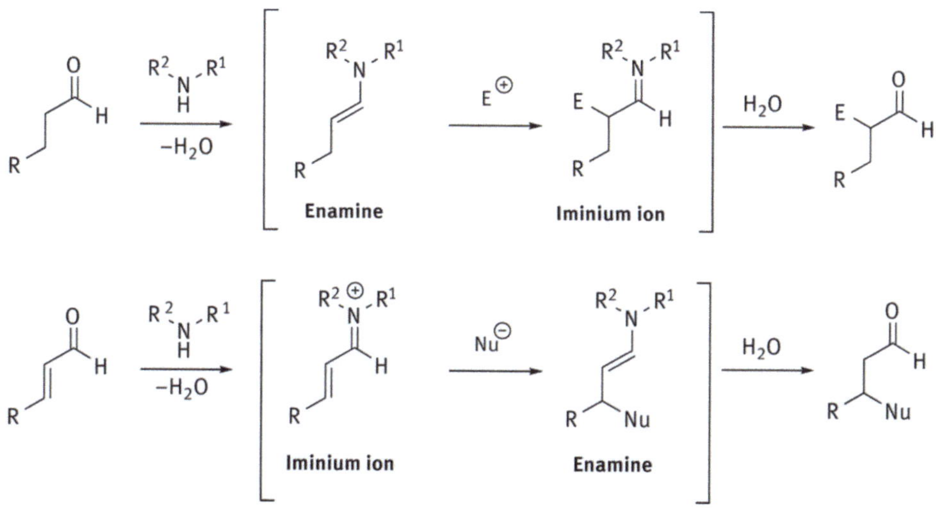

Figure 1.2: General reaction model of secondary and primary amines.

When diphenylprolinol silyl ether **1** reacts with aldehyde, enamine is formed, in which one of the enantiofaces of enamine is completely shielded by the bulky diphenyl(trimethylsilyloxy)methyl moiety (Figure 1.3). However, when diphenylprolinol silyl ether **1** reacts with α, β-unsaturated aldehyde, iminium ion will be formed, in

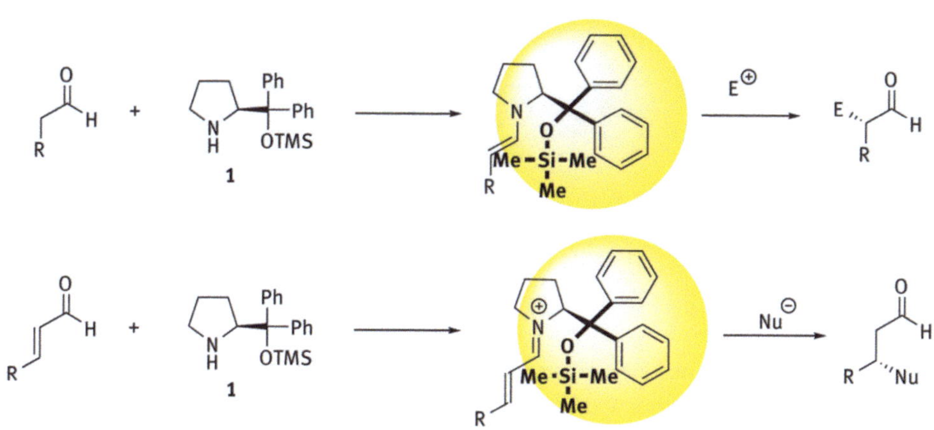

Figure 1.3: The enamine and iminium ions of diphenylprolinol silyl ether **1**.

which one of the enantiofaces of iminium ion is also completely shielded by the bulky diphenyl(trimethylsilyloxy)methyl moiety. Thus, a high enantioselectivity is generally expected when this catalyst is employed and many asymmetric catalytic reactions using this catalyst have been developed [12].

1.2.2 Effect of the silyl substituent on diarylprolinol silyl ether

There are usually two reaction categories of the secondary amine-type reactions, which are based on either enamine or iminium ions. However, we would like to define three types of reactions catalyzed by diphenylprolinol silyl ethers (Figure 1.4) [13]. Type A is a Michael-type reaction of α, β-unsaturated aldehydes involving iminium ion intermediates, in which a higher enantioselectivity is realized when the catalyst with a bulkier silyl moiety is employed except for the Michael reaction of an α, β-unsaturated aldehyde with nitromethane. One of the best substituent patterns on the silyl atom in the diphenylprolinol silyl ether catalyst is the diphenylmethyl group [14]. Type B is a cycloaddition reaction via iminium ion intermediates, in which small substituents on the silyl atom, such as the trimethyl silyl group, affords excellent enantioselectivity. Type C is a reaction involving enamine intermediates derived from an aldehyde as a nucleophile. In this reaction, small substituents on the silyl group provide excellent enantioselectivity. In general, a bulky silyl group on the diphenylprolinol silyl ether catalyst retards the reaction for steric reasons. In certain cases, however, a better yield results because the bulky silyl substituents suppress unproductive side reactions or destruction of the catalyst molecule which become more prevalent with more nucleophilic diphenylprolinol silyl ether catalysts.

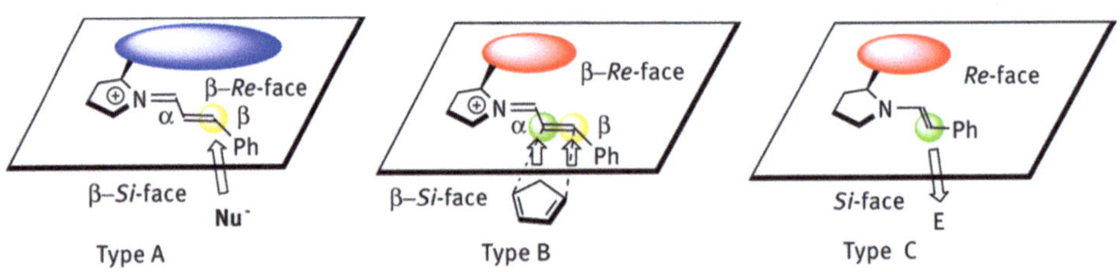

Figure 1.4: Schematic approach of reactants for types A, B, and C processes.

1.2.3 Effect of the aryl substituents on diarylprolinol silyl ether [15]

As for the diarylprolinol silyl ether catalyst, there are two major catalysts such as diphenylprolinol silyl ether **1** [16], and 2-[bis[3,5-bis(trifluoromethyl)phenyl][(trimethylsilyl)oxy]methyl]pyrrolidine **2** (Figure 1.1).

In the reaction of α, β-unsaturated aldehydes using diarylprolinol silyl ether as an amine organocatalyst, we proposed in the previous section that reactions involving iminium ions as intermediates can be classified into two types: Michael-type reaction (type A) or cycloaddition (type B). The LUMO of the iminium ion derived from trifluoromethyl-substituted diarylprolinol silyl ether **2** is lower than that derived from diphenylprolinol silyl ether **1** because of the electron-withdrawing trifluoromethyl groups, which was confirmed by *ab-initio* calculations. Thus, the iminium ion of the former catalyst **2** is more reactive and the catalyst **2** with trifluoromethyl groups is better suited for Diels–Alder reaction type B. The generation of iminium ions from α, β-unsaturated aldehydes is faster when the relatively electron-rich diphenylprolinol silyl ether **1** is employed compared to trifluoromethyl-substituted diarylprolinol silyl ether **2.** In the Michael reaction (type A), the use of diphenylprolinol silyl ether catalyst **1** is preferable. Because the generation of iminium ion is promoted by acid, acid accelerates the reaction. However, acid also reduces the concentration of anionic nucleophiles by protonation in the Michael reaction (type A). Thus, an appropriate selection of acid is advisable. Although there are exceptions, in general, diphenylprolinol silyl ether **1** is a superior catalyst in type A reactions, while trifluoromethyl-substituted diarylprolinol silyl ether **2** is preferable in type B reactions.

1.3 Syntheses of baclofen

Several asymmetric Michael reactions are catalyzed by organocatalysts, such as secondary amine, thiourea, and phase transfer catalysts, which afford the products with excellent enantioselectivities. In this section, the synthesis of baclofen will be explained using the asymmetric Michael reaction as a key step.

Baclofen is a drug currently used clinically in its racemic form as a muscle relaxant with a γ-amino acid structure, while the (*R*)-isomer is biologically active [17]. Therefore, development of industrially applicable synthetic method for optically active baclofen is desirable. In 2007, the synthesis of baclofen using diphenylprolinol silyl ether catalyst **1** was independently reported by Hayashi [18] and Wang et al [19]., in which the key step is the asymmetric Michael reaction of nitromethane using α, β-unsaturated aldehyde (Figure 1.5). This reaction proceeds in MeOH in the presence of catalyst **1** to afford the corresponding Michael adduct **4** in high yield with excellent enantioselectivity. Stereocontrol in the catalytic reaction is explained by approaching the nucleophilic nitronate to the optically active iminium ion intermediate formed from the catalyst and aldehyde to avoid the bulky substituents. The oxidation and reduction of the Michael adduct **4** provided baclofen HCl salt (**5**).

In 2016, we investigated the one-pot synthesis of chiral baclofen from commercially available compounds using just one reaction vessel (Figure 1.6) [20]. While *p*-chlorobenzaldehyde **6** is commercially available, α, β-unsaturated aldehydes, such as 3-(*p*-chlorophenyl)prop-1-enal **3** are not available. Thus, 3-(*p*-chlorophenyl)prop-

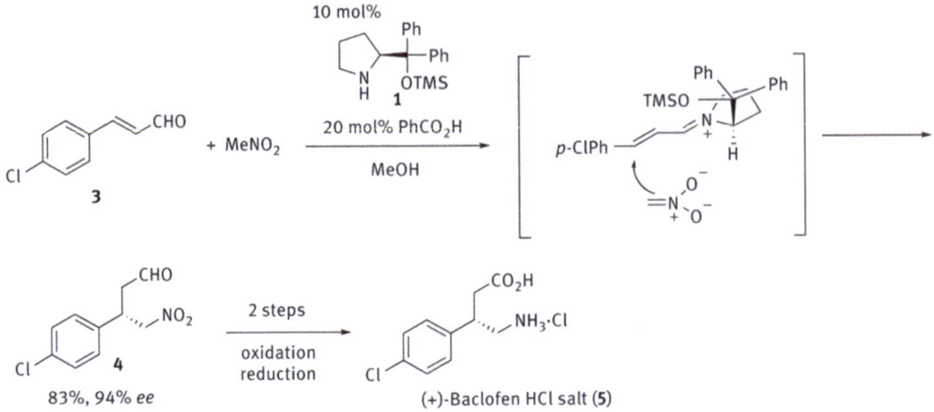

Figure 1.5: Synthesis of baclofen.

1-enal **3** must be prepared from *p*-chlorobenzaldehyde **6** in such a manner that the reaction should be suitable for the successive one-pot reactions including a catalytic asymmetric reaction. Usually α, β-unsaturated aldehydes are synthesized by the Wittig reaction from aldehyde to afford α, β-unsaturated esters, which are reduced to α, β-unsaturated aldehydes via DIBAL (diisobutylaluminium hydride). But these reactions are not be suitable for one-pot operations, because the successive asymmetric catalytic reaction of diphenylprolinol silyl ether catalyst **1** does not proceed well. As we have previously developed an asymmetric aldol reaction of acetaldehyde [21], we thought that the aldol condensation of acetaldehyde would afford α, β-unsaturated aldehydes with the generation of water as a by-product, which would not interfere with the successive asymmetric Michael reaction. Aldol condensation of *p*-chlorobenzaldehyde **6** and acetaldehyde was found to be catalyzed by DBU to give α, β-unsaturated aldehyde **3**, which was treated with diphenylprolinol silyl ether **1** with a combination of HCO_2H to afford Michael product **4**. Oxidation and reduction proceeded in the same vessel to afford baclofen **5** in a total yield of 31 % from *p*-chlorobenzaldehyde **6** in a single pot.

Asymmetric synthesis of baclofen was also reported by Takemoto et al. [22]. The key step is an asymmetric Michael reaction of nitroalkene **8** and diethylmalonate catalyzed by Takemoto's thiourea catalyst **9** to afford the Michael adduct **10** in good yield with excellent enantioselectivity (94 % ee) (Figure 1.7) [22]. The efficient enantioselectivity is attributed to the hydrogen bond between the amino group and the thiourea moiety as shown in Figure 1.7. Subsequent reduction of the nitro group afforded lactam **11** after cyclization reaction. Conversion from lactam **11** to baclofen (**5**) was achieved by hydrolysis, followed by decarboxylation.

Bernardi and Adamo (2009) reported the asymmetric synthesis of baclofen by a phase transfer catalyst **13** derived from cinchona alkaloid (Figure 1.8) [23]. Using 3-methyl-4-nitro-5-styrylisoxazole **12** as a Michael acceptor, the asymmetric Michael

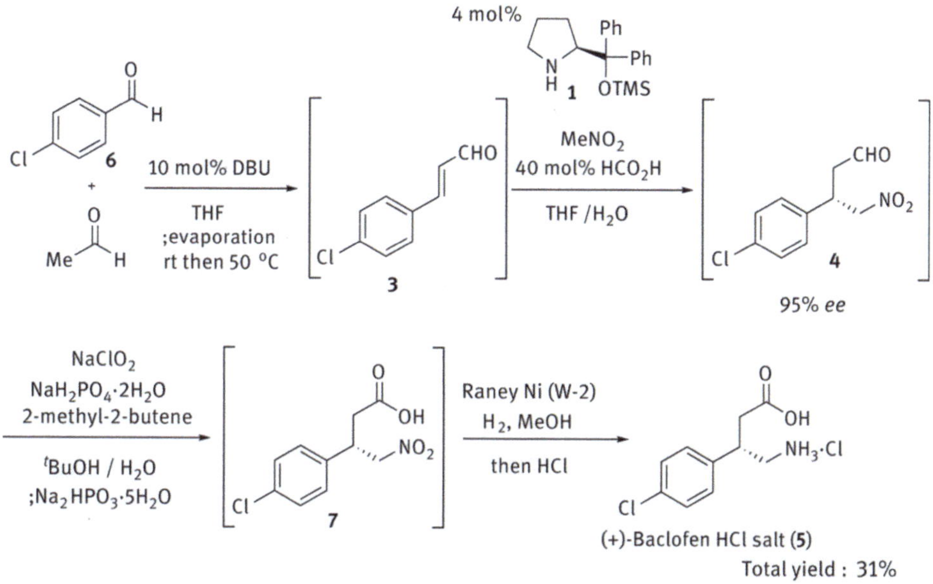

Figure 1.6: One-pot synthesis of baclofen.

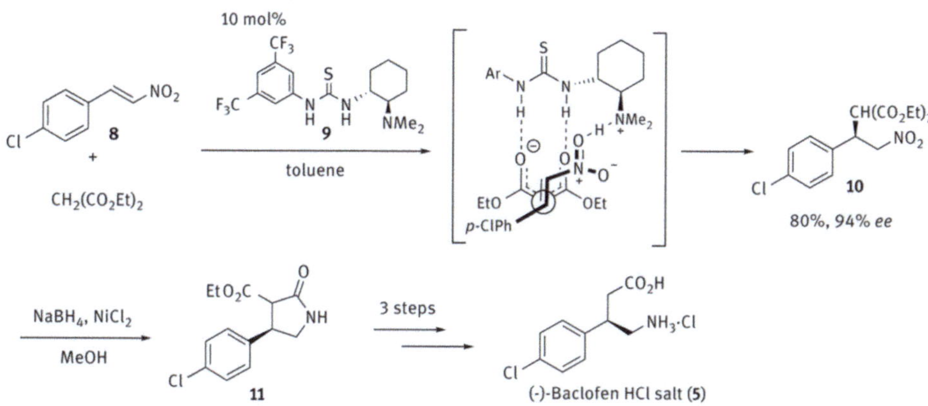

Figure 1.7: Takemoto's synthesis of baclofen.

reaction of nitromethane proceeds in a highly enantioselective manner to afford the product **14** in 91% ee. Hydrolysis afforded the nitro carboxylic acid **7**, which is a known precursor for baclofen. As the same target molecule can be synthesized using various organocatalysts, the synthesis of baclofen is a good example for understanding the stereocontrol of organocatalysts.

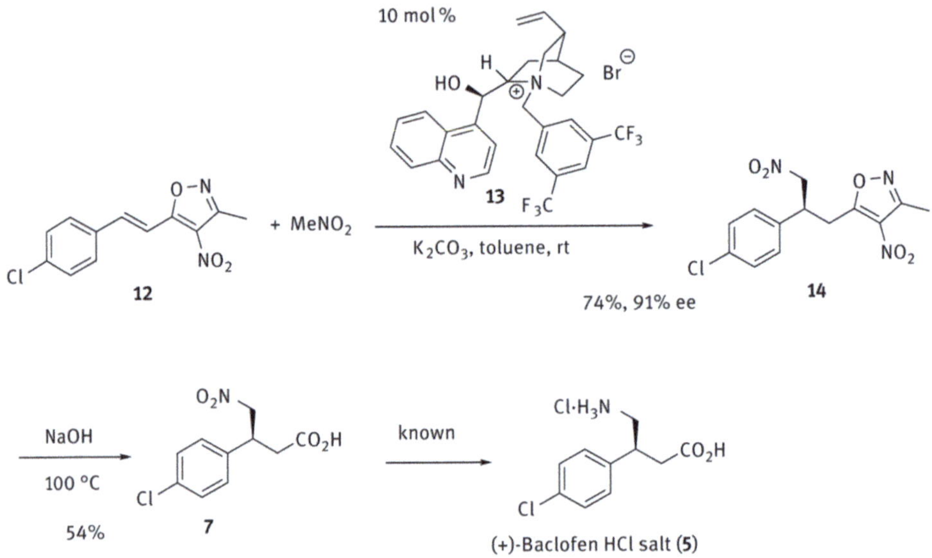

Figure 1.8: Synthesis of baclofen by Bernardi and Adamo.

1.4 Synthesis of telcagepant

The asymmetric Michael reaction of nitromethane and α, β-unsaturated aldehyde catalyzed by diphenylprolinol silyl ether was successfully employed as one of the key steps in the synthesis of telcagepant developed by Merck (Figure 1.9) [24]. Telcagepant is an antagonist of calcitonin gene-related peptide receptor and is an

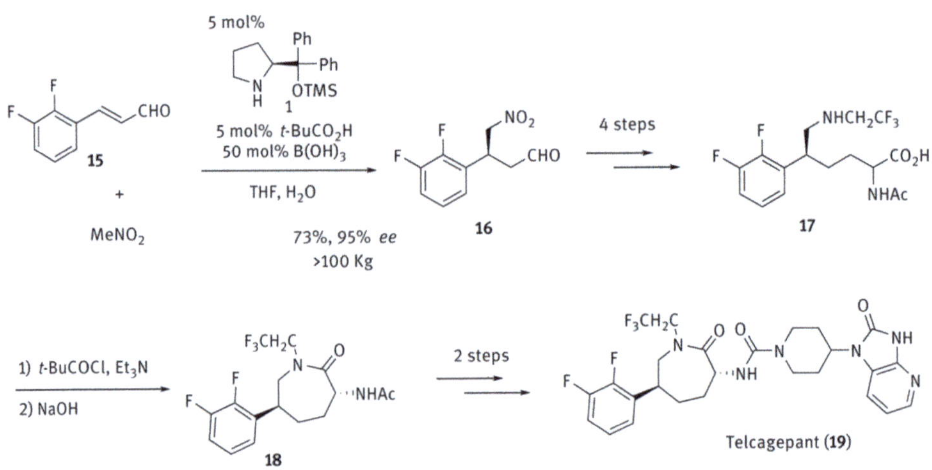

Figure 1.9: The total synthesis of telcagepant by Merck.

expected candidate for treatment of migraine [25]. The asymmetric construction of an optically active disubstituted seven-membered ring lactam is a synthetic challenge. Merck researchers reported an asymmetric Michael reaction with nitromethane and cinnamaldehyde derivative **15** as a substrate at 100 Kg scale to afford γ-nitroaldehyde **16** in high yield with excellent enantioselectivity. Aldehyde **16** was converted into a seven-membered ring lactam **18**. From lactam, hydrolysis of the acetamide group and introduction of the amine side chain resulted in an efficient total synthesis of telcagepant (**19**). This industrial synthesis indicates that asymmetric Michael reaction catalyzed by diphenylprolinol silyl ether catalyst is practical and useful, even at an industrial scale.

1.5 Synthesis of oseltamivir

1.5.1 Total synthesis of oseltamivir in 60 min

Tamiflu has been extensively used worldwide for the treatment of influenza. As a result of the intense need for this life-saving medicine, many synthetic organic chemists have investigated its effective preparation and many syntheses have been reported [26]. Even though Tamiflu is currently still effective, the recent emergence of Tamiflu-resistant virus has prompted efforts by the chemical community to develop medicines active against this mutated virus. This has led to the need for simple syntheses capable of rapidly producing a wide and diverse range of derivatives.

We have been interested in the synthesis of Tamiflu for a long time. We reported a three-pot synthesis in 2009 [7], two-pot synthesis in 2010 [27], and one-pot synthesis in 2013 [28]. Moreover, a 60-min synthesis in a single pot was developed in 2016 [29]. A single flow, multistep synthesis of Tamiflu was reported in 2017 [30]. In this section, we will describe the 60-min synthesis and a multistep flow synthesis. Ma [31], Sebesta [32], and Lu [33] also reported the synthesis of Tamiflu using diphenylprolinol silyl ether as an asymmetric organocatalyst.

The key reaction involving organocatalyst is an asymmetric Michael reaction of aldehyde and nitroalkene catalyzed by diphenylprolinol silyl ether **1**, which we developed in 2005 [10]. In 2011, we found that this Michael reaction was greatly accelerated by an addition of acid (eq. 2) [34]. We investigated the role of acid, and we proposed the following reaction mechanism (Figure 1.10). First, catalyst **1** reacts with aldehyde to generate enamine **20**. When an equimolar amount of enamine **20** and nitroalkene were mixed, immediate generation of cyclobutene **22** and oxazine oxide derivatives **23** were observed. A $CDCl_3$ solution of these two intermediates are heated at 60 °C and formation of enamine **20** and nitroalkene were observed with cyclobutene **22** and oxazine oxide derivatives **23**. The $CDCl_3$ solution of this mixture was cooled down, and enamine **20** and nitroalkene were converted into cyclobutene **22** and oxazine oxide derivatives **23**. Thus, the reaction is under equilibrium. When an acid is added to the mixture of cyclobutene **22** and oxazine oxide derivatives **23**, an immediate formation of the Michael product **25** was observed. When cyclobutene **22** and oxazine oxide derivatives **23** were generated, these two species capture the diphenylprolinol silyl ether catalyst **1** and there is no available catalyst to continue the reaction circle. Thus, cyclobutene **22** and oxazine oxide derivatives **23** are regarded as parasites. The major role of the acid is to decompose these species and regenerate the catalyst, and to return the catalyst into the reaction circle. Although this is our proposal, there is still a lingering controversy about the role of acid [35].

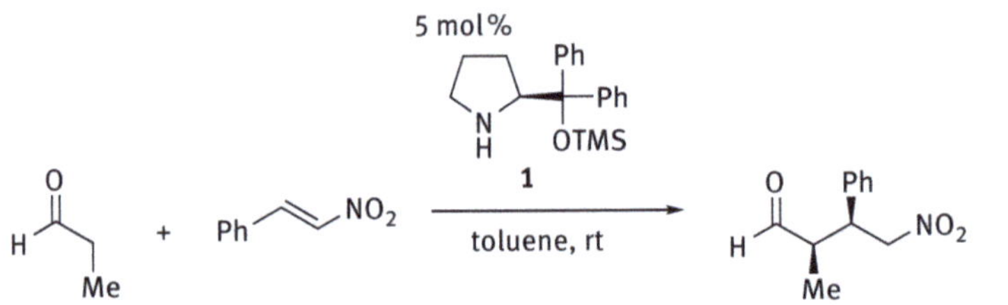

(1.2)

without additive 6 h, 98%, syn:anti = 14:1, 99% ee
with 4-nitrophenol 15 min, 98%, syn:anti = 15:1, 99% ee

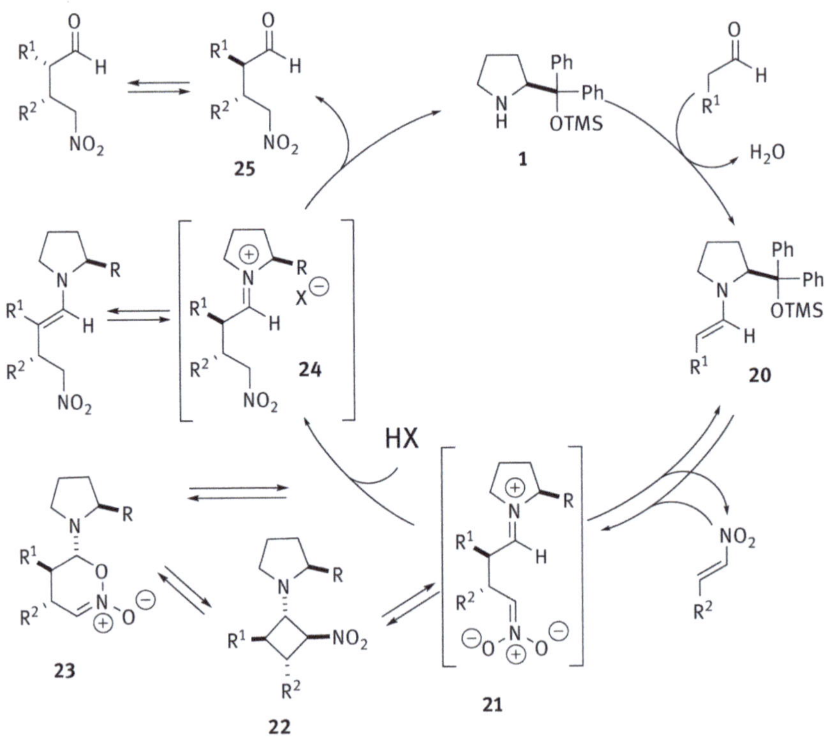

Figure 1.10: The mechanism of the Michael reaction of aldehyde and nitroalkene.

The first step for the synthesis of Tamiflu is a Michael reaction of α-alkoxyacetaldehyde **27** and (*Z*)-*N*-2-nitroethynylacetamide **26** catalyzed by diphenylprolinol silyl ether **28** with a combination of HCO_2H and Schreiner's thiourea **29** (Figure 1.11). This Michael reaction was completed within 30 min. Without thiourea, it took 90 min with lower yield. The effect of thiourea **29** is to activate the nitroalkene **26** via hydrogen bond activation [36]. In this reaction, both cis- and trans-enamines are generated from aldehydes and catalyst, and there is a fast equilibrium between the cis and trans-enamines. cis-Enamine reacts with nitroalkene much faster compared with the trans-enamine because of the steric hindrance at the transition state.

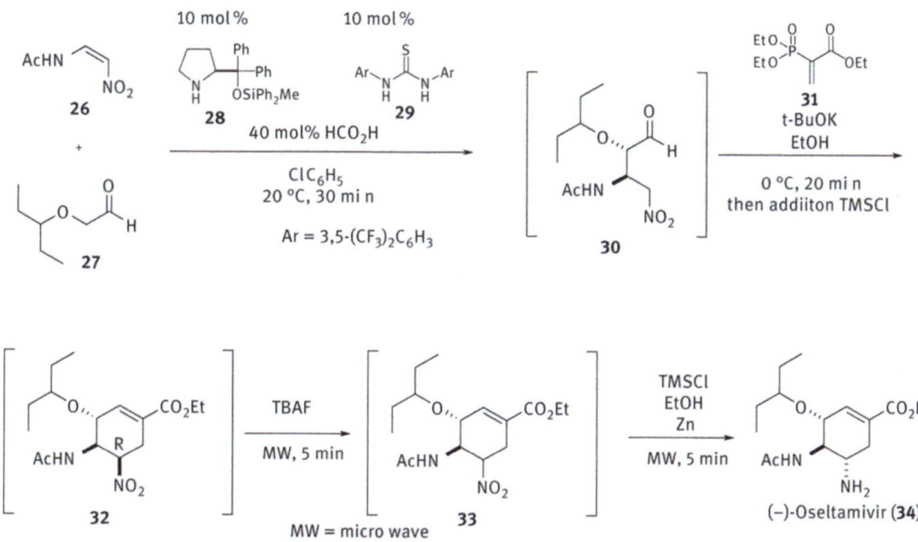

Figure 1.11: One-pot 60-min total synthesis of (-)-oseltamivir.

Michael product **30** was treated with phosphoric ester derivative **31** and t-BuOK with the addition of EtOH to afford cyclohexene derivative **32** via Michael reaction, followed by the Horner–Wadsworth–Emmons reaction. After addition of TMSCl, partial isomerization at α-position of the nitro group occurred by the reaction with tetrabutylammonium fluoride (TBAF). Reduction of nitro group to amine was accomplished by the addition of Zn to afford (-)-oseltamivir (**34**). Both the isomerization using TBAF and the reduction of nitro group using Zn were carried out within 5 min under microwave irradiation. All reactions can be conducted in a same reaction vessel successively, and this is a one-pot synthesis. All we need to do is to add the reagents successively without solvent swap nor solvent exchange. The total yield is 15 %. The total reaction time is 60 min using microwave irradiation and 170 min with normal heating using oil bath.

1.5.2 Multistep continuous flow synthesis for (–)-oseltamivir

Flow synthesis has attracted considerable attention recently because of its efficiency, productivity, safety, and reproducibility [37]. Moreover, a continuous flow technique has been developed to combine a multistep synthesis with a single, continuous, and uninterrupted reactor network without isolation of the intermediate [38].

In our one-pot synthesis of (–)-oseltamivir [29], we simply added the reagents sequentially without evaporation and solvent swap. Thus, we considered that we could apply flow techniques to the synthesis of (–)-oseltamivir. However, under the reaction conditions of the 60-min synthesis, the starting material, nitroalkene, is not completely soluble in toluene. To apply the flow synthesis, all the reagents must be soluble from the beginning. Thus, the concentration of the reagents and other parameters must be optimized again. After several trials, we could finally establish the single flow synthesis of (-)-oseltamivir as shown in Figure 1.12 [30].

The first Michael reaction using organocatalyst proceeds in 71 min (l = 4.2 m, flow 1), and the next step is a domino Michael/Horner–Wadsworth–Emmons reaction in 35 min (l = 4.0 m, flow 2). Protonation is within 7 min (l = 1.0 m, flow 3) and isomerization takes 67 min (l = 10.0 m, flow 4). Final reduction of nitro group to amine was conducted using a cartridge packed with Zn and celite, which takes 120 min (flow 5). Total residence time is 310 min and total yield is 13 %. Thus, a continuous flow synthesis of (–)-oseltamivir with five flow units was accomplished. It should be noted that all reagents except for Zn were dissolved in solvent. The

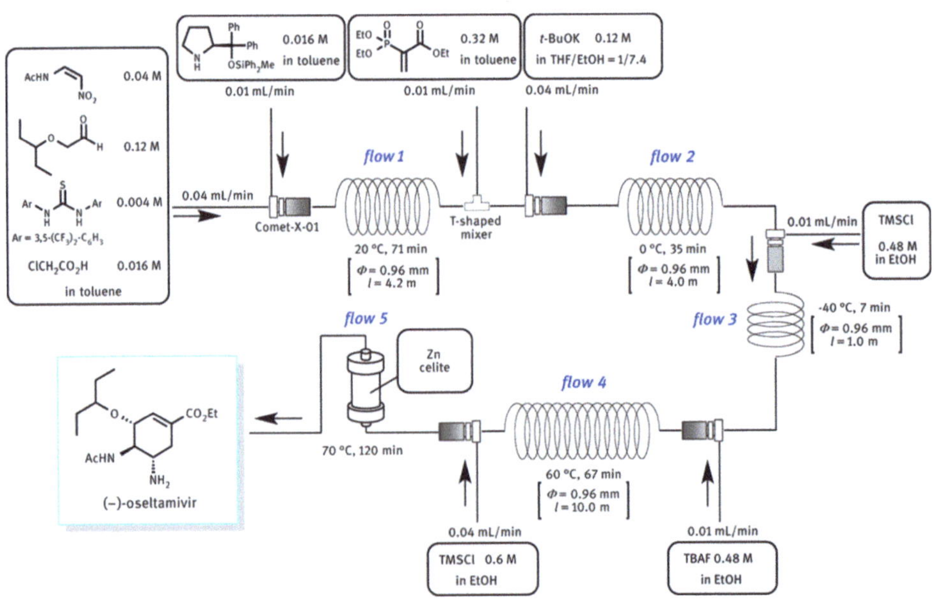

Figure 1.12: Multistep continuous flow synthesis for (–)-oseltamivir.

multistep continuous flow synthesis for (–)-oseltamivir with three continuous chiral centers was successfully realized without isolating any intermediates via a single flow.

1.6 Synthesis of ABT-341

ABT-341 is a potent, and orally bioavailable DPP4 (dipeptidyl peptidase IV) inhibitor, which is a drug candidate for type 2 diabetes being developed by Abbott Laboratories [39]. As ABT-341 is 3,4-disubstituted ethyl cyclohexenecarboxylate, which has a similar structure as that of Tamiflu, we thought that the synthetic methodology developed for Tamiflu would also be applicable to the synthesis of ABT-341. A one-pot synthesis of ATB-341 is based on our organocatalyst-mediated Michael reaction as a key step (Figure 1.13). In 2008, we reported a Michael reaction of acetaldehyde and nitroalkene [40]. This Michael reaction of **35** proceeded smoothly to afford **36** as a first step, followed by the Michael/Horner–Wadsworth–Emmons reactions, the sequence of which has been developed for Tamiflu

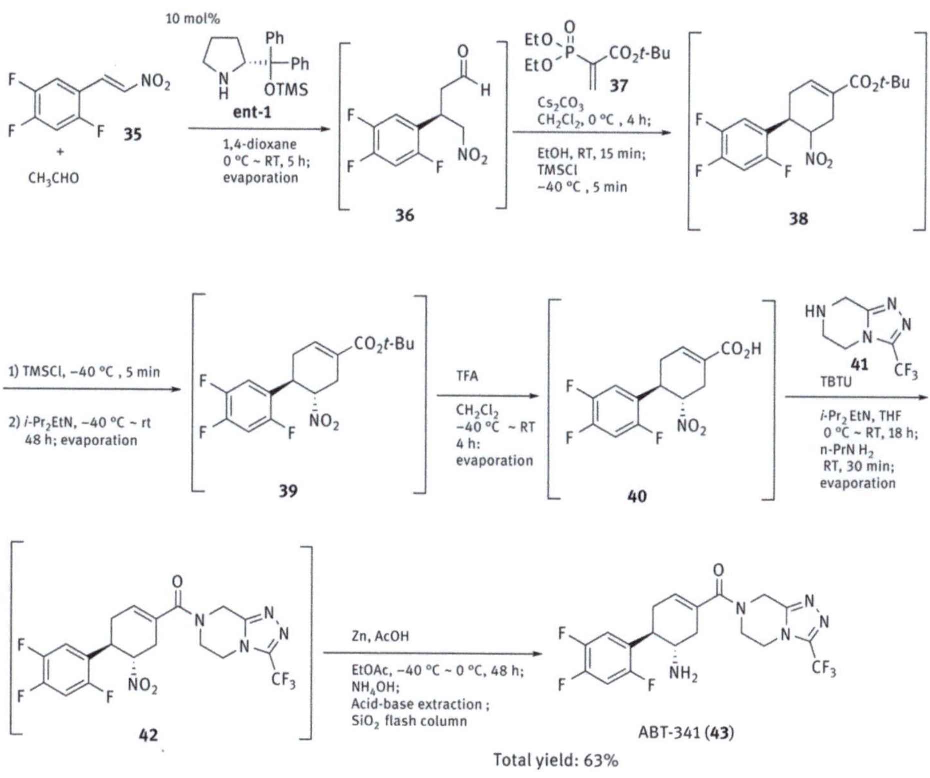

Figure 1.13: One-pot synthesis of ABT-341.

synthesis, to afford the cyclohexene carboxylate **38** in good yield. Isomerization at the α-position of nitro group using *i*-Pr_2EtN provided the trans isomer **39** stereoselectively. tert-Butyl ester **39** can be converted into carboxylic acid **40** by treatment with TFA, which can be coupled with secondary amine **41** using TBTU to afford amide **42**. Nitro group was reduced to amine by treatment with Zn to afford ABT-341 (**43**). All transformations can be conducted in a single reaction vessel and the total yield is 63 % [41].

1.7 Synthesis of prostaglandins

The prostaglandins are known to act as local hormones, controlling a multitude of important physiological properties in only trace amounts, and some of their derivatives are important to human as useful medicines [42]. Because of their biological importance and scarce availability from natural sources, the scientific community has put a great deal of effort and ingenuity into their efficient synthesis [43]. Beginning with Corey's landmark synthesis [44] and continuing through the more than 40 instructive syntheses that have followed, these molecules have prompted the chemical community to devise many different synthetic strategies, although all the previous syntheses require many operations. Thus, it is still a synthetic challenge to synthesize a molecule of this complexity with three contiguous chiral centers in a short number of steps and via a sustainable process.

Proline is known to promote asymmetric aldol reaction, and recently, Aggarwal and coworkers achieved a short step synthesis of PGF2α using proline-mediated aldol dimerization reaction of succinaldehyde **45** as a key step (Figure 1.14) [45]. The proline mediated asymmetric aldol reaction of succinaldehyde **45**, prepared from 2,5-dimethoxytetrahydrofuran **44**, proceeded, followed by aldol dehydration and acetalization to provide a bicyclic compound **46**, which was converted into methyl acetal **47** in 14 % yield with 98 % ee. Although the yield of this acetal **47** in 2012 is 14 % [45], it was improved to 29 % in 2018 [46]. The addition of the side chain **48** via Michael reaction, and oxidative cleavage of the alkene afforded bicyclic compound **50**. The conversion of methyl acetal to acetal, followed by the Wittig reaction provided PGF2α (**50**) in seven steps.

However, in 2013, we reported a three-pot synthesis of PGE_1 methyl ester using a domino asymmetric Michael/intramolecular Henry reaction as a key step (Figure 1.15) [47]. An asymmetric Michael reaction of succinaldehyde **45** and nitroalkene **51** proceeded in the presence of 5 mol% of diphenylprolinol silyl ether catalyst **ent-1** with *p*-nitrophenol, followed by Henry reaction to provide aldehyde **52**. In the same vessel, Horner–Wadsworth–Emmons reaction with phosphate ester **53** proceeded to afford three-component coupling product **54** in good yield, which possesses all the necessary carbon framework for prostaglandin E_1 methyl ester. The reduction of α, β-unsaturated ketone moiety with diisopinocamphenyl chloroborane provided allyl alcohol **55**

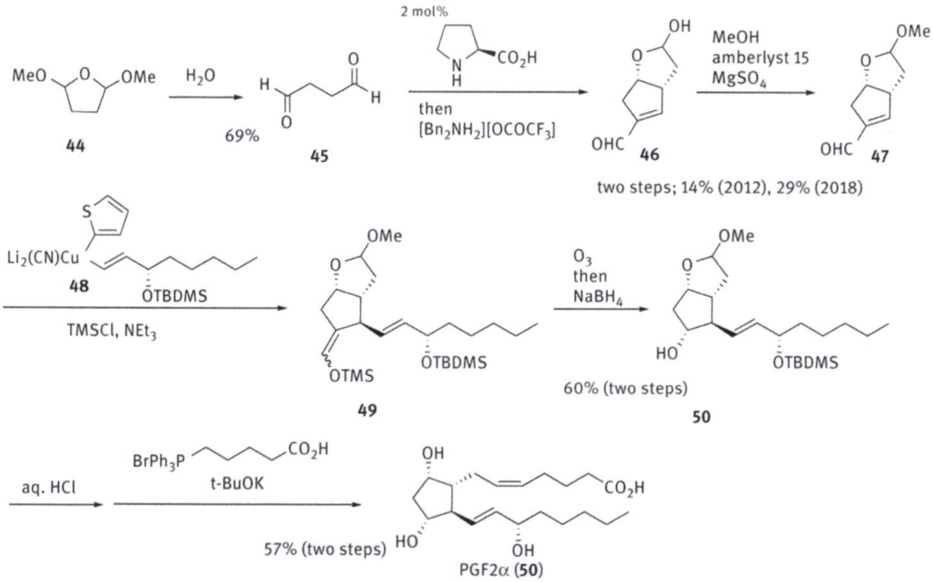

Figure 1.14: Aggarwal's total synthesis of PGF2α.

with high diastereoselectivity. Dehydration afforded nitroalkene, followed by oxidative Nef reaction, which was developed by our group [48] to afford the cyclopentenone. Diastereoselective epoxidation, followed by reductive opening of the epoxide with Zn, provided PGE_1 methyl ester (**56**). The enantioselective total synthesis of PGE_1 methyl ester (**56**) was accomplished in 14 % total yield, using inexpensive and simple starting materials, and via a pot-economical process.

1.8 Synthesis of estradiol

Steroids with a tetracyclic molecular framework have versatile and important biological activities [49]. Given their importance in medicine and biology, many synthetic methods have been developed for the efficient synthesis of steroids [50]. Estrogen is a member of the steroid family and is a female sex hormone. There are three estrogen compounds that have estrogenic hormonal activity in females; namely, estrone, estradiol, and estriol. Given their important biological activities, many total syntheses of these molecules have been developed [51]. Racemic estrone was synthesized by cobalt-catalyzed reaction by Vollhardt [52], and Torgov reported a short synthesis of racemic estrone through acid-catalyzed reaction as a key step [53]. Corey reported an excellent short synthesis of chiral estrone by using CBS reduction [54]. Organocatalysts have also been used in the synthesis of steroids.

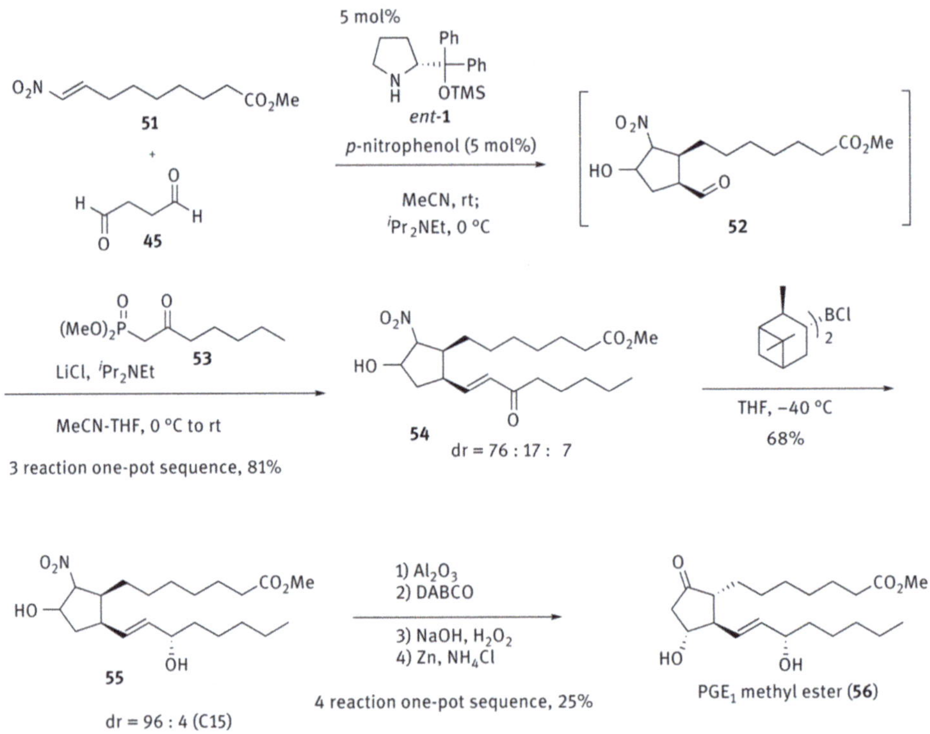

Figure 1.15: Three-pot total synthesis of PGE_1 methyl ester.

The proline-mediated intramolecular aldol reaction reported in 1971 by Hajos and Parrish at Hoffmann-La Roche [55] and by Eder, Sauer, Wiechert at Schering AG [56] independently, is a landmark discovery in organocatalysis (eq. 3). This approach is effective for the synthesis of Wieland–Miescher ketone, which is a key intermediate in steroid synthesis [57].

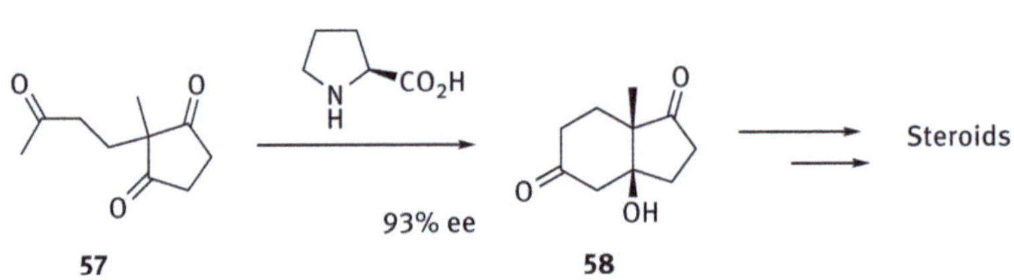

(1.3)

Recently, Hong reported the one-pot synthesis of a steroid framework through organocatalyst-mediated Michael/Michael/aldol/Henry and Michael/Michael/Henry reactions [58]. Jørgensen reported the construction of the steroid structure by using Torgov's diene via a dienamine intermediate (Figure 1.16) [59]. The Diels–Alder reaction of α, β-unsaturated aldehyde **59** and diketone **60** proceeded in the presence of 20 mol% of diphenylprolinol silyl ether **1** with benzoic acid to afford tetracyclic

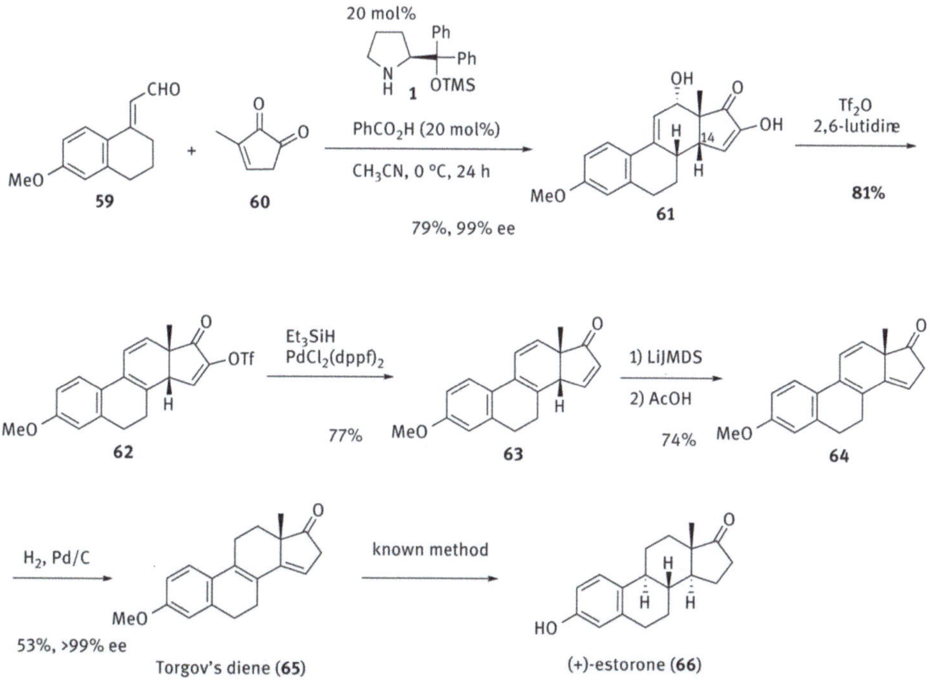

Figure 1.16: Jørgensen' total synthesis of Torgov's diene.

compound **61** in good yield with excellent enantioselectivity. As the stereochemistry at C14 was opposite, it must be changed to the desired stereochemistry. In five steps, the Diels–Alder product **61** was converted into the Torgov's diene **65**, which is known to be converted into (+)-estrone **66**.

Torgov cyclization is an effective method for the synthesis of steroids. Recently, List reported that Torgov cyclization of **67**, which was easily prepared from 6-methoxy-1-tetralone, proceeded in the presence of chiral disulfon imide **68** to afford Torgov diene **65**, the steroid core skeleton, in good yield with excellent enantioselectivity (Figure 1.17) [60]. (+)-Estrone **66** was synthesized in two steps from Torgov's diene **65** by a known procedure reported by Corey.

Recently our group synthesized (+)-estradiol methyl ether (**74**) via a pot-economical manner [51, 61]. In this study, we developed an efficient synthetic method for the steroid framework construction using an organocatalyst-mediated domino reaction. Our strategy is to synthesize a key intermediate **73** with A, C, and D rings of steroids from readily available compounds, and then construct the B-ring of steroid. Our idea is as follows (Figure 1.18). When the asymmetric Michael reaction of nitroalkane **70** having a diketone moiety and α, β-unsaturated aldehyde **71** proceeds, an optically active enamine **72** will be produced. If the successive intramolecular aldol reaction proceeds, it is expected that the compound **73** having steroid

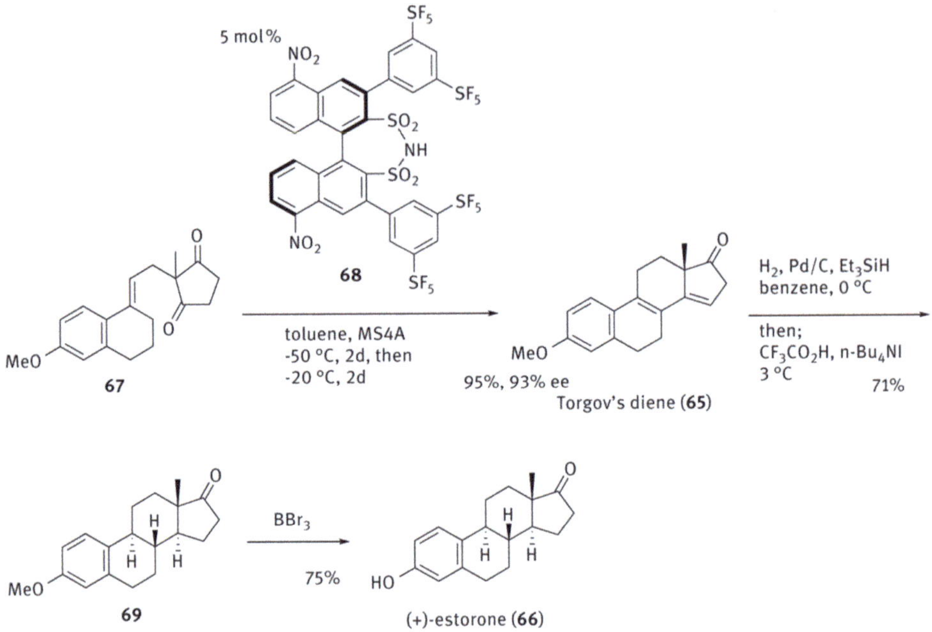

Figure 1.17: List's total synthesis of (+)-estrone.

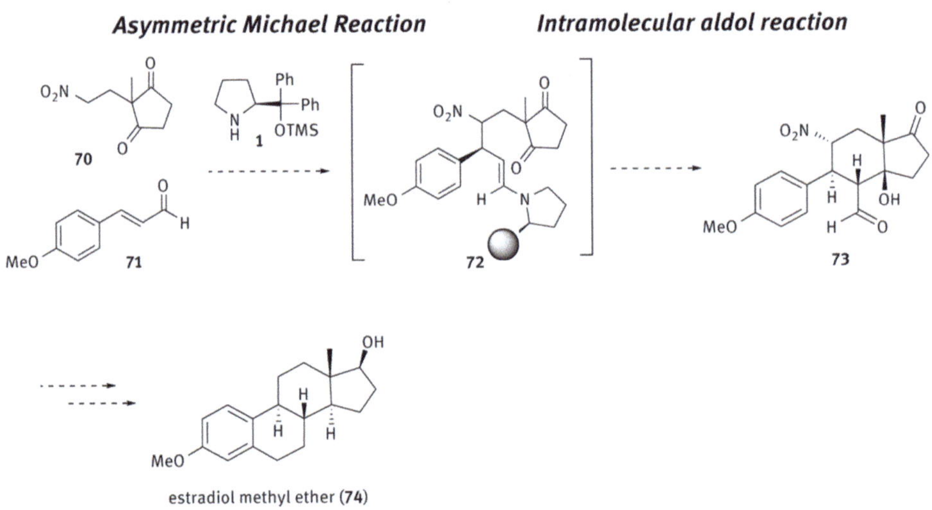

Figure 1.18: Our idea for a domino reaction in the synthesis of a steroid intermediate with A, C, and D rings.

A, C, and D rings would be obtained in a single pot via a domino reaction. The key issues in this reaction are whether the reaction proceeds smoothly, whether the stereoselectivity of five consecutive asymmetric centers is high or whether a compound having the desired stereochemistry can be obtained.

After several trials, we identified suitable reaction conditions: The desired bicyclo compound **73** was obtained in good yield with excellent diastereo- and enantioselectivities when a mixture of **70** and **71** were treated with the catalyst **1** and benzoic acid in the presence of water (eq. 4). The stereochemistry of five contiguous stereocenters are the same as that of estradiol.

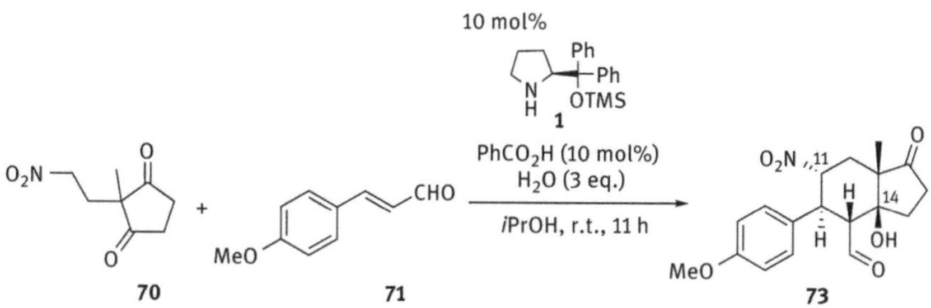

89%, >99% *ee*, single isomer (1.4)

As we obtained the key intermediate of the estradiol synthesis, we investigated the total synthesis based on pot economy (Figure 1.19). In the synthesis of estradiol, it is not necessary to isolate compound **73**. Addition of KCN to the same reaction vessel of **73** afforded cyanohydrin, which is further converted into xanthate **74** by the reaction with CS_2 and MeI. Then, dehydration using $SOCl_2$ gave **75** in a single reaction vessel from **70** and **71** in 78 % yield.

Both the nitro group and xanthate of **75** were reduced under radical conditions (n-Bu_3SnH, cat. AIBN) to afford **76** in 59 %. It was found that the yield of the simultaneous reduction was higher than that of the stepwise reduction. Diastereoselective ketone reduction **76** with $LiBHEt_3$, followed by the reductive conversion of cyano group via DIBAL gave aldehyde **77**. Protection of alcohol **77** gave silyl ether **78**.

The transformation from **78** to estradiol methyl ether (**74**) can be carried out in a single reaction vessel including six reaction steps: Kraus–Pinnick oxidation of aldehyde **78** gave carboxylic acid **79**, in which the remaining oxidant was destroyed by the addition of CH_3CHO. The second reaction was diastereoselective hydrogenation from **79** to **80**. The third reaction is a conversion of carboxylic acid **80** into acid chloride **81** by the treatment with oxalyl chloride. The fourth reaction was the Friedel–Crafts acylation from **81** to **82**, which proceeded well by the addition of $AlCl_3$. The addition of methanol, which reacts with remaining $AlCl_3$ to generate HCl, cleaves silyl ether to generate alcohol **83** as a fifth reaction. The sixth reaction is the hydrogenolysis of **83** employing $Pd(OH)_2$ in AcOH, which gave estradiol methyl ether (**74**).

It should be noted that the metal-catalyzed hydrogenolysis reaction proceeded well even after five reactions in the presence of many by-products and side products.

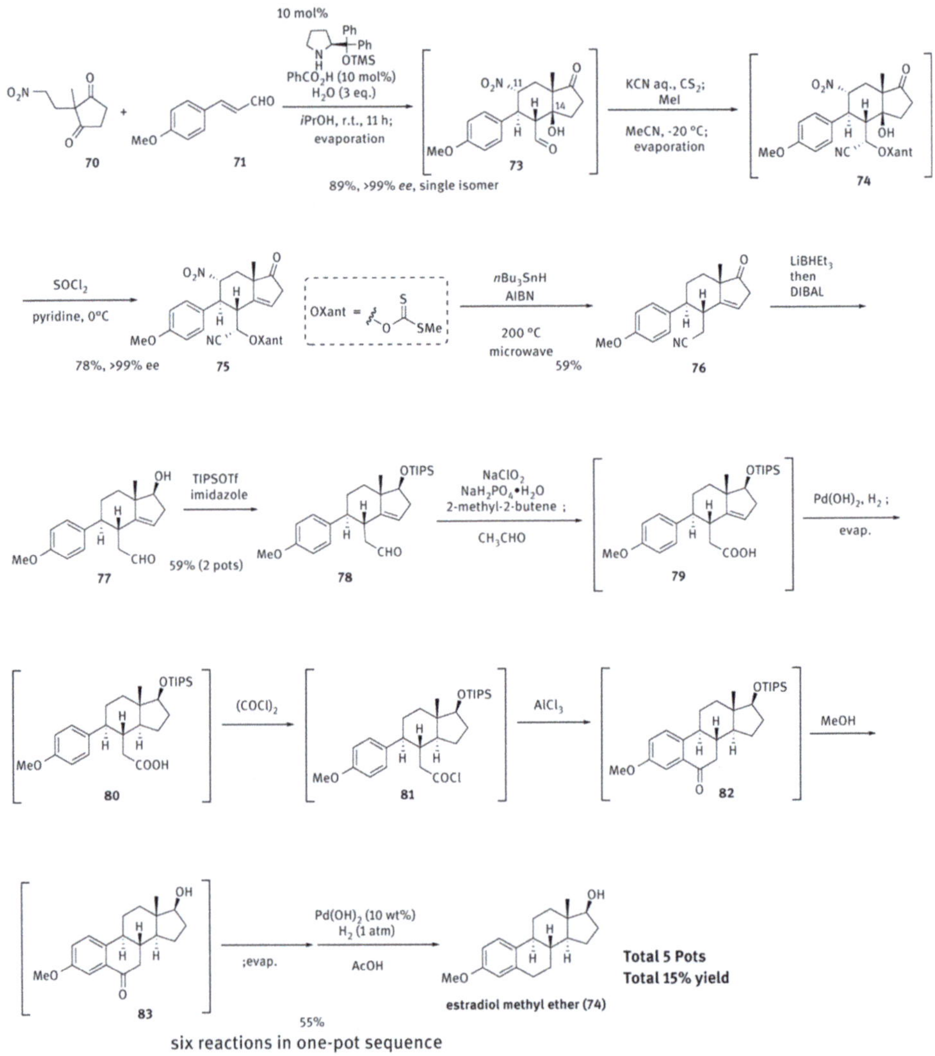

Figure 1.19: Five-pot total synthesis of estradiol methyl ether.

1.9 Horsfiline and coerulescine

We have reported the Michael reaction of α, β-unsaturated aldehyde and nitromethane. We also found that unreactive β, β-disubstituted α, β-unsaturated aldehyde is also a useful Michael acceptor for the construction of quaternary chiral centers with excellent enantioselectivity (eq. 5). Using this reaction as a key step, we have synthesized (−)-horsfiline (**84**) and (−)-coerulescine (**85**) in three pots (Figure 1.20) [62].

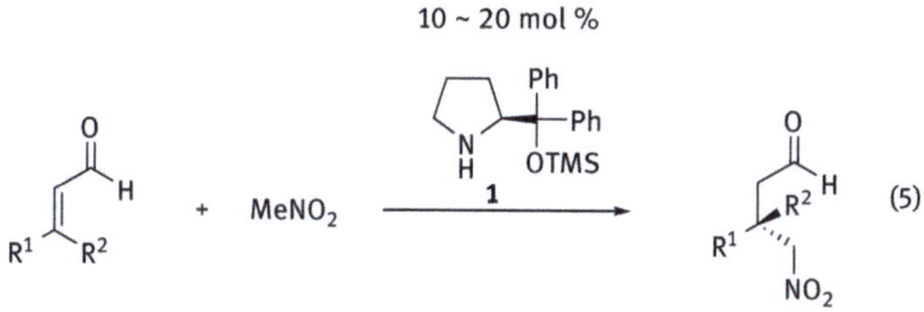

(1.5)

(−)-Horsfiline (**84**) and (−)-coerulescine (**85**) are spirooxyindole alkaloids that have been isolated from *Horsfieldia superba* in 1991 by Bodo's group [63] and from *Pharalis coerulescens* in 1998 by Colegate's group [64], respectively. The first reaction is an aldol condensation reaction of acetaldehyde and isatin derivatives **86, 87**. Aldol condensation reaction of acetaldehyde using DBU was developed by our group, which is described in Section 3 of this paper. β, β-Disubstituted α, β-unsaturated aldehydes **90, 91** were obtained in good yield. The next reaction is a key asymmetric reaction catalyzed by diarylprolinol silyl ether **2** of β, β-disubstituted α, β-unsaturated aldehydes **90, 91** with nitromethane to afford the Michael products **92, 93**. By treatment of **92, 93**

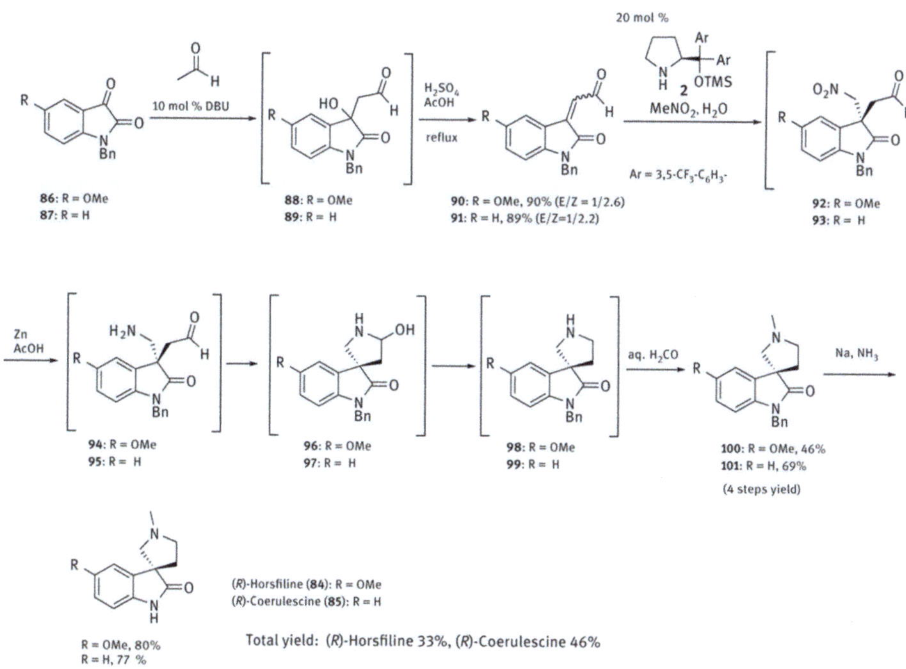

Figure 1.20: Three-pot total synthesis of (−)-horsfiline (**84**) and (−)-coerulescine (**85**).

with Zn and CH_3CO_2H, a domino reaction comprising the reduction of nitro group and reductive amination proceeded to afford **98, 99.** Second reductive amination occurred by the further addition of formaldehyde in the same reaction vessel to provide *N*-methyl amines **100, 101**. Deprotection of Bn-protecting group provided (–)-horsfiline (**84**) and (–)-coerulescine (**85**) in good yield.

1.10 Summary

In this paper, we have described the successful application of diphenylprolinol silyl ether, which is one of the widely used organocatalysts, to the synthesis of natural products and drugs, mostly focusing on the author's results. As an organocatalyst is a suitable catalyst in domino and one-pot reactions, complex molecular structures can be created in short steps. As many kinds of reactions have already been developed using enamine and iminium ions as intermediates, organocatalysts will soon be more frequently applied to the synthesis of biologically active molecules.

References

[1] Selected reviews on organocatalysis: (a) List B, eds. Asymmetric organocatalysis 1; lewis base and acid catalysts, Thieme, Stuttgart, 2012. (b) Dalko PI, editors. Comprehensive Enantioselective Organocatalysis: Catalysts, Reactions, and Applications. Weinheim: Wiley-VCH, 2013.

[2] (a) Wender PA, Croatt MP, Witulski B. New reactions and step economy: the total synthesis of (±)-salsolene oxide based on the type II transition metal-catalyzed intramolecular [4+4] cycloaddition. Tetrahedron 2006, 62, 7505–11. (b) Wender PA, Verma VA, Paxton TJ, Pillow TH. Function-oriented synthesis, step economy, and drug design. Acc Chem Res. 2008;41:40–9. (c) Wender PA. Toward the ideal synthesis and molecular function through synthesis-informed design. Nat Prod Rep. 2014;31:433–40.

[3] Gaich T, Baran PS. Aiming for the ideal synthesis. J Org Chem. 2010;75:4657–73.

[4] (a) Tietze LF. Domino reactions in organic synthesis. Chem Rev 1996, 96, 115–36. (b) Tietze LF, editors. Domino Reactions. Weinheim: Wiley-VCH, 2014.

[5] Enders D, Huttl MR, Grondal C, Raabe G. Control of four stereocentres in a triple cascade organocatalytic reaction. Nature. 2006;441:861–3.

[6] Review; (a) Grondal C, Jeanty M, Enders D. Organocatalytic cascade reactions as a new tool in total synthesis. Nature Chem 2010, 2, 167–78. (b) Albrecht L, Jiang H, Jørgensen KA. A simple recipe for sophisticated cocktails: organocatalytic one-pot reactions–concept, nomenclature, and future perspectives. Angew Chem Int Ed. 2011;50:8492–82509. (c) Pellissier H. Recent Developments in asymmetric organocatalytic domino reactions. Adv Synth Catal. 2012;354:237–94. (d) Volla CM, Atodiresei I, Rueping M. Catalytic C–C. Bond-forming multi-component cascade or domino reactions: Pushing the boundaries of complexity in asymmetric organocatalysis. Chem Rev. 2014;114:2390–431. (e) Hong B-C, Raja A, Sheth VM.

Asymmetric synthesis of natural products and medicinal drugs through one-pot-reaction strategies. Synthesis. 2015;47:3257–85.
[7] Ishikawa H, Suzuki T, Hayashi Y. High-yielding synthesis of the anti-influenza neuramidase inhibitor (-)-oseltamivir by three “one-pot” operations. Angew Chem Int Ed. 2009;48:1304–7.
[8] Hayashi Y. Pot economy and one-pot synthesis. Chem Sci. 2016;7:866–80.
[9] Reviews, (a) Marques-Lopez E, Herrera RP, Christmann M. Asymmetric organocatalysis in total synthesis – a trial by fire. Nat Prod Rep 2010, 27, 1138–67. (b) Dibello E, Gamenara D, Seoane G. Organocatalysis in the Synthesis of Natural Products: Recent development in aldol and mannich reactions, and 1,4-conjugated additions. Current Organocatal. 2015;2:124–49. (c) Sun B-F. Total synthesis of natural and pharmaceutical products powered by organocatalytic reactions. Tetrahedron Lett. 2015;56:2133–40.
[10] Hayashi Y, Gotoh H, Hayashi T, Shoji M. Diphenylprolinol silyl ethers as efficient organocatalysts for the asymmetric Michael reaction of aldehydes and nitroalkenes. Angew Chem Int Ed. 2005;44:4212–15.
[11] Marigo M, Wabnitz TC, Fielenbach D, Jørgensen KA. Enantioselective organocatalyzed α sulfenylation of aldehydes. Angew Chem Int Ed. 2005;44:794–7.
[12] For reviews, see: (a) Palomo C, Mielgo A. Diarylprolinol ethers: expanding the potential of enamine/iminium-ion catalysis. Angew Chem Int Ed 2006, 45, 7876–80. (b) Mielgo A, Palomo C. α,α-Diarylprolinol Ethers: New tools for functionalization of carbonyl compounds. Chem Asian J. 2008;3:922–48. (c) Xu LW, Li L, Shi ZH. Asymmetric synthesis with silicon-based bulky amino organocatalysts. Adv Synth Catal. 2010;352:243–79. (d) Jensen KL, Dickmeiss G, Jiang H, Albrecht L, Jørgensen KA. The diarylprolinol silyl ether system: A general organocatalyst. Acc Chem Res. 2012;45:248–64. (e) Gotoh H, Hayashi Y. in Sustainable catalysis, Dunn J, Hii KK, Krische M J, Williams M T. editor. Hoboken: Wiley, 2013:287–316. (f) Donslund BS, Johansen TK, Poulsen PH, Halskov KS, Jørgensen KA. The diarylprolinol silyl ethers: Ten years after. Angew Chem Int Ed. 2015;54:13860–74.
[13] Hayashi Y, Okamura D, Yamazaki T, Ameda Y, Gotoh H, Tsuzuki S, et al. A theoretical and experimental study of the effects of silyl substituents in enantioselective reactions catalyzed by diphenylprolinol silyl ether. Chem Eur J. 2014;20:17077–88.
[14] (a) Seebach D, Grošelj U, Badine D M, Schweizer W B, Beck A K. Isolation and x-ray structures of reactive intermediates of organocatalysis with diphenylprolinol ethers and with imidazolidinones: a survey and comparison with computed structures and with 1-acyl-imidazolidinones: the 1,5-repulsion and the geminal-diaryl effect at work. Helv Chem Acta 2008, 91, 1999–2034. (b) Grošelj U, Seebach D, Badine DM, Schweizer WB, Beck AK, Krossing I, Klose P, Hayashi Y. Uchimaru T. Structures of the reactive intermediates in organocatalysis with diarylprolinol ethers. Helv Chem Acta. 2009;92:1225–59.
[15] Gotoh H, Uchimaru T, Hayashi Y. Two reaction mechanisms via iminium ion intermediates: the different reactivities of diphenylprolinol silyl ether and trifluoromethyl-substituted diarylprolinol silyl ether. Chem Eur J. 2015;21:12337–46.
[16] 2-[diphenyl[(trimethylsilyl)oxy]methyl]pyrrolidine
[17] Olpe HR, Demieville H, Baltzer V, Bencze WL, Koella WP, Wolf P, et al. The biological activity of d-baclofen (Lipresal®). Eur J Pharmacol. 1978;52:133–6.
[18] Gotoh H, Ishikawa H, Hayashi Y. Diphenylprolinol silyl ether as catalyst of an asymmetric, catalytic, and direct michael reaction of nitroalkanes with α,β-unsaturated aldehydes. Org Lett. 2007;9:5307–9.
[19] Zu L, Xie H, Li H, Wang J, Wang W. Highly enantioselective organocatalytic conjugate addition of nitromethane to α,β-unsaturated aldehydes: three-step synthesis of optically active baclofen. Add Synth Catal. 2007;349:2660–4.

[20] Hayashi Y, Sakamoto D, Okamura D. One-pot synthesis of (*S*)-baclofen via aldol condensation of acetaldehyde with diphenylprolinol silyl ether mediated asymmetric michael reaction as a key step. Org Lett. 2016;18:4–7.
[21] (a) Hayashi Y, Itoh T, Aratake S, Ishikawa H. A diarylprolinol in an asymmetric, catalytic, and direct crossed-aldol reaction of acetaldehyde. Angew Chem Int Ed 2008, 47, 2082–4. (b) Hayashi Y, Samanta S, Itoh T, Ishikawa H. Asymmetric, Catalytic, and direct self-aldol reaction of acetaldehyde catalyzed by diarylprolinol. Org Lett. 2008;10:5581–3.
[22] Okino T, Hoashi Y, Furukawa T, Xu X, Takemoto Y. Enantio- and diastereoselective michael reaction of 1,3-dicarbonyl compounds to nitroolefins catalyzed by a bifunctional thiourea. J Am Chem Soc. 2005;127:119–25.
[23] Baschieri A, Bernardi L, Ricci A, Suresh S, Adamo MFA. Catalytic asymmetric conjugate addition of nitroalkanes to 4-nitro-5-styrylisoxazoles. Angew Chem Int Ed. 2009;48:9342–5.
[24] Xu F, Zacuto M, Yoshikawa N, Desmond R, Hoerrner S, Itoh T, et al. Asymmetric synthesis of telcagepant, a CGRP receptor antagonist for the treatment of migraine. J Org Chem. 2010;75:7829–41.
[25] Paone DV, Shaw AW, Nguyen DN, Burgey CS, Deng JZ, Kane SA, et al. Potent, orally bioavailable calcitonin gene-related peptide receptor antagonists for the treatment of migraine: discovery of *N*-[(3*R*,6*S*)-6-(2,3-Difluorophenyl)-2-oxo-1-(2,2,2-trifluoroethyl)azepan-3-yl]-4-(2-oxo-2,3-dihydro-1*H*-imidazo[4,5-*b*]pyridin-1-yl)piperidine-1-carboxamide (MK-0974). J Med Chem. 2007;50:5564–7.
[26] Reviews, see (a) Farina V, Brown JD Tamiflu: the supply problem. Angew Chem Int Ed 2006, 45, 7330–4. (b) Shibasaki M, Kanai M. Synthetic strategies for oseltamivir phosphate. Eur J Org Chem. 2008;1839–50. (c) Magano J. Synthetic approaches to the neuraminidase inhibitors zanamivir (relenza) and oseltamivir phosphate (tamiflu) for the treatment of influenza. Chem Rev. 2009;109:4398–438. (d) Andraos J. Global green chemistry metrics analysis algorithm and spreadsheets: evaluation of the material efficiency performances of synthesis plans for oseltamivir phosphate (tamiflu) as a test case. Org Process Res Dev. 2009;13:161–85. (e) Magano J. Recent synthetic approaches to oseltamivir phosphate (Tamiflu™) for the treatment of influenza. Tetrahedron. 2011;67:7875–99. (f) Shibasaki M, Kanai M, Yamatsugu K. Recent development in synthetic strategies for oseltamivir phosphate. Isr J Chem. 2011;51:316–28. (g) Li N-G, Shi Z-H, Tang Y-P, Shi Q-P, Zhang W, Zhang P-X, Dong Z-X, Li W, Duan J-A. Recent progress on the total synthesis of (–)-oseltamivir phosphate (tamiflu) for the treatment of influenza disease. Curr Org Chem. 2014;18:2125–38.
[27] Ishikawa H, Suzuki T, Orita H, Uchimaru T, Hayashi Y. High-yielding synthesis of the anti-influenza neuramidase inhibitor (-)-oseltamivir by two "one-pot" sequences. Chem Eur J. 2010;16:12616–26.
[28] Mukaiyama T, Ishikawa H, Koshino H, Hayashi Y. One-pot synthesis of (-)-oseltamivir and mechanistic insights into the organocatalyzed michael reaction. Chem Eur J. 2013;19:17789–800.
[29] Hayashi Y, Ogasawara S. Time economical total synthesis of (-)-oseltamivir. Org Lett. 2016;18:3426–9.
[30] Ogasawara S, Hayashi Y. Multistep continuous-flow synthesis of (-)-oseltamivir. Synthesis. 2017;49:424–8.
[31] Zhu S, Yu S, Wang Y, Ma D. Organocatalytic michael addition of aldehydes to protected 2-amino-1-nitroethenes: the practical syntheses of oseltamivir (tamiflu) and substituted 3-aminopyrrolidines. Angew Chem Int Ed. 2010;49:4656–60.
[32] (a) Rehak J, Hutka M, Latika A, Brath H, Almassy A, Hajzer V, et al. Thiol-free synthesis of oseltamivir and its analogues via organocatalytic michael additions of oxyacetaldehydes to

2-acylaminonitroalkenes. Synthesis 2012, 44, 2424–30. (b) Hajzer V, Latika A, Durmis J, Sebesta R. Enantioselective *Michael* addition of the 2-(1-ethylpropoxy)acetaldehyde to *N*-[(1*Z*)-2-nitroethenyl]acetamide – optimization of the key step in the organocatalytic oseltamivir synthesis. Helv Chem Acta. 2012;95:2421–8.

[33] Weng J, Li Y-B, Wang R-B, Lu G. Organocatalytic michael reaction of nitroenamine derivatives with aldehydes: short and efficient asymmetric synthesis of (−)-oseltamivir. Chem Cat Chem. 2012;4:1007–12.

[34] Patora-Komisarska K, Benohoud M, Ishikawa H, Seebach D, Hayashi Y. Organocatalyzed *Michael* addition of aldehydes to nitro alkenes – generally accepted mechanism revisited and revised. Helv Chem Acta. 2011;94:719–45.

[35] (a) Burés J, Armstrong A, Blackmond DG Mechanistic rationalization of organocatalyzed conjugate addition of linear aldehydes to nitro-olefins. J Am Che Soc 2011, 133, 8822–5. (b) Burés J, Armstrong A, Blackmond D G. Curtin–Hammett paradigm for stereocontrol in organocatalysis by diarylprolinol ether catalysts. J Am Che Soc. 2012;134:6741–50. (c) Bächle F, Duschmalé J, Ebner C, Pfaltz A, Wennemers H. Organocatalytic asymmetric conjugate addition of aldehydes to nitroolefins: identification of catalytic intermediates and the stereoselectivity-determining step by ESI-MS. Angew Chem Int Ed. 2013;52:12619–23. (d) Isenegger PG, Pfaltz A. Mass spectrometric back reaction screening of quasi-enantiomeric products as a mechanistic tool. Chem Rev. 2016;16:2534–43. (e) Földes T, Madarász Á, Révész Á, Dobi Z, Varga Sz, Hamza A, Nagy P R, Pihko P M, Pápai I. Stereocontrol in diphenylprolinol silyl ether catalyzed michael additions: steric shielding or curtin-hammett scenario? J Am Chem Soc. 2017;139:17052–63.

[36] (a) Schreiner PR Metal-free organocatalysis through explicit hydrogen bonding interactions. Chem Soc Rev 2003, 32, 289–96. (b) Zhang Z, Schreiner P R. Chem Soc Rev. 2009;38:1187–98.

[37] (a) Yoshida J, Nagaki A, Yamada T Flash chemistry: fast chemical synthesis by using microreactors. Chem Eur J 2008, 14, 7450–9. (b) Webb D, Jamison T F. Continuous flow multi-step organic synthesis. Chem Sci. 2010;1, 675–80. (c) Wiles C, Watts P. Continuous flow reactors: a perspective. Green Chem. 2012;14:38–54. (d) Poechlauer P, Manley J, Broxterman R, Gregertsen B, Ridemark M. Continuous processing in the manufacture of active pharmaceutical ingredients and finished dosage forms: an industry perspective. Org Process Res Dev. 2012;16:1586–190. (e) Hartman R L, McMullen J P, Jensen K F. Deciding whether to go with the flow: evaluating the merits of flow reactors for synthesis. Angew Chem Int Ed. 2011;50:7502–19. (f) Newman S G, Jensen K F. The role of flow in green chemistry and engineering. Green Chem. 2013;15:1456–72. (g) Tsubogo T, Ishikawa T, Kobayashi S. Asymmetric carbon-carbon bond formation under continuous-flow conditions with chiral heterogeneous catalysts. Angew Chem Int Ed. 2013;52:6590–604.

[38] Hessel V. Novel process windows – gate to maximizing process intensification via flow chemistry. Chem Eng Technol. 2009;32:1655–81.

[39] Pei Z, Li X, von Geldern TW, Madar DJ, Longenecker K, Yong H, et al. Discovery of ((4*R*,5*S*)-5-Amino-4-(2,4,5- trifluorophenyl)cyclohex-1-enyl)-(3- (trifluoromethyl)-5,6-dihydro- [1,2,4] triazolo[4,3-a]pyrazin-7(8 *H*)-yl)methanone (ABT-341), a highly potent, selective, orally efficacious, and safe dipeptidyl peptidase IV inhibitor for the treatment of type 2 diabetes. J Med Chem. 2006;49:6439–42.

[40] Hayashi Y, Itoh T, Ohkubo M, Ishikawa H. Asymmetric michael reaction of acetaldehyde catalyzed by diphenylprolinol silyl ether. Angew Chem Int Ed. 2008;47:4722–4.

[41] Ishikawa H, Honma M, Hayashi Y. One-pot high-yielding synthesis of the DPP4-selective Inhibitor ABT-341 by a four-component coupling mediated by a diphenylprolinol silyl ether. Angew Chem Int Ed. 2011;50:2824–7.

[42] Funk CD. Prostaglandins and leukotrienes: advances in eicosanoid biology. Science. 2001;294:1871–5.

[43] Das D, Chandrasekhar S, Yadav JS, Gree R. Recent developments in the synthesis of prostaglandins and analogues. Chem Rev. 2007;107:3286–337.

[44] (a) Corey EJ, Weinshenker NM, Schaaf TK, Huber W Stereo-controlled synthesis of dl-prostaglandins $F_{2\alpha}$ and E2. J Am Chem Soc 1969, 91, 5675–7. *(b)* Corey E J, Cheng X M. The Logic of chemical synthesis, New York: Wiley, 1995.

[45] Coulthard G, Erb W, Aggarwal VK. Stereocontrolled organocatalytic synthesis of prostaglandin $PGF_{2\alpha}$ in seven steps. Nature. 2012;489:278–81.

[46] Pelss A, Gandhamsetty N, Smith JR, Mailhol D, Silvi M, Watson A, et al. Re-optimization of the organocatalyzed double aldol domino process to a key enal intermediate and its application to the total synthesis of Δ^{12}-prostaglandin J_3. Chem Eur J. 2018;24:9542–5.

[47] Hayashi Y, Umemiya S. Pot economy in the synthesis of prostaglandin a_1 and e_1 methyl esters. Angew Chem Int Ed. 2013;52:3450–2.

[48] Umemiya S, Nishino K, Sato I, Hayashi Y. Nef reaction with molecular oxygen in the absence of metal additives, and mechanistic insights. Chem Eur J. 2014;20:15753–9.

[49] Biellmann J-F. Enantiomeric steroids: synthesis, physical, and biological properties. Chem Rev. 2003;103:2019–34.

[50] (a) Kametani T, Nemoto H Recent advances in the total synthesis of steroids via intramolecular cycloaddition reactions. Tetrahedron 1981, 37, 3–16. (b) Chapelon A S, Moralëda D, Rodriguez R, Ollivier C, Santelli M. Enantioselective synthesis of steroids. Tetrahedron. 2007;63:11511–616. (c) Gupta P, Panda G. Asymmetric assembly of steroidal tetracyclic skeletons. Eur J Org Chem. 2014;8004–19.

[51] See the references cited therein Koshino S, Kwon E, Hayashi Y. Total synthesis of estradiol methyl ether and its five-pot synthesis with an organocatalyst. Eur J Org Chem. 2018;5629–38.

[52] Funk RL, Vollhardt KPC. The cobalt way to dl-estrone, a highly regiospecific functionalization of 2,3-bis(trimethylsilyl)estratrien-17-one. J Am Chem Soc. 1979;101:215–17.

[53] Ananchenko SN, Torgov IV. New syntheses of estrone, d,1-8-iso-oestrone and d,1-19-nortestosterone. Tetrahedron Lett. 1963;4:1553–8.

[54] Yeung Y-Y, Chein R-J, Corey EJ. Conversion of torgov's synthesis of estrone into a highly enantioselective and efficient process. J Am Chem So. 2007;129:10346–7.

[55] (a) Hajos ZG, Parrish DR German Patent DE 2102623. July 29, 1971. (b) Hajos Z G, Parrish D R. Asymmetric synthesis of bicyclic intermediates of natural product chemistry. J Org Chem. 1974;39:1615–21.

[56] (a) Eder U, Sauer G, Wiechert R German Patent DE 2014757, Oct 7, 1971 (b) Eder U, Sauer G, Wiechert R. New Type of asymmetric cyclization to optically active steroid CD partial structures. Angew Chem Int Ed Engl. 1971;10:496–7.

[57] Bradshaw B, Bonjoch J. The wieland–miescher ketone: a journey from organocatalysis to natural product synthesis. Synlett. 2012;23:337–56.

[58] Jhuo D-H, Hong B-C, Chang C-W, Lee G-H. One-pot organocatalytic enantioselective domino double-michael reaction and pictet-spengler–lactamization reaction. a facile entry to the "inside yohimbane" system with five contiguous stereogenic centers. Org Lett. 2014;16:2724–7.

[59] Halskov KS, Donslund BS, Barfüsser S, Jørgensen KA. Organocatalytic asymmetric formation of steroids. Angew Chem Int Ed. 2014;53:4137–41.

[60] Prévost S, Dupré N, Leutzsch M, Wang Q, Wakchaure V, List B. Catalytic asymmetric Torgov cyclization: a concise total synthesis of (+)-estrone. Angew Chem Int Ed. 2014;53:8770–3.

[61] Hayashi Y, Koshino S, Ojima K, Kwon E. Pot economy in the total synthesis of estradiol methyl ether by using an organocatalyst. Angew Chem Int Ed. 2017;56:11812–15.

[62] Mukaiyama T, Ogata K, Sato I, Hayashi Y. Asymmetric organocatalyzed michael addition of nitromethane to a 2-oxoindoline-3-ylidene acetaldehyde and the three one-pot sequential synthesis of (-)-horsfiline and (-)-coerulescine. Chem Eur J. 2014;20:13583.
[63] Jossang A, Jossang P, Hadi HA, Sévenet T, Bodo B. Horsfiline, an oxindole alkaloid from horsfieldia superba. J Org Chem. 1991;56:6527–30.
[64] Anderton N, Cockrum PA, Colegate SM, Edgar JA, Flower K, Vit I, et al. Oxindoles from *Phalaris Coerulescens*. Phytochemistry. 1998;48:437–9.

Ierasia Triandafillidi, Errika Voutyritsa
and Christoforos G. Kokotos

2 Recent advances in reactions promoted by amino acids and oligopeptides

Abstract: During the last 20 years, Organocatalysis has become one of the major fields of Catalysis. Herein, we provide a recent overview on reactions where the use of amino acids and peptides as the organocatalysts was employed. All aspects regarding aldol reactions, Michael reactions, epoxidation, Henry reactions and many others that are crucial for the reaction conditions and reaction mechanisms are discussed.

Keywords: organocatalysis, proline, amino acids, peptides

2.1 Introduction

Traditionally, asymmetric catalysis is directly linked to transition metal catalysis, which has the lion's share on literature regarding novel reactions. Another unique branch of asymmetric catalysis is biocatalysis, the use of enzymes to promote asymmetric transformations. Although transition metal-catalysis and biocatalysis are well established, the use of "pure" organic molecules as catalysts was neglected for long, although it turned out to be an additional efficient tool for the synthesis of chiral building blocks. Nowadays, asymmetric organocatalysis is one of the most powerful tools for the synthesis of enantiopure molecules [1]. The first organocatalytic reaction was reported about one century ago, but till the end of last century, the field of organocatalysis did not flourish. In 2000, MacMillan named this new field [2], which became quickly the hottest research area in organic synthesis. In the last few years, organocatalysis has experienced an explosive growth with many organocatalysts being discovered [3]. Amino acids possess one of the most important place in this field, employed in a variety of organic synthetic reactions, and are very popular and well-explored catalysts [4]. Amino acids have the advantage of being cheap and can be used without any specialized approaches to perform catalytic reactions in a highly enantioselective manner. They can be available in both enantiomeric forms and different derivatives can provide complimentary selectivities. In this overview, the authors would like to put forward the importance of

This article has previously been published in the journal Physical Sciences Reviews. Please cite as: Ierasia Triandafillidi, I., Voutyritsa, E., Kokotos, C. Recent advances in reactions promoted by amino acids and oligopeptides *Physical Sciences Reviews* [Online] 2020, 11. DOI: 10.1515/psr-2018-0086

https://doi.org/10.1515/9783110590050-002

using such metal-free organic catalysts in organic chemistry. Numerous reaction types for the formation of C–C bonds are presented, as well as reactions with potential applications in chemical industry.

2.1.1 α-Functionalization of carbonyl compounds

2.1.1.1 Asymmetric aldol reaction

The asymmetric aldol reaction is one of the most important topics in modern asymmetric synthesis that has been used for carbon–carbon bond formation and has been developed widely in the last few decades [5]. In 2000, List, Barbas and Lerner reported the first direct aldol reaction of aldehydes with acetone, utilizing (*L*)-proline as the organocatalyst (Figure 2.1) [6].

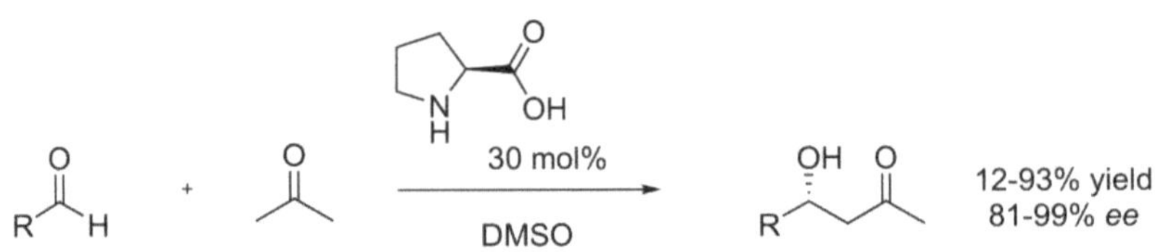

Figure 2.1: The first asymmetric organocatalytic aldol reaction.

Afterwards, a variety of organocatalysts, based on amino-acids' skeleton, has been synthesized for the aldol reaction [7]. Proline, being an amino acid that contains a secondary amine, immediately surfaced as a potential catalyst that can couple with the ketone forming the desired enamine, which acts as the nucleophile and attacks the electrophilic aldehyde (Figure 2.2).

In most cases, the acidic group of proline can be replaced by other moieties, which provide additional interactions with the substrates, providing the desired levels of stereocontrol in the reaction. A number of groups have been successfully combined with the proline scaffold and most of these functionalities are capable of acting as hydrogen-bond donors. This notion is highlighted by the high catalytic efficiency of prolinamides. Examples of prolinamides that have an additional hydroxy-group for additional hydrogen bond have been developed (Figure 2.3) [8].

In 2011, Kokotos and co-workers have developed a series of prolinamides carrying a (thi)ourea group [9]. The aim of that study was to combine either a prolinamide unit or an α-amino acid moiety with a chiral unit bearing the thiourea group, because it is a well-known double hydrogen-bond donor and chiral thioureas are known to be an important class of organocatalysts. The presence of an acidic additive improved both the yield and the enantioselectivity, since it stabilized the transition state with three hydrogen bonds, leading to excellent yields and *ee*s in the

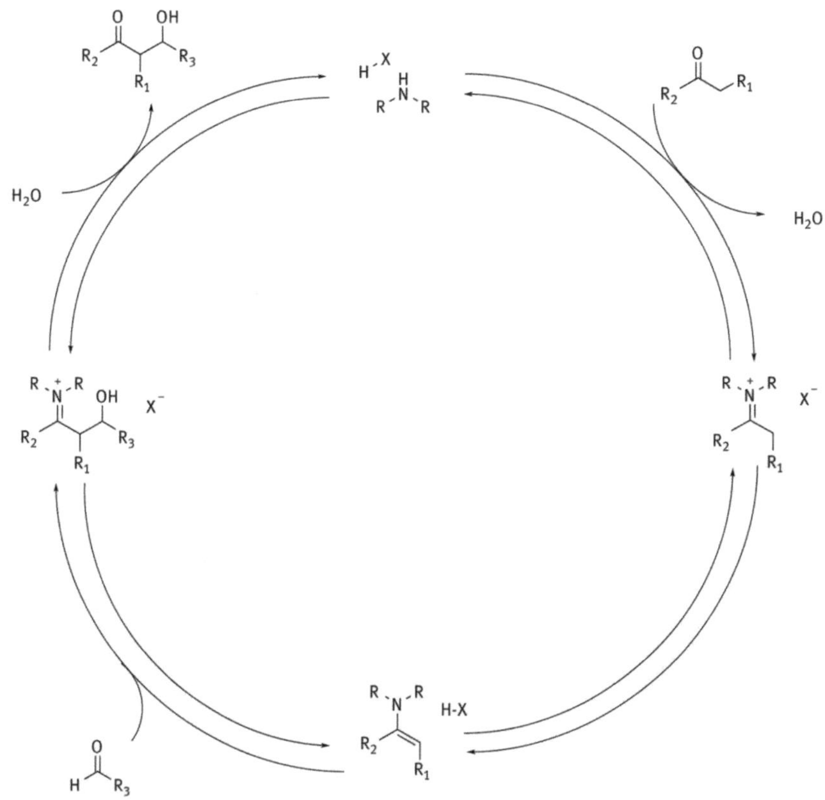

Figure 2.2: Enamine-catalytic cycle with secondary amines.

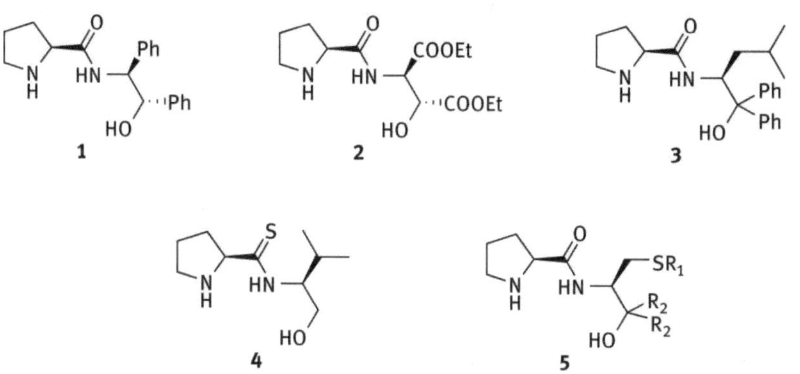

Figure 2.3: A number prolinamides as organocatalysts.

aldol reaction (up to 99 % *ee*) (Figure 2.4). Selected examples of the substrate scope are presented below (Figure 2.5).

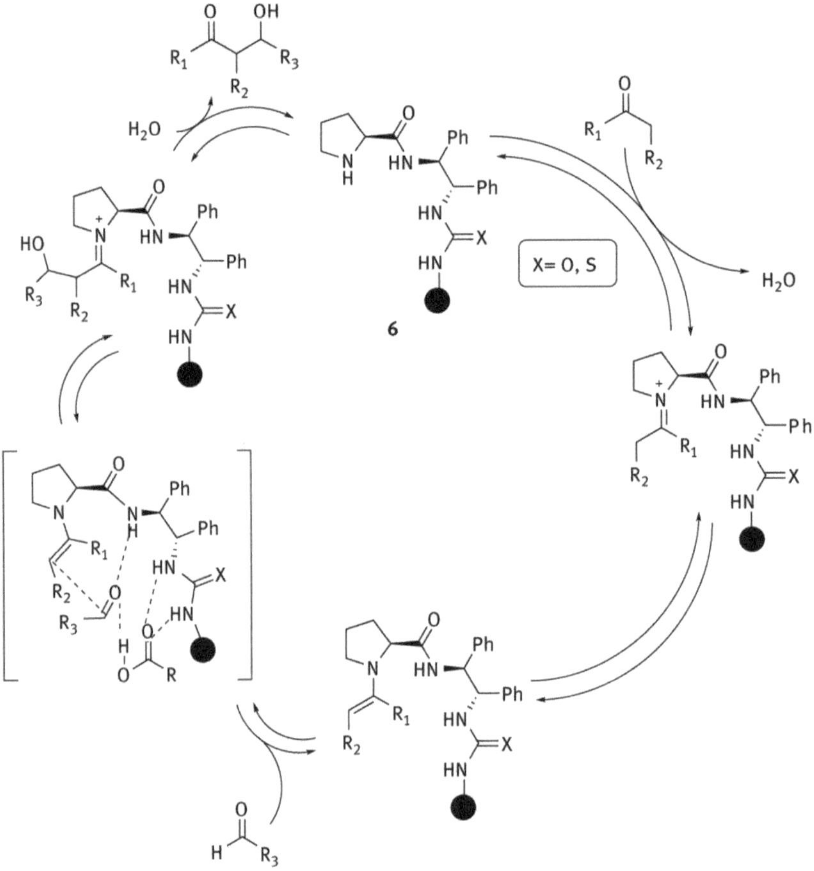

Figure 2.4: Kokotos' prolinamides and proposed mechanism.

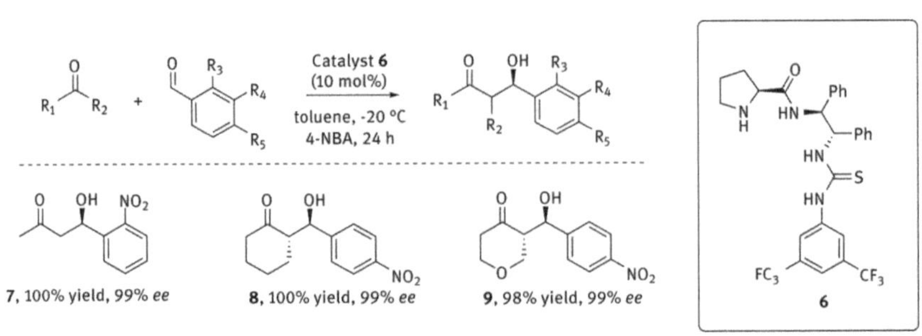

Figure 2.5: Selected examples of the substrate scope.

Juaristi and co-workers became interested in the preparation and use of derivatives that would be anchored to a resin in order to improve the reaction conditions, and to allow reuse of the catalyst, fulfilling one of the requirements of sustainable chemistry. Therefore, they synthesized a number of proline-derived organocatalysts, bound to resins, and used them in the asymmetric aldol reaction, affording the desired products in very good enantiomeric excesses (Figure 2.6) [10]. A simple hydrogen bond from the amide of the catalyst with the electrophile guides the addition for the stereoselectivity. Along the same lines, Fulop and co-workers used a resin-based proline oligopeptide for the aldol reaction [11].

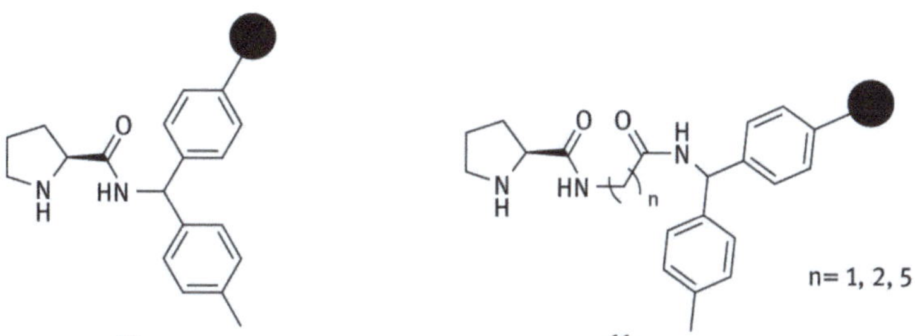

10 11

Figure 2.6: Organocatalysts anchored to resin and the transition state.

Replacement of the carboxylic acid moiety of proline with suitably positioned chelated Lewis acids offers the opportunity to access alternative product stereochemistries, and could additionally provide improved reactivity over monofunctional catalalysts. At the basis of this data, Dockendorff and co-workers replaced proline's carboxylic group with multifunctional heterocyclic scaffolds, developing a new category of organocatalysts, in order to provide better affinity with a Lewis acid additive (Figure 2.7) [12].

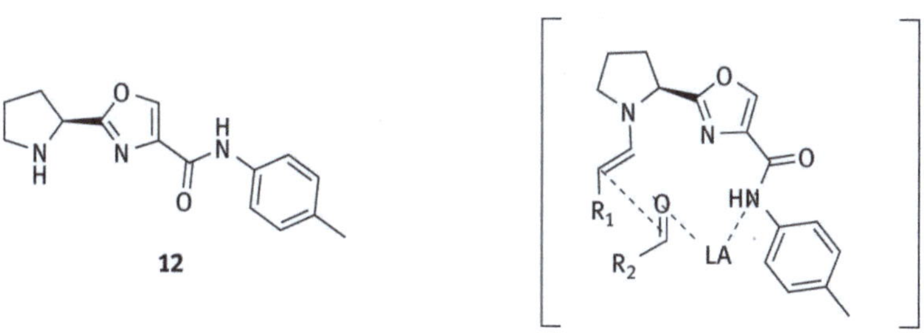

Figure 2.7: Organocatalysts bearing heterocyclic scaffolds.

In 2013, Benaglia synthesized a variety of effective organocatalysts, based on proline, affording the desired products in very good yields and excellent *ees* (Figure 2.8) [13]. These catalysts, containing a heterocyclic NH group, offer the possibility of forming hydrogen bonds not only by the NH of the amide group, but also by the NH+ of the protonated pyridine ring. The enamine was formed by the reaction of cyclohexanone with the secondary amine of the pyrrolidine ring, while the protonated part of the catalyst can interact with the electrophile in two different manners, and thus, can adopt two different conformations when coordinated with the aldehyde. The proposed mechanism is presented in Figure 2.9.

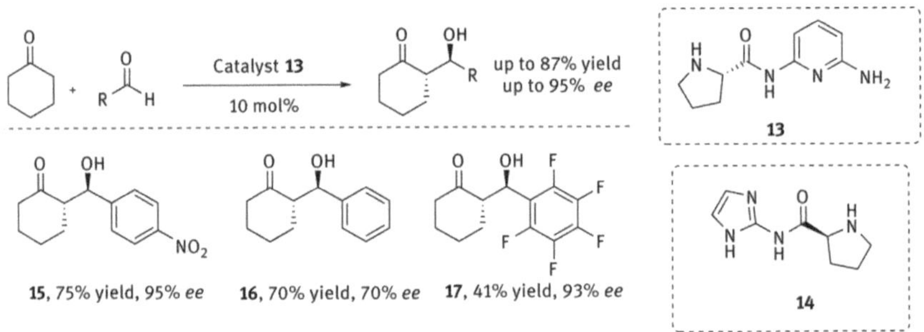

Figure 2.8: Organocatalysts bearing heterocyclic moieties.

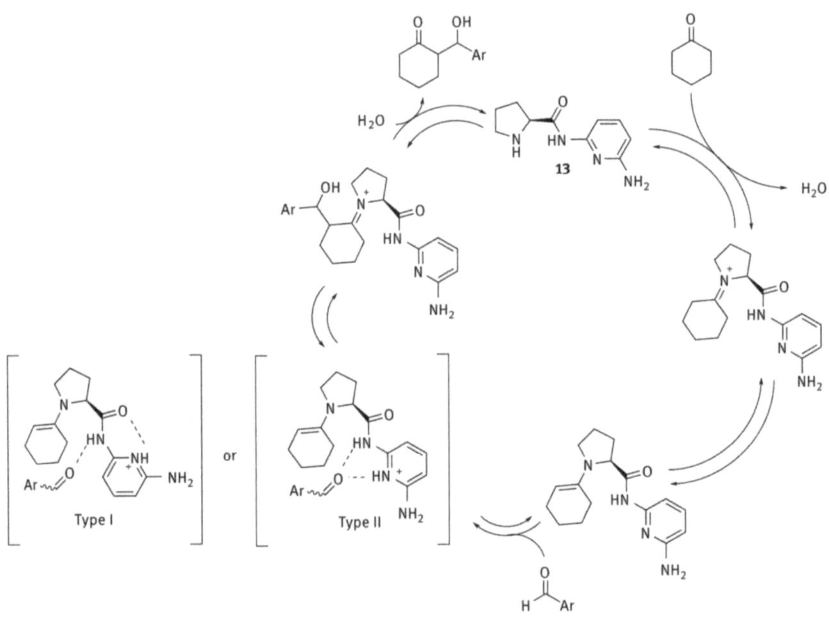

Figure 2.9: Proposed mechanism.

Very recently, Illa and co-workers synthesized unnatural cyclobutyl-based oligopeptides as catalysts. The remarkable conformation stability of their structure leads to the formation of intra- and inter-residue hydrogen bonds in peptides that incorporate either cis-cyclobutane β-amino acids or cis-cyclobutane γ-amino acids. They have employed cyclobutyl-based amino acids in proline-based catalysts, aiming to provide a specific spatial arrangement of the catalyst, in order to achieve high yields and stereoselectivities (Figure 2.10). The proposed reaction mechanism is presented in Figure 2.11 [14].

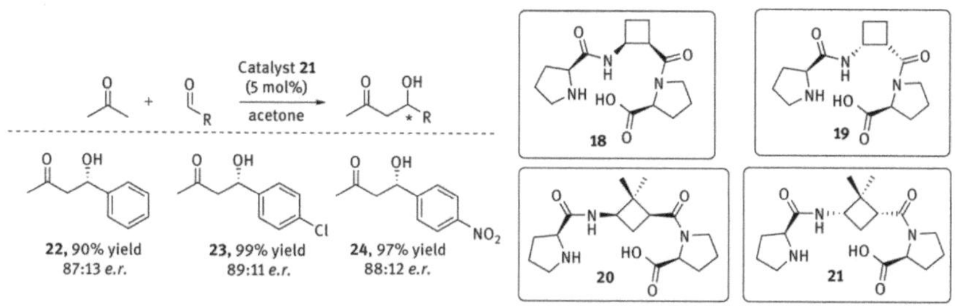

Figure 2.10: Organocatalysts bearing a cyclobutyl-group.

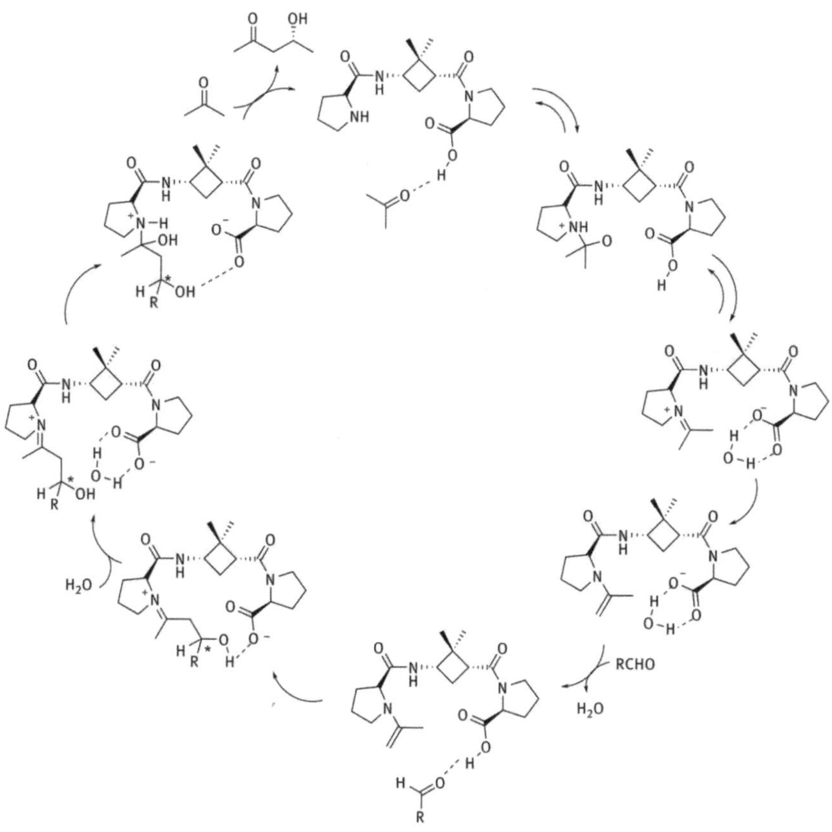

Figure 2.11: Proposed mechanism.

Phosphoramides are known to be active catalysts for different transformations. Based on this, the Benaglia group proposed (*S*)-proline-derived phosphoramides as the organocatalysts for the stereoselective aldol reaction of activated thioesters, aiming to improve the versatility and chemical and stereochemical efficiency of previous methodologies in the aldol reaction. According to the authors, a chiral cationic hypervalent hexacoordinated silicon species was formed, with the two chlorine atoms in the apical positions. Both enolate and aldehyde are coordinated to the silicon atom. The parallel-offset orientation of the phenyl rings of the aldehyde and of the enolate, and the T-shaped arrangement of the phenyls of the enolate and of the phosphoroamide seem to play the major role in determining the reaction stereoselection (Figure 2.12) [15].

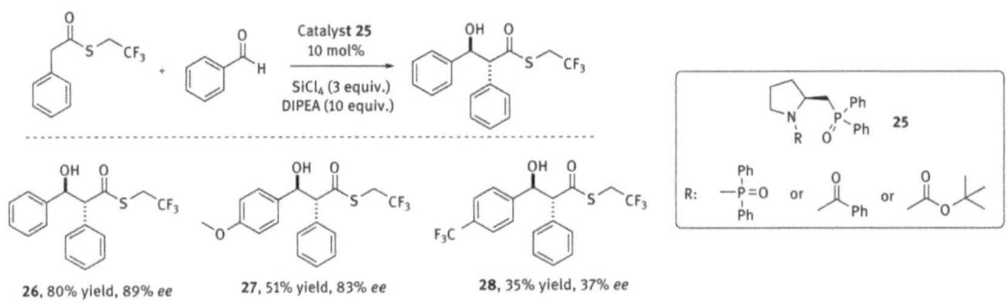

Figure 2.12: (*S*)-Proline-derived phosphoramide as the organocatalyst.

In 2018, Juaristi and co-workers, trying to generate a chiral *C2*-symmetric, helix-type conformation in the catalyst, inserted a chiral phosphoramide group, which creates a big blocking area. These structural characteristics were anticipated to induce efficient enantiodiscrimination, by selectively blocking one of the enantiotopic faces of the prochiral carbonyl group in isatin or in aldehyde leading to high *ee*s (Figure 2.13) [16]. Plausible transition states for the enamine-catalyzed addition of cyclohexanone to isatin are presented in Figure 2.14.

Additionally, proline has been used as a linker for the synthesis of oligopeptides, which can be used in the aldol reaction providing excellent results [17]. Also, in 2016, Benaglia studied extensively the kinetics and thermodynamics effects for the aldol reaction catalyzed by proline [18]. In that study, the reversibility of the reaction was proved using Reaction Progress Kinetic Analysis (RPKA). The obtained kinetic law, being applicable to a wider range of substrates, could be implemented even when the reaction does not provide quantitative conversions. Moreover, Linear Free Energy Relationships (LFERs) was used to investigate the equilibrating phenomenon and rationalize its effect on the aldol reaction outcome.

When 4-hydroxy-proline was used as the starting material for the synthesis of catalysts, this new chiral center plays a different role in the reaction. Zlotin's and Liebscher's groups protected the hydroxy group with alkyl chains, which had heterocyclic rings [19]. The two groups made no mention of the opposite enantiomer of 4-OH proline that could lead to new potential catalysts for the aldol reaction.

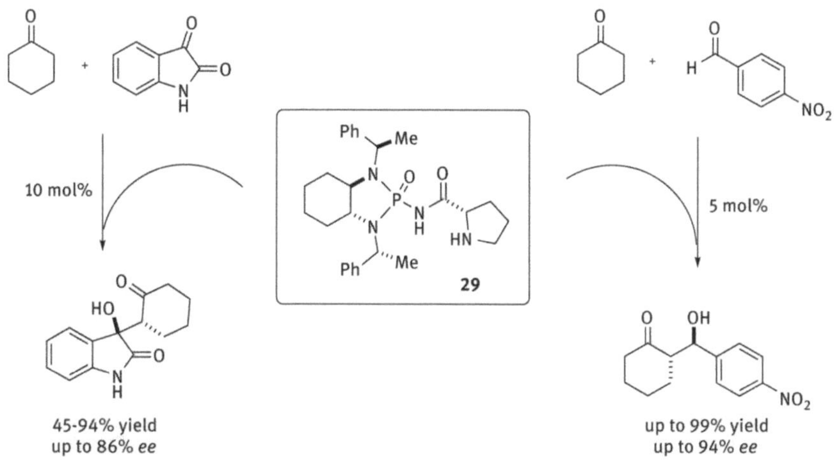

Figure 2.13: Organocatalysts bearing a phosphoramide group.

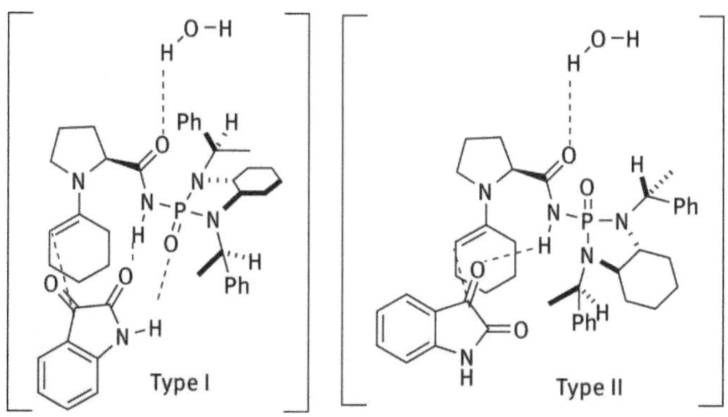

Figure 2.14: Proposed transition states.

Besides proline-based catalysts, a variety of organocatalysts have been synthesized derived from other amino acids. In 2005, Córdova utilized alanine, one of the simplest amino acids, in order to achieve high stereoselectivity in intermolecular aldol reactions [20].

Some years later, the Miura group achieved the synthesis of *anti*-aldol products in excellent yields, using β-aminosulfonamide **30** as the organocatalyst (Figure 2.15) [21]. The proposed transition state is presented in Figure 2.16. The primary amino group is employed to form the nucleophilic enamine, while the acidic NH of the triflate ensures a hydrogen-bond interaction with the electrophile, providing high levels of selectivity.

3-Substituted-3-hydroxy-2-oxindoles are heterocyclic organic compounds that constitute desirable synthetic targets as these structures form the core units of many natural products and pharmaceuticals. The most direct method to 3-substituted-3-hydroxy

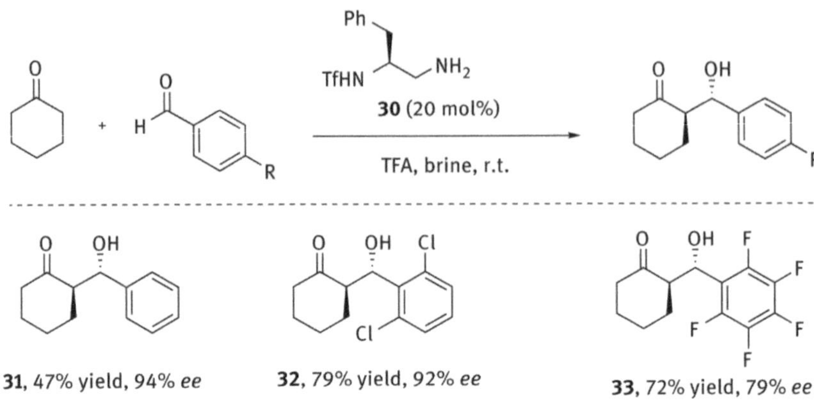

Figure 2.15: Phenylalanine-derivative as the organocatalyst for the aldol reaction.

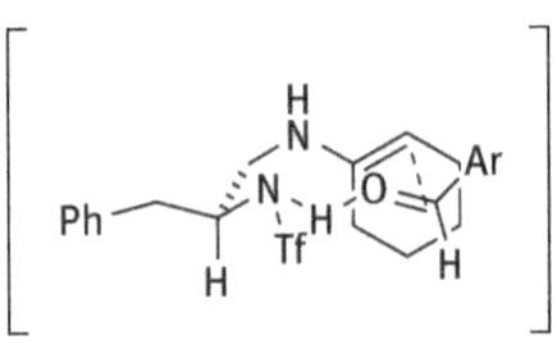

Figure 2.16: Proposed transition state.

oxindoles is a nucleophilic addition of appropriate nucleophiles to isatins, such as the aldol reaction. In 2013, Liu and Xie utilized arginine as the organocatalyst for the aldol reaction of *α*,*β*-unsaturated ketones and isatins, leading to the oxindole-skeleton in excellent yields (Figure 2.17) [22]. The proposed reaction mechanism is presented in Figure 2.18.

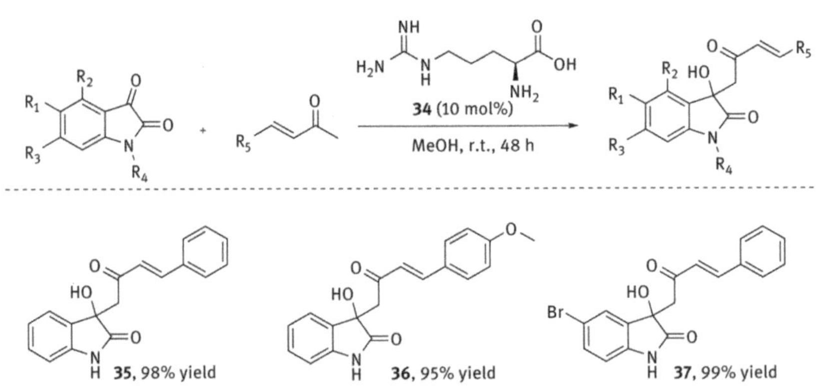

Figure 2.17: Arginine as the organocatalyst for the aldol reaction.

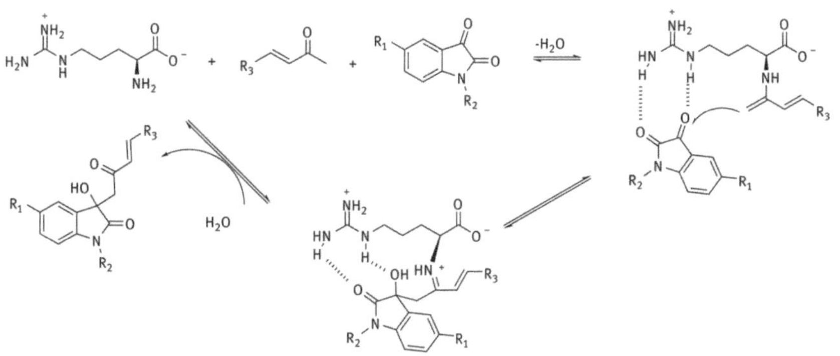

Figure 2.18: Proposed mechanism.

One year later, Acevedo investigated the aldol reaction between benzaldehyde and acetone using QM/MM Monte Carlo calculations, in order to clarify the enamine mechanism and the intermolecular interactions responsible for the amine catalysis and enhanced enantioselectivity using water conditions and the aldolase antibody 33F12 [23]. In the key-step of the mechanism, lysine-skeleton played decisive role for the enamine formation (Figure 2.19).

Furthermore, valine has been extensively studied as the organocatalyst [24]. In this case, the *iso*-propyl side chain plays an important role for the catalysis in the aldol reaction whereas electron withdrawing moieties were introduced to enhancing the amide NH acidity in the hope of improving the stereocontrol of the aldol reaction.

A different category in aldol organocatalysis is the use of amino acids that possess nucleophilic side-chains. Thus, serine with protected side-chain catalyzes the aldol reaction providing very good results [25].

Fu and co-workers took advantage of the organosulfur side-chain of cysteine and they synthesized an amino-acid-catalyst for the aldol reaction [26]. This characteristic of cysteine's side-chain has been used as an advantage for the synthesis of thiazolidine-based organocatalysts (Figure 2.20) [27]. The proposed transition state is presented in Figure 2.21.

A variety of threonine-derived catalysts have been also synthesized, anchored to several materials and polymers [28]. Leucine, *iso*-leucine and *tert*-leucine, possessing an hydrophobic moiety, provide catalysts with different features [29].

2.1.1.2 Asymmetric Mannich reaction

The Mannich reaction is an extensively studied and significant tool in the synthesis of β-amino ketones. Organocatalyzed asymmetric variants of this reaction were developed in the early 2000s, primarily utilizing proline and its derivatives as the organocatalysts [30].

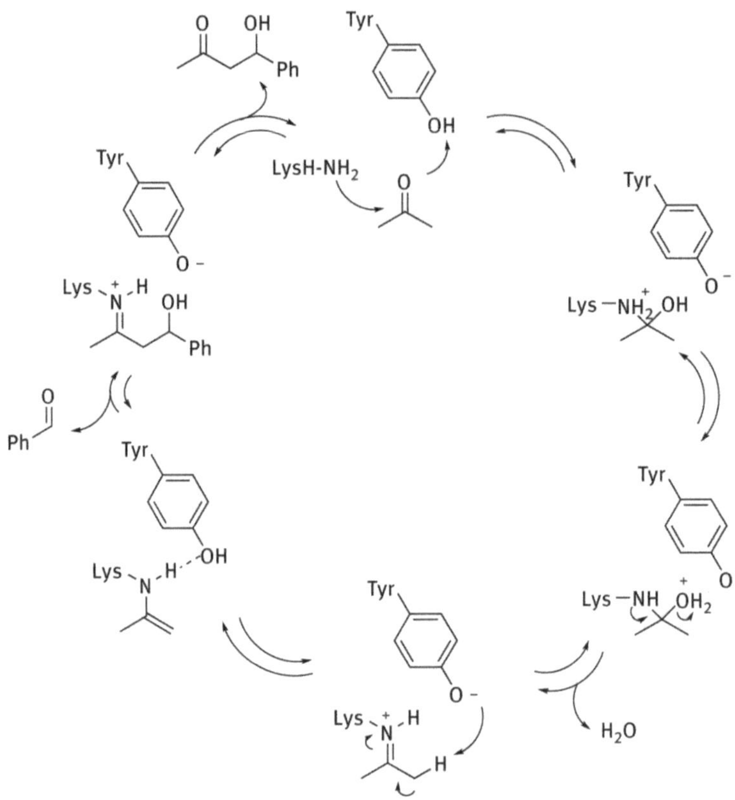

Figure 2.19: Proposed mechanism.

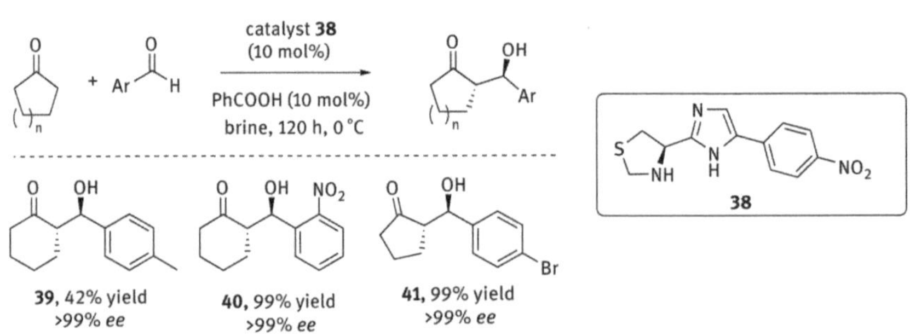

Figure 2.20: Cysteine-based organocatalysts for the aldol reaction.

Tao and co-workers utilized 4-hydroxy-proline as the starting material for the synthesis of isosteviol–proline conjugates. Aiming to develop a highly enantioselective Mannich reaction, this group synthesized catalysts based on proline, containing the chiral cavity of isosteviol, and tested them in the Mannich reaction [31].

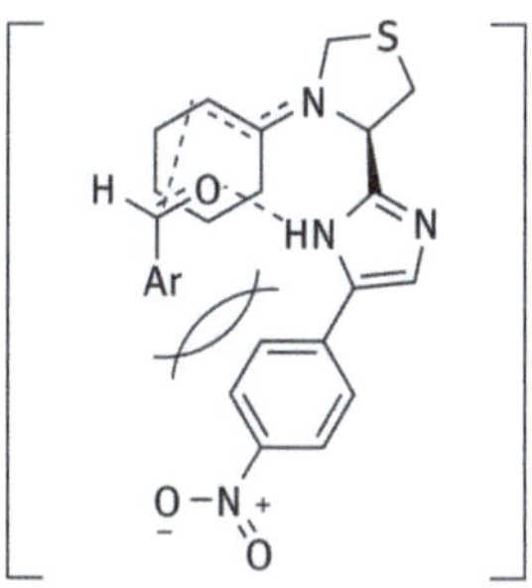

Figure 2.21: Proposed transition state.

These molecules catalyzed the Mannich reaction, in the presence of water, besides the bulky hydrophobic group (Figure 2.22). The proposed transition state is presented in Figure 2.23. The secondary amine of proline forms the nucleophilic enamine, while the carboxylic group forms the required hydrogen bond with the electrophile. The role of the steviol side chain is to embed the aryl moiety of the imine providing the appropriate orientation.

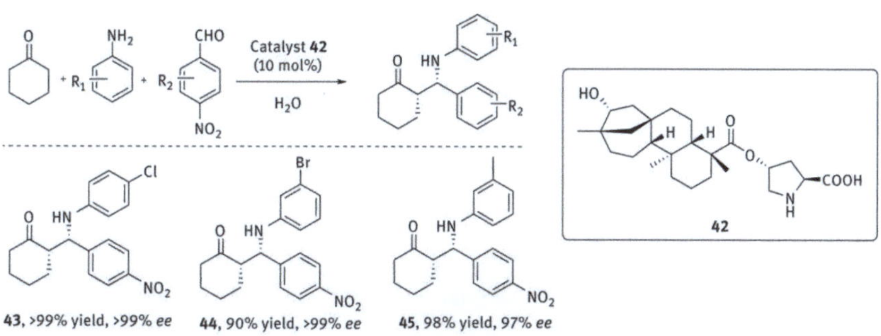

Figure 2.22: Mannich reaction catalyzed by substituted-4-hydroxy-proline.

Also, a variety of primary amino-catalysts have been tested. Córdova and his group reported an asymmetric Mannich reaction, utilizing α-amino acids as the organocatalysts [32]. Based on this research, alanine, valine and serine catalyzed efficiently the Mannich reaction, affording the desired β-amino ketones in excellent enantioselectivity.

Amino acids with other side-chains have also been extensively researched (Ser, Cys, Thr). Thus, siloxy-protected serine [33], cysteine derivative bound to ferrite magnetic nanoparticles [34] and threonine either protected [35] or immobilized on polymer support [36] have been employed in the Mannich reaction providing excellent results. Amino acid derivatives as chiral stuctures and hydrogen bonding donors can lead to a variety of structurally variable chiral ammonium salts, containing multiple hydrogen-bonding donors. These novel quaternary ammonium salts can be used as

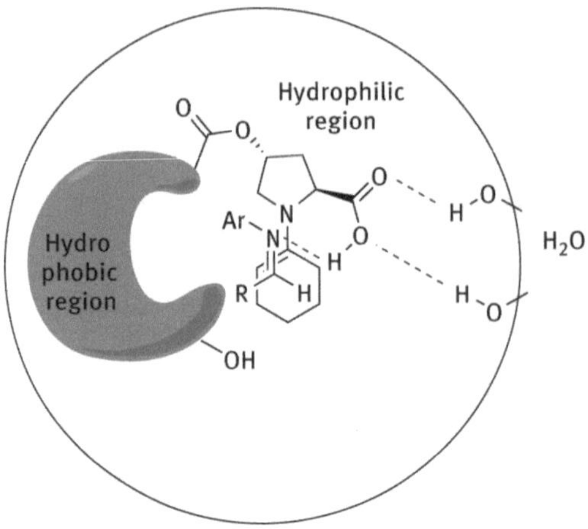

Figure 2.23: Proposed transition state.

effective phase transfer catalyst for some conventional and challenging asymmetric reactions. At the basis of these data, Duan and co-workers presented a variety of α-amino-acid-derived phase transfer catalysts for the nitro-Mannich reaction [37]. Thus, starting from *α*-amino alcohols, a series of bulky catalysts have been synthesized and catalyzed the desired reaction, leading to very good selectivities (Figure 2.24). Mechanistic experiments proposed that there is a synergistic catalysis and indicated that both the hydroxyl group on the phenylglycinol moiety and the quaternary ammonium center are crucial to achieve excellent catalytic activity and stereoselectivity in this asymmetric nitro-Mannich reaction.

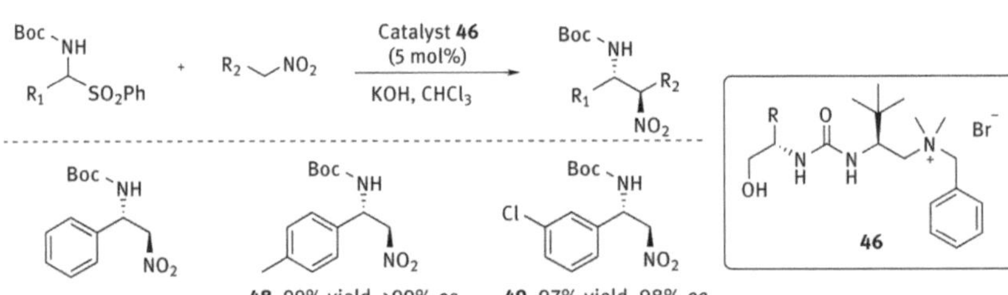

Figure 2.24: Nitro-Mannich reaction catalyzed by an *α*-amino alcohol-based catalyst.

Finally, Bhadury and co-workers synthesized *β*-amino acids, in order to catalyze the Mannich reaction [38]. *β*-Amino esters have been isolated in excellent yield and enantioselectivity utilizing *β*-amino acid **50** (Figure 2.25). It has to be noted that these types of catalysts provide the opposite diastereoselectivity than proline-derived catalysts [methyl group blocks the enamine formation towards the opposite side of the carboxylic group (enamine formed in the same side as the carboxylic group), vs what happens in proline (enamine *E*- to the carboxylic group)].

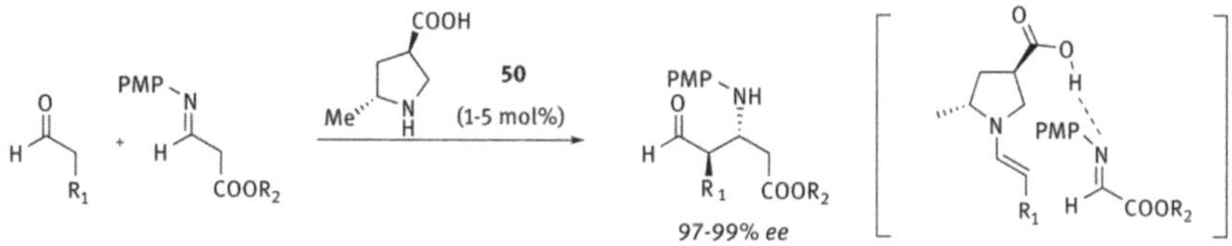

Figure 2.25: Mannich reaction with β-amino acid as the catalyst.

2.1.1.3 Asymmetric *α*-fluorination reaction

Fluorine occupies a very special position as an element in organic synthesis. Fluorine as a substituent has highly advantageous effects; thus, there is a truly intense effort into the synthesis and characterization of novel fluorine-containing compounds. Furthermore, the introduction of fluorine in a diastereo- and enantio-controlled manner presents enormous challenges, especially under catalytic conditions. Organocatalytic methods for the *α*-fluorination of carbonyl compounds have enjoyed an explosive growth in recent years. In 2005, Enders and Huttl published the first example of a purely organocatalytic fluorination process [39]. Utilizing Selectfluor as the fluorine source and proline-derivatives as the organocatalyst succeeded in the *α*-fluorination of aldehydes in good yields, but with very low enantioselectivity.

By the time that work was published, three more organocatalytic enantioselective *α*-fluorination reactions were published. Jørgensen [40], Barbas [41] and MacMillan [42] described highly enantioselective *α*-fluorination of aldehydes, using a variety amines or amines salts as catalysts and *N*-fluorobenzenesulfonimide (NFSI) as the fluorine source.

In 2014, Toste and co-workers merged two separate chiral catalytic cycles: a chiral anion phase-transfer catalytic cycle to activate Selectfluor and an amino acid-based catalytic cycle via enamine intermediate [43]. Given that the combination of transition metal catalysts or Lewis acids with organocatalysts has been already used as a promising strategy for developing new and synthetically useful reactions, the Xu group investigated whether the enamine intermediate can be activated by Lewis acids, Brønsted acids, metal complexes, Lewis bases or other additives in order to change the electronic and bulky effect of β-ketoester in fluorination. At the basis of their data, the combinational use of L-leucine and a cinchona-based catalyst constitutes a novel dual organocatalyst with promising catalytic activity for the Selectfluor activation [44] (Figure 2.26). The proposed reaction mechanism is proposed in Figure 2.27.

Yamamoto [45] and Shibatomi [46] also utilized proline-derivatives in order to insert fluorine in *α*-chloroaldehydes utilizing NFSI as reagent.

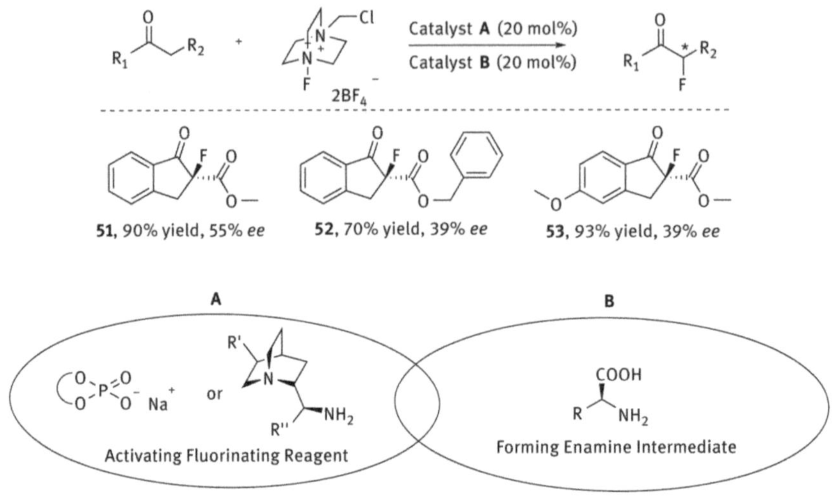

Figure 2.26: α-Fluorination of carbonyls merging two catalytic cycles.

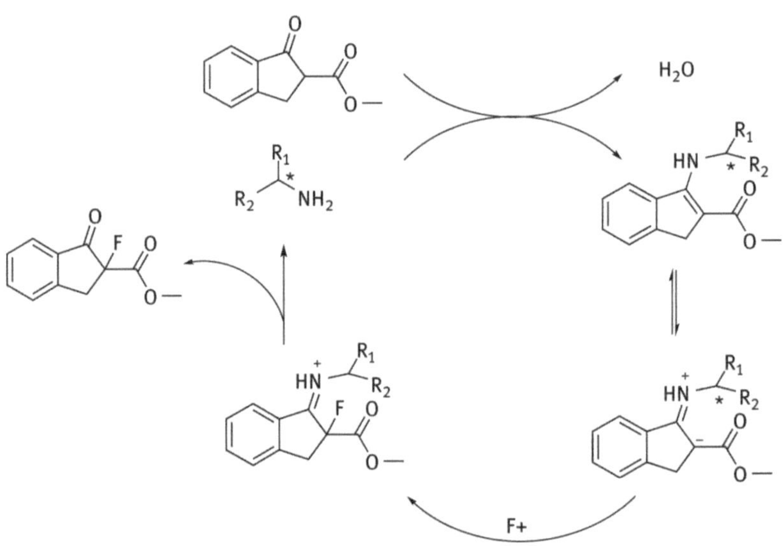

Figure 2.27: Proposed mechanism.

2.1.1.4 Other α-functionalization reactions

The direct enantioselective introduction of a stereogenic carbon–heteroatom bond adjacent to α-carbonyl functionality has been extensively studied. Azodicarboxylate esters have been used as the heteroatom source, in order to form a new C–N bond. Proline is one of the most popular organocatalysts for this reaction, leading to excellent yields and enantioselectivities [47].

In 2013, Kokotos utilized *β*-tert-butyl aspartate as the organocatalyst for the *α*-heteroatom functionalization of α,α-disubstituted aldehydes ([48]a), enhancing the knowledge that primary amine can form the desired enamine, better than secondary amines (Figure 2.28). Before that contribution, three publications from different groups had already reported the use of primary amine catalysts that efficiently catalyzed the reaction between α,α-disubstituted aldehydes and azodicarboxylates ([48]b–d). The proposed transition state is presented in Figure 2.29.

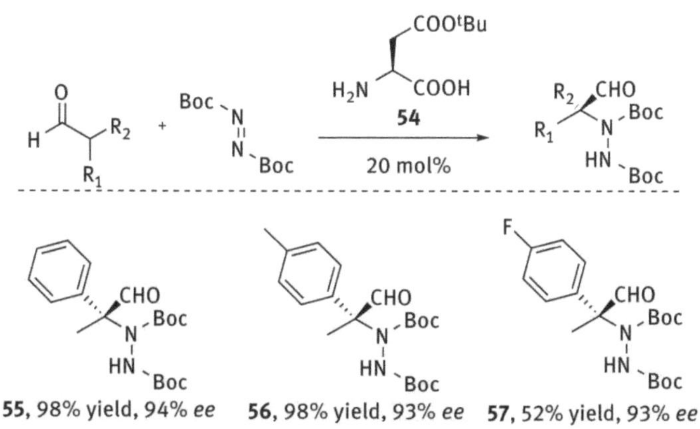

Figure 2.28: *a*-Functionalization of *a*,*a*-disubstituted aldehydes catalyzed by primary amino acids.

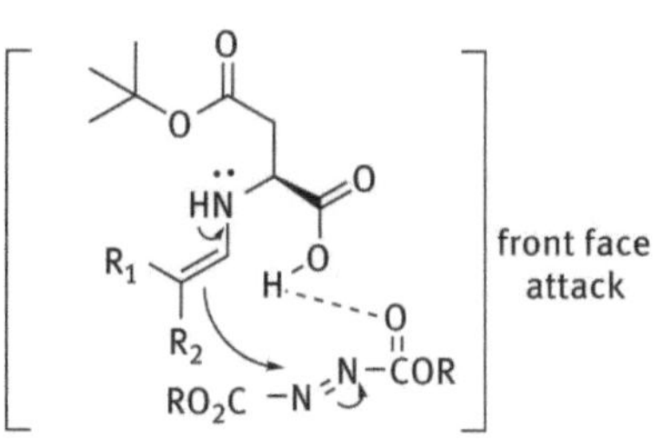

Figure 2.29: Proposed transition state.

In 2016, Wennemers synthesized a tripeptide organocatalyst, in order to achieve the Michael addition between aldehydes and maleimides (Figure 2.30, **A**) [49], and one year later achieved the additional reaction of acetophenones to dicyanoolefins, utilizing the same catalyst's skeleton (Figure 2.30, **B**) [50]. In both cases, enamine is formed between the carbonyl group and the pyrrolidine nitrogen, while the Michael acceptor is activated via hydrogen bonding.

Additionally, the Morita-Baylis-Hillman reaction can be defined as an *a*-functionalization reaction of *α*,*β*-unsaturated carbonyl compounds. In 2009, Wu and co-workers synthesized a new type of chiral bifunctional phospinothioureas derived

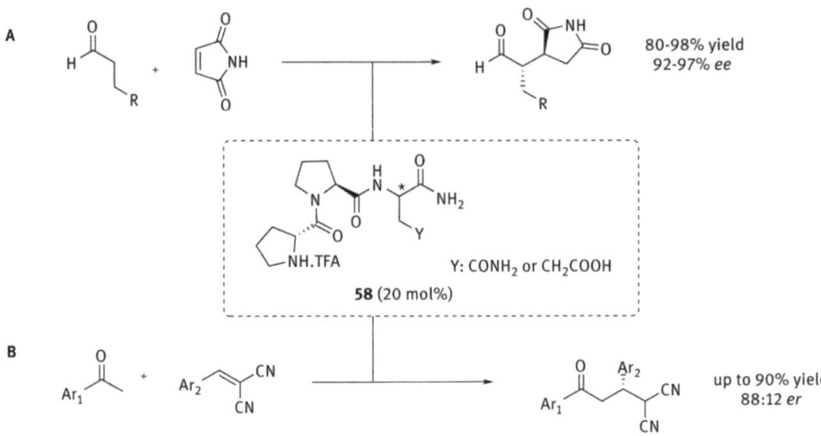

Figure 2.30: α-Functionalization of carbonyls catalyzed by oligopeptides.

from *L*-valine [51]. Thus, aromatic aldehydes reacted with acrylates, affording the desired products in very good selectivity. In 2015, Vesely and co-workers synthesized a bifunctional (thio)urea-phosphine organocatalyst derived from *D*-glycose and *α*-amino acids, catalyzing the same reaction (Figure 2.31) [52].

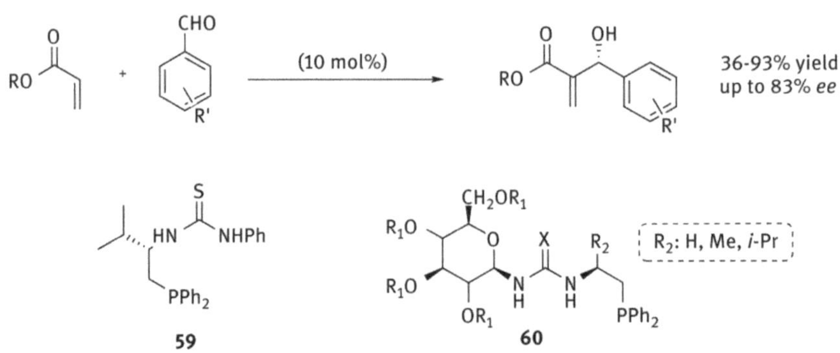

Figure 2.31: Morita-Baylis-Hillman reaction catalyzed by valine-based catalyst.

(L)-Threonine-derived phosphine sulfonamide **61** accomplished excellent aza-Morita-Baylis-Hillman reaction, affording the desired products in high yields with excellent enantioselectivities [53]. For the *aza*-Morita-Baylis-Hillman reaction, a chiral phosphine-phenylalanine-derived catalyst has been synthesized from the Shi group (Figure 2.32, catalyst **62**) [54].

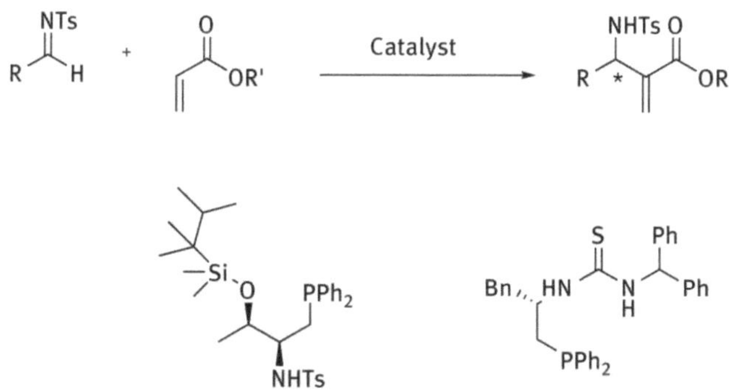

61 (20 mol%)
76-96% yield
88-97% *ee*

62 (20 mol%)
59-92% yield
59-91% *ee*

Figure 2.32: aza-Morita-Baylis-Hillman reaction catalyzed by amino acid-based catalyst.

2.1.2 Olefins as electron acceptors

2.1.2.1 Asymmetric Michael reaction

The asymmetric Michael addition or 1,4-conjugate addition is considered one of the fundamental C–C bond-forming reactions for the construction of chiral *β*-nitro, *β*-carbonyl and several other important structural motifs in organic synthesis. The last two decades have witnessed an explosive growth of asymmetric organocatalysis in the asymmetric Michael addition reaction [5]c, [55]. Among the natural amino acids, proline is the most popular to catalyze the Michael reaction. Starting from (*L*)-proline and inserting a variety of groups on the acid edge, a plethora of organocatalysts have been synthesized. In 2005, Barbas utilized (*L*)-proline to synthesize a catalyst with an additional hydrophobic chain [56], which afforded excellent enantioselectivities in the addition of cyclohexanone to *β*-nitrostyrene. In 2011, Kokotos synthesized a proline-derivative bearing a cyclic thiourea group [57]. This catalyst was used for the enantioselective Michael addition of cyclohexanone to *β*-nitrostyrene, ([57]a) phenyl nitrodiene ([57]b) or changing the nucleophile to 1,4-cyclohexanodione, where a bicyclic product was isolated ([57]c).

Previous studies have already shown that the C6 epimeric catalyst *epi*-PTU can successfully be applied to many asymmetric reactions with good yields and stereoselectivities ([58]a–i). At the basis of this data, Ryu synthesized a series of proline-based organocatalysts, which carried a (thio)urea group, in order to investigate the role of the C6 stereocenter on the catalytic activity of these catalysts in Michael reaction. The results indicated that the relative stereochemistry on the C2 and C6 chiral centers in the organocatalysts are important for successful catalysis (Figure 2.33) ([58]j). All possible diastereomeric catalysts afforded high yields of the product, but the difference was observed in the enantioselection of the product. Then, this study

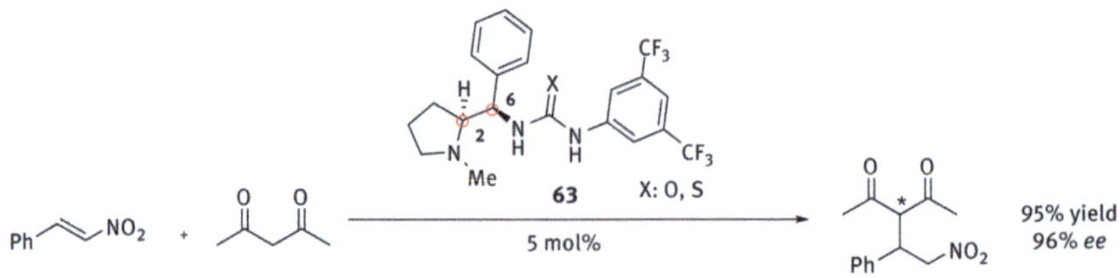

Figure 2.33: Configuration at C2, C6 for the catalytic activity of the proline-derived catalyst.

proves the matching abilities of the correct choice of stereochemistry, which allows the catalyst to adopt the confirmation that leads to high *ees*.

Based on proline, Chimni synthesized a triamine catalyst [59] and Oriyama inserted an acridinium group on the amino acid backbone allowing the formation of hydrogen bonds (Figure 2.34) [60]. Szollosi and co-workers achieved the asymmetric Michael addition catalyzed by amino acids adsorbed on Laponite [61]. The aqueous suspension of Lap RD is basic and hydrogen-bond donor silanol groups are available on the surface. Adsorption of the amino acid may occur both through the secondary amino group and by the carboxylic acid group. The latter may also interact with the charge compensating cations. Surface anchored enamine intermediate is formed by reaction with the aldehyde. These surface intermediates react with nitrostyrene, which may also be adsorbed on the surface.

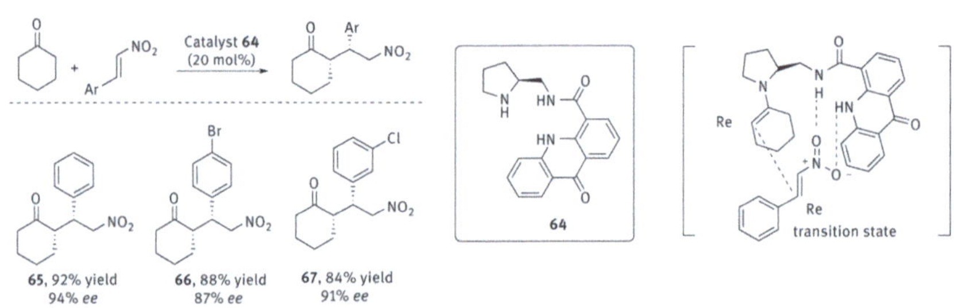

Figure 2.34: Acridinium-based organocatalyst.

Recently, the fluorine effect has been studied in organocatalysis. Fluorine-containing molecules are found to improve the catalytic properties, without any major alterations to the reaction conditions. At the basis of these data, substituting the hydroxy-group of 4-hydroxy-proline, Kokotos synthesized a new organocatalyst with a fluoro-group that was applied in the Michael reaction (Figure 2.35) [62]. It is interesting to note that this catalyst employs brine as the solvent. The fact that this catalyst provided the product in aqueous environment could be explained by hydrogen-bonding interaction between the catalyst and molecules of water.

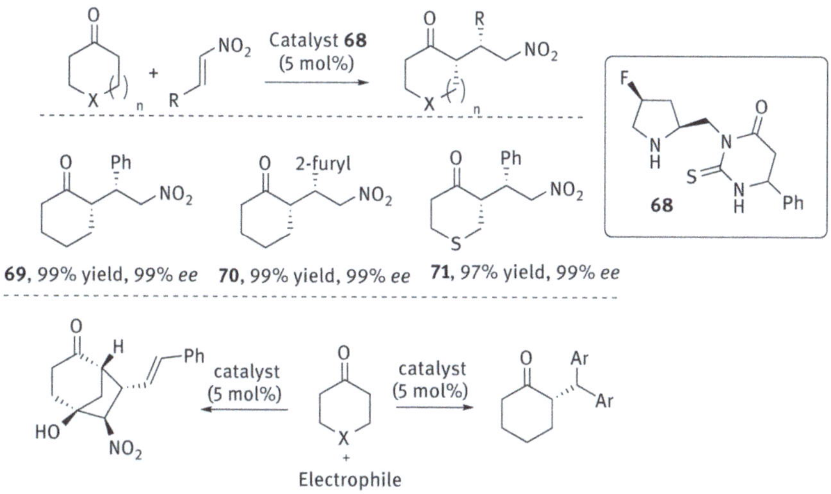

Figure 2.35: 4-Fluoro-proline-based organocatalyst.

Simple acyclic secondary amine derivatives of amino acids could very effectively catalyze asymmetric conjugated additions via iminium ion activation, often giving stereoselectivities as high as those obtained with commercial cyclic organocatalysts, albeit requiring longer reaction times. Such acyclic chiral moieties can be readily incorporated into a rotaxane thread. In 2014, a very interesting work by the Leigh group showed that a phenylalanine-catalytic center in a rotaxane-based switchable asymmetric organocatalyst plays an important role in the position of the macrocycle, affording the desired products in excellent stereoselectivities (Figure 2.36) [63]. The mechanism of the rotaxane is a switching mechanism. The chiral secondary ammonium group in the protonated form is preferred by the macrocycle as a better binding site than the triazolium ring. The macrocycle, in this position, blocks access of reactants to the

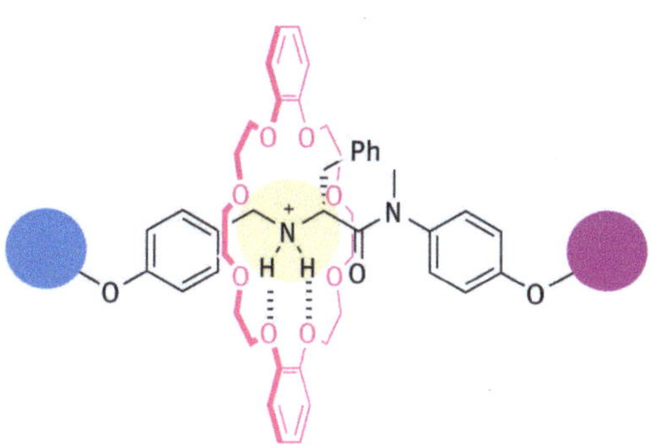

72

Figure 2.36: Rotaxane-based organocatalyst.

catalytic site. When the secondary amine of the rotaxane is not protonated, the triazolium group is the preferred binding site for the macrocycle, leaving open the secondary amine to perform the catalysis.

Pedrosa and co-workers synthesized fullerene-thioureas, derived from natural amino acids, such as valine phenylalanine and *tert*-leucine [64]. The additional products were obtained in excellent yields and selectivities. Additionally, cysteine-derived catalysts have been used in the Michael addition [65] or in intramolecular variants [66]. Isoleucine and *tert*-leucine are widely used as starting materials for the synthesis of organocatalysts for the direct Michael reaction [67], the 1,6-Michael addition [68] and the sulfa-Michael addition [69].

Utilizing amino acids as the starting materials, a variety of oligopeptides have been synthesized, in order to catalyze the Michael addition. In 2011, Wennemers synthesized an oligopeptide, providing the desired addition to nitro-olefins by a variety of aldehydes ([70]a). This reaction's intermediates have been studied by the same group, utilizing ESI-MS as the mechanistic tool ([70]b). One year later, a similar oligopeptide catalyzed the addition of substituted nitro-olefins [71]. In 2013, the same group achieved the formation of tertiary stereocenters with high diastereoselectivities and enantioselectivities, starting from β,β-disubstituted nitro-olefins [72].

Very recently, Wennemers studied the effect of γ-substituted proline derivatives on the performance of the peptidic catalyst **73** (Figure 2.37) [73]. A recent crystal structure of the catalyst shows that the middle proline residue prefers a Cγ-*endo* pucker confirmation. In contrast, the pucker of the (4*R*)-configured proline derivatives with electron-withdrawing groups adopts a Cγ-*exo* pucker confirmation. In the case of 4R-configured proline derivatives, the H-bonded β-turn structure of the peptide is stronger than $n \rightarrow \pi^*$ interaction and cannot stabilize the *trans* conformer of the peptide.

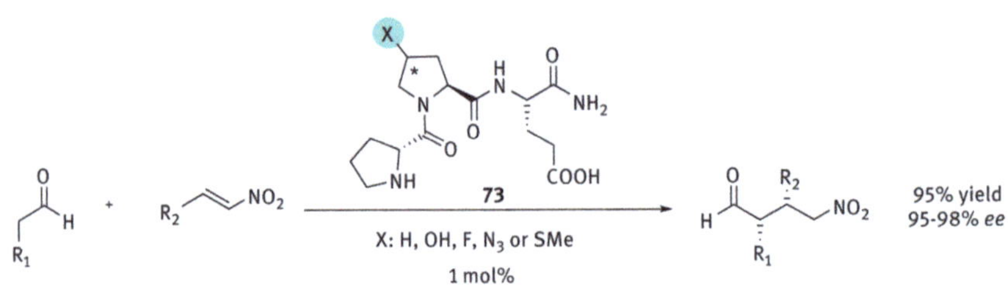

Figure 2.37: 4-Substituted-proline derivatives as the organocatalysts for addition reactions.

Besides the natural amino acids, a variety of catalysts have been synthesized based on β-amino acids acting as peptidomimics [74]. β-Proline-derivatives [75] and β-phenylalanine [76] have been used as catalysts for Michael addition.

2.1.2.2 Epoxidation of double bonds

Epoxidation of double bonds is one of the most studied chapters in organic synthesis. A variety of catalysts, oxidants and substrates have been reported, but the activation of hydrogen peroxide, a cheap but poor oxidant, has been less investigated. In 2005, Jørgensen achieved the asymmetric organocatalytic epoxidation of *α*, *β*-unsaturated aldehydes, utilizing hydrogen peroxide as the oxidant and an proline-derivative as the catalyst [77], while Córdova utilized the same oxidant and similar proline derivatives [78].

In 2007, Miller and co-workers provided an alternative approach in the asymmetric epoxidation, utilizing an aspartate-catalyst with unprotected acid group, achieving the activation of hydrogen peroxide via an hyperoxy-intermediate and the epoxidation of double bond in excellent yields and enantiomeric excesses (Figure 2.38) [79].

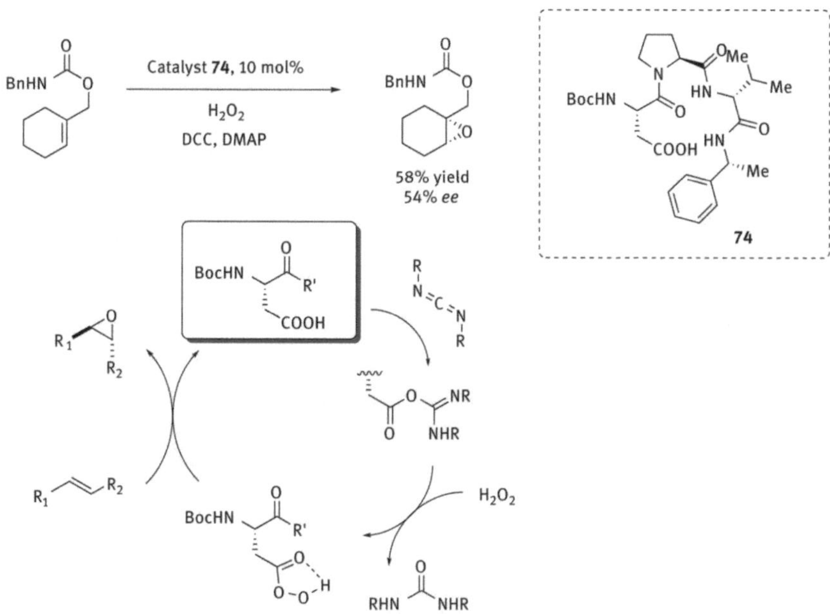

Figure 2.38: Catalytic cycle via hyperoxy-intermediate.

Based on that idea, Miller synthesized aspartate-analogues, in order to confirm the reaction mechanism ([80]a) and achieve the selective epoxidation of polyene substrates ([80]b) (Figure 2.39).

Kudo and his group presented the epoxidation of *α*,*β*-unsaturated aldehydes using a resin-supported peptide catalyst. Although long reaction time was required,

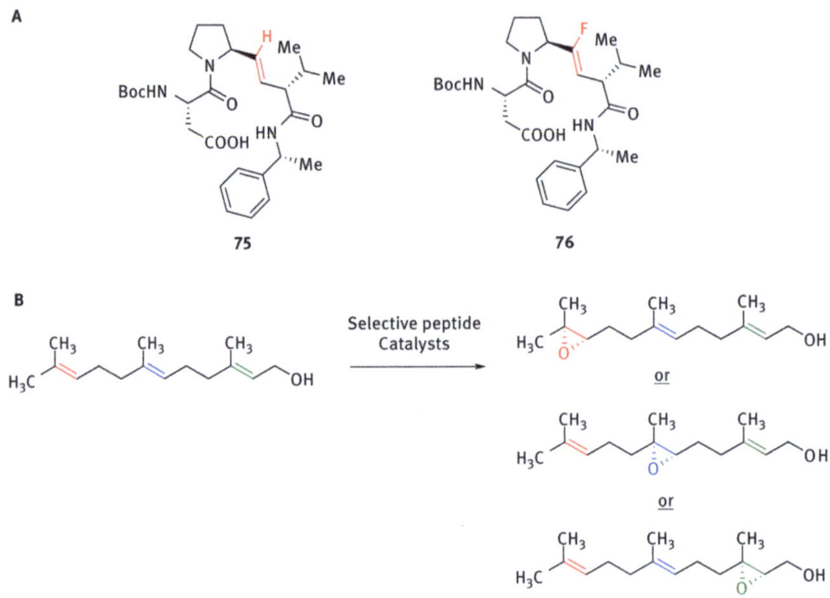

Figure 2.39: **(a)**. Analogues of catalyst 67. **(b)**. Selective epoxidation of polyene.

nevertheless, the peptide catalysis has potential for improving and a resin-supported peptides offer an easy way of screening [81].

Eantioselective organocatalytic epoxidation of *a*,*β*-unsaturated ketones has been achieved by Tang, utilizing poly-(*L*)-leucine as the catalyst [82] and Kurihara utilized a stabilized short helical peptide as the organocatalyst [83]. On the other hand, Park [84] and Yao [85] utilized simple amino acids as organocatalysts to epoxide *a*,*β*-unsaturated carbonyl compounds.

2.1.3 Reactions including carbon–nitrogen bond

2.1.3.1 Asymmetric Henry reaction

The Henry reaction has been established as a powerful methodology for the formation of carbon–carbon bonds between nitroalkenes and carbonyl compounds. Although the Henry reaction was discovered in 1896, a chiral variant was not developed until the 1990. Researchers have sought to expand the reaction scope, testing a variety of chiral organocatalysts. In 2005, Nagasawa and co-workers achieved the asymmetric Henry reaction, utilizing guanidine-thiourea bifunctional organocatalyst, leading to excellent yields and stereoselectivities [86]. In 2013, Karadeniz and Astley utilized amino acids and their derivatives, in order to catalyze the Henry reaction [87], while Yin achieved a domino aza-Michael-Henry reaction for the synthesis of 3-nitro-1,2-dihydroquinolines utilizing 4-hydroxy-proline-derivatives (Figure 2.40) [88].

In the later protocol, an electron-withdrawing group was added on the amino moiety of 2-aminobenzaldehyde in order to increase the N–H acidity, thus enabling deprotonation and subsequently improving nucleophilicity. The proposed reaction mechanism is presented in Figure 2.41.

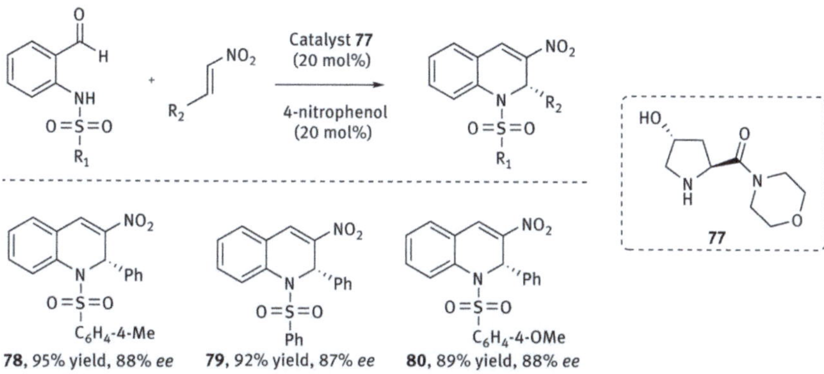

Figure 2.40: Synthesis of dihydroquinolines via the Henry reaction.

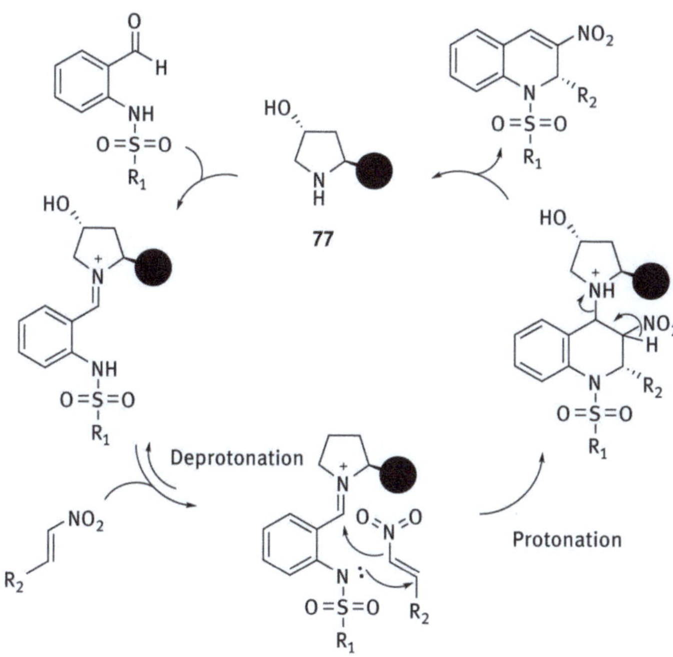

Figure 2.41: Proposed mechanism.

2.1.3.2 Asymmetric Strecker reaction

The hydrocyanation of carbonyl compounds (Strecker reaction) is a different methodology to form a new carbon–carbon bond, providing direct access to a diverse range of unnatural substituted amino acid precursors. In 2000, Jacobsen reported a new organocatalyst based on (*L*)-*tert*-leucine, in order to catalyze the hydrocyanation of imines [89]. The desired products were isolated in excellent yields and enantioselectivities. Two years later, the same group analyzed and optimized the reaction conditions, utilizing computational chemical data [90].

In 2012, Khan and co-workers achieved the asymmetric Strecker reaction of *N*-benzydrylimines with ethyl cyanoformate as the cyanide source, utilizing a (*S*)-phenylalanine derivative **78** as the organocatalyst (Figure 2.42) [91]. The proposed reaction mechanism is presented in Figure 2.43.

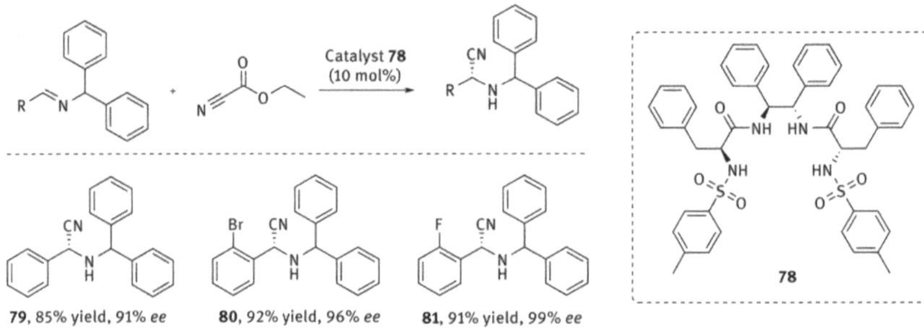

Figure 2.42: (*S*)-Phenylalanine derivative as catalyst for the Strecker reaction.

In 2013, Nasreen utilized (*L*)-proline to catalyze the synthesis of *α*-aminonitriles from aldehydes, affording the racemic products in good yields [92].

2.1.3.3 Reduction of ketimines

Chiral amines are very important structural components of biologically important compounds, such as natural products and agrochemicals. Organocatalysts catalyzing the enantioselective reduction of prochiral imines with trichlorosilane ($HSiCl_3$) represent one of the most important methods for preparing chiral amines. In 2004, Malkov and Kocovsky achieved the asymmetric reduction of ketimines with $HSiCl_3$ as the reductant and a valine-derivative as the organocatalyst to insert the desired chirality [93]. In the next years, the same group inserted new valine-analogues with bulky groups, in order to increase the reaction yield and selectivity [94]. Given the importance of organocatalytic asymmetric hydrosilylation, it is desirable to search for catalysts that could provide high enantioselectivity for the hydrosilylation of 1,4-benzooxazine. In 2013, the Sun group utilized an (*L*)-phenylalanine-derived catalyst with a sulfur chiral center, in order to hydrosilylate 1,4-benzooxazines, affording the desired chiral products in good yields and excellent *ee* (Figure 2.44) [95].

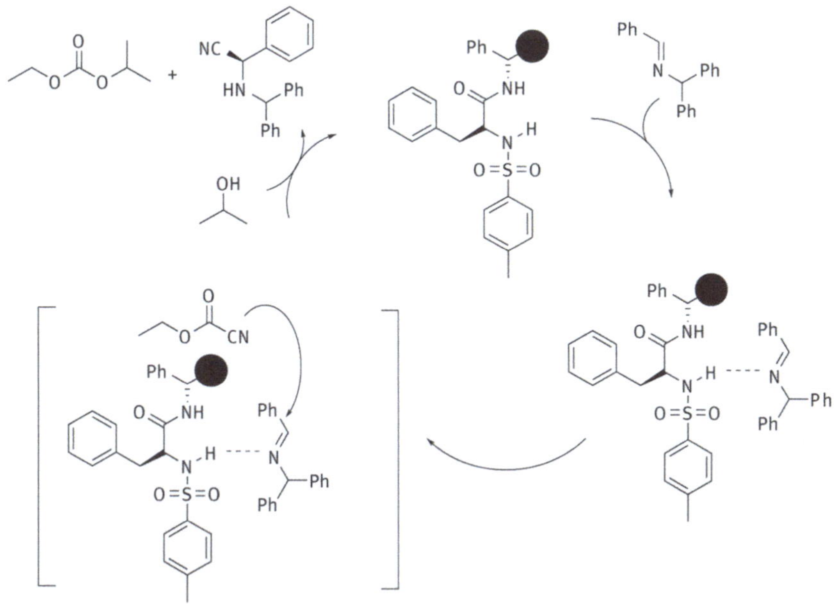

Figure 2.43: Proposed mechanism.

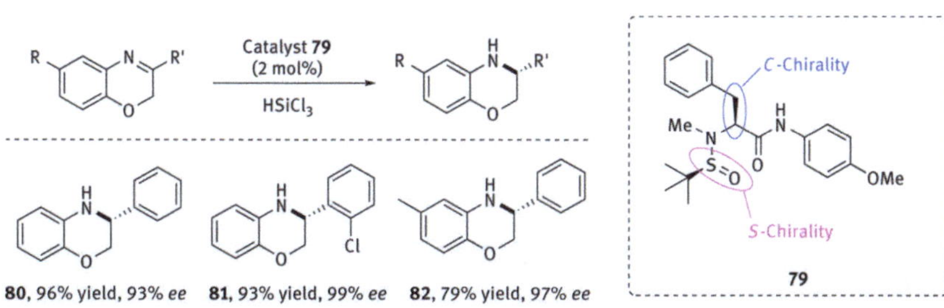

Figure 2.44: Reduction of 1,4-benzooxazines.

The proposed transition state is a compact hexacordinating silicon atom-based and is presented in Figure 2.45.

One year later, Qian and co-workers synthesized a bulky organocatalyst for ketimines' reduction, based on valine and *D*-glucose groups (Figure 2.46, Catalyst **83**) [96], while the Chen group stabilized valine-derived formamide onto the surface of Fe_3O_4 magnetic nanoparticles and used it as the catalyst for the asymmetric reduction (Figure 2.47, Catalyst **84**) [97]. The proposed transition state is similar as before and is presented in Figure 2.48.

Except valine and phenylalanine, proline has also been employed as the skeleton for the synthesis of organocatalysts suitable for asymmetric ketimines' reduction. In 2013, Benaglia utilized *N*-protected proline as the organocatalyst for the reduction of

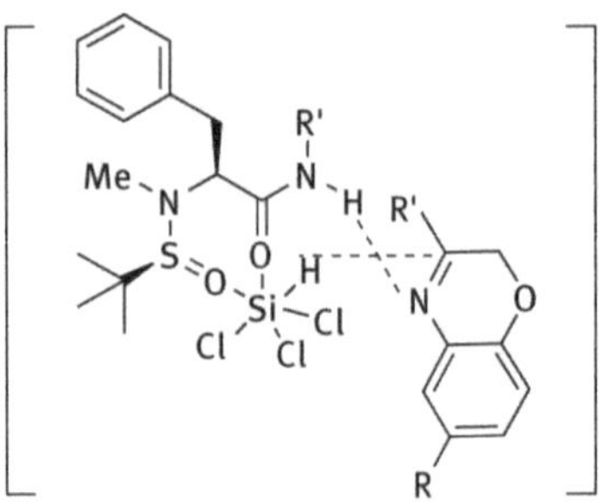

Figure 2.45: Proposed transition state.

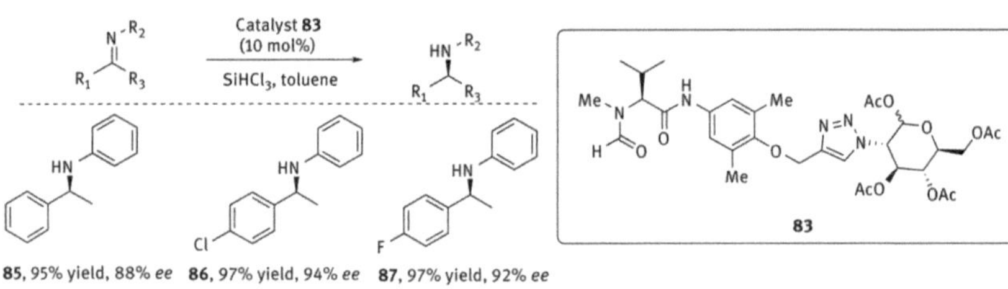

Figure 2.46: Valine-derived organocatalyst with *D*-glycose group.

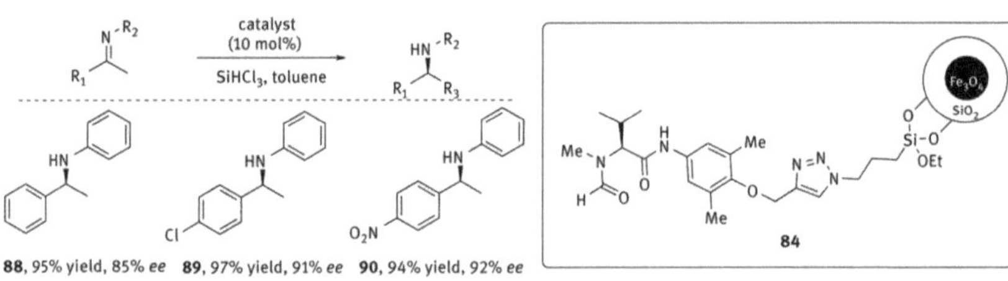

Figure 2.47: Valine-derived organocatalyst stabilized onto the surface of Fe_3O_4 magnetic nanoparticles.

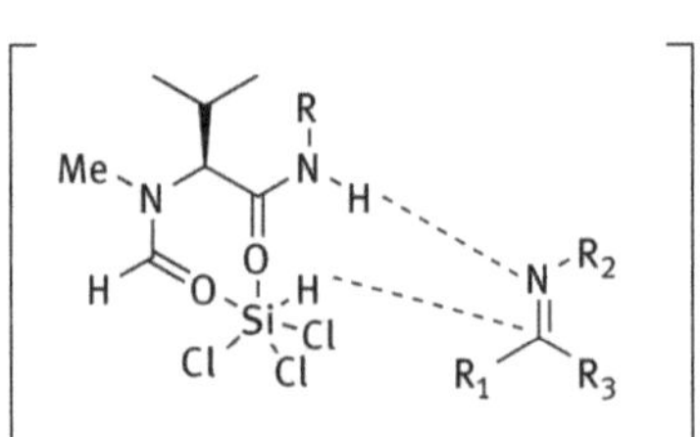

Figure 2.48: Proposed transition state.

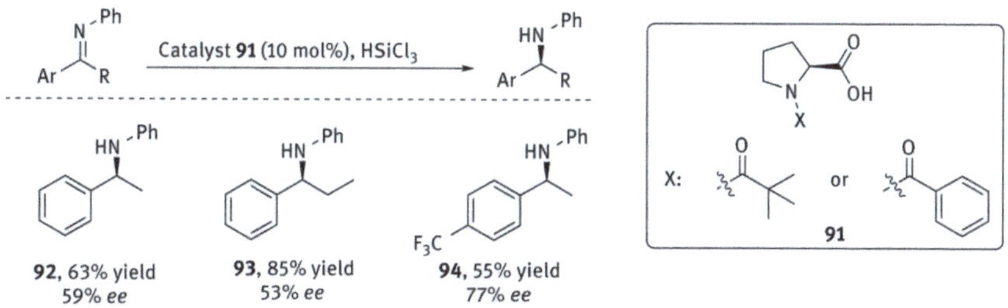

Figure 2.49: *N*-Protected-proline-derived organocatalysts.

ketimines, leading to very good results (Figure 2.49) [98]. The similar proposed transition state is presented in Figure 2.50.

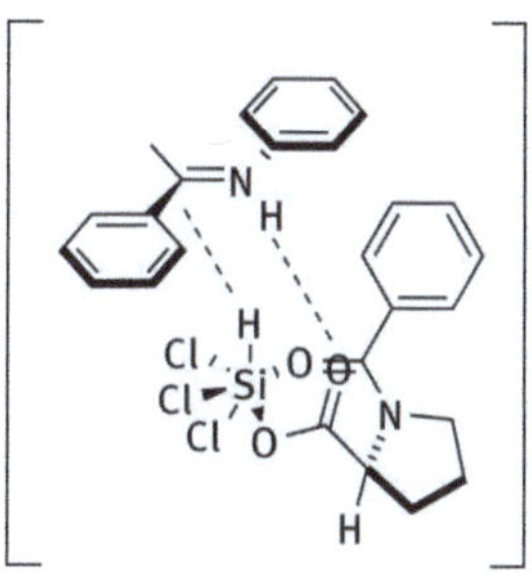

Figure 2.50: Proposed transition state.

Very recently, Yang, Bai and co-workers, aiming to use an organocatalyst which can be recycled, developed a homogeneous organocatalyst based on proline, which can be reused for more than seven times (Figure 2.51) [99], based on host–guest molecular recognition using cyclodextrinmodified Fe3O4@SiO2 magnetic nanoparticles (MNPs). The proposed transition state is presented in Figure 2.52.

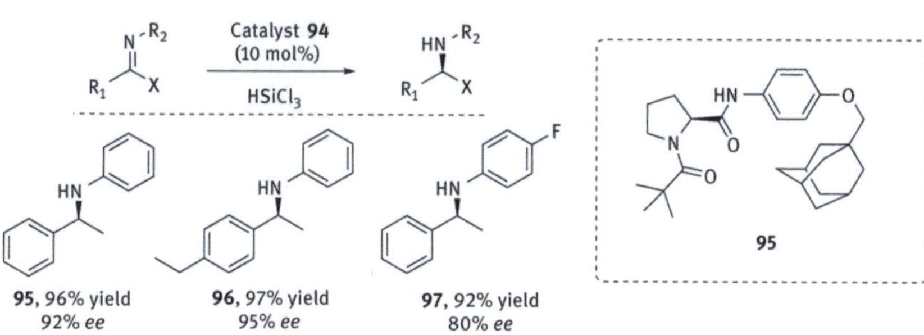

Figure 2.51: Proline-derived organocatalyst for the ketimines' reduction.

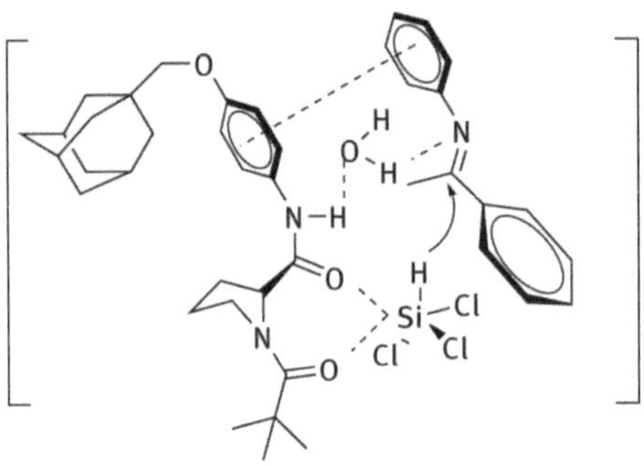

Figure 2.52: Proposed transition state.

2.1.4 Addition to aromatic rings

2.1.4.1 Friedel-crafts reaction

Iminium activation is one of the most common methods for the enantioselective formation of carbon–carbon bonds. In 2001, MacMillan developed an enantioselective Friedel-Crafts alkylation of *N*-methyl-pyrrole, utilizing a phenylalanine-based imidazolinone in very good selectivities ([100]a). One year later, the same group utilized a more bulky imidazolinone, in order to improve the reaction results ([100]b). A variety of studies have been developed in order to confirm the results and decode the reaction mechanism [101].

In 2007, Melchiorre and co-workers presented a very interesting alternative. A Friedel-Crafts alkylation of indoles with α,β-unsaturated ketones took place, in the presence of a new catalyst amine salt, which both the cation and the anion are chiral, which exhibited high reactivity and selectivity for iminium ion catalysis [102]. The cation is a Cinchona alkaloid derivative, while the anion is based on the phenylglycine skeleton.

2.1.4.2 Asymmetric bromination

A very interesting field of organic synthesis is introducing axial chirality. The dynamic kinetic resolution (DKR) is one of the most important techniques to transform racemic axially chiral biaryls into enantiopure atropisomers. This strategy was successfully applied by Miller and co-workers using the peptide-catalyzed asymmetric bromination of carboxylic acids [103]. Racemic biarylic benzoic acids were converted into chiral analogues, using a simple bromination strategy employing *N*-bromopthalimide (NBP) as the brominating agent. It was envisaged that peptidic-based organocatalysts could insert excellent axial chirality in the products (Figure 2.53). The transition state is presented in Figure 2.54.

In 2013, the same group confirmed the usefulness of the peptidic framework as the catalyst for the DKR of racemic benzamides, utilizing dibromodimethylhydantoin

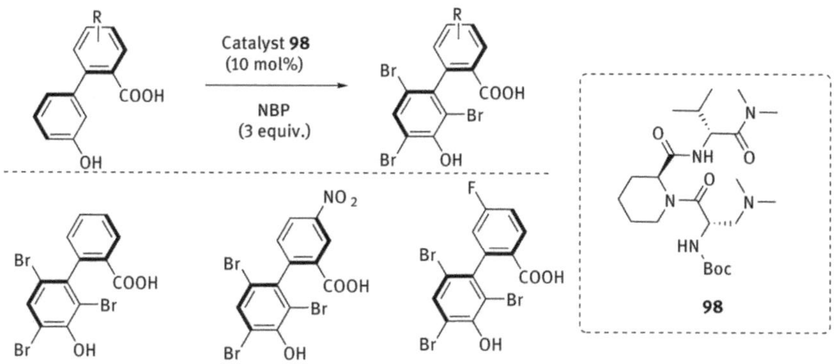

99, 80% yield, 97:3 *e.r.* 100, 85% yield, 97:3 *e.r.* 101, 70% yield, 97:3 *e.r.*

Figure 2.53: Peptide-catalyst for the asymmetric bromination.

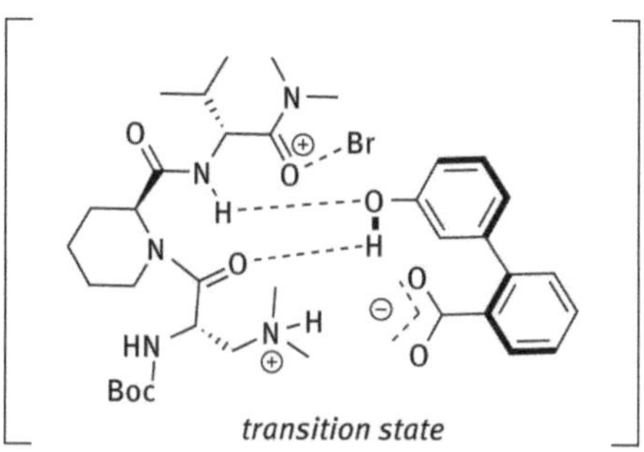

Figure 2.54: Proposed transition state.

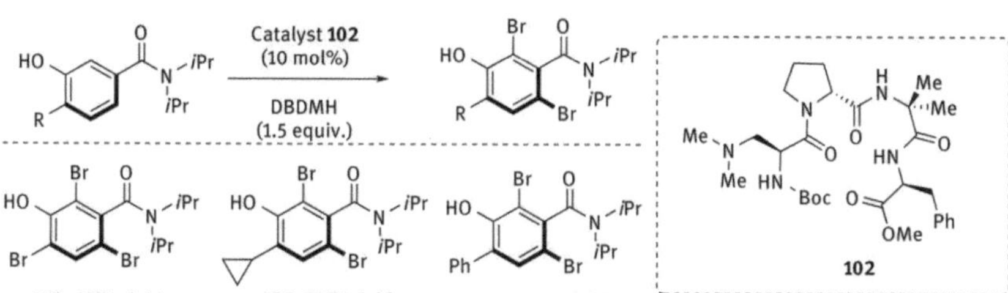

103, 89% yield
94:6 *e.r.*

104, 79% yield
96:4 *e.r.*

105, 86% yield
93:7*e.r.*

Figure 2.55: Peptide-catalyst for the asymmetric bromination of benzamides.

(DBDMH) as the brominating agent (Figure 2.55) [104]. The mechanism and intermediate transition states have been studied extensively [105].

In 2015, the same group, in order to extend the limits of this methodology, achieved the peptide-catalyzed atroposelective bromination of quinazolinones to access enantioenriched cyclic products (Figure 2.56) [106]. This work was followed by structural studies and X-ray crystallography, in order to confirm the 3-D structure of molecules [107]. The highly organized proposed transition state is presented in Figure 2.57.

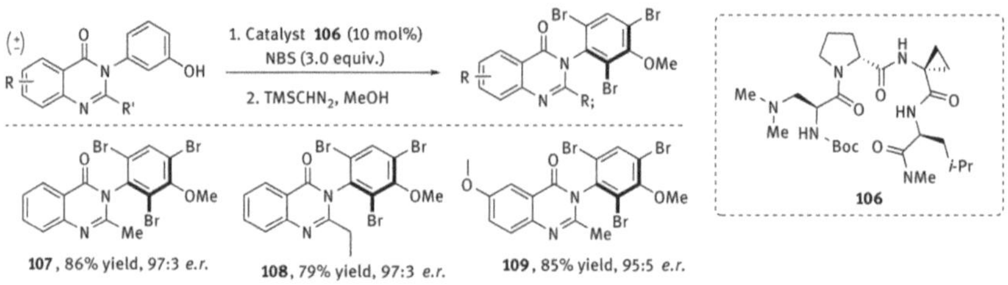

Figure 2.56: Peptide-catalyst for the asymmetric bromination of quinazolinones.

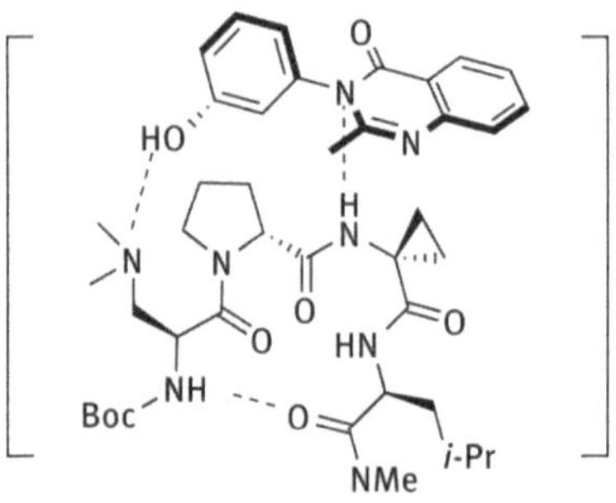

Figure 2.57: Proposed transition state.

2.1.5 Other reactions and applications

2.1.5.1 Synthesis of polycyclic products

Amino acids have a widespread use in catalysis leading to polycyclic products, which can be used as intermediates for further synthetic pathways. First of all, glycine, the smallest, simplest and the sole non-chiral amino acid, has been used in a variety of reactions. Pasha used glycine for the synthesis of pyranopyrazoles via one-pot multicomponent reactions (Figure 2.58, reaction **A**) ([108]a). Changing one of the reagent, Singh achieved to synthesize a plethora of furopyranes, benzochromenes and benzoxanthenes in very good yields (Figure 2.58, reaction **B**) ([108]b). Finally, glycine has been employed as the catalyst in the synthesis of polyhydroquinolines, under controlled microwave conditions (Figure 2.58, reaction **C**) ([108]c).

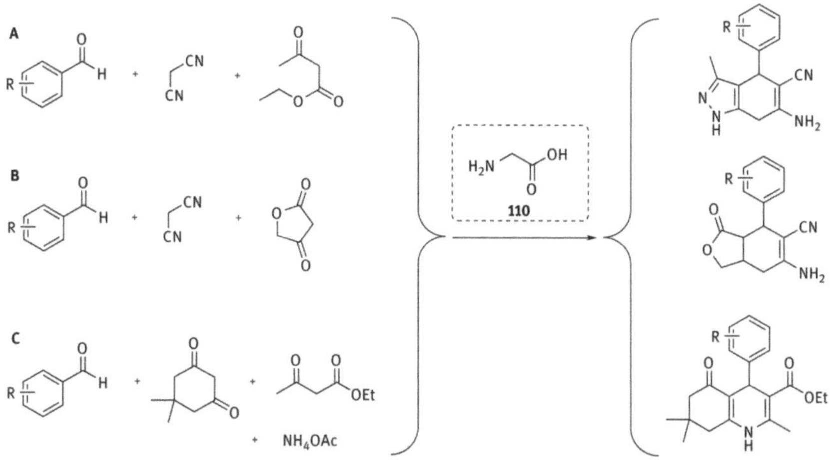

Figure 2.58: Reactions catalyzed by glycine.

Taking advantage of the unique side chain of cysteine, a variety of studies have been introduced, utilizing cysteine combined with magnetic nanoparticles (LCMNP) as the catalyst for the synthesis of polycyclic compounds (Figure 2.59) [109].

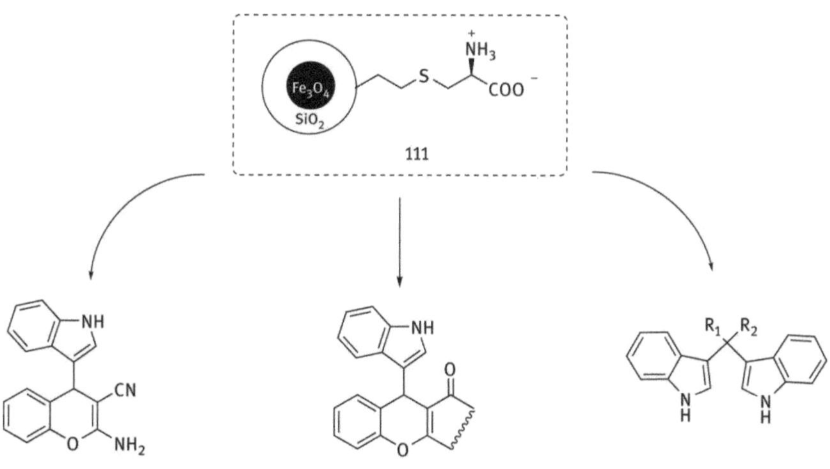

Figure 2.59: Cysteine-derivative for the synthesis of polycyclic compounds.

In 2013, Roy utilized (*L*)-cysteine as the organocatalyst for the synthesis of trisubstituted imidazoles from benzyl, which was condensed with aldehydes and ammonium acetate (Figure 2.60) [110]. The proposed reaction mechanism has two potential routes that finally end up in the same proposed aminal intermediate, which cyclizes and aromatizes and is presented in Figure 2.61.

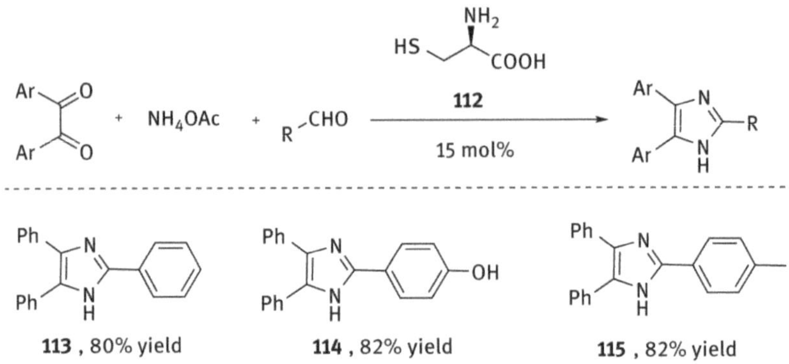

Figure 2.60: Cysteine as the catalyst for the synthesis of substituted imidazoles.

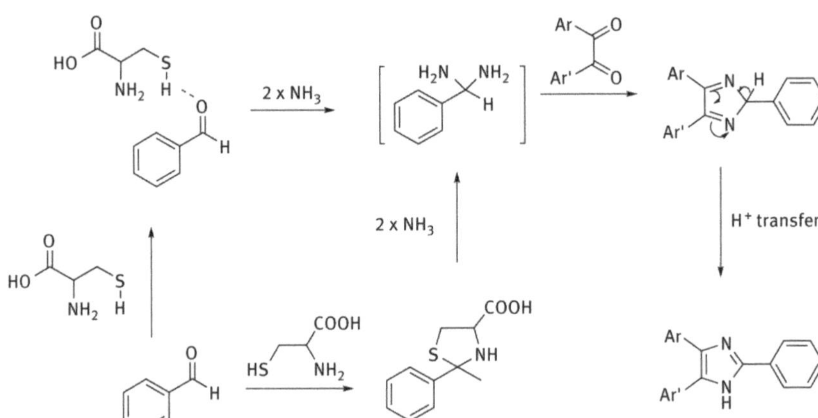

Figure 2.61: Proposed reaction mechanism.

(*L*)-Valine has been used for the synthesis of aminocyanopyrans starting from aromatic aldehydes, malononitrile and a diverse array of electron-rich enolizable carbonyl compounds (Figure 2.62) [111]. The proposed reaction mechanism is presented in Figure 2.63.

In 2011, Maas utilized proline or proline-derived organocatalysts to achieve the cyclopropanation of *α*-methylacrolein with *α*-diazobenzylphosphonate, leading to excellent results (Figure 2.64) [112]. The proposed transition states are presented in Figure 2.65.

(*L*)-Proline is also an excellent organocatalyst for the synthesis of a plethora of cyclic compounds (Figure 2.66) [113].

Amino acids bearing an aromatic moiety in their side chain have been studied extensively for the synthesis of cyclic compounds in high enantioselectivity. In 2015, Spring utilized a peptidic phosphane catalyst, based on (*L*)-phenylalanine, in

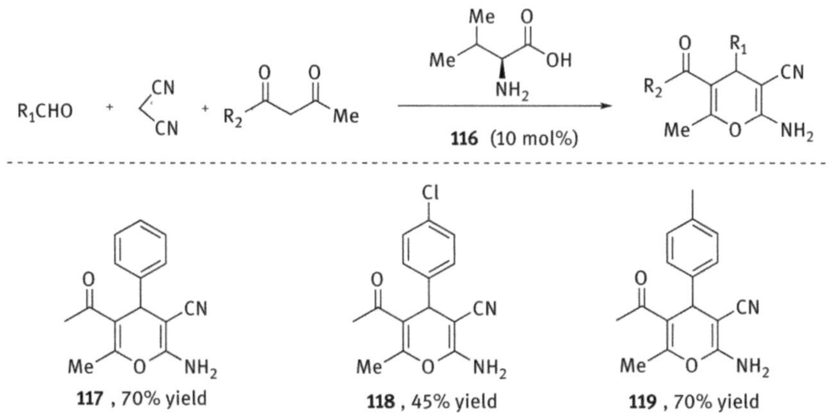

Figure 2.62: (*L*)-Valine as the catalyst for the synthesis of cyclic compounds.

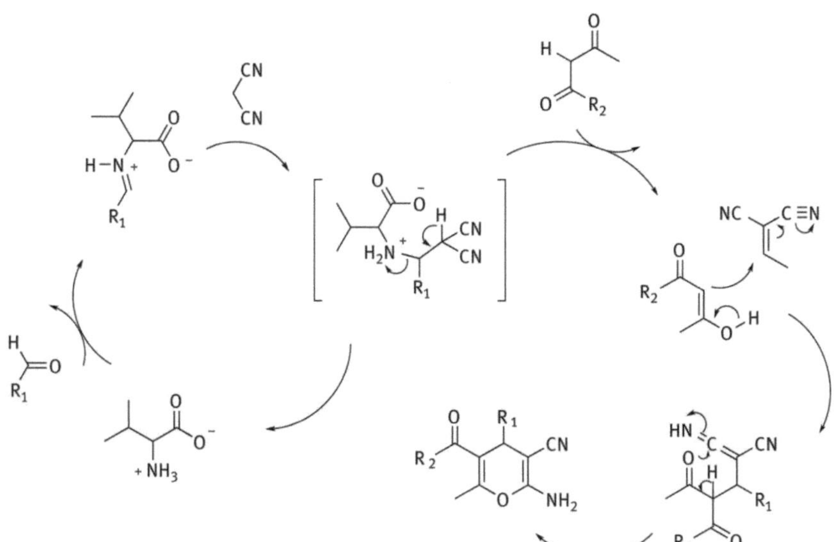

Figure 2.63: Proposed reaction mechanism.

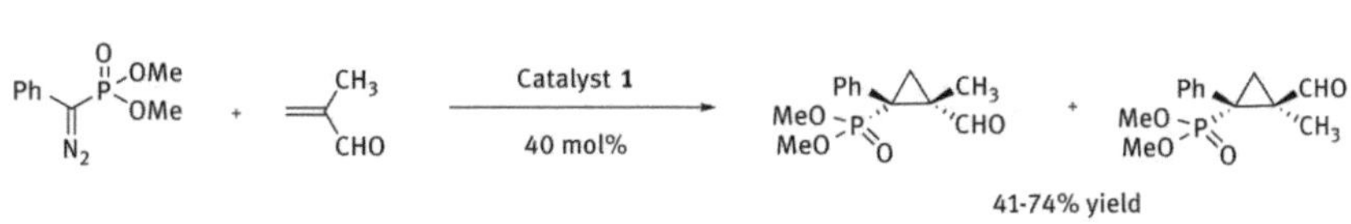

Figure 2.64: (*L*)-Proline as the catalyst for the synthesis of cyclopropyl compounds.

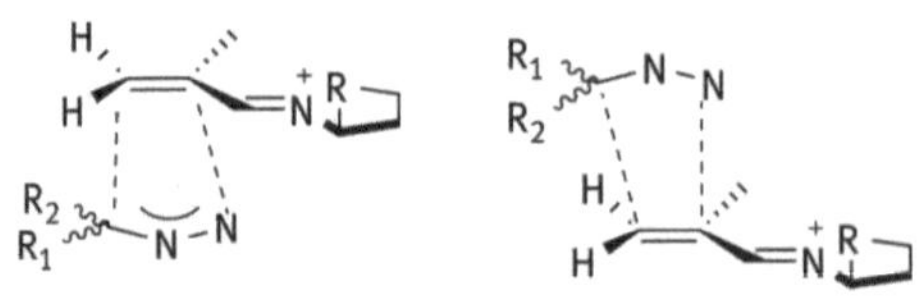

Figure 2.65: Possible transition states.

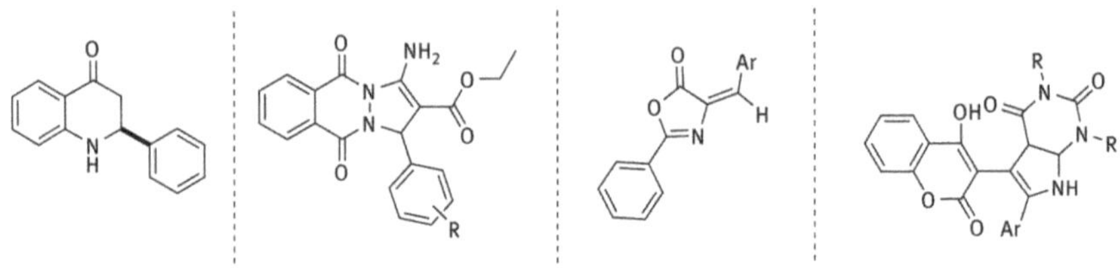

Figure 2.66: (*L*)-Proline as the catalyst for the synthesis of cyclic compounds.

order to achieve an intramolecular cyclization (Figure 2.67) [114]. The proposed reaction mechanism is presented in Figure 2.68.

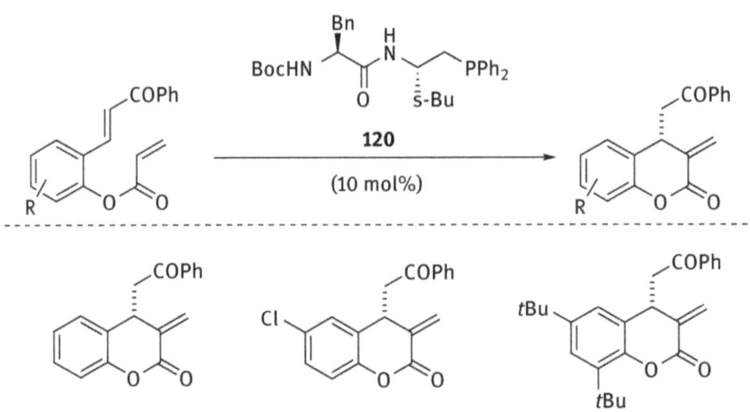

121, 93% yield, 97% *ee* **122**, 78% yield, 99% *ee* **123**, 99% yield, 94% *ee*

Figure 2.67: (*L*)-Phenylalanine-derivative as the catalyst for the intramolecular cyclization.

Very recently, Wang and co-workers developed an asymmetric protocol for the [3+3] annulation of pyridine *N*-oxides with acetone, utilizing (*L*)-phenylalanine (salt with potassium) as the catalyst (Figure 2.69, **A**) [115], while Li utilized the same amino acid for the synthesis of imidazopyridines and 1,4-diazepane derivatives (Figure 2.65, **B** and **C**) [116].

(*L*)-Tyrosine has been used for the synthesis of dihydropyrimidinones in excellent yields, taking advantage of the hydroxy-group of the side-chain for additional hydrogen bonding (Figure 2.70) [117].

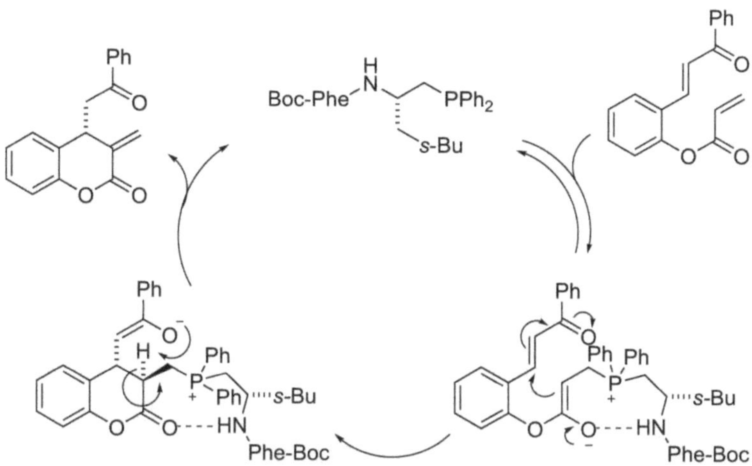

Figure 2.68: Proposed reaction mechanism.

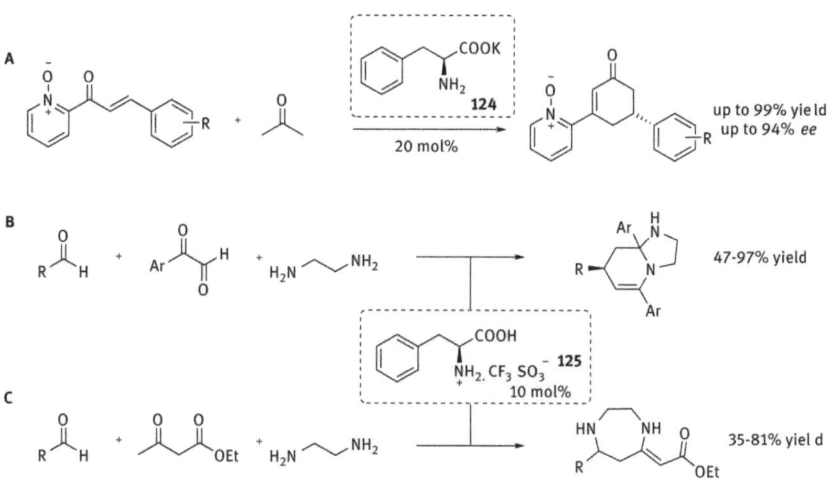

Figure 2.69: (*L*)-Phenylalanine as catalyst for cyclic compounds.

(*L*)-Aspartic acid has been employed for the diastereoselective synthesis of *trans*-isoquinolonic acids and poly-substituted imidazoles (Figure 2.71) [118].

Finally, taking advantage of the functional side chain of arginine, Heydari and co-workers synthesized arginine-functionalized magnetic nanoparticles, in order to catalyze the efficient synthesis of chromenes (Figure 2.72) [119]. The proposed mechanism is presented in Figure 2.73.

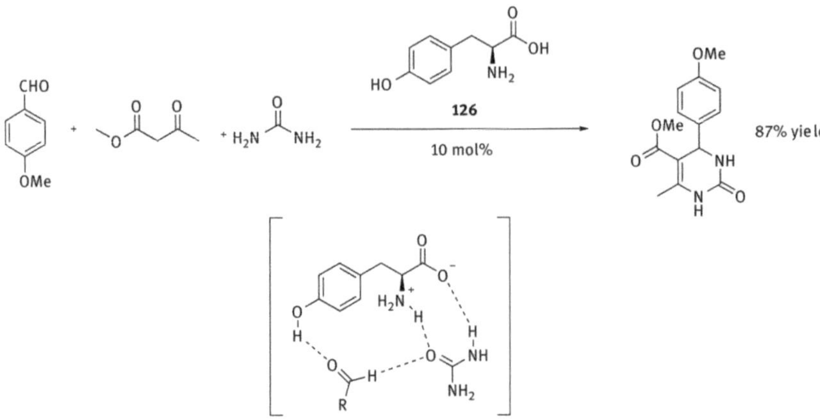

Figure 2.70: (*L*)-Tyrosine as the organocatalyst for the synthesis of cyclic compounds.

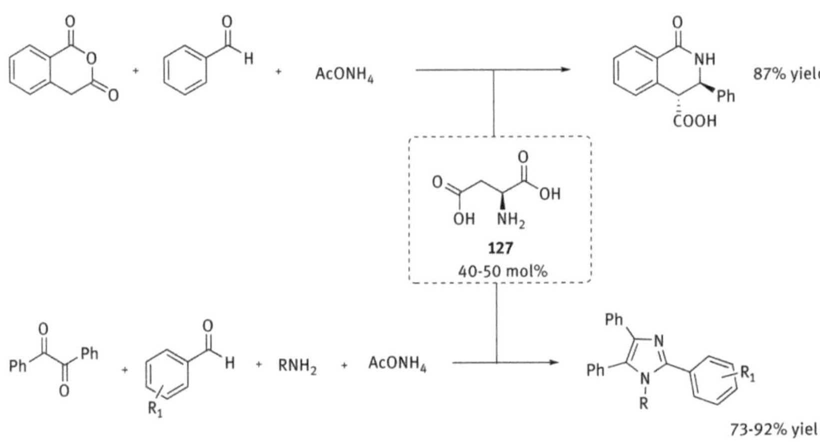

Figure 2.71: (*L*)-Aspartic acid as the organocatalyst for the synthesis of cyclic compounds.

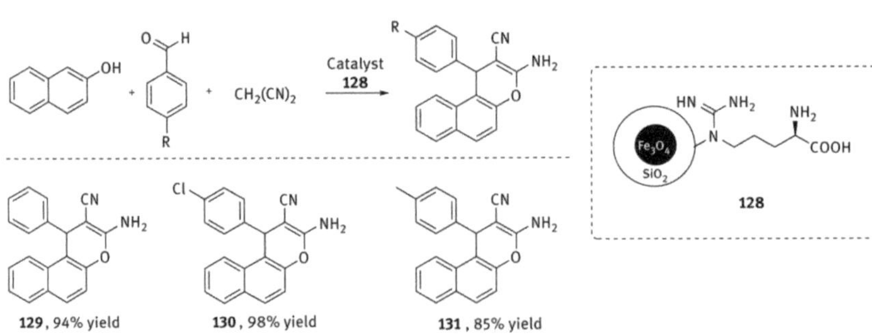

Figure 2.72: (*L*)-Arginine as the organocatalyst for the synthesis of cyclic compounds.

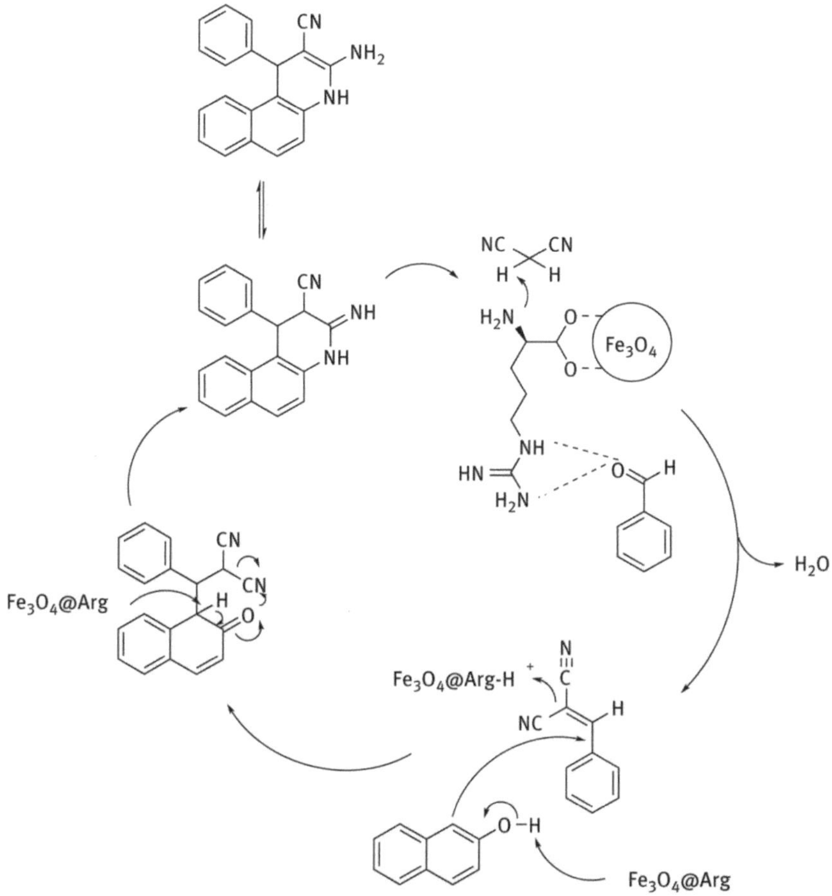

Figure 2.73: Proposed reaction mechanism.

2.1.5.2 Other reactions and potential applications in the chemical industry

Alkylation of aromatics has attracted increasing interest in organic chemistry, being widely used in the synthesis of petrochemicals, fine chemicals and pharmaceuticals ([120]a–c). The esterification process is very common in the preparation of fine chemicals used in the synthesis of drugs, food preservatives, solvents, perfumes, pharmaceuticals, plasticizers and cosmetics ([120]d–e). In 2010, Pitchumani achieved the chemoselective *O*-methylation of phenols and the esterification of carboxylic acids, utilizing (*L*)-methionine onto layered double hydroxides (LDH) as the organocatalyst (Figure 2.74) ([120]f). The proposed reaction mechanism is presented in Figure 2.75.

Allylation reactions have been widely used in total synthesis of biologically active compounds ([121]a–j). In 2013, Abdi, Ganguly and co-workers synthesized a

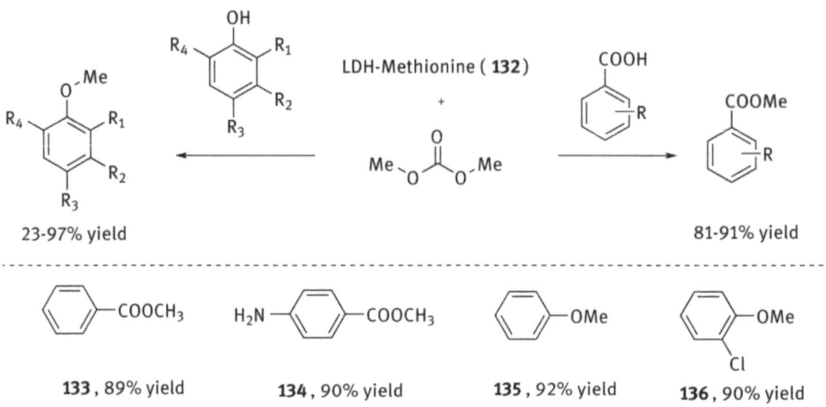

Figure 2.74: (*L*)-Methionine as the organocatalyst for the *O*-methylation and esterification of carboxylic acids.

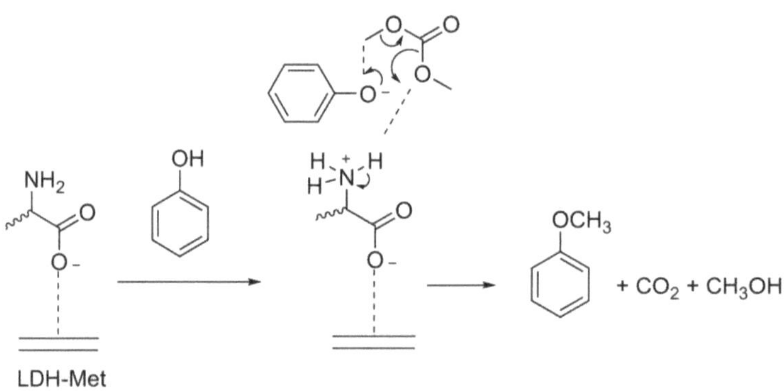

Figure 2.75: Proposed reaction mechanism.

variety of organocatalysts based on amino acids, in order to catalyze the asymmetric reaction of aldehydes with allyltrichlorosilane, affording allylic alcohols. The (*L*)-phenylalanine derivative **137** afforded the desired products in excellent yields and enantioselective excesses (Figure 2.76) ([121]k). The proposed mechanism, which is assumed to proceed via a Zimmerman-Traxler transition state, involving silicon activation is presented in Figure 2.77.

L-Tryptophan is a naturally occurring amino acid essentially used as a psychotherapeutic drug for mood regulation and also as a food supplement ([122]a). Due to its pharmaceutical usefulness, a broad range of methods have been already been applied for its production ([122]b). In 2016, Balavar utilized (*L*)-tryptophan as the skeleton for the synthesis of a catalyst for the Paal-Knorr pyrrole cyclocondensation ([122]c). The amino acid led to excellent results, that could be recycled and re-used (Figure 2.78). The proposed reaction mechanism is presented in Figure 2.79.

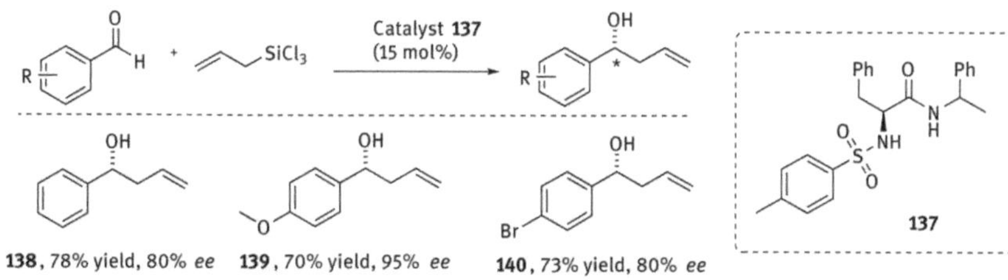

Figure 2.76: (*L*)-Phenylalanine-derivative as the organocatalyst for the synthesis of chiral allylic alcohols.

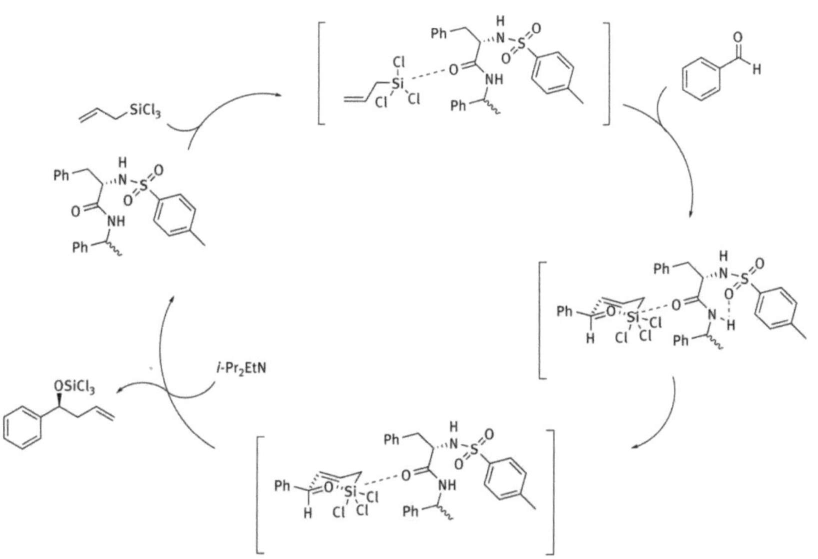

Figure 2.77: Proposed reaction mechanism.

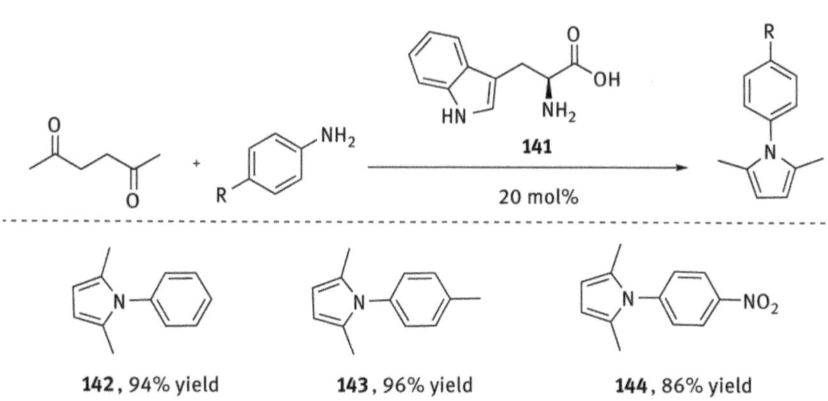

Figure 2.78: (*L*)-Tryptophan as the organocatalyst for the Paal-Knorr synthesis.

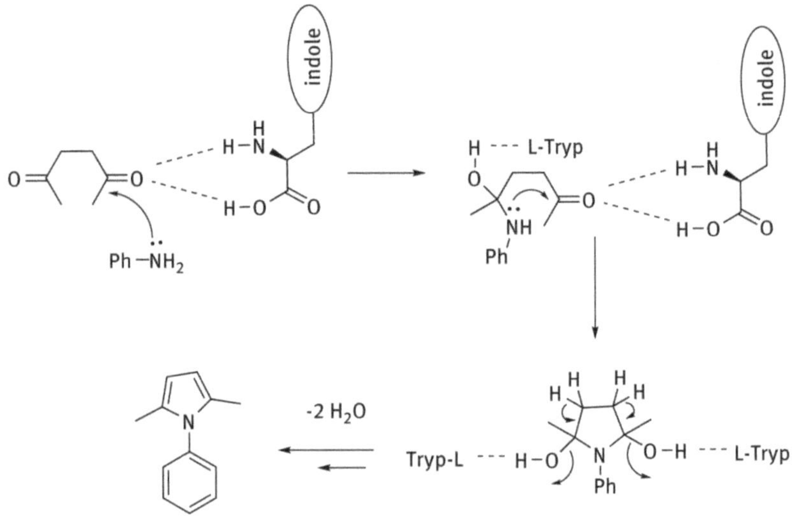

Figure 2.79: Proposed reaction mechanism.

Darzens reaction provides an efficient method for the synthesis of several biologically useful compounds. For example, biologically active vitamin A products were produced by the application of the Darzens synthesis to β-ionone ([123]a). In 2017, Siva achieved the synthesis of *α,β*-epoxy carbonyl compounds from *α*-halo carbonyl moieties, utilizing proline-based organocatalysts as the starting point for the formation of chiral quaternary ammonium salts, that can promote the Darzens reaction both asymmetrically, but also as the leaving group (Figure 2.80, **A**), ([123]b), while Wagenknecht, utilizing the same amino acid, achieved the first successful example for the introduction of secondary-structured peptides with photoredox catalysis (Figure 2.80, **B**) [124].

Fixation of CO_2 with amines, which combines both reduction of CO_2 and C–N bond construction, produces versatile and useful chemicals and energy-storage materials, such as formamides, aminals and methylamines that are usually derived from petroleum feedstocks ([125]a–c). Recently, He and co-workers reported that glycine betaine (GB) is an excellent and sustainable organocatalyst for the reductive functionalization of CO_2 with various amines and diphenylsilane (Figure 2.81) ([125]d). The proposed reaction mechanism is presented in Figure 2.82.

Amino acids, due to their easy use and commercially availability, have been used as catalysts in large-scale reactions, in order to find useful applications in chemical industry. For example, proline tetrazole has been used as the catalyst for the aldol, the Mannich and the *o*-nitroso aldol reaction in a column-flow system, keeping up the excellent yields, enantio- and chemoselectivity values [126]. Additionally, (*L*)-tyrosine-derivative have been used excellently as the filling material for a “homemade” HPLC column, in order to achieve continuous-flow

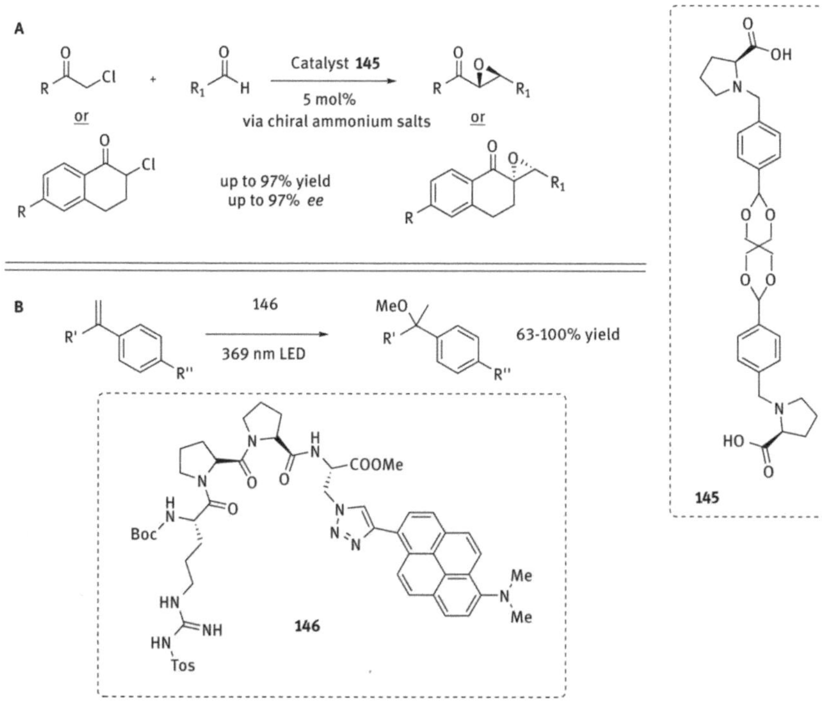

Figure 2.80: (*L*)-Proline-derivatives as the organocatalysts.

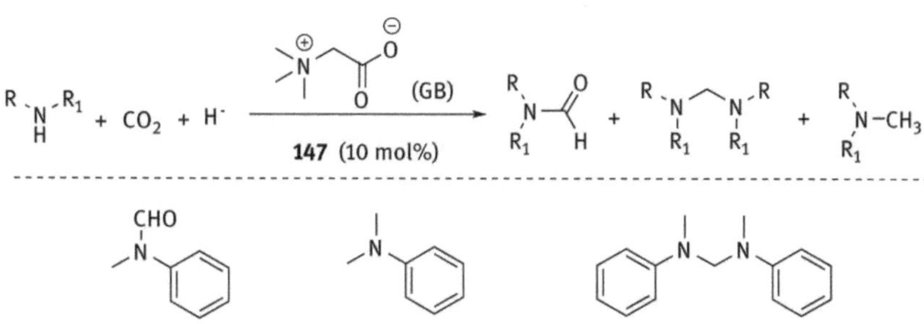

Figure 2.81: (*L*)-Glycine betaine as the organocatalyst.

organocatalyzed addition reactions with very good results [127]. Finally, Fu has demonstrated the use of threonine-based organocatalysts for the large-scale aldol and Mannich reactions providing the products in excellent results [128].

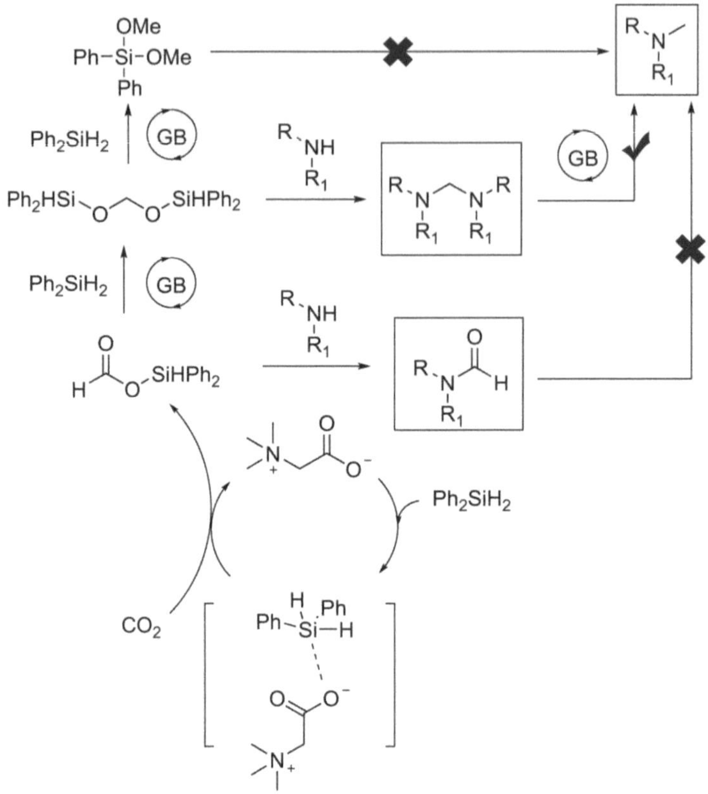

Figure 2.82: Proposed reaction mechanism.

2.2 Conclusion

In this chapter, we present the usefulness and simplicity of amino acids as organocatalysts for the successful synthesis of numerous and various motifs in organic synthesis. We analyzed extensively the recent advances in reactions, which were promoted by amino acids, emphasizing the important role of the skeleton of the catalysts. Reactions such as the aldol, the Mannich, the Michael and a variety of others have been studied, presenting a plethora of organocatalysts for each one. It goes without saying that amino acids have a special and different role in modern asymmetric synthesis and have a lot more to give in the near future.

References

[1] For selected books on asymmetric organocatalysis, see: (a) Berkessel A, Groger H. Metal free organic catalysis in asymmetric synthesis, Wiley-VCH:Weinheim, 2004; (b) Berkessel A, Groger H. Asymmetric organocatalysis – from biomimetic concepts to applications in asymmetric synthesis. Weinheim: Wiley-VCH, 2005; (c) Dalko PI. Enantioselective organocatalysis. Weinheim: Wiley-VCH, 2007; (d) List B. Asymmetric organocatalysis. Berlin, Heidelberg: Springer, 2009; (e) Pellissier H. Recent developments in asymmetric organocatalysis. Cambridge: RSC Publishing, 2012; (f) Dalko PI. Comprehensive stereoselective organocatalysis, vols 1–3, Weinheim: Wiley-VCH, 2013; (g) Rios Torres R. Stereoselective organocatalysis-bond formation methodologies and activation modes. Hoboken, New Jersey: Wiley, 2013. For reviews, see: (h) Dalko PI, Moisan L. In the golden age of organocatalysis. Angew Chem Int Ed. 2004;43:5138–75; (i) Mukherjee S, Yang JW, Hoffmann S, List B. Asymmetric enamine catalysis. Chem Rev. 2007;107:5471–569; (j) Dondoni A, Massi A. Asymmetric organocatalysis: from infancy to adolescence. Angew Chem Int Ed. 2008;47:4638–60; (k) Melchiorre P, Marigo M, Carlone A, Bartoli G. Asymmetric aminocatalysis-gold rush in organic chemistry. Angew Chem Int Ed. 2008;47:6138–71.

[2] (a) Ahrendt KA, Borths CJ, MacMillan DW. New strategies for organic catalysis: the first highly enantioselective organocatalytic Diels-Alder reaction. J Am Chem Soc. 2000;122:4243–4; (b) MacMillan DWC. The advent and development of organocatalysis. Nature. 2008;455:304–8.

[3] For selected reviews, see: (a) Wang YB, Tan B. Construction of axially chiral compounds via asymmetric organocatalysis. Acc Chem Res. 2018, 51, 534–47; (b) Zhu Y, Wang Q, Cornwall RG, Shi Y. Organocatalytic asymmetric epoxidation and aziridination of olefins and their synthetic applications. Chem Rev. 2014;114:8199–256; (c) Kano T, Maruoka K. Unique properties of chiral biaryl-based secondary amine catalysts for asymmetric enamine catalysis. Chem Sci. 2013;4:907–15.

[4] For recent reviews, see: (a) Liu J, Wang L. Recent advances in asymmetric reactions catalyzed by proline and its derivatives. Synthesis. 2017;49:960–72; (b) Wieczorek R, Adamala K, Gasperi T, Polticelli F, Stano P. Small and random peptides: an unexplored reservoir of potentially functional primitive organocatalysts. The case of seryl-histidine. Life. 2017;7: 19–43; (c) Kumar P, Sharmaa BM. Proline-catalyzed asymmetric α-amination in the synthesis of bioactive molecules. Synlett. 2018;29:1944–56.

[5] For recent reviews, see: (a) Trost BM, Brindle CS. The direct catalytic asymmetric aldol reaction. Chem Soc Rev. 2010;39:1600–32; (b) Heravi MM, Zadsirjan V, Dehghani M, Hosseintash N. Current applications of organocatalysts in asymmetric aldol reactions: an update. Tetrahedron: Asymmetry. 2017;28:587–707; (c) Kaur J, Chauhan P, Singh S, Chimni SS. Journey heading towards enantioselective synthesis assisted by organocatalysis. Chem Rec. 2018;18:137–53; (d) Evans CS, Davis LO. Recent advances in organocatalyzed domino C–C bond-forming reactions. Molecules. 2018;23:33–46.

[6] List B, Lerner RA, Barbas CF. Proline-catalyzed direct asymmetric aldol reactions. J Am Chem Soc. 2000;122:2395–6.

[7] (a) Sakthivel K, Notz W, Bui T, Barbas CF. Amino acid catalyzed direct asymmetric aldol reactions: a bioorganic approach to catalytic asymmetric carbon–carbon bond-forming reactions. J Am Chem Soc. 2001;123:5260–7; (b) Tang Z, Jiang F, Yu LT, et al. Novel small organic molecules for a highly enantioselective direct aldol reaction. J Am Chem Soc. 2003;125:5262–3; (c) Ostrowski KA, Lichte D, Stuck M, Vorholt AJ. A comprehensive investigation and optimisation on the proteinogenic amino acid catalysed homo aldol condensation. Tetrahedron. 2016;72:592–8.

[8] (a)Tang Z, Yang ZH, Chen XH, Cun LF, Mi AQ, Jiang YZ, et al. A highly efficient organocatalyst for direct aldol reactions of ketones with aldehydes. J Am Chem Soc. 2005;127:9285–9; (b) Raj M, Ginotra SK, Singh VK. Highly enantioselective direct aldol reaction catalyzed by organic molecules. Org Lett. 2006;8:4097–9; (c) Gandhi S, Singh V K. Synthesis of chiral organocatalysts derived from aziridines: application in asymmetric aldol reaction. J Org Chem. 2008;73:9411–416; (d) Jia YN, Wu FC, Ma X, Zhu GJ, Da CS. Highly efficient prolinamide-based organocatalysts for the direct asymmetric aldol reaction in brine. Tetrahedron Lett 2009;50:3059–62; (e) Wang B, Chen G, Liu L, Chang W, Li J. A novel proline-valinol thioamide small organic molecule for a highly enantioselective direct aldol reaction. Adv Synth Catal. 2009;351:2441–28; (f) Schwab RS, Galetto FZ, Azeredo GB, Braga AL, Lüdtke DS, Paixão MW. Organocatalytic asymmetric aldol reactions mediated by a cysteine-derived prolinamide, Tetrahedron Lett 2008;49:5094–7.

[9] (a)Fotaras S, Kokotos CG, Tsandi E, Kokotos G. Prolinamides bearing thiourea groups as catalysts for asymmetric aldol reactions. Eur J Org Chem. 2011:1310–7; (b) Fotaras S, Kokotos CG, Kokotos G. A tripeptide-like prolinamide-thiourea as an aldol reaction catalyst. Org Biomol Chem. 2012;10:5613–9; (c) Revelou P, Kokotos CG, Moutevelis-Minakakis P. Novel prolinamide-ureas as organocatalysts for the asymmetric aldol reaction. Tetrahedron. 2012;68:8732–8; (d) Kokotos CG. Construction of tertiary alcohols bearing perfluoroalkyl chains catalyzed by prolinamide-thioureas. J Org Chem. 2012;77:1131–5.

[10] Machuca E, Granados G, Hinojosa B, Juaristi E. Synthesis and evaluation of (*S*)-proline-containing dipeptidic organocatalysts bound to MBHA resin in asymmetric aldol reactions. Tetrahedron Lett. 2015;56:6047–51.

[11] (a)Otvos SB, Mandity IM, Fulop F. Highly efficient 1,4-addition of aldehydes to nitroolefins: organocatalysis in continuous flow by solid-supported peptidic catalysts. ChemSusChem. 2012;5:266–9; (b) Otvos SB, Mandity IM, Fulop F. Asymmetric aldol reaction in a continuous-flow reactor catalyzed by a highly reusable heterogeneous peptide. J Catal. 2012;295:179–85.

[12] Wiedenhoeft D, Benoit AR, Wu Y, Porter JD, Meyle E. Multifunctional heterocyclic scaffolds for hybrid Lewis acid/Lewis base catalysis of carbonecarbon bond formation. Tetrahedron. 2016;72:3905–16.

[13] Orlandi M, Benaglia M, Raimondi L, Celentano G. 2-aminoimidazolyl and 2-aminopyridyl (*S*)-prolinamides as versatile multifunctional organic catalysts for aldol, Michael, and Diels-Alder reactions. Eur J Org Chem. 2013;2346–54.

[14] Illa O, Porcar-Tost O, Robledillo C, et al. Stereoselectivity of proline/cyclobutane amino acid-containing peptide organocatalysts for asymmetric aldol additions: a rationale. J Org Chem. 2018;83:350–63.

[15] Bonsignore M, Benaglia M, Cozzi F, Genoni A, Rossi S, Raimondi L. Readily available (*S*)-proline-derived organocatalysts for the Lewis acid-mediated Lewis base-catalyzed stereoselective aldol reactions of activated thioesters. Tetrahedron. 2012;68:8251–5.

[16] (a)Cruz-Hernández C, Hernández-González PE, Juaristi E. (*R*)- and (*S*)-Proline-derived chiral phosphoramides as organocatalysts for the enantiodivergent aldol reaction of isatins with cyclohexanone in the presence of water. Synthesis 2018, 50, 1827–40; (b) Cruz-Hernández C, Hernández-González PE, Juaristi E. Proline-glycine dipeptidic derivatives of chiral phosphoramides as organocatalysts for the enantiodivergent aldol reaction of aryl aldehydes and isatins with cyclohexanone in the presence of water. Synthesis. 2018;50:3445–59.

[17] (a)Wu FC, Da C-S, Du Z-X, et al. *N*-Primary-amine-terminal β-turn tetrapeptides as organocatalysts for highly enantioselective aldol reaction. J Org Chem. 2009;74:4812–8; (b) Chandrasekhar S, Johny K, Reddy CR. Proline-threonine dipeptide as an organocatalyst for the direct asymmetric aldol reaction. Tetrahedron: Asymmetry. 2009;20:1742–5;

(c) Hernandez JG, Juaristi E. Efficient ball-mill procedure in the 'green' asymmetric aldol reaction organocatalyzed by (*S*)-proline-containing dipeptides in the presence of water. Tetrahedron. 2011;67:6953–9; (d) Paradowska J, Pasternak M, Gut B, Gryzło B, Mlynarski J. Direct asymmetric aldol reactions inspired by two types of natural aldolases: water-compatible organocatalysts and Zn^{II} complexes. J Org Chem. 2012;77:173–87; (e) Liu L, Wang B, Zhu Y, Chang W, Li J. Additive-controlled regioselective direct asymmetric aldol reaction of hydroxyacetone and aldehyde. Tetrahedron: Asymmetry. 2013;24:533–42; (f) Bayat S, Tejo BA, Salleh AB, Abdmalek E, Normi YM, Rahman MB. Various polar tripeptides as asymmetric organocatalyst in direct aldol reactions in aqueous media. Chirality. 2013;25:726–34; (g) Popik O, Pasternak-Suder M, Leśniak K, Jawiczuk M, Gorecki M, Frelek J, Mlynarski J. Amine-catalyzed direct aldol reactions of hydroxy- and dihydroxyacetone: biomimetic synthesis of carbohydrates. J Org Chem. 2014;79:5728–39; (h) Psarra A, Kokotos CG, Moutevelis-Minakakis P. *tert*-Butyl esters of tripeptides based on Pro-Phe as organocatalysts for the asymmetric aldol reaction in aqueous or organic medium. Tetrahedron. 2014;70:608–15; (i) Triandafillidi I, Bisticha A, Voutyritsa E, Galiatsatou G, Kokotos CG. *tert*-Butyl ester or benzylamide of the dipeptide Pro-Gly as organocatalysts for the asymmetric aldol reaction. Tetrahedron. 2015;71:932–40; (j) Vlasserou I, Sfetsa M, Gerokonstantis D-T, Kokotos CG, Moutevelis-Minakakis P. Combining prolinamides with 2-pyrrolidinone: novel organocatalysts for the asymmetric aldol reaction. Tetrahedron. 2018;74:2338–49; (k) Messeser M, Wennemers H. Reversing the enantioselectivity of a peptidic catalyst by changing the solvent. Synlett. 2011;4:499–502.

[18] Orlandi M, Ceotto M, Benaglia M. Kinetics versus thermodynamics in the proline catalyzed aldol reaction. Chem Sci. 2016;7:5421–7.

[19] (a)Siyutkin DE, Kucherenko AS, Zlotin SG. Hydroxy-α-amino acids modified by ionic liquid moieties: recoverable organocatalysts for asymmetric aldol reactions in the presence of water. Tetrahedron. 2009;65:1366–72; (b) Khan SS, Shah J, Liebscher J. Synthesis of new ionic-liquid-tagged organocatalysts and their application in stereoselective direct aldol reactions. Tetrahedron. 2010;66:5082–88.

[20] Córdova A, Zou W, Ibrahem I, Reyes E, Engqvist M, Liao W-W. Acyclic amino acid-catalyzed direct asymmetric aldol reactions: alanine, the simplest stereoselective organocatalyst. Chem Commun. 2005:3586–8.

[21] Miura T, Ina M, Imai K, et al. β-Aminosulfonamide-catalyzed direct asymmetric aldol reaction in brine. Synlett. 2011;3:410–4.

[22] Yan T, Wang X, Sun H, Liu J, Xie Y. Facile creation of 3-substituted-3-hydroxy-2-oxindoles by arginine-catalyzed aldol reactions of α,β-unsaturated ketones with isatins. Mol. 2013;18:14505–18.

[23] Armacost K, Acevedo O. Exploring the aldol reaction using catalytic antibodies and "on water" organocatalysts from QM/MM calculations. J Am Chem Soc. 2014;136:147–56.

[24] (a)Huang W, Tian H, Xu H, Zheng L, Liu Q, Zhang S. *L*-Valine dipeptide organocatalysts with two amide units for the direct asymmetric aldol reaction in brine. Catal Lett. 2011;141:872–6; (b) Gerasimchuk VV, Kucherenko AS, Fakhrutdinov AN, Medvedev MG, Nelyubina YV, Zlotin SG. Towards sustainable amino acid derived organocatalysts for asymmetric *syn*-aldol reactions. Eur J Org Chem. 2017:2540–4.

[25] (a)Teo Y-C. Direct asymmetric aldol reactions catalyzed by a siloxy serine organocatalyst in water. Tetrahedron Asymmetry. 2007;18:1155–8; (b) Teo Y-C, Chua G-L, Ong C-Y, Poh C-Y. Organocatalytic asymmetric *syn*-selective direct aldol reactions in water. Tetrahedron Lett. 2009;50:4854–6; (c) Teo Y-C, Lee P P-F. Organic solvent-free direct asymmetric aldol reactions catalyzed by a siloxy serine organocatalyst in brine. Synth Commun. 2009;39:3081–91; (d) Yong F-F, Poh C-Y, Chua G-L, Teo Y-C. Organocatalytic asymmetric *syn*-

selective direct aldol reaction in ionic liquid. Chem Lett. 2010;39:490–2; (e) Jiang Z, Lu Y. Direct asymmetric aldol reaction of acetone with *α*-ketoesters catalyzed by primary-tertiary diamine organocatalysts. Tetrahedron Lett. 2010;51:1884–6; (f) Wu C, Fu X, Li S. New simple and recyclable *O*-acylation serine derivatives as highly enantioselective catalysts for the large-scale asymmetric direct aldol reactions in the presence of water. Tetrahedron. 2011;67:4283–90.

[26] Li S, Wu C, Fu X, Miao Q. Cysteine-based organocatalysts for the highly efficient direct stoichiometric *anti*- and *syn*-aldol reactions. Ind Eng Chem Res. 2011;50:13711–6.

[27] (a)Rambo RS, Schneider PH. Thiazolidine-based organocatalysts for a highly enantioselective direct aldol reaction. Tetrahedron Asymmetry. 2010;21:2254–7; (b) Jacoby CG, Vontobel PH, Bacha MF, Schneider PH. Highly efficient organocatalysts for the asymmetric aldol reaction. New J Chem. 2018;42:7416–21.

[28] (a)Wu C, Fu X, Li S. A highly efficient, large-scale, asymmetric direct aldol reaction employing simple threonine derivatives as recoverable organocatalysts in the presence of water. Eur J Org Chem. 2011:1291–9; (b) Larionova NA, Kucherenko AS, Siyutkin DE, Zlotin SG. (*S*)-Threonine/α,α-(*S*)-diphenylvalinol-derived chiral ionic liquid: an immobilized organocatalyst for asymmetric *syn*-aldol reactions. Tetrahedron. 2011;67:1948–54; (c) Henseler AH, Ayats C, Pericas MA. An enantioselective recyclable polystyrene-supported threonine-derived organocatalyst for aldol reactions. Adv Synth Catal. 2014;356:1795–1802; (d) Deng J, Zhao B, Deng J. Optically active helical polyacetylene bearing ferrocenyl amino-acid derivative in pendants. Preparation and application as chiral organocatalyst for asymmetric aldol reaction. Ind Eng Chem Res 0.2016;55:7328–37; (e) Gerasimchuk VV, Romanov RR, Woo G H-T, Dmitriev IA, Kucherenko AS, Zlotin SG. Novel *L*-threonine-based ionic liquid supported organocatalyst for asymmetric *syn*-aldol reactions: activity and recyclability design. Arkivoc. 2017:241–9.

[29] (a)Umehara A, Kanemitsu T, Nagata K, Itoh T. Stereoselective synthesis of *vic*-halohydrins via *L*-*tert*-leucine-catalyzed *syn*-selective aldol reaction. Synlett. 2012;23:453–7; (b) Fanjul-Mosteirín N, Concellón C, del Amo V. *L*-Isoleucine in a choline chloride/ethylene glycol deep eutectic solvent: a reusable reaction kit for the asymmetric cross-aldol carboligation. Org Lett. 2016;18:4266–9; (c) Owolabi IA, Reddy UV, Chennapuram M, et al. A new type of amino amide organocatalyzed enantioselective crossed aldol reaction of ketones with aromatic aldehydes. Tetrahedron. 2018;74:4705–11.

[30] Verkade JM, van Hemert LJ, Quaedflieg PJ, Rutjes FP. Organocatalyzed asymmetric Mannich reactions. Chem Soc Rev. 2008;37:29–41.

[31] An Y-J, Wang C-C, Liu Z-P, Tao J-C. Isosteviol-proline conjugates as highly efficient amphiphilic organocatalysts for asymmetric three-component Mannich reactions in the presence of water. Helv Chim Acta. 2012;95:43–51.

[32] Ibrahem I, Zou W, Engqvist M, Xu Y, Córdova A. Acyclic chiral amines and amino acids as inexpensive and readily tunable catalysts for the direct asymmetric three-component Mannich reaction. Chem Eur J. 2005;11:7024–9.

[33] (a)Teo Y-C, Lau J-J, Wu M-C. Direct asymmetric three-component Mannich reactions catalyzed by a siloxy serine organocatalyst in water. Tetrahedron Asymmetry. 2008;19:186–90; (b) Yong F-F, Teo Y-C. Recyclable siloxy serine organocatalyst for the direct asymmetric Mannich reactions in ionic liquids. Synth Commun. 2011;41:1293–1300.

[34] (a)Gawande MB, Velhinho A, Nogueira ID, Ghumman CA, Teodoro OM, Branco PS. A facile synthesis of cysteine-ferrite magnetic nanoparticles for application in multicomponent reactions – a sustainable protocol. RSC Adv. 2012;2:6144–9; (b) Huang Y, Guan D, Wang L. One-pot synthesis of *β*-amino esters by recyclable magnetic organocatalyst-catalysed Mannich-type reaction. J Chem Res. 2014;38:297–9.

[35] (a)Wu C, Fu X, Ma X, Li X, Li C. Threonine-surfactant organocatalysts for the highly diastereo- and enantioselective direct anti-Mannich reactions of hydroxyacetone. Tetrahedron Lett. 2010;51:5775–7; (b) Nugent TC, Sadiq A, Bibi A, et al. Noncovalent bifunctional organocatalysts: powerful tools for contiguous quaternary-tertiary stereogenic carbon formation, scope, and origin of enantioselectivity. Chem Eur J. 2012;18:4088–98.

[36] Ayats C, Henseler H, Dibello E, Pericas MA. Continuous flow enantioselective three-component *anti*-Mannich reactions catalyzed by a polymer-supported threonine derivative. ACS Catal. 2014;4:3027–33.

[37] Liu Y, Wei Z, Liu Y, et al. Novel α-amino acid-derived phase-transfer catalyst application to a highly enantio- and diastereoselective nitro-Mannich reaction. Org Biomol Chem. 2017;15:9234–42.

[38] Bhadury PS, Yang S, Song B-A. Catalytic synthesis of optically active β-amino acid derivatives. Curr Org Synth. 2012;9:695–726.

[39] Enders D. Hüttl M R M. Direct organocatalytic a-fluorination of aldehydes and ketones. Synlett. 2005:991–3.

[40] Marigo M, Fielenbach D, Braunton A, Kjærsgaard A, Jørgensen KA. Enantioselective formation of stereogenic carbon-fluorine centers by a simple catalytic method. Angew Chem Int Ed. 2005;44:3703–6.

[41] Steiner DD, Mase N, Barbas CF. Direct asymmetric a-fluorination of aldehydes. Angew Chem Int Ed. 2005;44:3706–10.

[42] Beeson TD, MacMillan DW. Enantioselective organocatalytic a-fluorination of aldehydes. J Am Chem Soc. 2005;127:8826–8.

[43] Yang X, Phipps RJ, Toste FD. Asymmetric fluorination of α-branched cyclohexanones enabled by a combination of chiral anion phase-transfer catalysis and enamine catalysis using protected amino acids. J Am Chem Soc. 2014;136:5225–8.

[44] Shang J-Y, Li L, Lu Y, Yang K-F, Xu L-W. Enantioselective fluorination reaction of β-ketoester-catalyzed chiral primary amine-based multifunctional catalyst systems. Synth Commun. 2014;44:101–14.

[45] Shibatomi K, Yamamoto Y. Stereoselective synthesis of a,a-chlorofluoro carbonyl compounds leading to the construction of fluorinated chiral quaternary carbon centers. Angew Chem Int Ed. 2008;47:5796–8.

[46] Shibatomi K, Okimi T, Abe Y, Narayama A, Nakamura N, Iwasa S. Organocatalytic asymmetric fluorination of α-chloroaldehydes involving kinetic resolution. Beilstein J Org Chem. 2014;10:323–31.

[47] (a)Marigo M, Jørgensen KA Organocatalytic direct asymmetric a-heteroatom functionalization of aldehydes and ketones. Chem Commun. 2006:2001–11; (b) Makino K, Kubota S, Hara S, Sakaguchi M, Hamajima A, Hamada Y. Reagent-controlled diastereoselective synthesis of (2*S*,3*R*)- and (2*R*,3*R*)-2,3-diaminobutanoic acid derivatives using proline-catalyzed a-hydrazination reaction. Tetrahedron. 2009;65:9468–73.

[48] (a)Theodorou A, Papadopoulos GN, Kokotos CG. β-*tert*-Butyl aspartate as an organocatalyst for the asymmetric a-amination of a,a-disubstituted aldehydes. Tetrahedron. 2013;69:5438–43; (b) Fu J-Y, Yang Q-C, Wang Q-L, Ming J-N, Wang F-Y, Xu X-Y, Wang L-X. Enantioselective α-amination of branched aldehydes promoted by simple chiral primary amino acids. Org Chem. 2011;76:4661–4; (c) Liu C, Zhu Q, Huang K-W, Lu Y. Primary amine/CSA ion pair: a powerful catalytic system for the asymmetric enamine catalysis. Org Lett. 2011;13:2638–41; (d) Desmarchelier A, Yalgin H, Coeffard V, Moreau X, Greck C. Primary amine catalyzed electrophilic amination of α,α-disubstituted aldehydes. Tetrahedron Lett. 2011;52:4430–2.

[49] Grunenfelder CE, Kisunzu JK, Wennemers H. Peptide-catalyzed stereoselective conjugate addition reactions of aldehydes to maleimide. Angew Chem Int Ed. 2016;55:8571–4.

[50] Schnitzer T, Wennemers H. Thieme chemistry journals awardees – where are they now? A stereoselective tripeptide catalyst for conjugate addition reactions of acetophenones to dicyanoolefins. Synlett. 2017;28:1282–6.

[51] Gong J-J, Yuan K, Wu X-Y. Valine-derived phosphinothiourea as organocatalyst in enantioselective Morita-Baylis-Hillman reactions of acrylates with aromatic aldehydes. Tetrahedron Asymmetry. 2009;20:2117–20.

[52] Gergelitsova I, Tauchman J, Cisarova I, Vesely J. Bifunctional (thio)urea-phosphine organocatalysts derived from *D*-glucose and *α*-amino acids and their application to the enantioselective Morita-Baylis-Hillman reaction. Synlett. 2015;26:2690–6.

[53] Zhong F, Wang Y, Han X, Huang K-W. *L*-Threonine-derived novel bifunctional phosphine-sulfonamide catalyst-promoted enantioselective aza-Morita-Baylis-Hillman reaction. Org Lett. 2011;13:1310–3.

[54] Gao Y, Xu Q, Shi M. Enantioselective synthesis of polycyclic indole derivatives based on aza-Morita-Baylis-Hillman reaction. ACS Catal. 2015;5:6608–14.

[55] For review see: (a)Mondal A, Bhowmick S, Ghosh A, Chanda T, Bhowmick KC. Advances on asymmetric organocatalytic 1,4-conjugate addition reactions in aqueous and semi-aqueous media. Tetrahedron Asymmetry. 2017;28:849–75; (b) Butler RN, Coyne AG. Water: nature's reaction enforcer – comparative effects for organic synthesis "in-water" and "on-water". Chem Rev. 2010;110:6302–37; (c) Li C-J. Organic reactions in aqueous media with a focus on carbon–carbon bond formations: a decade update. Chem Rev. 2005;105:3095–3165; (d) Fuxin C, Pin G, Yongfei G, Huikuan Z, Anning Z. Asymmetric organocatalytic Michael additions in aqueous media. Mini-Rev Org Chem. 2013;10: 207–16; (e) Hong L, Sun W, Yang D, Li G, Wang R. Additive effects on asymmetric catalysis. Chem Rev. 2016;116:4006–4123.

[56] Mase N, Watanabe K, Yoda H, Takabe K, Tanaka F, Barbas CF. Organocatalytic direct Michael reaction of ketones and aldehydes with *β*-nitrostyrene in brine. J Am Chem Soc. 2006;128:4966–7.

[57] (a)Kokotos CG, Limnios D, Triggidou D, Trifonidou M, Kokotos G. Novel pyrrolidine-thiohydantoins/thioxotetrahydropyrimidinones as highly effective catalysts for the asymmetric Michael addition. Org Biomol Chem. 2011;9:3386–95; (b) Tsakos M, Trifonidou M, Kokotos CG, Tetrahedron. 2012;68:8630–5; (c) Tsakos M, Elsegood M R, Kokotos CG, Organocatalytic asymmetric domino Michael-Henry reaction for the synthesis of substituted bicyclo[3.2.1]octan-2-ones. Chem Commun. 2013;49:2219–21.

[58] (a)Vinayagam P, Vishwanath M, Kesavan V. New class of bifunctional thioureas from l-proline: highly enantioselective Michael addition of 1,3-dicarbonyls to nitroolefins. Tetrahedron Asymmetry. 2014:25:568–77; (b) Pratap Reddy Gajulapalli V, Vinayagam P, Kesavan V. Organocatalytic asymmetric decarboxylative cyanomethylation of isatins using *L*-proline derived bifunctional thiourea. Org Biomol Chem. 2014;12:4186–91; (c) Reddy Gajulapalli VP, Vinayagam P, Kesavan V. Enantioselective assembly of functionalized carbocyclic spirooxindoles using an L-proline derived thiourea organocatalyst. RSC Adv. 2015;5:7370–9; (d) Pratap Reddy Gajulapalli V, Lokesh K, Vishwanath M, Kesavan V. Organocatalytic construction of spirooxindole naphthoquinones through Michael/ hemiketalization using L-proline derived bifunctional thiourea. RSC Adv. 2016;6:12180–4; (e) Vishwanath M, Prakash M, Vinayagam P, Kesavan V. Construction of polycyclic fused pyrrolidines with three contiguous stereocenters via Michael addition of vinylmalononitriles to nitrostyrenes – using l-proline-derived thiourea. Synthesis. 2016;48:2671–8; (f) Muthusamy S, Prakash M, Ramakrishnan C, Gromiha MM, Kesavan V. Organocatalytic enantioselective assembly of spirooxindole-naphthopyrans through Tandem Friedel-Crafts Type/Hemiketalization. Chem Cat Chem. 2016;8:1708–12; (g) Kumarswamyreddy N, Kesavan

V. Enantioselective synthesis of dihydrospiro[indoline-3,4′-pyrano[2,3-*c*]pyrazole] derivatives via Michael/Hemiketalization reaction. Org Lett. 2016;18:1354–7; (h) Vishwanath M, Sivamuthuraman K, Kesavan V. An efficient construction of *N,N*-bicyclic pyrazolidinones comprising enaminonitriles via asymmetric [3+2] cycloaddition. Chem Commun. 2016;52:12314–7; (i) Vishwanath M, Vinayagam P, Gajulapalli V, Kesavan V. Asymmetric organocatalytic assembly of oxindoles fused with spiro-3,4-dihydropyrans with three contiguous stereocenters consisting of vicinal quaternary centers. Asian J Org Chem. 2016;5:613–6; (j) Jin H, Cho SM, Lee J, Ryu DH. Role of configuration at C6 in catalytic activity of *L*-proline-derived bifunctional organocatalysts. Org Lett. 2017;19:2434–7.

[59] Singh S, Chimni SS. Chiral amine catalyzed enantio- and diastereoselective Michael reaction in brine. Tetrahedron Asymmetry. 2012;23:1068–79.

[60] Wang H-W, Oriyama T. A novel proline-based acridone bifunctional organocatalyst for the asymmetric Michael addition of cyclohexanone. Chem Lett. 2018;47:472–4.

[61] Szollosi G, Gombkoto D, Mogyoros AZ, Fulop F. Surface-improved asymmetric Michael addition catalyzed by amino acids adsorbed on laponite. Adv Synth Catal. 2018;360:1992–2004.

[62] Kaplaneris N, Koutoulogenis G, Raftopoulou M, Kokotos CG. 4-Fluoro and 4-hydroxy pyrrolidine-thioxotetrahydropyrimidinones: organocatalysts for green asymmetric transformations in brine. J Org Chem. 2015;80:5464–73.

[63] Blanco V, Leigh DA, Marcos V, Morales-Serna JA, Nussbaumer AL. A switchable [2]rotaxane asymmetric organocatalyst that utilizes an acyclic chiral secondary amine. J Am Chem Soc. 2014;136:4905–8.

[64] Andres JM, Gonzalez M, Maestro A, Naharro D, Pedrosa R. Recyclable chiral bifunctional thioureas derived from [60]fullerene and their use as highly efficient organocatalysts for the asymmetric nitro-Michael reaction. Eur J Org Chem. 2017:2683–91.

[65] Wang H, Wang Z, Li S, Qiu Y, Liu B, Song Z, et al. Synthesis of novel thiazoline catalysts and their application in Michael addition reaction. Chem Res Chin Univ. 2016;32:373–9.

[66] Hayashi Y, Gotoh H, Tamura T, Yamaguchi H, Masui R, Shoji M. Cysteine-derived organocatalyst in a highly enantioselective intramolecular Michael reaction. J Am Chem Soc. 2005;127:16028–9.

[67] (a)Huang H, Zhu K, Wu W, Jin Z, Ye J. Highly diastereoselective and enantioselective Michael addition of 5*H*-oxazol-4-ones to α,β-unsaturated ketones catalyzed by a new bifunctional organocatalyst with broad substrate scope and applicability. Chem Commun. 2012;48: 461–3; (b) Bera K, Satam NS, Namboothiri IN. Enantioselective synthesis of quaternary α-amino acids via *L-tert*-leucine-derived squaramide-catalyzed conjugate addition of α-nitrocarboxylates to enones. J Org Chem. 2016;81:5670–80; (c) Bera K, Namboothiri IN. Enantioselective synthesis of γ-tetrasubstituted nitrosulfonyl carboxylates and amides via *L-tert*-leucine-derived-squaramide catalyzed conjugate addition of nitrosulfones to acrylates and acrylamides. Org Biomol Chem. 2014;12:6425–31.

[68] Uraguchi D, Shibazaki R, Tanaka N, Yamada K, Yoshioka K, Ooi T. Catalyst-enabled site-divergent stereoselective Michael reactions: overriding intrinsic reactivity of enynyl carbonyl acceptors. Angew Chem Int Ed. 2018;57:4732–6.

[69] Farley AJ, Sandford C, Dixon DJ. Bifunctional iminophosphorane catalyzed enantioselective sulfa-Michael addition to unactivated α-substituted acrylate esters. J Am Chem Soc. 2015;137:15992–5.

[70] (a)Arakawa Y, Wiesner M, Wennemers H. Efficient recovery and reuse of an immobilized peptidic organocatalyst. Adv Synth Catal. 2011;353:1201–6; (b) Bachle F, Duschmale J, Ebner C, Pfaltz A, Wennemers H. Organocatalytic asymmetric conjugate addition of aldehydes to nitroolefins: identification of catalytic intermediates and the stereoselectivity-determining step by ESI-MS. Angew Chem Int Ed. 2013;52:12619–23.

[71] Duschmale J, Wennemers H. Adapting to substrate challenges: peptides as catalysts for conjugate addition reactions of aldehydes to *α,β*-disubstituted nitroolefins. Chem Eur J. 2012;18:1111–20.

[72] Kastl R, Wennemers H. Peptide-catalyzed stereoselective conjugate addition reactions generating all-carbon quaternary stereogenic centers. Angew Chem Int Ed. 2013;52:7228–32.

[73] Schnitzer T, Wennemers H. Effect of *γ*-substituted proline derivatives on the performance of the peptidic catalyst H-*D*-Pro-Pro-Glu-NH_2. Synthesis. 2018;50:4377–82.

[74] Durini M, Sahr FA, Kuhn M, Civera M, Gennari C, Piarulli U. Bifunctional 2,5-diketopiperazines as efficient organocatalysts for the enantioselective conjugate addition of aldehydes to nitroolefins. Eur J Org Chem. 2011;5599–607.

[75] Yang H, Wong MW. *β*-Amino acid catalyzed asymmetric Michael additions: design of organocatalysts with catalytic acid/base dyad inspired by serine proteases. J Org Chem. 2011;76:7399–405.

[76] (a)Kokotos CG. An asymmetric Michael addition of *α,α*-disubstituted aldehydes to maleimides leading to a one-pot enantioselective synthesis of lactones catalyzed by amino acids. Org Lett. 2013;15:2406–9; (b) Schiza A, Spiliopoulou N, Shahu A, Kokotos C G. Combining organocatalysis with photoorganocatalysis: photocatalytic hydroacylation of asymmetric organocatalytic Michael addition products. New J Chem. 2018;42:18844–9.

[77] Marigo M, Franzen J, Poulsen TB, Zhuang W, Jørgensen KA. Asymmetric organocatalytic epoxidation of *α,β*-unsaturated aldehydes with hydrogen peroxide. J Am Chem Soc. 2005;127:6964–5.

[78] (a)Sunden H, Ibrahem I, Córdova A. Direct organocatalytic asymmetric epoxidation of *α,β*-unsaturated aldehydes. Tetrahedron Lett. 2006;46:99–103; (b) Zhao G-L, Ibrahem I, Sunden H, Córdova A. Amine-catalyzed asymmetric epoxidation of *α,β*-unsaturated aldehydes. Adv Synth Catal. 2007;349:1210–24.

[79] Peris G, Jakobsche CE, Miller SJ. Aspartate-catalyzed asymmetric epoxidation reactions. J Am Chem Soc. 2007;129:8710–1.

[80] (a)Jakobsche CE, Peris G, Miller SJ. Functional analysis of an aspartate-based epoxidation catalyst with amide-to-alkene peptidomimetic catalyst analogues. Angew Chem Int Ed. 2008, 47, 6707–11; (b) Lichtor PA, Miller SJ. Combinatorial evolution of site- and enantioselective catalysts for polyene epoxidation. Nature Chem. 2012;4:990–5.

[81] (a)Akagawa K, Kudo K. Asymmetric epoxidation of *α,β*-unsaturated aldehydes in aqueous media catalyzed by resin-supported peptide-containing unnatural amino acids. Adv Synth Catal. 2011;353:843–7; (b) Akagawa K, Hirata T, Kudo K. Asymmetric epoxidation of enones by peptide-based catalyst: a strategy inverting Julia-Colonna stereoselectivity. Synlett. 2016;27:1217–22.

[82] Yi H, Zou G, Li Q, Chen Q, Tang J, He M. Asymmetric epoxidation of *α,β*-unsaturated ketones catalyzed by silica-grafted poly-(*L*)-leucine catalysts. Tetrahedron Lett. 2005;46:5665–8.

[83] (a) Demizu Y, Yamagata N, Nagoya S, Sato Y, Doi M, Tanaka M, et al. Enantioselective epoxidation of *α,β*-unsaturated ketones catalyzed by stapled helical *L*-Leu-based peptides. Tetrahedron. 2011;67:6155–65; (b) Yamagata N, Demizu Y, Sato Y, Doi M, Tanaka M, Nagasawa K, Okuda H, Kurihara M. Design of a stabilized short helical peptide and its application to catalytic enantioselective epoxidation of (*E*)-chalcone. Tetrahedron Lett. 2011;52:798–801.

[84] Kim J, Jung S, Park S, Park S. Amino-acid-mediated epoxidation of *α,β*-unsaturated ketones by hydrogen peroxide in aqueous media. Tetrahedron Lett. 2011;52:2866–8.

[85] Huang L, Yao X, An Y, Peng C. Hydrolysis of cinnamaldehyde in aqueous solution at room temperature catalyzed by amino acids. Adv Mater Res. 2012;512:2361–5.

[86] Sohtome Y, Hashimoto Y, Nagasawa K. Guanidine-thiourea bifunctional organocatalyst for the asymmetric Henry (nitroaldol) reaction. Adv Synth Catal. 2005;347:1643–8.

[87] Karadeniz L, Astley ST. Catalyst functional group cooperativity in the amino acid-catalysed nitroaldol condensation reaction. Res Chem Intermed. 2013;39:3407–15.

[88] Luo H, Yan X, Chen L, Li Y, Liu N, Yin G. Enantioselective catalytic domino aza-Michael-Henry reactions: one-pot asymmetric synthesis of 3-nitro-1,2-dihydroquinolines via iminium activation. Eur J Org Chem. 1707;2016:1702–.

[89] Signman MS, Vachal P, Jacobsen EN. A general catalyst for the asymmetric Strecker reaction. Angew Chem Int Ed. 2000;39:1279–81.

[90] Vachal P, Jacobsen EN. Structure-based analysis and optimization of a highly enantioselective catalyst for the Strecker reaction. J Am Chem Soc. 2002;124:10012–4.

[91] Saravanan S, Sadhukhan A, Khan N, Kureshy RI, Abdi SH, Bajaj HC. C_2-Symmetric recyclable organocatalyst for enantioselective Strecker reaction for the synthesis of α-amino acid and chiral diamine – an intermediate for APN inhibitor. J Org Chem. 2012;77:4375–84.

[92] Nasreen A. *L*-Proline catalyzed one pot synthesis of α-aminonitriles. Tetrahedron Lett. 2013;54:3797–800.

[93] Malkov AV, Mariani A, MacDougall KN, Kocovsky P. Role of noncovalent interactions in the enantioselective reduction of aromatic ketimines with trichlorosilane. Org Lett. 2004;6:2253–6.

[94] (a)Malkov AV, Stoncius S, MacDougall KN, Mariani A, McGeoch GD, Kocovsky P. Formamides derived from *N*-methyl amino acids serve as new chiral organocatalysts in the enantioselective reduction of aromatic ketimines with trichlorosilane. Tetrahedron. 2006;62:264–84; (b) Malkov AV, Vrankova K, Sigerson RC, Stoncius S, Kocovsky P. New organocatalysts for the asymmetric reduction of imines with trichlorosilane. Tetrahedron. 2009;65:9481–6.

[95] (a)Liu X-W, Wang C, Yan Y, Wang Y-Q, Sun J. An organocatalyst bearing stereogenic carbon and sulfur centers as an efficient promoter for enantioselective hydrosilylation of 1,4-benzooxazines. J Org Chem. 2013;78:6276–80; (b) Wang C, Wu X, Zhou L, Sun J. *L*-Valine derived chiral *N*-sulfinamides as effective organocatalysts for the asymmetric hydrosilylation of *N*-alkyl and *N*-aryl protected ketimines. Org Biomol Chem. 2015;13:577–82.

[96] Ge X, Qian C, Chen X. Synthesis of novel carbohydrate-based valine-derived formamide organocatalysts by CuAAC click chemistry and their application in asymmetric reduction of imines with trichlorosilane. Tetrahedron Asymmetry. 2014;25:1450–5.

[97] Ge X, Qian C, Ye X, Chen X. Asymmetric reduction of imines with trichlorosilane catalyzed by valine-derived formamide immobilized onto magnetic nano-Fe_3O_4. RSC Adv. 2015;5:65402–7.

[98] Bonsignore M, Benaglia M, Raimondi L, Orlandi M, Celentano G. Enantioselective reduction of ketoimines promoted by easily available (*S*)-proline derivatives. Beilstein J Org Chem. 2013;9:633–40.

[99] Li Q, Li Y, Wang J, Lin Y, Wei Z, Duan H, et al. An efficient proline-based homogeneous organocatalyst with recyclability. New J Chem. 2018;42:827–31.

[100] (a)Paras NA, MacMillan DW. New strategies in organic catalysis: the first enantioselective organocatalytic Friedel-Crafts alkylation. J Am Chem Soc. 2001;123:4370–1; (b) Paras NA, MacMillan DW. The enantioselective organocatalytic 1,4-addition of electron-rich benzenes to α,β-unsaturated aldehydes. J Am Chem Soc. 2002;124:7894–5.

[101] (a)Sparr C, Gilmour R Fluoro-organocatalysts: conformer equivalents as a tool for mechanistic studies. Angew Chem Int Ed. 2010;49:6520–3; (b) Lakhdar S, Mayr H. Counterion effects in iminium-activated electrophilic aromatic substitutions of pyrroles. Chem Commun. 2011;47:1866–8; (c) Holland MC, Paul S, Schweizer WB, Bergander K, Muck-Lichtenfeld C, Lakhdar S, Mayr H, Gilmour R. Noncovalent interactions in oganocatalysis: modulating conformational diversity and reactivity in the MacMillan catalyst. Angew Chem Int Ed. 2013;52:7967–71.

[102] Bartoli G, Bosco M, Carlone A, Pesciaioli F, Sambri L, Melchiorre P. Organocatalytic asymmetric Friedel-Crafts alkylation of indoles with simple *α,β*-unsaturated ketones. Org Lett. 2007;9:1403–5.

[103] Gustafson JL, Lim D, Miller SJ. Dynamic kinetic resolution of biaryl atropisomers via peptide-catalyzed asymmetric bromination. Science. 2010;328:1251–5.

[104] Barrett KT, Miller SJ. Enantioselective synthesis of atropisomeric benzamides through peptide-catalyzed bromination. J Am Chem Soc. 2013;135:2963–6.

[105] Barrett KT, Metrano AJ, Rablen PR, Miller SJ. Spontaneous transfer of chirality in an atropisomerically enriched two-axis system. Nature. 2014;509:71–5.

[106] Diener ME, Metrano AJ, Kusano S, Miller SJ. Enantioselective synthesis of 3-arylquinazolin-4 (3*H*)-ones via peptide-catalyzed atroposelective bromination. J Am Chem Soc. 2015;137:12369–77.

[107] (a)Metrano AJ, Abascal NC, Mercado BQ, Paulson EK, Miller SJ. Structural studies of *β*-turn-containing peptide catalysts for atroposelective quinazolinone bromination. Chem Commun. 2016;52:4816–9; (b) Metrano AJ, Abascal NC, Mercado BQ, Paulson EK, Hurtley AE, Miller SJ. Diversity of secondary structure in catalytic peptides with *β*-turn-biased sequences. J Am Chem Soc. 2017;139:492–516.

[108] For examples with glycine, see: (a) Reddy MB, Jayashankara VP, Pasha MA Glycine-catalyzed efficient synthesis of pyranopyrazoles via one-pot multicomponent reaction. Synth Commun. 2010;40:2930–4; (b) Singh S, Tiwari J, Jaiswal D, Sharma AK, Singh J, Singh V, Singh J. Organocatalyst mediated one pot synthesis of 4*H*-furo[3,4-b]pyran, 4*H*-benzo[g]chromene and 1*H*-benzo[b]xanthene derivatives in aqueous medium: a green approach. Curr Organocatalysis. 2018;5:51–7; (c) Singh SK, Singh KN. Glycine-catalyzed easy and efficient one-pot synthesis of polyhydroquinolines through Hantzsch multicomponent condensation under controlled microwave. J Heterocyclic Chem. 2010;47:194–8.

[109] (a)Khalafi-Nezhad A, Nourisefat M, Panahi F. *L*-Cysteine functionalized magnetic nanoparticles (LCMNP): a novel magnetically separable organocatalyst for one-pot synthesis of 2-amino-4*H*-chromene-3-carbonitriles in water. Org Biomol Chem. 2015;13:7772–9; (b) Nourisefat M, Panahi F, Nabipour M, Heidari S, Khalafi-Nezhad A. *L*-Cysteine-functionalized magnetic nanoparticles (LCMNP): as a magnetic reusable organocatalyst for one-pot synthesis of 9-(1*H*-indol-3-yl) xanthen-4-(9*H*)-ones. J Iran Chem Soc. 2016;13:1853–65; (c) Bankar SR. Nano-Fe_3O_4 @ *L*-cysteine as an efficient recyclable organocatalyst for the green synthesis of bis (indolyl) methanes under microwave irradiation. Curr Organocatalysis. 2018;5:42–50.

[110] Roy HN, Rahman MM, Pramanick PK. Rapid access of some trisubstituted imidazoles from benzyl condensed with aldehydes and ammonium acetate catalyzed by *L*-cysteine. Ind J Chem Sec. 2013;52:153–9.

[111] Tiwari J, Singh S, Saquid M, Tufail F, Sharma AK, Singh S, et al. Organocatalytic mediated green approach: a versatile new *L*-valine promoted synthesis of diverse and densely functionalized 2-amino-3-cyano-4*H*-pyrans. Synth Commun. 2018;48:188–96.

[112] El-Gokha A, Maas G. (2-Formyl-1-phenylcyclopropyl)phosphonates as building blocks for (2-aminomethyl-cyclopropyl)phosphonates. Tetrahedron. 2011;67:2849–57.

[113] (a)Zheng H, Liu Q, Wen S, Yang H, Luo Y. One-pot asymmetric synthesis of 2-aryl-2,3-dihydro-4-quinolones catalyzed by amino acid-derived sulfonamides. Tetrahedron Asymmetry. 2013;24:875–82; (b) Roy HN, Rana M, Munsur AZ A, Lee K-I, Sarker AK. Efficient and convenient synthesis of *1H*-pyrazolo[1,2-b]phthalazine-5,10-dione derivatives mediated by *L*-proline. Synth Commun. 2016;46:1370–6; (c) Bhandari S, Kasana V. Synthesis and herbicidal activity of 4-benzylidene-2-phenyl-oxazol-5(4*H*)-one derivatives using *L*-proline as catalyst. Asian J Chem. 2018;30:1717–22; (d) Javahershenas R, Khalafy J. Heterocycl Commun. 2018;24:37–41.

[114] Scanes RJ, Grossmann O, Grossmann A, Spring DR. Enantioselective synthesis of chromanones via a peptidic phosphane catalyzed Rauhut-Currier reaction. Org Lett. 2015;17:2462–5.

[115] Xu Y, Zhang S, Li L, Wang Y, Zha Z, Wang Z. *L*-Phenylalanine potassium catalyzed asymmetric formal [3+3] annulation of 2-enoyl-pyridine *N*-oxides with acetone. Org Chem Front. 2018;5:376–9.

[116] Jiang J, Zhang M, Wu W-B, Lu H-B, Shi Y-L, Li J-J. *L*-Phenylalanine triflate as organocatalyst for divergent approaches to trisubstituted hexahydroimidazo[1,2-*a*]pyridine and 1,4-diazepane derivatives. Synlett. 2018;29:246–50.

[117] Khaskel A, Gogoi P, Barman P, Bandyopadhyay B. Grindstone chemistry: a highly efficient and green method for synthesis of 3,4-dihydropyrimidin-2-(*1H*)-ones by *L*-tyrosine as an organocatalyst: a combined experimental and DFT study. RSC Adv. 2014;4:35559–67.

[118] (a)Ghorbani-Choghmarani A, Hajjami M, Norouzi M, Abbasityula Y, Eigner V, Dusek M. Diastereoselective and one-pot synthesis of trans-isoquinolonic acids via three-component condensation of homophthalic anhydride, aldehydes, and ammonium acetate catalyzed by aspartic acid. Tetrahedron. 2013;69:6541–4; (b) Ghorbani-Choghmarani A, Hajjami M, Gholamian F, Azadi G. Aspartic acid as a highly efficient and nontoxic organocatalyst for the one-pot synthesis of tri- and tetrasubstituted imidazoles under solvent-free conditions. Russ J Org Chem. 2015;51:352–6.

[119] Azizi K, Karimi M, Shaterian HR, Heydari A. Ultrasound irradiation for the green synthesis of chromenes using arginine-functionalized magnetic nanoparticles as a recyclable organocatalyst. RSC Adv. 2014;4:42220–5.

[120] (a)Cejka J, Wichterlova B. Acid-catalyzed synthesis of mono-and dialkyl benzenes over zeolites: active sites, zeolite topology, and reaction mechanisms. Catal Rev Sci Eng. 2002;44:375–421; (b) Sreekumar K, Sugunan S. A comparison on the catalytic activity of $Zn_{1-x}Co_xFe_2O_4$ (x=0, 0.2, 0.5, 0.8 and 1.0)-type ferrospinels prepared via a low temperature route for the alkylation of aniline and phenol using methanol as the alkylating agentAppl. Catal A. 2002;230:245–51; (c) Perego C, Ingallina P. Combining alkylation and transalkylation for alkylaromatic production. Green Chem. 2004;6:274–9; (d) Izci A, Bodur F. Liquid-phase esterification of acetic acid with isobutanol catalyzed by ion-exchange resins. React Funct Polym. 2007;67:1458–64; (e) Aracil J, Martinez M, Sanchez N, Corma A. Formation of a jojoba oil analog by esterification of oleic acid using zeolites as catalyst. Zeolites. 1992;12:233–6; (f) Dhakshinamoorthy A, Sharmila A, Pitchumani K. Layered double hydroxide-supported *L*-methionine-catalyzed chemoselective *O*-methylation of phenols and esterification of carboxylic acids with dimethyl carbonate: a "green" protocol. Chem Eur J. 2010;16:1128–32.

[121] (a)Fürstner A, Langemann K. Total syntheses of (+)-ricinelaidic acid lactone and of (−)-gloeosporone based on transition-metal-catalyzed C–C bond formations. J Am Chem Soc. 1997;119:9130–6; (b) Meng D, Bertinato P, Balog A, Su D-S, Kamenecka T, Sorensen EJ, Danishefsky SJ. Total syntheses of epothilones A and B. J Am Chem Soc. 1997;119:10073–92; (c) Smith III AB, Zhu W, Shirakami S, Sfouggatakis C, Doughty VA, Bennett CS, Sakamoto Y. Total synthesis of (+)-spongistatin 1. An effective second-generation construction of an advanced EF Wittig salt, fragment union, and final elaboration. Org Lett. 2003;5:761–4; (d) Evans PA, Cui J, Gharpure SJ. Stereoselective construction of *cis*-2,6-disubstituted tetrahydropyrans *via* the reductive etherification of *δ*-trialkylsilyloxy substituted ketones: total synthesis of (-)-centrolobine. Org Lett. 2003;5:3883–5; (e) Keck GE, Truong AP. Synthetic studies on the bryostatins: synthetic routes to analogues containing the tricyclic macrolactone core. Org Lett. 2005;7:2153–6; (f) Denmark SE, Fu J. Asymmetric construction of quaternary centers by enantioselective allylation: application to the synthesis of the serotonin antagonist LY426965. Org Lett. 2002;4:1951–3; (g) Denmark SE, Regens CS, Kobayashi T. Total synthesis of papulacandin D. J Am Chem Soc. 2007;129:2774–6; (h) Maki K, Motoki R, Fujii K, Kanai M,

Kobayashi T, Tamura S, Shibasaki M. Catalyst-controlled asymmetric synthesis of fostriecin and 8-*epi*-fostriecin. J Am Chem Soc. 2005;127:17111–7; (i) Rauniyar V, Hall DG. Rationally improved chiral Brønsted acid for catalytic enantioselective allylboration of aldehydes with an expanded reagent scope. J Org Chem. 2009;74:4236–41; (j) Han SB, Hassan A, Kim IS, Krische MJ. Total synthesis of (+)-roxaticin via C–C bond forming transfer hydrogenation: a departure from stoichiometric chiral reagents, auxiliaries, and premetalated nucleophiles in polyketide construction. J Am Chem Soc. 2010;132:15559–61; (k) Ghosh D, Sahu D, Saravanan S, Abdi SH, Ganguly B, Khan NH, Kureshy RI, Bajaj HC. Synthetically amenable amide derivatives of tosylatedamino acids as organocatalysts for enantioselective allylation of aldehydes: computational rationale for enantioselectivity. Org Biomol Chem. 2013;11:3451–60.

[122] (a)Gibson EL, Vargas K, Hogan E, Holmes A, Rogers PJ, Wittwer J, et al. Effects of acute treatment with a tryptophan-rich protein hydrolysate on plasma amino acids, mood and emotional functioning in older women. Psychopharmacol. 2014;231:4595–610; (b) Nyeste L, Pécs M, Sevella B, Holló J. Production of *L*-tryptophan by microbial processes. Adv Biochem Eng Biotech. 1983;26:175–202; (c) Aghapoor K, Mohsenzadeh F, Darabi HR, Sayahi H, Balavar Y. *L*-Tryptophan-catalyzed Paal-Knorr pyrrole cyclocondensation: an efficient, clean and recyclable organocatalyst. Res Chem Intermed. 2016;42:407–15.

[123] (a)Cheeseman GW, Sir IH, Jones ER, Sondheimer F, Weedon BC. Studies in the polyene series. Part XXVIII. The structure of the C_{14} aldehyde derived from β-ionone and its use for the synthesis of norvitamin A and isovitamin A derivatives. J Chem Soc. 1949:1516–24; (b) Ashokkumar V, Siva A, Chidambaram RR. A highly enantioselective asymmetric Darzens reaction catalysed by proline based efficient organocatalysts for the synthesis of di- and tri-substituted epoxides. Chem Commun. 2017;53:10926–9.

[124] Hermann S, Sack D, Wagenknecht H-A. Proline-rich short peptides with photocatalytic activity for the nucleophilic addition of methanol to phenylethylenes. Eur J Org Chem. 2018:2204–7.

[125] Reviews on reductive functionalization of CO_2 with amines: (a)Tlili A, Blondiaux E, Frogneux X, Cantat T. Reductive functionalization of CO_2 with amines: an entry to formamide, formamidine and methylamine derivatives. Green Chem. 2015;17:157–68; (b) Li Y, Cui X, Dong K, Junge K, Beller M. Utilization of CO_2 as a C_1 building block for catalytic methylation reactions. ACS Catal. 2017;7:1077–86; (c) Klankermayer J, Wesselbaum S, Beydoun K, Leitner W. Selective catalytic synthesis using the combination of carbon dioxide and hydrogen: catalytic chess at the interface of energy and chemistry. Angew Chem Int Ed. 2016;55:7296–343; (d) Liu X-F, Li X-Y, Qiao C, Fu H-C, He L-N. Betaine catalysis for hierarchical reduction of CO_2 with amines and hydrosilane to form formamides, aminals, and methylamines. Angew Chem Int Ed. 2017;56:7425–9.

[126] Nakashima E, Yamamoto H. Process catalyst mass efficiency by using proline tetrazole column-flow system. Chem Eur J. 2018;24:1076–9.

[127] Chiroli V, Benaglia M, Cozzi F, Puglisi A, Annunziata R, Celentano G. Continuous-flow stereoselective organocatalyzed Diels-Alder reactions in a chiral catalytic "homemade" HPLC column. Org Lett. 2013;15:3590–3.

[128] (a)Wu C, Fu X, Li S. Simple and inexpensive threonine-based organocatalysts for the highly diastereo- and enantioselective direct large-scale *syn*-aldol and *anti*-Mannich reactions of *α*-hydroxyacetone. Tetrahedron Asymmetry. 2011;22:1063–73; (b) Wu C, Long X, Li S, Fu X. Simple and inexpensive threonine-based organocatalysts as highly active and recoverable catalysts for large-scale asymmetric direct stoichiometric aldol reactions on water. Tetrahedron: Asymmetry. 2012;23:315–28.

Anthony J. Burke and Gesine J. Hermann

3 Amino-cinchona derivatives

Abstract: In this chapter attention is given to the use of amino-cinchonas and their derivatives (including quaternary ammonium salts) for a whole host of very interesting chemical transformations. The focus is on the stereoselective (enantioselective and diastereoselective) of the target compounds. Much attention is given to the synthesis of medicinal compounds, and other biologically active compounds, as well as the use of sequential catalytic protocols to form the compounds in a highly sustainable manner. Enabling technologies such as flow-chemistry and catalyst immobilization are also reviewed. The inexhaustible array of product types accessed by these catalysts is clearly highlighted.

Keywords: amino-cinchona, organocatalysis, squramide-cinchona, stereoseletive reactions, biologically active compounds

3.1 Introduction

Amino-cinchona alkaloids and their derivatives have gained a lot of interest over the last couple of decades as asymmetric organocatalysts in the synthesis of optically pure compounds. Since the pioneering work of *Chen* [1] and *Melchiorre* [2] numerous research groups in academia and industry have embarked on exploring this challenging field and have contributed to the wealth of inspiring examples and wide applications to date. A number of comprehensive reviews give an excellent overview on the history and developments in the area [3]. In this chapter we will focus on the latest research achievements published since 2015.

Amino-cinchona organocatalysts are bifunctional catalysts bearing a Lewis/Brønsted base functionality, the quinuclidine moiety, and a hydrogen bond donating group, the primary amine group (Figure 3.1). The catalysts are very versatile, they can be conveniently prepared from natural sources [4] and further derivatised on the primary amine group and also the quinuclidine nitrogen. Amino-cinchona compounds can be easily converted into amide, thioamide, sulfonamide, urea, thiourea or squaramide analogues by reaction of the primary amine functionality. This means catalysts can be individually modified and tuned to cover a wide range of catalytic reactions, such as asymmetric conjugate additions, aldol reactions, Friedel–Crafts alkylations, cycloadditions, Mannich reactions, halogenations or cascade reactions, to name a few.

This article has previously been published in the journal Physical Sciences Reviews. Please cite as: Burke, A., Hermann, G. J. Amino-Cinchona Derivatives *Physical Sciences Reviews* [Online] 2021, 2. DOI: 10.1515/psr-2018-0089

https://doi.org/10.1515/9783110590050-003

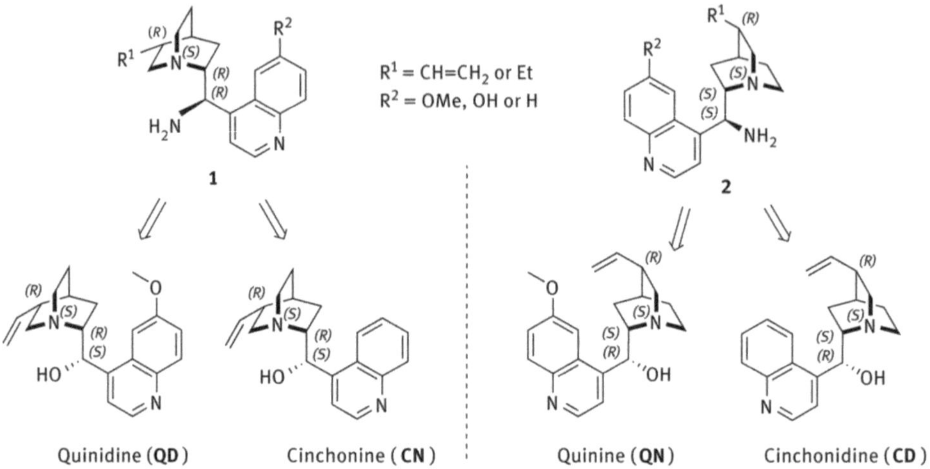

Figure 3.1: Chemical structures of cinchona-based primary amines **1** and **2** derived from quinidine (**QD**)/cinchonine (**CN**) and quinine (**QN**)/cinchonidine (**CD**).

Additionally, the quinuclidine nitrogen can be alkylated giving access to quaternized ammonium salts to produce phase transfer catalyst (PTC) type systems, which adds further beneficial features to certain catalytic processes. Reaction conditions are generally mild with reaction temperatures ranging predominantly from sub-zero to ambient, and occasionally to slightly elevated or reflux temperatures. The catalytic systems are typically metal free (apart from a few exceptions), and they are stable and of low toxicity. This makes them extremely attractive for the development of efficient and environmentally benign processes, in small and large scale, as well as in batch and continuous flow operations. The reaction mechanisms, which are in many cases not fully understood, will not be discussed in detail in this chapter, but references are given where appropriate. The reader can get an insight into some recent elucidation attempts, including computational work by studying reports from *Tanvier et al.* [5], *Hong* [6] and *Houk* [7].

In principal amino-cinchona systems can be categorised into two groups, namely:
1. Quinidine (**QD**)/cinchonine (**CN**)-derived compounds of structure **1** (Figure 3.1)
2. Quinine (**QN**)/cinchonidine (**CD**)-derived compounds of structure **2** (Figure 3.1).

In general, cinchonine (**CN**) and cinchonidine (**CD**) systems (with $R^2 = H$, Figure 3.1) are less frequently employed in asymmetric organocatalysis compared to their quinine (**QN**) and quinidine (**QD**) analogues (with $R^2 = OMe$ or OH).

In the following section (Section 3.2) we will first be discussing simple amino-cinchona organocatalysts of type **1** and **2**. This will then lead us to a range of primary amine derivatised analogues:

- Amides (Section 3.3)
- Thioamides (Section 3.4)
- Sulfonamides (Section 3.5)
- Ureas (Section 3.6)
- Thioureas (Section 3.7)
- Other related derivatives (Section 3.8)
- Squaramides (Section 3.9)

Additionally, quinuclidine substituted quaternary ammonium salts will be included. Due to the wealth of information available for each catalyst type only a selection of recently (since 2015) developed catalysts and their applications will be given.

3.2 Amino-cinchona catalysed reactions

3.2.1 Conjugate addition reactions

3.2.1.1 Michael additions

Amino-cinchona organocatalysts have been extensively studied over the last few years and most successfully been used in conjugate addition reactions, both inter- and intramolecularly.

In 2016, Benaglia and Capriati described asymmetric conjugate additions utilising amino-cinchona catalysts in deep eutectic solvents (DESs) [8]. This is most intriguing as DESs are considered as new environmentally friendly solvents and have not yet been explored much in the field of organocatalysis. The authors screened three different types of eutectic mixtures (DES **A** (choline chloride/urea 1:1), DES **B** (choline chloride/fructose/water 1:1:1) and DES **C** (choline chloride/glycerol 1:2)) with different model reactions. All reactions delivered high conversions and high enantioselectivities within short reaction times. Figure 3.2 shows the conjugate addition of isobutyraldehyde to β-nitrostyrene employing 20 mol% organocatalyst **3** in DES **B**. After 5 h at 25 °C product **4a** was isolated in 89% yield and 95% ee. The reaction was reported to undergo an enamine-activation mechanism between the primary amino group of alkaloid catalyst **3** and the aldehyde substrate. When toluene was used as a solvent only 15% conversion was observed at 25 °C after 5 h with an enantioselectivity of 97% ee. Performing the reaction in water resulted in just 21% yield and 61% ee.

A second example given by the same authors describes the Michael addition of 4-hydroxycoumarin to benzalacetone to afford the pharmaceutically active compound (*S*)-(-)-*Warfarin*, an anticoagulant for the treatment and prevention of blood clots (Figure 3.3). In this reaction the carbonyl functionality required activation with an acidic co-catalyst (TFA) to form an iminium intermediate with organocatalyst **3**.

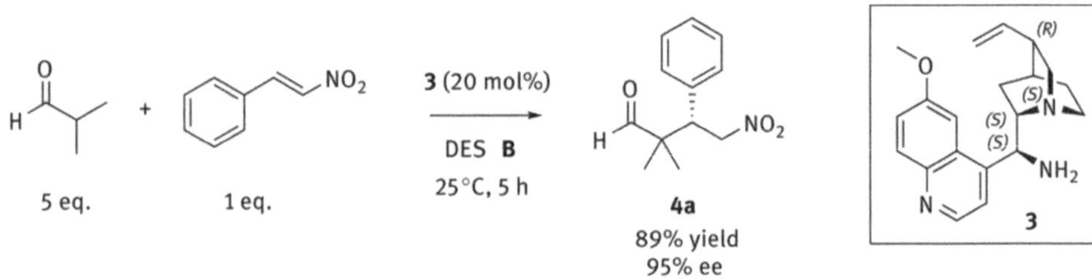

Figure 3.2: Conjugate addition of isobutyraldehyde to β-nitrostyrene in DES **B** catalysed by amino-cinchona organocatalyst **3** as described by Benaglia and Capriati.

The reaction was carried out with 20 mol% amino-cinchona catalyst **3** and 40 mol% TFA in DES **A** at 25 °C in 24 h. A yield of 70% was reported with 87% ee for the product (*S*)-(-)-*Warfarin*.

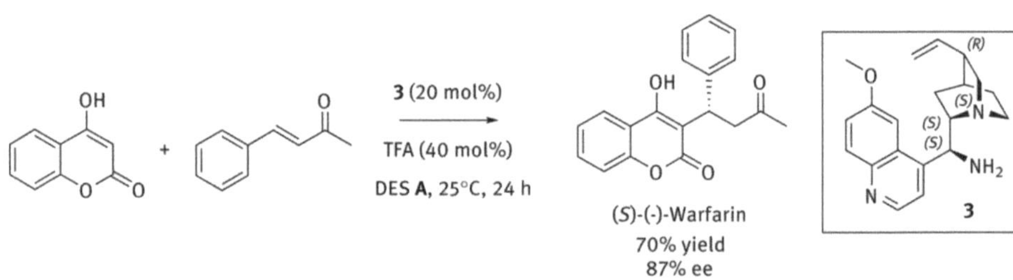

Figure 3.3: Amino-cinchona catalysed conjugate addition of 4-hydroxycoumarin to benzalacetone in the synthesis of (*S*)-(-)-*Warfarin* described by Benaglia and Capriati.

In the third example Benaglia employed organocatalyst **3** to conduct a conjugate addition reaction with concomitant cyclisation (Figure 3.4). In this cyclo-addition reaction (double Michael addition) *E*-3-methyl-3-nitroacrylate was reacted with benzalacetone to afford cyclohexanones, *cis*- and *trans*-product **5** and **6** with 76:24 dr. The reaction was carried out with 20 mol% catalyst **3** and 30 mol% salicylic acid co-catalyst in DES **A** at 50 °C. After a 2 h reaction time the product **5** was obtained in 50% yield with 94% ee and 72% ee for **6**. When toluene was used as a solvent a significantly longer reaction time (20 h) was required affording a 66:34 (*cis*/*trans*)-mixture of the two isomers, with an enantiopurity of 92% ee for **5** and 68% ee for **6**.

Benaglia and co-workers also established a process for attaching the very same 9-amino-9-deoxy-*epi*-quinine catalyst **3** to a polystyrene support to conduct for the first time an enantioselective Michael addition under continuous flow conditions (Figure 3.5) [9]. For this purpose a stainless steel column was packed with the polymer-supported organocatalyst **7**, and a mixture of isobutyraldehyde, β-nitrostyrene and benzoic acid in toluene was passed at room temperature through the catalyst-loaded column (Figure 3.5). This continuous flow set-up was suitable to deliver the conjugate

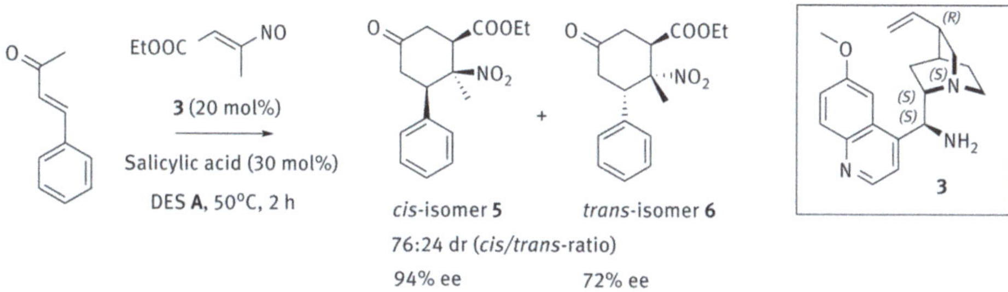

Figure 3.4: Cycloaddition reaction of *E*-3-methyl-3-nitroacrylate with benzalacetone in DES **A** in the presence of organocatalyst **3** and salicylic acid co-catalyst as described by Benaglia and Capriati.

addition product **4a** in 95% yield and 90% ee. The catalyst was reported to be recyclable and the method being applicable to a range of different carbonyl compounds and nitro-olefins; Figure 3.6 shows a cross-selection of reaction products (**4b-4e**) obtained from this process utilising different carbonyl compounds and nitro-olefins.

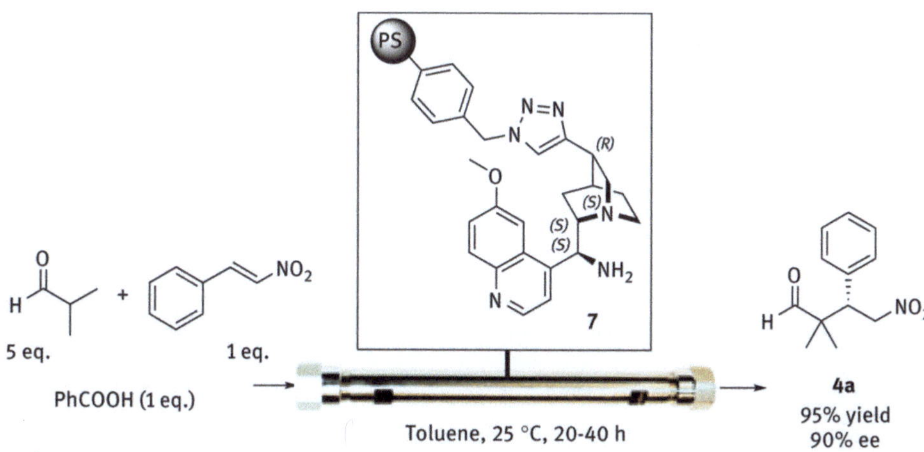

Figure 3.5: Enantioselective Michael addition of isobutyraldehyde to β-nitrostyrene carried out under continuous flow conditions with polymer-bound amino cinchona catalyst **7** as described by Benaglia and co-workers.

The same immobilized cinchona catalyst **7** as shown in Figure 3.5 was also successfully used by Pericàs's group, under batch and continuous flow conditions to promote asymmetric Michael additions with α,β-unsaturated ketones and nitro-acetates [10]. With 10 mol% catalyst **7** and 20 mol% benzoic acid additive the 1,4-conjugate addition reaction between benzalacetone and ethyl nitroacetate gave the product **8a** in 98% yield in chloroform at room temperature within 5 h (Figure 3.7). Product **8a** was obtained with a dr of 1.1:1 and 96/97% ee. The method was also applicable to a number of different enones and nucleophiles with catalyst **7** showing a high level of robustness

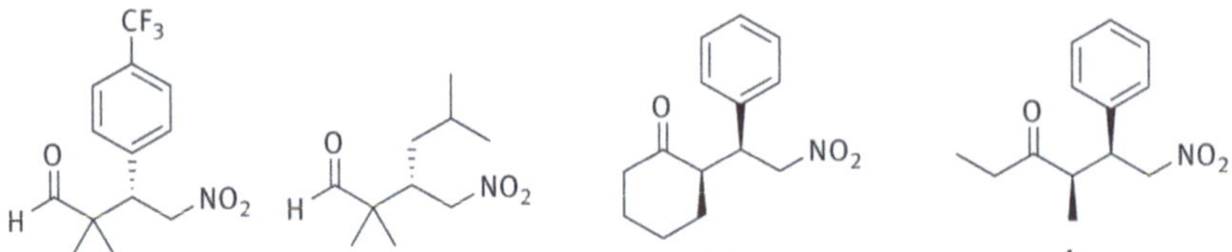

4b 78% yield 97% ee

4c 64% yield 97%ee

4d 65% yield 87:13 *syn:anti* 83% ee

4e 17% yield 95:5 *syn:anti* 98% ee

Figure 3.6: A cross-selection of conjugate addition reaction products (**4b**-**4e**) obtained from different carbonyl compounds and nitro-olefins with organocatalyst **7** under continuous flow conditions.

and recyclability. (For a selection of further examples (**8b**-**8g**) see Figure 3.8). The diastereoselectivity was generally low.

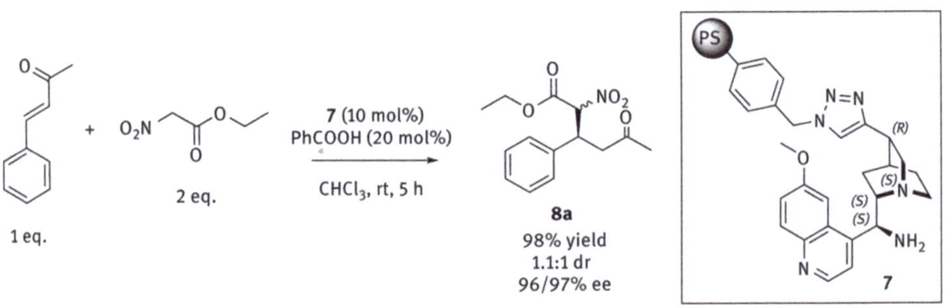

Figure 3.7: Asymmetric Michael addition of benzalacetone and ethyl nitroacetate employing polystyrene-supported amino-cinchona catalyst **7** as described by Pericàs and co-workers.

In 2017 Kim and co-workers carried out an asymmetric synthesis of 2,3-disubstituted indoles, which represent attractive building blocks for medicinal chemistry applications [11]. These compounds could be constructed *via* an organocatalytic intramolecular Michael addition employing amino-cinchona catalyst **9**. With 10 mol% catalyst **9**, 20 mol% 2-nitro-benzoic acid (as additive) in ethyl acetate at 0 °C, compound **10** was converted into the chiral indoline **11** in 95% yield with 62:38 dr (*cis*/*trans*), 98% ee (*cis*-isomer) and 94% ee (*trans*-isomer) (Figure 3.9). The absolute configuration of the major product (*cis*-product) was determined by X-ray crystallography and was shown to have the 2*S*,3*R* absolute configuration.

(-)-*Thelepamide* is a natural product isolated from tidal zone-derived annelid *Thelephus crispus* and shows activity against leukaemia cells. Rodríguez and Christmann [12] recently described a synthetic route towards (-)-*Thelepamide* which includes an amino-cinchona catalysed asymmetric thia-Michael addition (Figure 3.10). Best results were achieved with a catalytic system comprised of the 1:2-adduct **12** of an

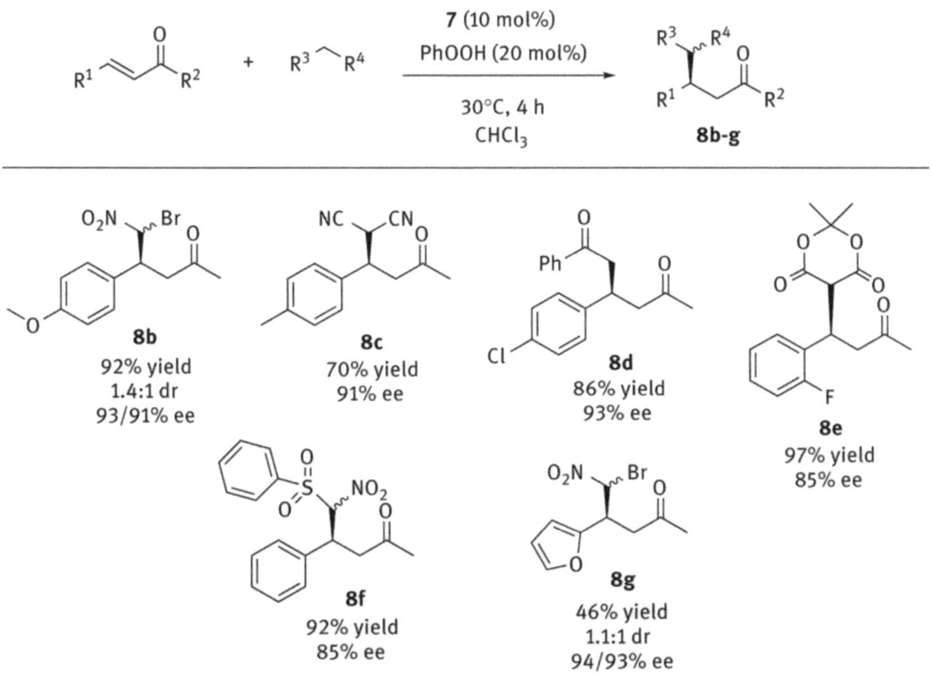

Figure 3.8: Further examples of the asymmetric Michael additions employing polystyrene-supported amino-cinchona catalyst **7** as described by Pericàs and co-workers.

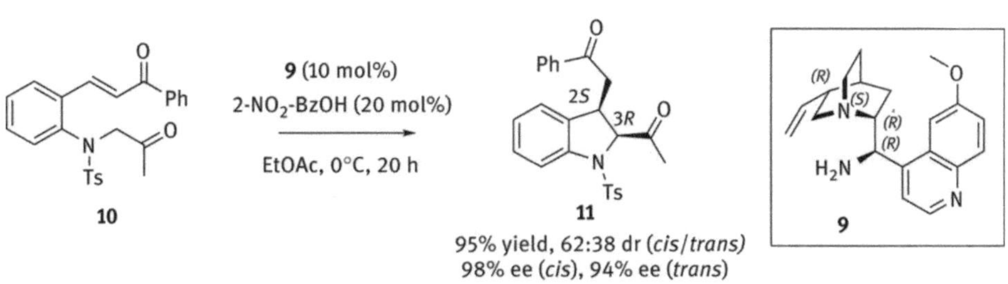

Figure 3.9: Amino-cinchona catalysed intramolecular Michael addition in the synthesis of chiral 2,3-disubstituted indole **11** as described by Kim and co-workers.

amino-cinchona alkaloid with *N*-Boc-phenyl glycine, which had previously been investigated by Melchiorre [13]. The key reaction in the sequence, the stereoselective 1,4-addition of *N*-Boc-protected *L*-cysteine methyl ester to (*E*)-5-octen-4-one was conducted in toluene at –28 °C to deliver the conjugate addition product in 98% yield and 24.5:1 dr. 20 mol% of organocatalyst **12** were used and the major product (*R*-isomer) could be successfully further converted into a diastereomeric mixture of *Thelepamide*. Unfortunately, the key Michael addition required 7 days.

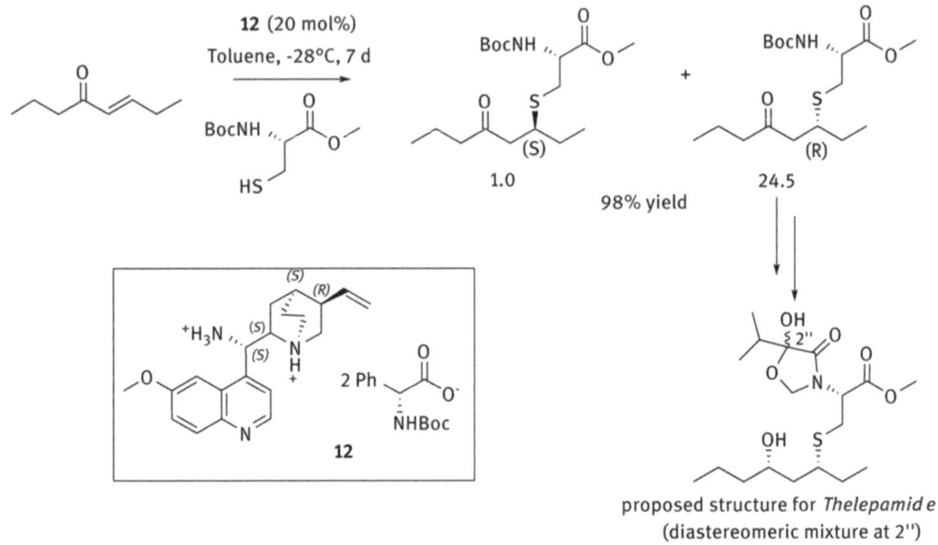

Figure 3.10: Amino-cinchona catalysed thia-Michael addition catalysed by amino-cinchona adduct **12** used in the synthesis of *Thelepamide* as described by Rodríguez and Christmann.

3.2.2 Oxa-Diels–Alder reactions

An example for an asymmetric inverse-electron demand (IED) oxa-Diels–Alder reaction was described by Chen and co-workers in 2016 [14]. The reaction of allyl-ketone **13** with α-cyano-α,β-unsaturated ketone **14** in the presence of 20 mol% of bifunctional amino-cinchona catalyst **15** and 40 mol% of salicylic acid co-catalyst delivered the oxa-Diels–Alder product **16a** in 92% yield with a dr of >19:1 and >99% ee (Figure 3.11). The reaction was conducted in toluene at room temperature over 20 h. The method was applicable to differently substituted allyl-ketones and cyano-ketones giving access to a wide range of highly substituted tetrahydropyrans. These compounds are very useful building blocks in medicinal chemistry and could be isolated in moderate to high yields with high enantioselectivity and moderate to high diastereoselectivity. Figure 3.12 shows a selection of tetra- and penta-substituted tetrahydropyrans (**16b**-**16g**) accessible *via* this route.

3.2.3 α-Fluorination reactions

In 2017 Higashi and co-workers employed amino-cinchona catalysts for carrying out highly enantioselective α-fluorination reactions on α-branched α,β-unsaturated aldehydes to construct allylic fluorides, which are compounds of significant interest for medicinal chemistry applications [15]. The quinuclidine nitrogen of the cinchona-

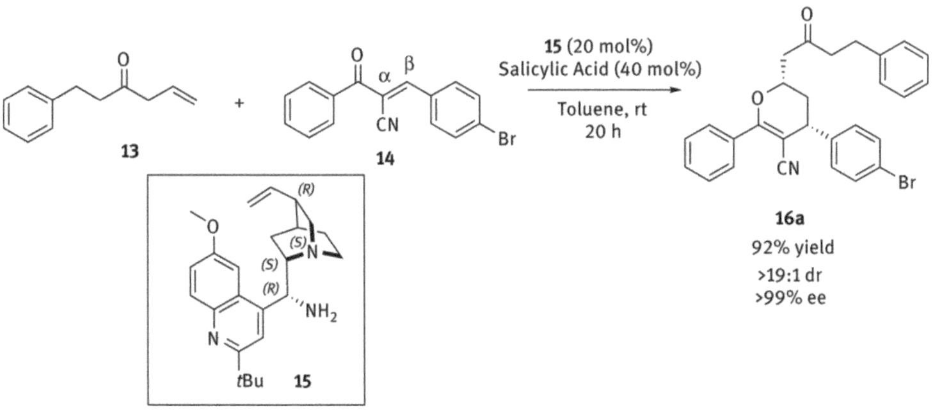

Figure 3.11: Amino-cinchona catalysed asymmetric inverse-electron demand (IED) oxa-Diels–Alder reaction catalysed by amino-cinchona catalyst **15** as developed by Chen's group.

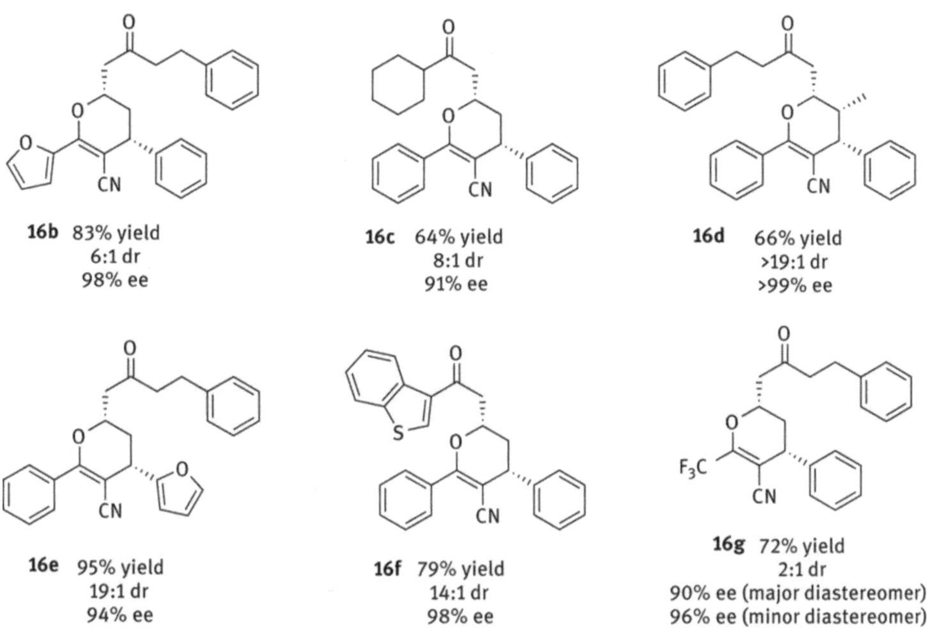

Figure 3.12: Selection of tetra- and penta-substituted tetrahydropyranes (**16b-16g**) synthesised *via* an amino-cinchona catalysed oxa-Diels–Alder reaction as developed by Chen's group.

amine catalyst **17** functions as a coordinating group to control selectivity, something that was demonstrated by density functional theory (DFT) calculations [16]. Best results were achieved with 30 mol% organocatalyst **17**, 60 mol% acid co-catalyst **18** and *N*-fluorobenzenesulfonimide (NFSI) to afford adduct **19** in 79% yield, > 20/1 *E*/*Z*-ratio and 90% ee (Figure 3.13). It was found that polar aprotic solvents, such as

DMF or NMP gave particularly good enantioselectivity; also the presence of Brønsted acid additives and water proved to be beneficial for reaction efficiency and stereoselectivity. To form the opposite enantiomer the analogous quinine derived amino-cinchona catalyst **20** could be used under the same conditions delivering the product with 76% yield, >20:1 dr and 91% ee.

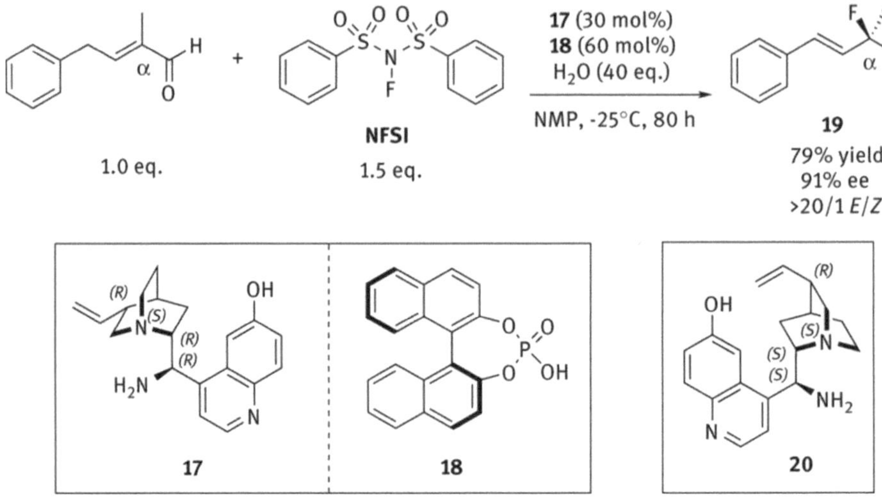

Figure 3.13: Amino-cinchona catalysed enantioselective α-fluorination of α-branched α,β-unsaturated aldehydes as reported by Higashi's group.

3.2.4 Peroxidation reactions

An unprecedented case of an asymmetric peroxidation reaction, catalysed by an amino-cinchona catalyst, was reported by Hu and Deng [17]. This reaction enabled the development of an enantioselective route towards the bicyclic 1,2-dioxane-tetrahydrofuran core structure **24** of stolonoxides, a family of natural products with anticancer properties. With 10 mol% of quinidine-based catalyst **22**, 20 mol% TFA and 2 eq. of peroxide in toluene at 0 °C starting material **21** was converted into a chiral peroxide intermediate. This intermediate was directly converted into the corresponding methyl ester **23** with 82% ee and 57% yield over three steps (Figure 3.14). It is important to note that the structure of the employed hydroperoxide had a major effect on the stereoselectivity of the reaction, with α-methoxydiphenyl hydroperoxide working best.

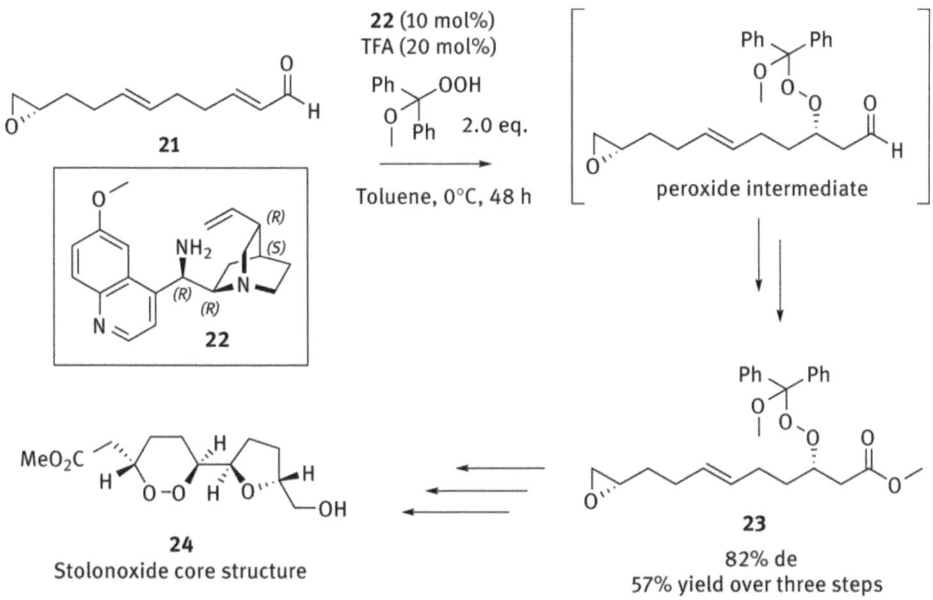

Figure 3.14: Amino-cinchona catalysed asymmetric peroxidation reaction in the synthesis of the stoloxonide core structure **24** as described by Hu and Deng.

3.2.5 Alkylation reactions

Amino-cinchona catalysts have also been reported to function as PTCs in alkylation reactions. Xia and Ma [18] developed a family of novel 9-amino-(9-deoxy) cinchona alkaloid ammonium salts of general structure **25** bearing different substituents on the quinuclidine nitrogen (Figure 3.15). These catalysts are a further development of the so-called third generation catalysts *N*-9-anthracenyl-methyl-*O*-allyl quaternary ammonium salts **26** (Figure 3.15) introduced by Lygo and Corey in 1997 [19].

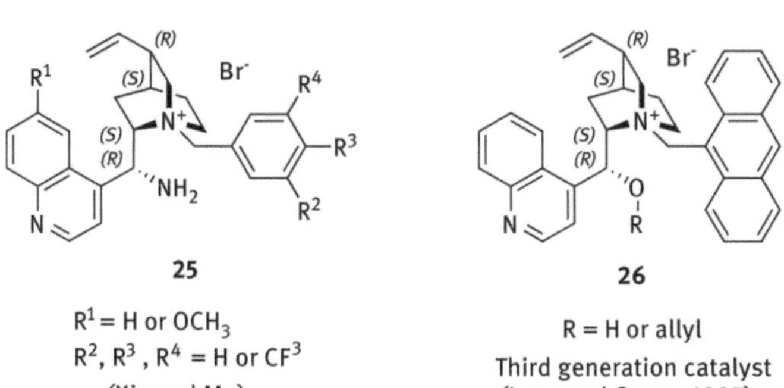

Figure 3.15: Structure of 9-amino-(9-deoxy) cinchona alkaloid ammonium salt PTCs **25** developed by Xie and Ma and third generation PTCs **26** previously developed by Lygo and Corey.

It could be demonstrated that some of the catalysts of general type **25** (Figure 3.15) delivered excellent enantioselectivities and high yields, for example, on using 10 mol% catalyst **27** the benzylation of *N*-(diphenylmethylene)-glycine *tert*-butyl ester as shown in Figure 3.16 worked very well, with enantioselectivities of up to 96% ee and yields of up to 99% having been reported, competing well with third generation PTCs of type **26** (Figure 3.15).

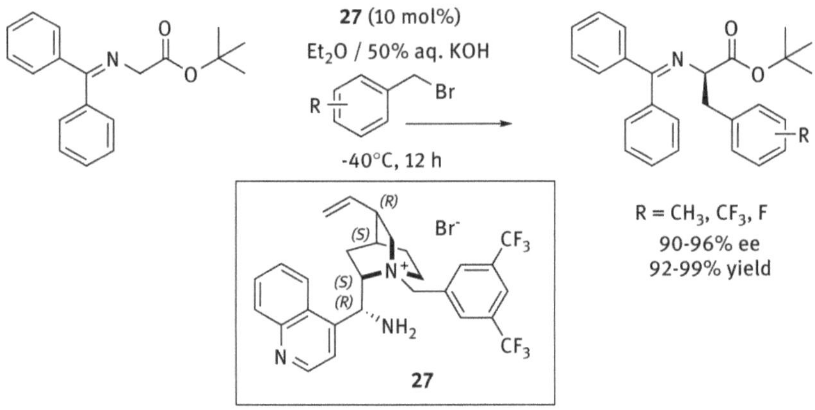

Figure 3.16: Enantioselective phase-transfer catalysed benzylation of *N*-(diphenyl-methylene)-glycine *tert*-butyl ester with amino-cinchona catalyst **27** as developed by Xie and Ma.

3.3 Amino-cinchona amide catalysed reactions

3.3.1 Conjugate addition reactions

3.3.1.1 Michael additions

Very recently Maddox *et al.* developed a novel atrop-selective addition of thiophenol nucleophiles to rapidly interconverting aryl-naphthoquinones to deliver stereochemically stable bisaryl sulphides under enantioselective control [20]. An amino-cinchona derived amide substituted catalyst promoted this reaction (see also Section 3.6.4 for atropisomeric cinchona urea catalysed halogenation reactions). In a typical procedure naphthoquinone **28** was reacted with thiophenol in the presence of 5 mol% cinchona amide catalyst **29** in toluene at 4 °C over 44 h. The conjugate addition intermediate was not isolated, but trapped *in situ* under reductive methylating conditions with $Na_2S_2O_4$ and dimethylsulfate to afford product **30** in 89% yield and 93:7 er (Figure 3.17). The absolute configuration of reaction product **30** was determined by X-ray crystallography. The reaction conditions were suitable for a range of substrates and thiophenol nucleophiles giving access to compounds that are of potential interest as chiral ligands or may function as a chiral scaffold for medicinal chemistry applications.

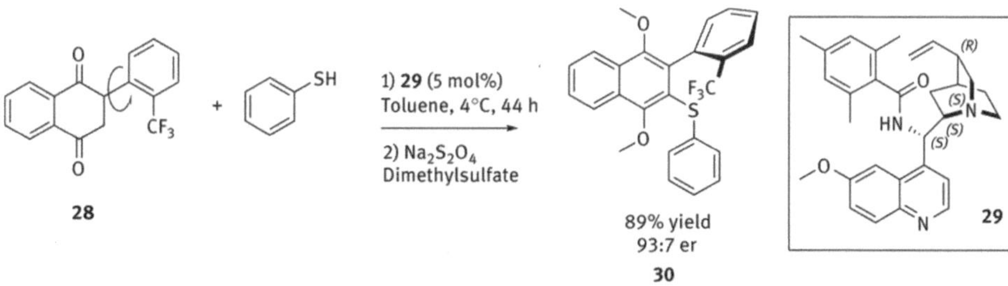

Figure 3.17: Enantioselective conjugate addition of thiophenol to naphthaquinones followed by reductive methylation in the synthesis of biaryl atropisomers as described by Maddox *et al.*

3.3.2 Aldol reactions

Ray and Mukherjee demonstrated the first successful use of a quinine derived bifunctional amide catalyst in enantioselective vinylogous aldol reactions of allyl ketones to pyrazole-4,5-diones [21]. When compound **31** was reacted with vinylketone **32** in the presence of 10 mol% of organocatalyst **33** in cyclopenyl methyl ether (CPME) aldol product **34a** was isolated in 82% yield and gave an er of 96:4 (Figure 3.18). The optical purity of the product could be further increased by re-crystallisation and the absolute configuration of the major product was determined by X-ray crystallography. In comparison, corresponding urea and thiourea catalysts performed less efficiently. The reaction proceeded exclusively in γ- and *E*-selective manner and was applicable to a range of different substrates. A selection of synthesised hydroxyl-pyrazolone aldol products (**34b-34g**) is depicted in Figure 3.19; these chiral compounds represent highly attractive synthons for medicinal chemistry use.

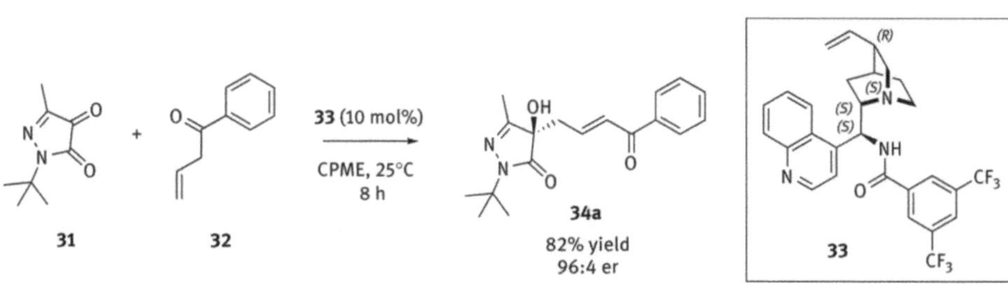

Figure 3.18: Example of an enantioselective vinylogous aldol reaction of allyl ketone **32** to pyrazole-4,5-dione **31** catalysed by cinchona amide catalyst **33** as described by Mukherjee and Ray.

3.3.3 Cycloaddition reactions

Oh and co-workers reported an interesting asymmetric intermolecular [3+2]-cycloaddition reaction between cyclopentenedione **35** and isocyanoacetate **36** [22] (with

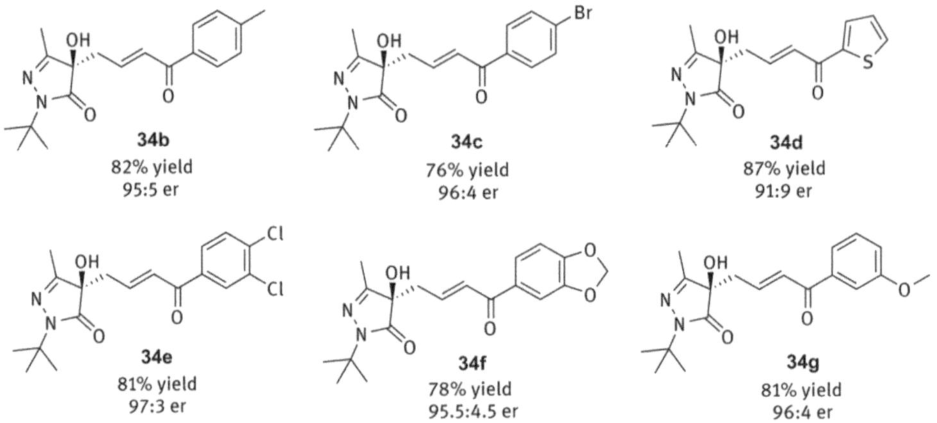

Figure 3.19: Selection of chiral hydroxyl-pyrazolone aldol products synthesised with amino-cinchona amide catalyst **33.**

regard to this reaction, please see Section 3.9.4 for a discussion on the application of cinchona-squaramide catalysts). This reaction was catalysed by an Ag(I)-complex bearing cinchona derived amide-phosphine ligand **37**. With 5 mol% catalyst 99% conversion was observed in ethyl acetate at 0 °C delivering the bicyclic product **38** in 92% ee (Figure 3.20). The chiral catalyst proved to be very air and moisture stable and gave moderate to high yields and high selectivities for a range of substrates.

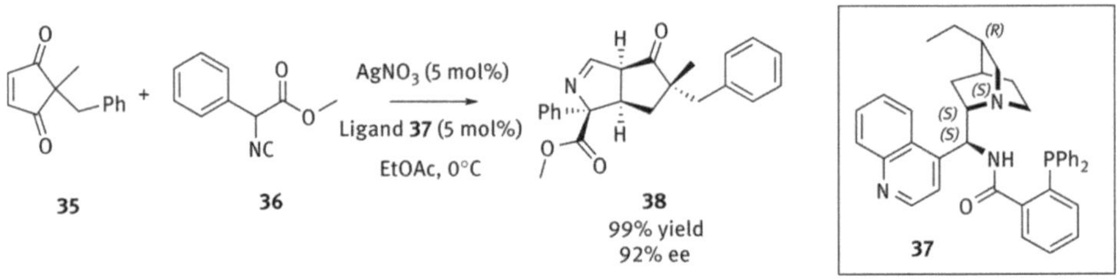

Figure 3.20: Example of a silver catalysed [3+2]-cycloaddition of cyclopentenedione **35** and isocyano-acetate **36** as developed by Oh *et al.*

3.3.4 Ring-opening reactions

Nakamura *et al.* developed a cinchona alkaloid amide catalyst suitable for enantio-selective oxidative ring-opening reactions of aziridines with α-nitro-esters [23]. The utilised organocatalyst **39** (which was just one of the several tested) is structurally characterised by a tetrazole group and a 2-picolinyl group (Figure 3.21). With 10 mol% catalyst, a Lewis acid ($NiBr_2$) and α-nitro-malonate the ring-opening reaction occurred in 95% yield and 97% ee within 4 h. The reaction was carried out in a mixture of toluene and 1,4-dioxane (9:1) at 50 °C. The method was applicable to a broad range of different aziridines giving access to valuable chiral α-amino ketones, which can be further converted under reductive conditions into very valuable optically pure 1,2-amino alcohols.

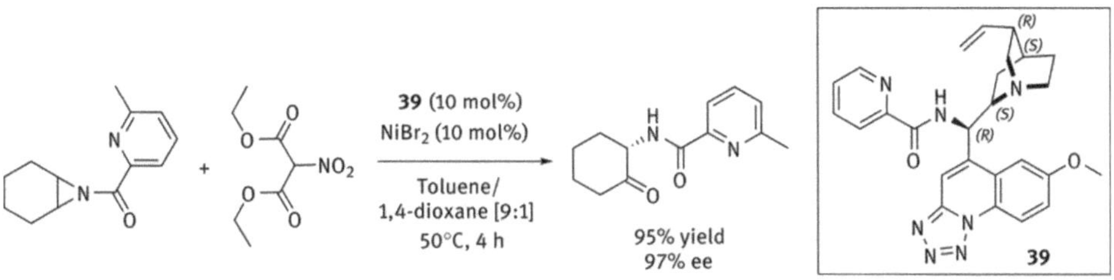

Figure 3.21: Example of an enantioselective oxidative ring-opening of an aziridine with α-nitromalonate catalysed by cinchona-amide catalyst **39** as described by Nakamura's group.

3.3.5 Hydrosilylation reactions

Benaglia and Burke reported a group of interesting cinchona amide catalysts, which proved to be very useful for enantioselective hydrosilylation reactions of ketimines [24]. These catalysts were successfully applied in the synthesis of an advanced precursor of *Rivastigmine*, an acetylcholinesterase inhibitor for the treatment of Alzheimer's disease. When compound **40** was reacted with trichlorosilane in dichloromethane at 0 °C in the presence of quaternized ammonium catalyst **41** hydrosilylation reduction product **42** was obtained in 81% yield and >96% ee (Figure 3.22). Compound **42** could then be further converted into *Rivastigmine*.

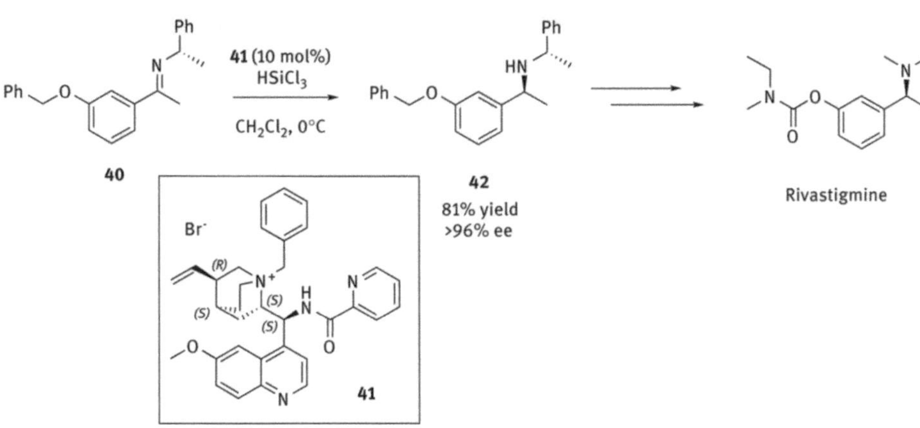

Figure 3.22: Amino-cinchona picolinamide **41** catalysed hydrosilylation reaction in the synthesis of *Rivastigamine* as described by Benaglia and Burke.

3.4 Amino-cinchona thioamide catalysed reactions

3.4.1 Decarboxylative Mannich and decarboxylative protonation reactions

Rouden and Baudoux reported a novel class of thioamide-substituted cinchona organocatalysts [25]; in fact, up to this point thioamide substituted amino-cinchona catalyst had not been investigated much. These catalysts showed activity in asymmetric decarboxylative Mannich and also decarboxylative protonation reactions of α-amido-substituted malonic acid half oxyesters (MAHOs) (see also Section 3.5.2 and Section 3.6.2 for cinchona-sulfonamide and -urea catalysed aldol reactions with MAHOs). When MAHO **43** was reacted with *N*-tosyl phenylimine **44** in the presence of 20 mol% thioamide catalyst **45** and molecular sieves in toluene at 20 °C Mannich products **46a** and **46b** were isolated in 78% yield with a dr of 66:34 (*anti*/*syn*); and an er of 68:32 was observed for the *anti*-product **46a** and an er of 88:12 for the *syn*-product **46b** (Figure 3.23**(A)**). In the case of decarboxylative protonation, benzyl-substituted MAHO **47** was treated with 20 mol% of catalyst **45** in THF at 30 °C to afford product **48** in 84% yield with an er of 92:8 (Figure 3.23**(B)**). It has to be noted that in this type of decarboxylative C–C and C–H bond-forming reactions, thioamide catalysts performed much more efficiently than their amide, thiourea or squaramide analogues.

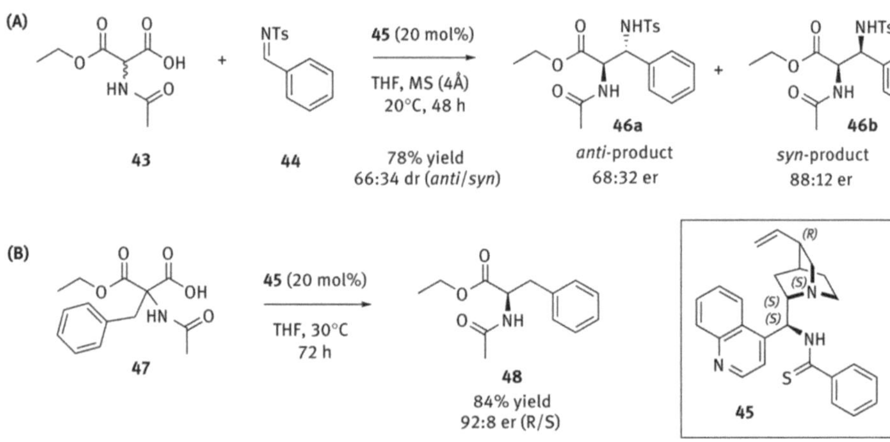

Figure 3.23: Asymmetric decarboxylative Mannich and decarboxylative protonation reactions catalysed by thioamide-substituted cinchona catalyst **45** developed by Rouden and Baudoux.

3.5 Amino-cinchona sulfonamide catalysed reactions

Like the above discussed urea and thiourea catalysts (Section 3.3 and Section 3.4), amino-cinchona sulfonamide organocatalysts can also conveniently be prepared from 9-amino-(9-deoxy) cinchona alkaloids, simply by reaction with sulfonyl chlorides.

These catalysts are also very powerful among the amino-cinchona family and find use in a range of different types of reactions, in monomeric or polymeric form, as exemplified below.

3.5.1 Conjugate addition reactions

3.5.1.1 1,2-Addition reactions

Nakamura *et al.* reported in 2018 the first highly enantioselective reaction of 2*H*-azirine **49** with sulphur nucleophiles [26]. It was demonstrated that a range of different thiols can be used for this addition reaction to deliver aziridines in high yield and very high ee. As an example, with *p*-bromobenzenethiol, 1 mol% catalyst **50** in dichloromethane at −78 °C, addition product **51** was isolated in 96% yield and 94% ee (Figure 3.24). The azirine intermediate **49** was generated *in situ* from an α-azidoacrylate, and the utilized catalyst **50** was the same as for the peroxidation of ketimines discussed in Section 3.4 (Figure 3.28).

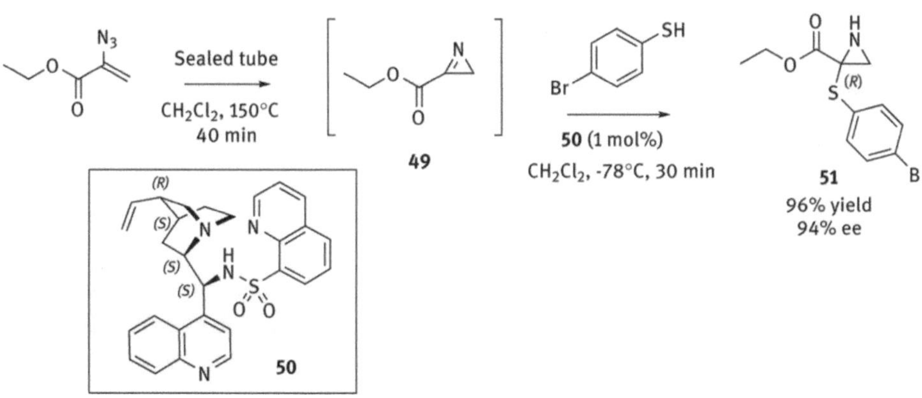

Figure 3.24: Example of an enantioselective reaction of azirine **49** with a thiol nucleophile catalysed by cinchona sulphonamide catalyst **50** as described by Nakamura *et al.*

3.5.1.2 Michael additions

Endo *et al.* demonstrated that polymeric cinchona sulphonamides are also suitable catalysts for asymmetric Michael additions of β-ketoesters on nitrostyrenes [27]. Despite their poor solubility in standard organic solvents these polymeric catalysts provide good diastereoselectivities, very high enantioselectivities and very high yields. With 10 mol% of catalyst **52** the reaction of methyl 2-oxocyclopentanecarboxylate and *trans*-β-nitrostyrene furnished the conjugate addition product **53** in 97% yield and 99% ee with a dr of 11:1 (Figure 3.25).

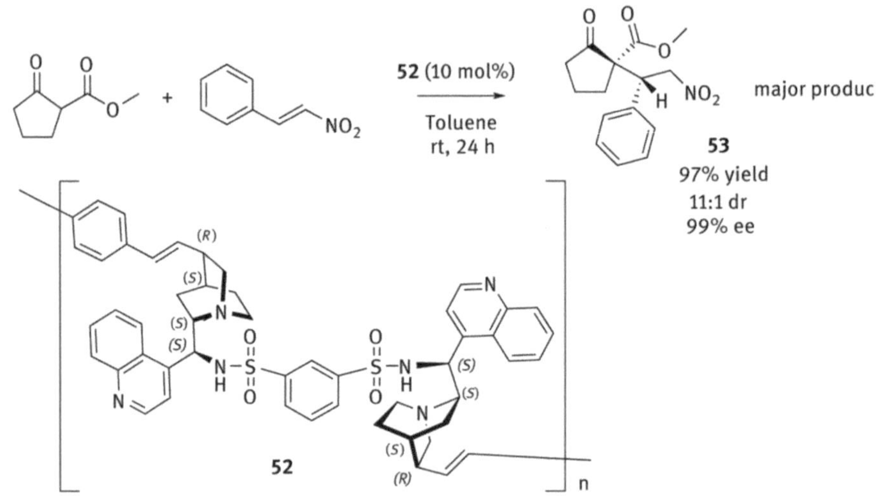

Figure 3.25: Example of an enantioselective Michael addition reaction of methyl 2-oxocyclopentanecarboxylate on *trans*-β-nitrostyrene catalysed by polymeric amino-cinchona sulfonamide catalyst **52** as described by Endo *et al.*

3.5.2 Decarboxylative aldol reactions

March *et al.* reported the first enantioselective and diastereoselective decarboxylative aldol addition of α-amino-substituted malonic half oxyesters (MAHOs) with electron-withdrawing aldehydes to provide *anti*-β-hydroxy-α-amino esters, which are very useful and versatile chiral synthons in organic synthesis [28] (see also Section 3.4.1 for amino-cinchona thioamide catalysed Mannich reactions and 6.2 for cinchona urea catalysed aldol reactions with MAHOs and F-MAHTs). A range of novel bifunctional amino-cinchona sulphonamide catalysts were found to be active delivering moderate to excellent yields and high selectivities for this type of reaction. Best results were obtained with phenyl/Fmoc protected MAHO **54** (Figure 3.26) as a substrate; when compound **54** was reacted with 10 eq. of *p*-nitrobenzaldehyde in CPME in the presence of 20 mol% catalyst **55** and 20 mol% pentafluorobenzoic acid aldol product **56a** was furnished in 95% yield with 88:12 *anti*/*syn* ratio and an er of 89:11. It was found that an *ortho*-substituent on the phenyl ring of the sulfonamide catalyst proved to be essential to achieve high enantioselectivity and diastereoselectivity with a wide range of electron-withdrawing aldehydes.

The method was applicable to a range of different aryl aldehydes. Figure 3.27 shows a selection of aldol products (**56b**-**56f**) that were obtained under these conditions in moderate to high yields and high diastereo- and enantioselectivities.

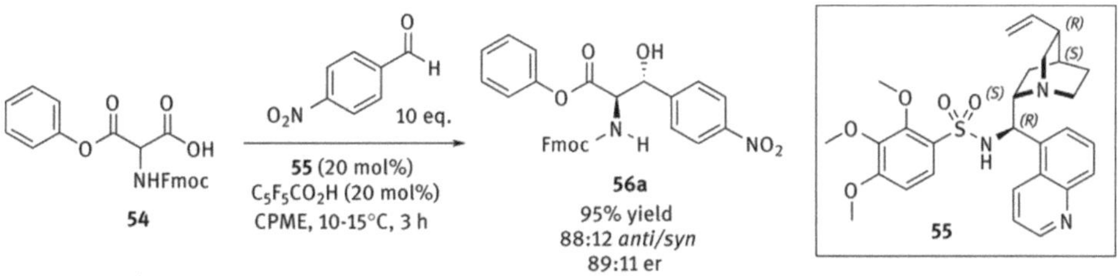

Figure 3.26: Example of a decarboxylative aldol reaction of phenyl/Fmoc-proteced MAHO **54** with *p*-nitrobenzaldehyde catalysed by sulfonamide cinchona catalyst **55** as published by March *et al.*

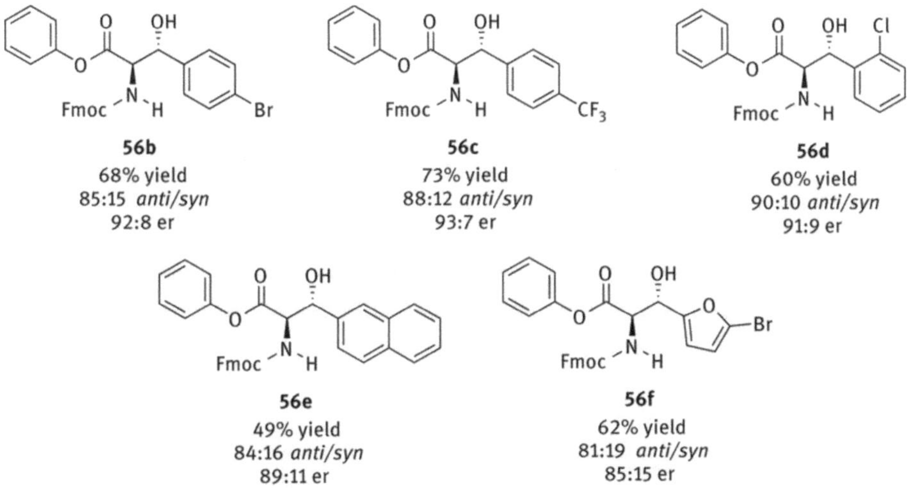

Figure 3.27: Selection of chiral decarboxylative aldol products **56b-56f** synthesised with amino-cinchona sulfonamide catalyst **55** as described by March *et al.*

3.5.3 Peroxidation reactions

In 2015 Nakamura and Takahashi reported the first enantioselective peroxidation reaction of ketimines, catalysed by a cinchona alkaloid sulphonamide catalyst [29]. The ketimines were derived from isatins and subsequently reacted with hydroperoxides to form α-amino peroxides, which are useful chiral building blocks for the preparation of novel compounds of pharmaceutically interest. As an example, with 5 mol% organocatalyst **50** in toluene at ambient temperature very high yields and excellent ees were obtained for the peroxidation of compound **57** with different peroxides (Figure 3.28). It should be noted that both enantiomers are accessible depending on the type of catalyst employed. Catalyst **50** was also successfully used in a 1,2-addition on 2*H*-azirines as shown in Section 3.5.1.1 (Figure 3.24).

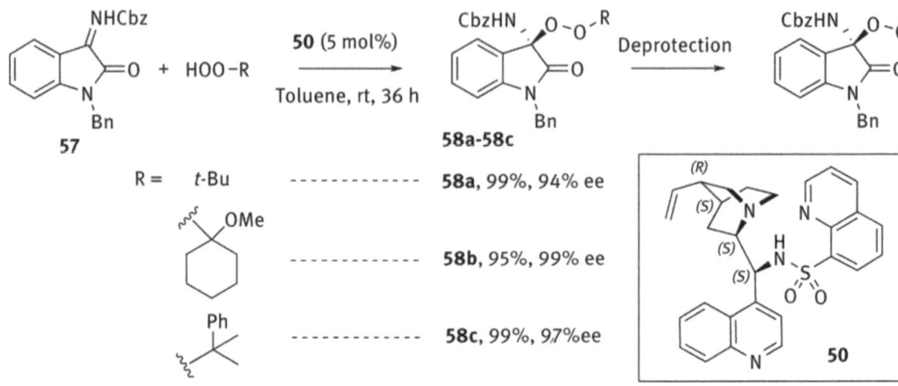

Figure 3.28: Enantioselective peroxidation reaction of ketimine **57** catalysed by cinchona sulphonamide catalyst **50** as described by Nakamura and Takahashi.

3.5.4 Desymmetrization reactions

When cinchona alkaloid sulphonamides are incorporated into polymeric structures, very stable catalysts with different solubility properties can be generated, which often deliver higher yields and improved selectivities compared to their monomeric analogues. The immobilisation makes them also very attractive systems for flow-chemistry applications. Takata *et al.* reported the synthesis of two different types of polymeric amino-cinchona sulfonamide catalysts [30] **59** (one-component catalyst), and **60** (two-component polymer) (Figure 3.29) and derivatives thereof.

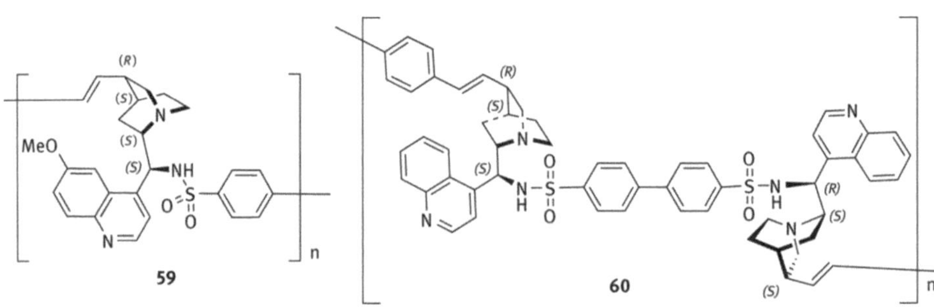

Figure 3.29: Polymeric amino-cinchona sulphonamide catalysts **59** and **60** developed by Takata *et al.*

Catalysts **59** and **60** were successfully used for organocatalytic asymmetric desymmetrization (ASD) reactions of cyclic anhydrides with different alcohols to provide very useful building blocks for the synthesis of biologically active compounds. With 10 mol% of polymeric organocatalyst **59** or **60** in THF at room temperature high yields and moderate to very high ees were obtained (Figure 3.30**(A)**–**(C)**) [31]. The catalysts could be easily recovered from the reaction mixture and were sufficiently stable to be re-used.

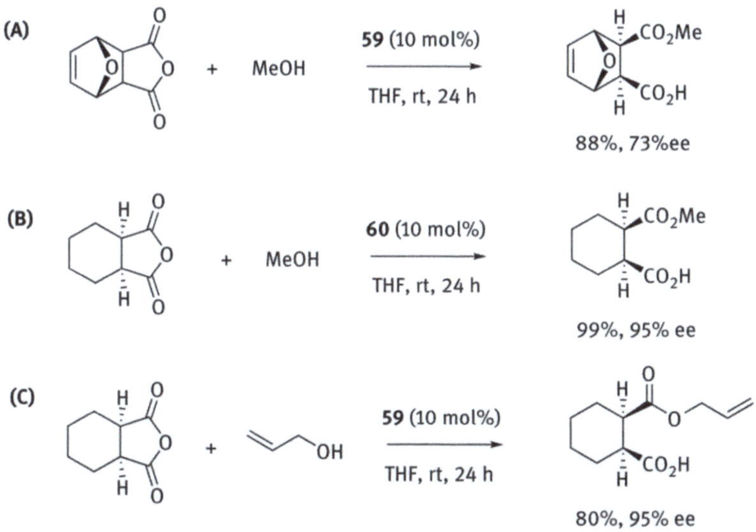

Figure 3.30: Examples of asymmetric desymmetrization reactions of cyclic anhydrides catalysed by polymeric amino-cinchona sulphonamide catalysts **59** and **60.**

3.5.5 Oxytrifluoromethylation reactions

Li *et al.* developed a class of amino-cinchona based sulphonamides to be used as ligands to form very potent catalysts with copper salts for asymmetric radical oxytrifluoromethylation reactions [32]. This method is of significant interest as the introduction of CF_3-groups is important in medicinal chemistry. When alkenyl oxime **61** was reacted with Togni's reagent **62** in the presence of $Cu(OAc)_2$ (10 mol%) and cinchona sulfonamide ligand **63** (15 mol%) in chloroform at −10 °C the chiral isoxazoline **64a** was isolated in 82% yield and 91% ee (Figure 3.31). The addition of molecular sieves (4 Å) proved to be beneficial for obtaining high ees in this reaction. The absolute configuration of the newly formed chiral centre in compound **64a** was determined by an X-ray crystal structure analysis. The obtained isoxazoline **64a** could be further converted into 1,3-aminoalcohols **65a** and **65b** in 49% and 47% yield, respectively with 90% ee. These products also represent very useful chiral building blocks for the synthesis of pharmaceutically active compounds.

The oxytrifluoromethylation method was applicable to a number of different substrates delivering isoxazolines in high yield and high ees. A selection of products (**64b**-**64i**) is shown in Figure 3.32.

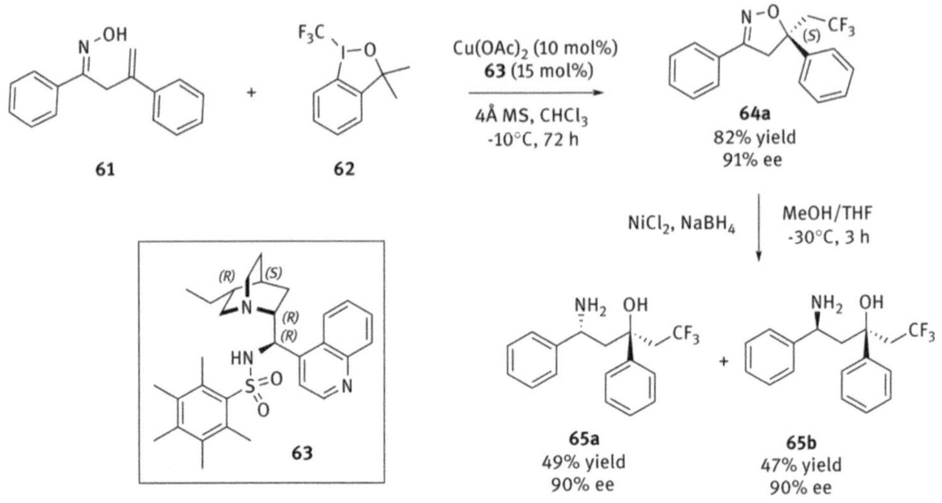

Figure 3.31: Example of an asymmetric radical oxytrifluoromethylation of alkenyl oxime **61** catalysed by an amino-cinchona sulphonamide **63** coordinated copper complex as described by Li *et al.*

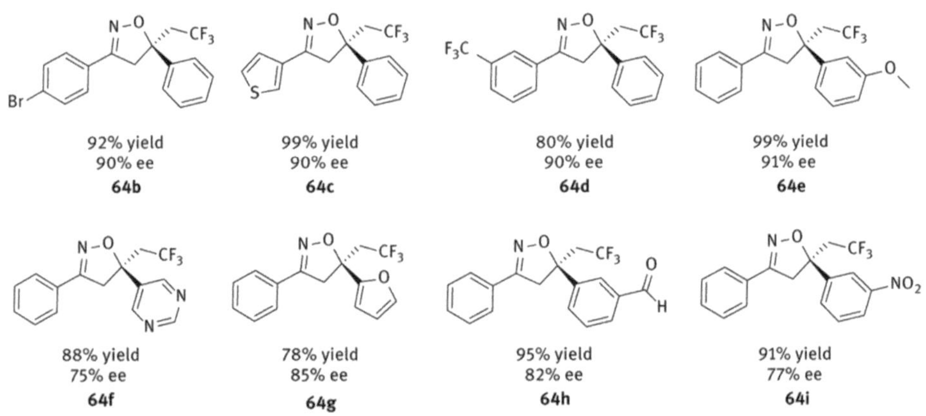

Figure 3.32: Selection of chiral trifluoromethyl isoxazolines (**64b-64i**) derived from amino-cinchona sulphonamide **63** complex catalysed radical oxytrifluoromethylation reactions.

3.6 Amino-cinchona urea catalysed reactions

Amino-cinchona urea organocatalysts are widely used for various types of asymmetric catalytic reactions.

3.6.1 Conjugate addition reactions

3.6.1.1 Michael additions

Wennemers' research group developed the conjugate addition reaction of fluorinated monothiomalonates (F-MTMs, the protected form of the below mentioned F-MAHTs, Section 3.6.2) with nitro-olefins [33]. The addition products represent very attractive synthons for the synthesis of pharmaceutically active compounds. When the PMP (*p*-methoxy phenyl) protected α-fluoro thiomalonate **66** was reacted with β-nitrostyrene **67** in the presence of just 1 mol% of an *epi*-cinchona urea (eCNU) catalyst **69** the addition product **70a** was isolated in 96% yield, with >20:1 dr and 99% ee (Figure 3.33). The reaction was best conducted in mesitylene **68** at −30 °C for 2 h. The resulting α-fluoro-γ-nitro thioester **70a** could be successfully further converted into the fluoro-analogue of *AC-264613*, a protease-activated receptor 2 (PAR2) agonist, an increasingly interesting pharmaceutical target for various diseases. The absolute configuration of the conjugate addition product was determined by X-ray crystal structure analysis.

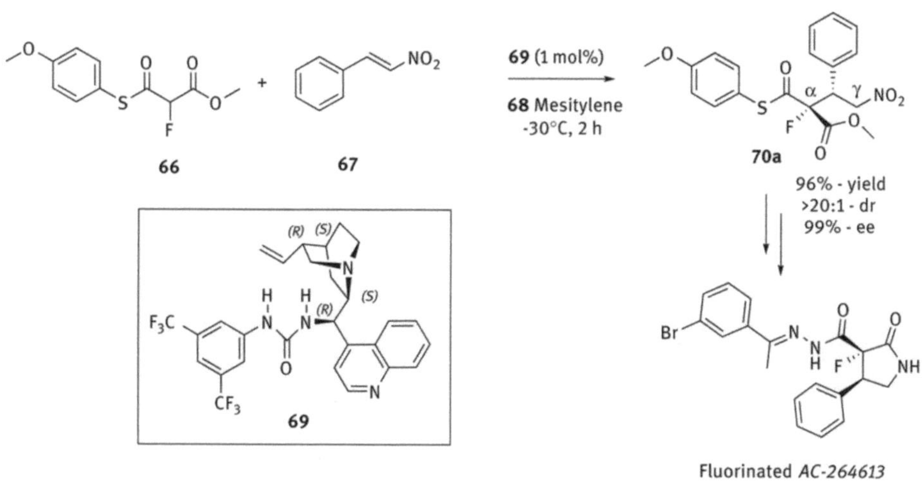

Figure 3.33: Example of the conjugate addition of α-fluoro thiomalonate (F-MTM) **68** with β-nitrostyrene, catalysed by *epi*-cinchona urea catalyst **69** in the synthesis of fluorinated *AC-264613* as described by Wennemers group.

The addition reaction of F-MTMs was applicable to a range of nitro-olefins delivering α-fluoro-γ-nitro thioesters in high yields and very high ees; a selection of successfully synthesised examples (**70b**-**70g**) is shown in Figure 3.34.

Phelan and Ellman developed a catalytic enantioselective addition reaction of pyrazolones to trisubstituted nitroalkenes utilising an *N*-sulfinylurea cinchona organocatalyst [34]. This conversion is special in the sense that the Michael addition

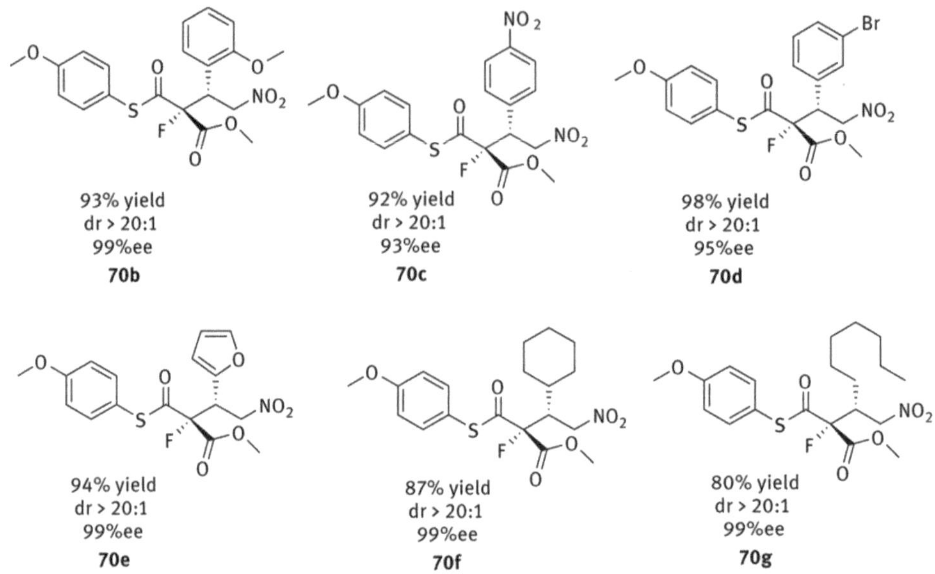

Figure 3.34: Selection of chiral α-fluoro-γ-nitro thioesters (**70b-70 g**) derived from amino-cinchona urea catalysed conjugate addition reactions with catalyst **69.**

occurs on a trisubstituted nitro-alkene, which is not very stable as it generates a product that is prone to undergo epimerisation (Figure 3.35). When compound **71** was reacted with nitro-olefin **72** in the presence of 10 mol% catalyst **73** in 1,4-dioxane at room temperature Michael addition product **74a** was isolated in 93% yield and with an er of 91:9.

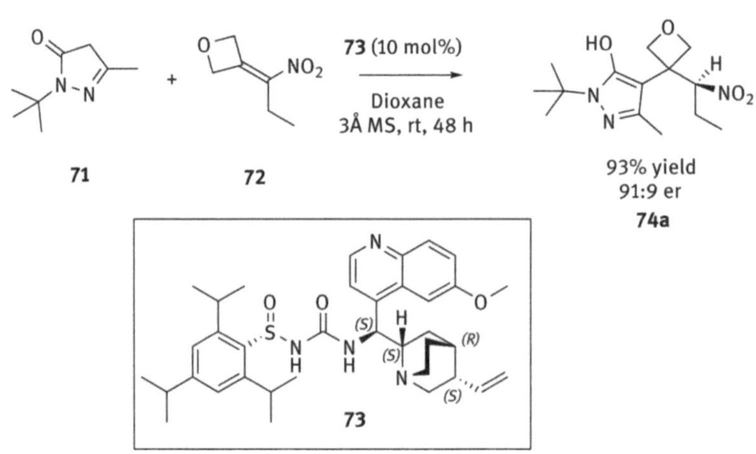

Figure 3.35: Example of an enantioselective Michael addition catalysed by cinchona-based *N*-sulfinylurea organocatalyst **73** as described by Phelan and Ellman.

The method was applicable to a range of different substrates giving access to novel building blocks of pharmaceutical interest. It has to be noted that besides nitro-oxetane Michael acceptors, nitro-azetidines were also suitable delivering moderate to high yields and high enantioselectivities. Figure 3.36 depicts a selection of synthesised conjugate addition products (**74b**-**74g**) employing catalyst **73**.

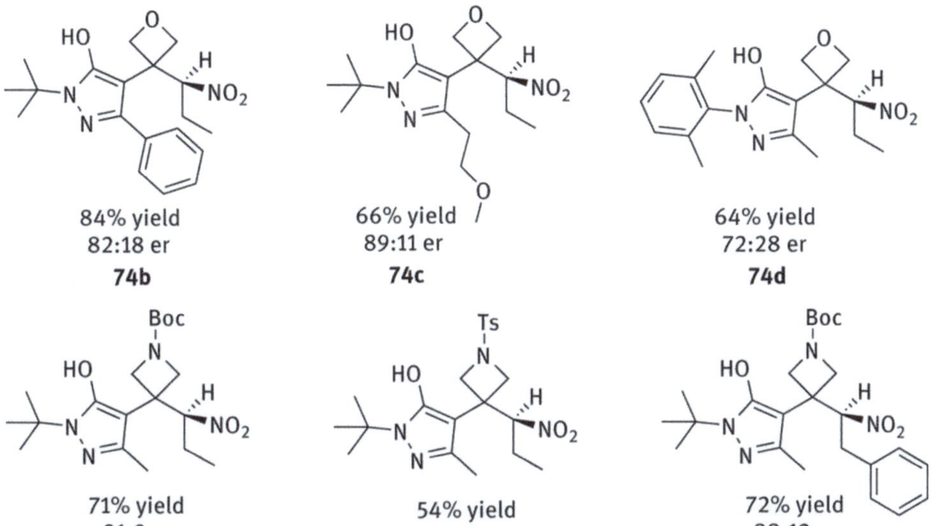

84% yield
82:18 er
74b

66% yield
89:11 er
74c

64% yield
72:28 er
74d

71% yield
91:9 er
74e

54% yield
90:10 er
74f

72% yield
88:12 er
74g

Figure 3.36: Selection of conjugate addition products (**74b**-**74g**) derived from pyrazolones and trisubstituted nitroalkenes using cinchona *N*-sulfinylurea substituted organocatalyst **73**.

3.6.1.2 Bromo-Lactonisation reactions

Very recently Jiang and Yeung established a catalytic asymmetric bromo-lactonisation reaction employing deactivated olefinic acids and *N*-bromosuccinimide (NBS) to furnish bromo-keto-lactones in high yield and high enantioselectivity [35]. These products showed significant anti–inflammatory properties on macrophage-like RAW 264.7 cells. When compound **75** was reacted with NBS in the presence of 15 mol% catalyst **76** under very mild conditions at 15 °C in toluene, bromo-lactone **77** was obtained in 96% yield and an er of 90:10 (Figure 3.37). To optimise the reaction a range of different catalysts were explored and very unexpectedly it was found that electron-donating urea catalysts showed a significant increase in enantioselectivity compared to more electron-deficient systems. The product enantiopode could also be efficiently obtained with the analogous quinine-derived organocatalyst **78** (Figure 3.37); however, no yield or er was reported for this reaction.

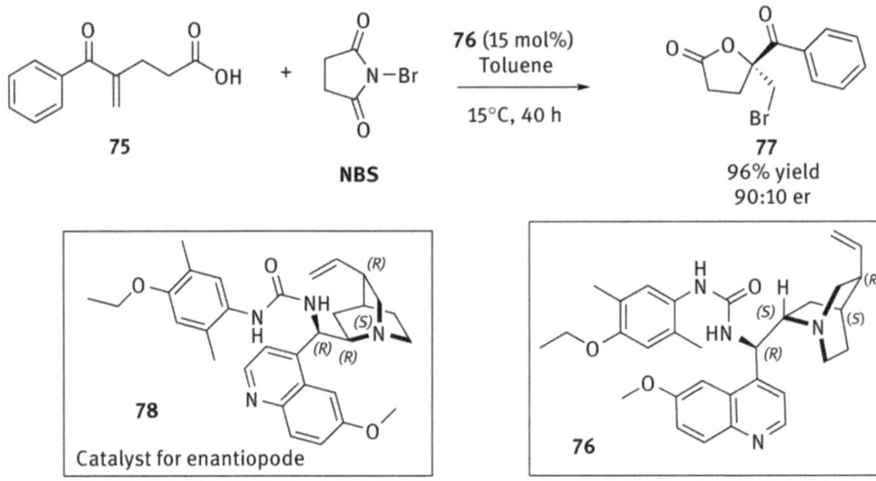

Figure 3.37: Example of an asymmetric bromo-lactonisation of a deactivated olefinic acid with *N*-bromosuccinimde (NBS) and electron-donating urea organocatalyst **76** as developed by Jiang and Yeung.

3.6.2 Aldol reactions

Wennemers and Saadi reported in 2016 enantioselective aldol reactions of aliphatic and aromatic aldehydes with fluoromalonic acid half thioesters (F-MAHTs) employing a cinchona urea catalyst [36] (see also Section 3.4.1 for cinchona thioamide catalysed Mannich reactions and Section 3.5.2 for cinchona sulphonamide catalysed aldol reactions with MAHOs). This method is of considerable interest as F-MAHTs are biomimetic fluoroacetate surrogates; they find use in medicinal chemistry applications for the preparation of fluorinated compounds with advanced pharmacological properties. The authors successfully applied this type of asymmetric aldol reaction to the synthesis of a fluorinated version of the API *Atorvastatin* (*Lipitor®*), the well-known lipid lowering drug for the regulation of cholesterol levels in the body (Figure 3.38). Compound **79** was reacted with F-MAHT **80** in the presence of 20 mol% of the urea cinchona organocatalyst **81**. Aldol product **82** was obtained in 98% yield with a dr of 9:1 (*anti*/*syn*) and 62% ee. By recrystallization the enantiopurity of the major product **82** could be increased to 99% with > 20:1 dr prior to final conversion to the calcium salt of a fluorinated *Atorvastatin* derivative.

3.6.3 Nitro-Mannich reaction

Duan and co-workers published a series of articles describing the synthesis of cinchona based urea PTCs and their application in nitro-Mannich reactions. In one example

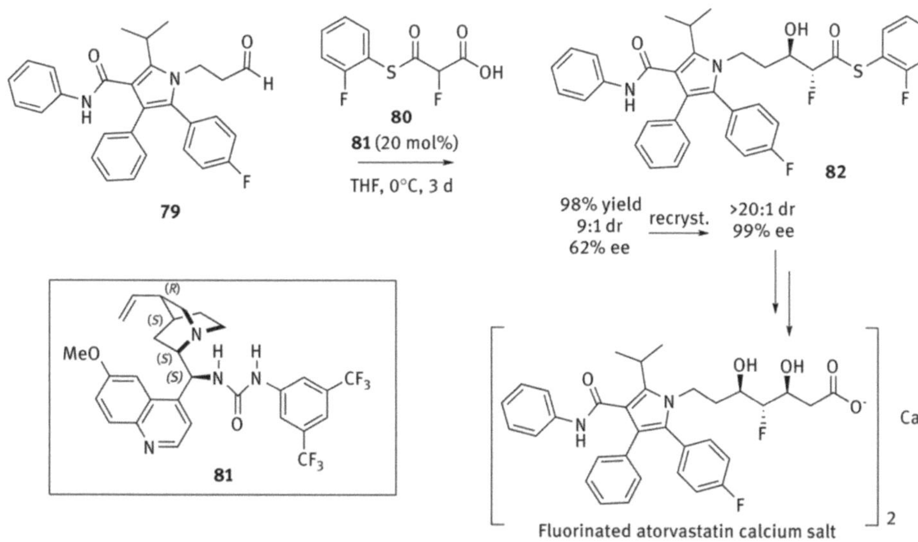

Figure 3.38: Example of an enantioselective aldol reaction of compound **79** with F-MAHT **80** catalysed by cinchona urea organocatalyst **81** in the synthesis of a fluorinated *Atorvastatin* derivative as described by Wennemers and Saadi.

organocatalyst **84** was used for the first time in nitro-Mannich reactions with isatin-derived *N*-Boc-protected ketimine **83** [37], (which is less common than reactions with aldimines). With 5 mol% catalyst **84**, lithium hydroxide monohydrate as the base and the conjugate base derived from nitromethane, it was possible to obtain the product **85a** in 99% yield and 88% ee after 20 h at −40 °C (Figure 3.39**(A)**). The method was found to be suitable for a range of differently substituted isatin ketimine substrates. It has to be noted that analogous urea-cinchona catalysts, which are not quaternized on the quinuclidine nitrogen performed poorly. In a second application less commonly used α-aryl nitromethanes were reacted in the same manner delivering products in high yield with high enantioselectivity and stereoselectivity [38]. With 10 mol% catalyst **84** and the conjugate base of α-nitrotoluene as the nucleophile, the nitro-Mannich product **85b** was isolated in 99% yield, 92% ee and 93:7 dr (Figure 3.39**(B)**). The conditions were also compatible for a range of differently substituted α-aryl nitromethanes. In a third case PTC **84** was employed in a highly efficient and selective nitro-Mannich reaction with amidosulfonates and α-aryl nitromethanes [39]. With 5 mol% organocatalyst **84** in the presence of lithium hydroxide monohydride compound **86** was reacted with α-nitrotoluene affording product **87** in 99% yield, 99% ee and 99:1 dr (Figure 3.39**(C)**). This asymmetric catalytic procedure gives access to optically pure 1,2-diarylethylenediamines, which are very useful building blocks for medicinal chemistry applications. Previously also analogous reactions of amidosulfonates with nitromethane had been described by the same authors [40].

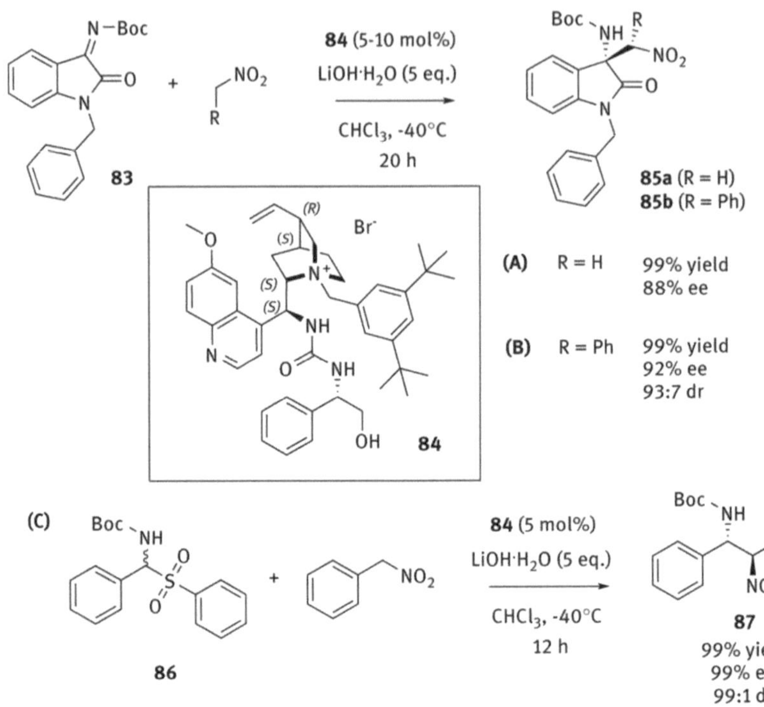

Figure 3.39: Cinchona alkaloid phase transfer catalysed nitro-Mannich reactions on ketimines and amidosulfonates with nitromethane and α-nitrotoluene in the presence of catalyst **84** as described by Lu *et al.*

3.6.4 Halogenation reactions

In 2017 Asano and Matsubara described a new route towards axially chiral 8-aryl-quinolines employing bifunctional cinchona urea catalysts [41]. The substrate 3-(quinolone-8-yl)phenol **88**, which has a low axial rotational barrier, successfully underwent atrop-selective bromination with *N*-bromoacetamide (NBA) in dichloromethane at 25 °C with 10 mol% urea organocatalyst **89** to deliver tri-brominated product **90** in 99% yield and 90% ee (Figure 3.40**(A)**). This product has in contrast to its starting material a high rotational barrier and makes it an interesting intermediate for further conversions. The analogous tri-iodination product **91** could also be prepared in 52% yield and 91% ee with *N*-iodosuccinimide (NIS) in ethyl acetate at −40 °C with the same organocatalyst **89** (Figure 3.40**(B)**).

Further it was demonstrated that *ortho*-brominated 8-aryl-quinoline **92** could be bis-iodinated to afford the mixed halogenated quinolone building block **93** in moderate 51% yield and 88% ee, the compound is a useful intermediates for further selective transformations (Figure 3.40**(C)**). The iodination was best conducted with NIS

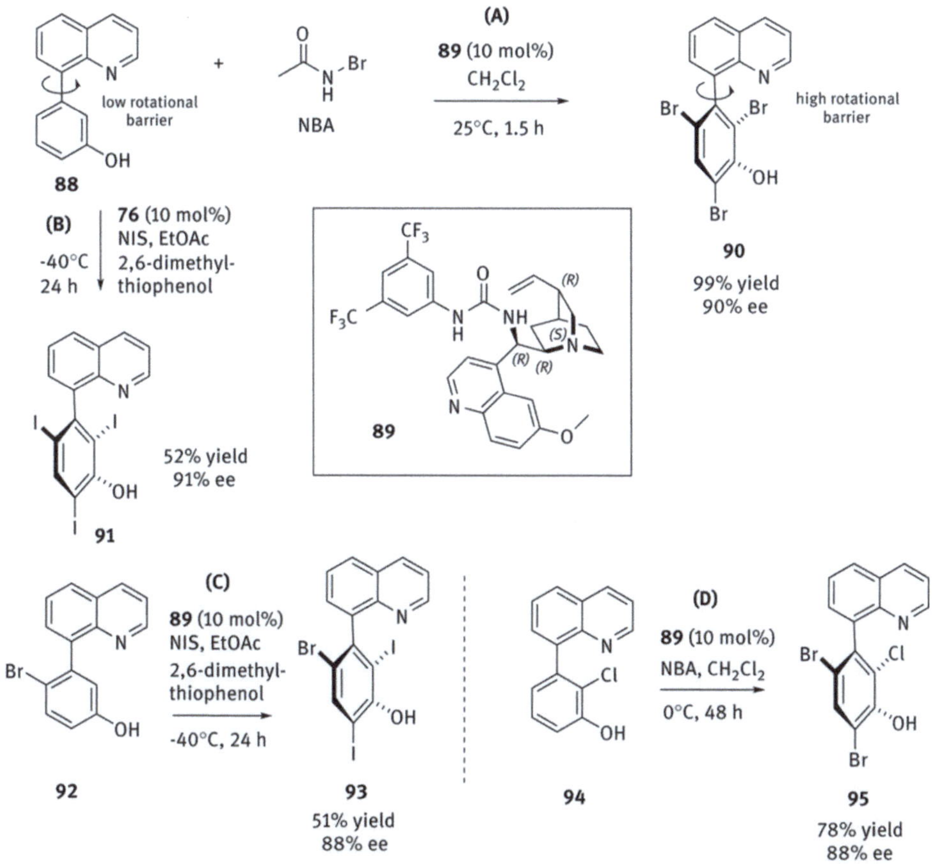

Figure 3.40: Atrop-selective halogenation of 8-aryl-quinoline derivatives catalysed by cinchona urea catalyst **89** for the synthesis of axially chiral compounds as developed by Asano and Matsubara.

at −40 °C in the presence of the additive 2,6-dimethyl-thiophenol and 10 mol% urea catalyst **89**. In an analogous manner *ortho*-chlorinated aryl-quinoline **94** could be bis-brominated in 78% yield and 88% ee also utilising cinchona urea organocatalyst **89** (Figure 3.40**(D)**). The mini-review published by Zilate *et al.* should be mentioned; as it features recent advances in the field of organocatalysed stereo-controlled atropisomer synthesis [42].

3.6.5 Alkylation reactions

Very recently Connon and co-workers reported enantioselective alkylation reactions of 3-substituted oxindoles with a cinchona urea derived PTC [43]. This reaction enabled the construction of a quaternary stereocentre in high yield and very high enantioselectivity. Using 5 mol% organocatalyst **98** the alkylation of oxindole **96** with alkylating agent **97** was carried out in dichloromethane at −60 °C in the presence of potassium carbonate to afford compound **99** in 74% yield and 91% ee (Figure 3.41). Product **99** could then be further converted into the marine alkaloid (-)-debromoflustramine B. It has to be noted that a very bulky cinchona urea catalyst was essential for the reaction to achieve high selectivity; less bulky systems or other analogues such as squaramides delivered poorer results. The method gave access to a range of differently C-3 alkylated and structurally very interesting quaternized oxindoles.

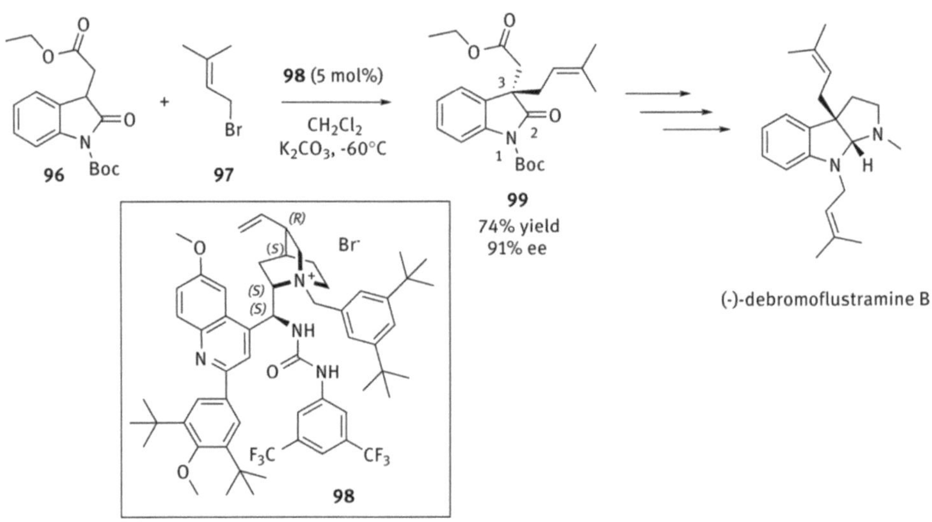

Figure 3.41: Enantioselective alkylation of 3-substituted oxindole **96** catalysed by cinchona urea phase transfer catalyst **98** as described by Connon and co-workers.

3.7 Amino-cinchona thiourea catalysed reactions

Amino-cinchona thiourea organocatalysts are much in demand for carrying out various types of asymmetric reactions. They are very powerful and particularly popular for conjugate addition reactions. Therefore an efficient and reliable synthesis of these catalysts is of importance. Soós and co-workers reported in 2005 a three-step process for the synthesis of such thiourea catalysts [44], which included two time-consuming purification steps by silica gel chromatography, proving to be impractical for producing large quantities of organocatalysts. In 2017 Wang *et al.* established a more scalable, robust and cost-effective three-step process that could be run at pilot plant scale and did not require any chromatographic purification procedures [45]. Starting from quinine (**QN**) a Mitsunobu reaction was carried out with triphenylphosphine, diisopropyl azodicarboxylate (DIAD) and diphenyl phosphoryl azide (DPPA) (Figure 3.42). The azide intermediate **100** was directly subjected to a

Staudinger reduction with triphenylphosphine to afford the 9-amino-9-deoxy-cinchona key intermediate **101**. Compound **101** was then reacted without purification with 3,5-bis(trifluoromethyl)phenyl isothio-cyanate to form the desired thiourea catalyst **102**. After recrystallization product **102** was isolated in 71% yield over three steps. The process was successfully run on an 11 kg scale and is also amenable to a range of other thiourea analogues.

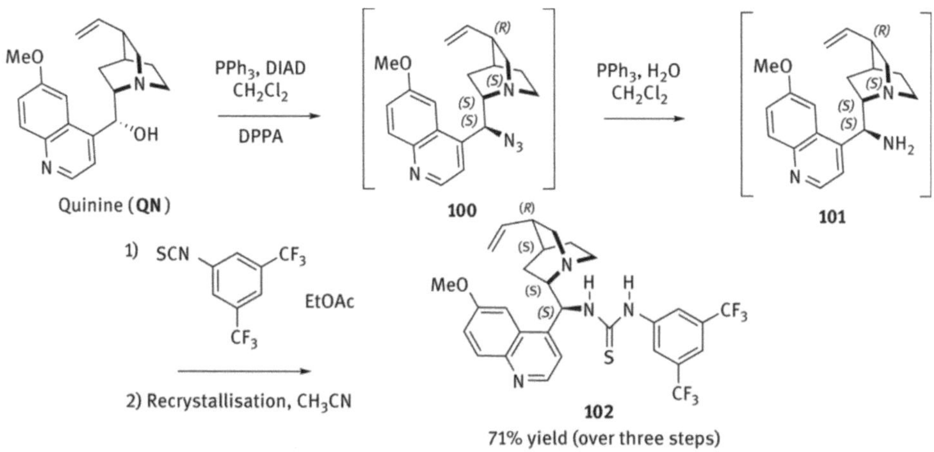

Figure 3.42: Scale-up route for the synthesis of an amino-cinchona thiourea catalysts by Wang *et al.*

3.7.1 Conjugate addition reactions

3.7.1.1 Michael additions

Liu *et al.* reported in 2017 the asymmetric Michael addition reaction of *S,S'*-diphenyl-dithiomalonate **103** with β-nitrostyrene [46]. With 10 mol% catalyst **104** in trifluoro-benzene at room temperature the conjugate addition product **106** ((*S*)-isomer) was isolated in 92% yield and 96% ee (Figure 3.43). The corresponding (*R*)-isomer could be obtained in 95% yield and 98% ee utilising the quasi-enantiomeric catalyst **105**.

Medina *et al.* discovered in 2016 the α-stereoselective glycosylation of 2-nitroga-lactals catalysed by bifunctional cinchona thiourea organocatalysts [47]. The glyco-sylation products are α-*O*-linked and represent useful intermediates for biologically relevant oligosaccharides and glycoconjugates. When perbenzylated 2-nitrogalactal **107** was reacted with compound **108** in acetonitrile under reflux conditions in the presence of 30 mol% of bis-cinchona alkaloid thiourea **109**, addition product **110** was obtained in 87% yield with (4:1) α:β-ratio (Figure 3.44). The reaction conditions were applicable to a range of different substrates.

Bhaskararao and Sunoji very recently described a dual-organocatalytic Michael addition reaction between a dicarbonyl compound and β-nitrostyrene in toluene at room temperature [48]. This conversion is a modularly designed organocatalytic

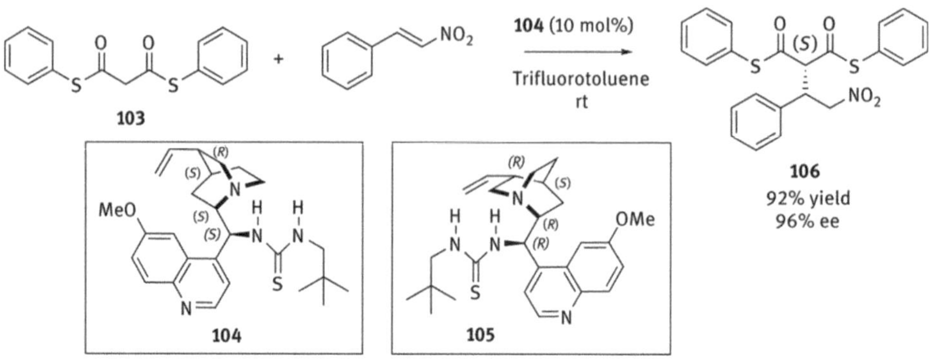

Figure 3.43: Example of an enantioselective Michael addition of *S,S'*-diphenyldithiomalonate **103** to β-nitrostyrene catalysed by cinchona thiourea catalyst **104** as reported by Liu *et al.*

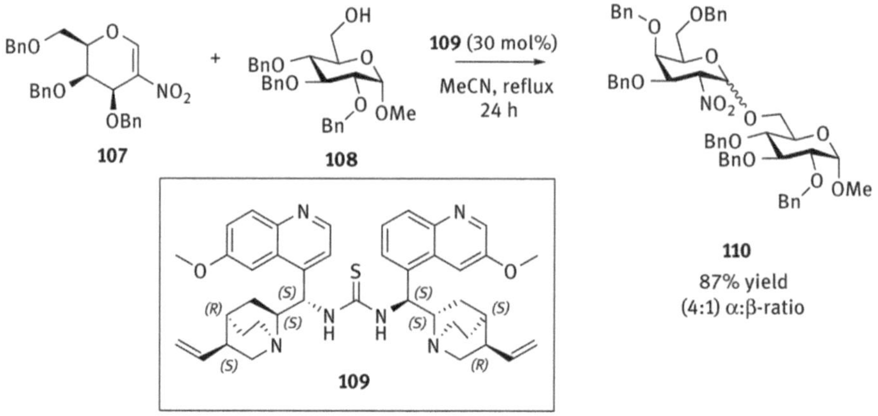

Figure 3.44: An example of an α-stereoselective glycosylation of 2-nitrogalactal **107**, catalysed by bis-cinchona alkaloid thiourea organocatalyst **109** as described by Medina *et al.*

reaction operated with two chiral organocatalysts (*L*-proline and quinidine derived thiourea organocatalyst **112**). When compound **111** was reacted with β-nitrostyrene in toluene at room temperature double-Michael addition product **113**, a tetra-substituted cyclohexane derivative, was obtained in 73% yield and 92% de (Figure 3.45). Unfortunately no information was given on the catalyst loading. The work was accompanied by theoretical studies to give an insight into a possible mechanism of this organo–organo catalysed tandem Michael–Michael reaction. The catalytic cycle is believed to be initiated by a proline/cinchona-thiourea enamine complex, followed by a first intermolecular conjugate addition and then a second intramolecular reaction. It should be noted that comprehensive mechanistic studies had previously been carried out by Dang and Zhou on trifunctional amino-cinchona-thiourea catalysed single Michael additions in the absence of a co-catalyst [49].

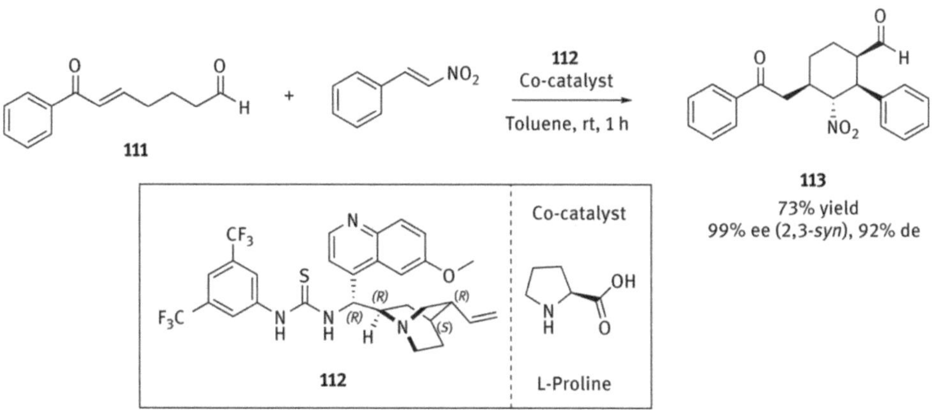

Figure 3.45: An example of a dual-organocatalytic double-Michael addition of the dicarbonyl compound **111** with β-nitrostyrene in the presence of thiourea catalyst **112** and *L*-proline as reported by Bhaskararao and Sunoji.

In fact, these newly emerged and very popular molecularly designed organocatalysts (MDOs) had also been studied by Ramachary *et al.* [50]. With the combination of two organocatalysts, cinchona thiourea **115** (5 mol%) and D-proline (5 mol%) a highly selective Michael addition between nitro-olefin **114** and cyclohexanone was conducted in dichloromethane at room temperature. The conjugate addition product **116a** was isolated in 99% yield, 98% ee and 99% de (Figure 3.46). The method was suitable for a range of different substrates and ketones giving access to novel building blocks of

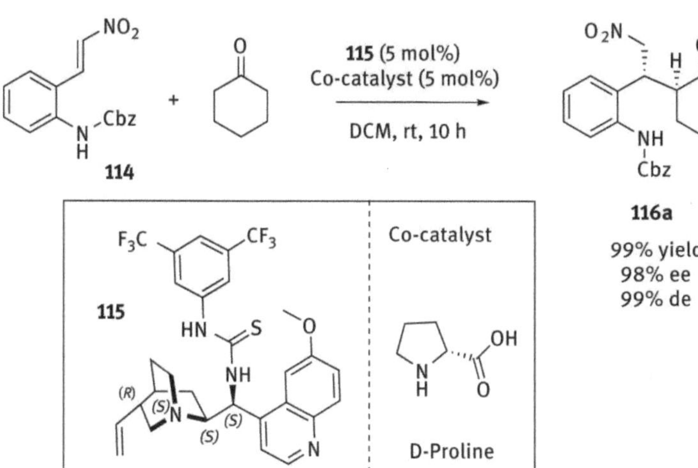

Figure 3.46: An example of an asymmetric Michael addition catalysed by a modularly designed supramolecular organocatalyst comprising the cinchona thiourea **115** and D-proline as described by Ramachary *et al.*

interest for medicinal chemistry applications; some examples of the Michael addition reaction products (**116b-116h**) that were obtained are shown in Figure 3.47.

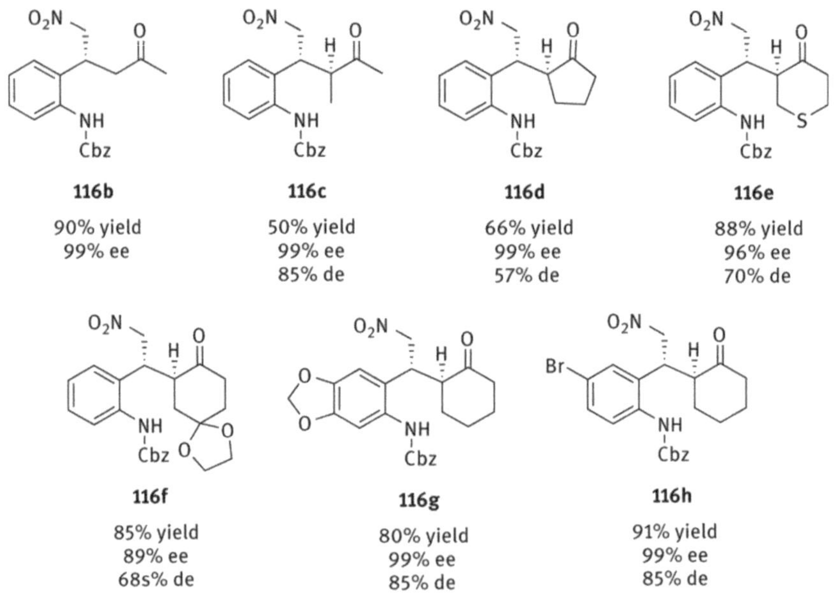

Figure 3.47: Some selected examples of the Michael addition products (**116b-116h**) obtained from reactions discussed above.

3.7.1.2 Robinson annulations

In 2018 Reddy and co-workers described the first enantioselective cinchona thiourea catalysed Robinson annulation (conjugate addition/cyclisation reaction) of 3-indolinone-2-carboxylate **117** with cyclohexenone employing 10 mol% quinidine derived thiourea catalyst **118** (Figure 3.48) [51]. The reaction was conducted in toluene at 55 °C to afford the chiral bridged hydrocarbazole **119** in 75% yield, >20:1 dr and 96% ee. The method was applicable to a range of different indolinone substrates.

3.7.1.3 Iodocyclisations

Suresh *et al.* reported in 2018 the first enantioselective synthesis of 5,6-dihydro-4*H*-1,2-oxazines bearing an oxygen-containing quaternary stereogenic centre [52]; the reaction was catalysed by an amino-cinchona thiourea organocatalyst. When the 1,4-diphenyl-substituted γ,δ-unsaturated (*E*)-oxime **120** was reacted with NIS in the presence of 10 mol% catalyst **121** at −80 °C the chiral iodo-oxazine **122** was formed in 99% yield and 97:3 er (Figure 3.49).

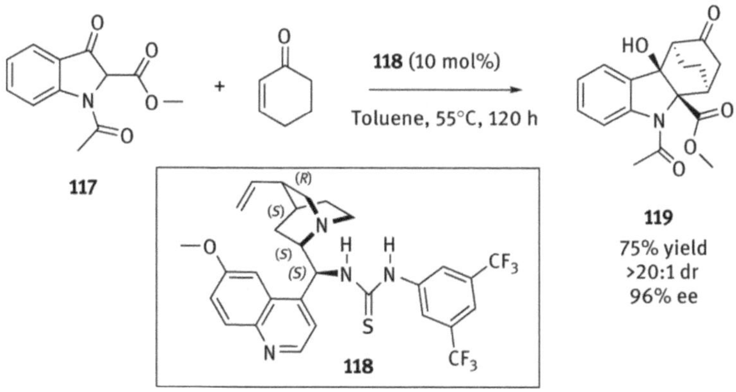

Figure 3.48: An example of an asymmetric Robinson annulation catalysed by quinidine derived thiourea organocatalyst **118** as described by Reddy and co-workers.

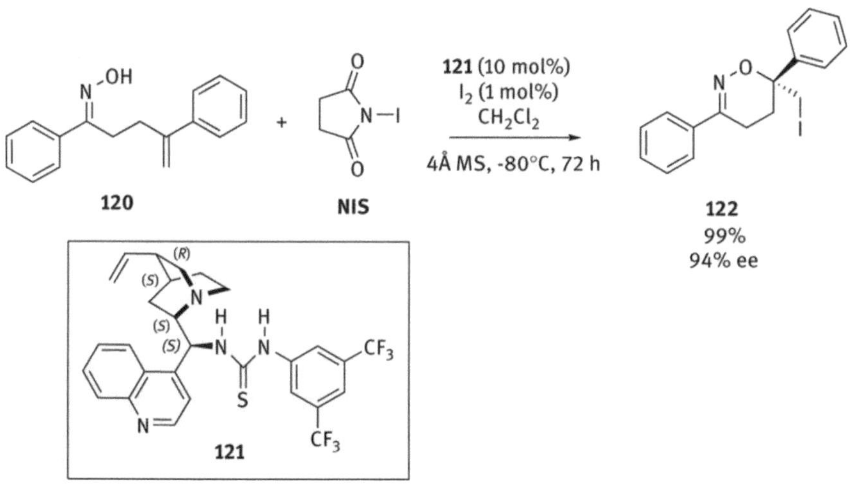

Figure 3.49: An example of an amino-cinchona thiourea **121** catalysed iodocyclisation in the synthesis of chiral dihydrooxazines by Suresh *et al.*

3.7.2 Mannich reactions

In 2018, Zhao and Konda succeeded in carrying out an *anti*-selective Mannich reaction, which is more difficult to achieve than the very commonly conducted *syn*-Mannich reactions [53]. Heptanal was reacted neat with (4-methoxyphenylimino) acetate **123** at room temperature in the presence of a modularly designed organocatalyst comprising of cinchona thiourea **124** (5 mol%) and L-proline (5 mol%). The *anti*-Mannich product **125** was then reduced to afford amino alcohol **126** in 90% yield, 99:1 dr and 99% ee (Figure 3.50). When the reaction was conducted in a solvent high selectivities were observed, however yields were reduced. The reaction conditions were successfully applied to a range of substrates providing access to various amino alcohols which were of interest for medicinal chemistry applications.

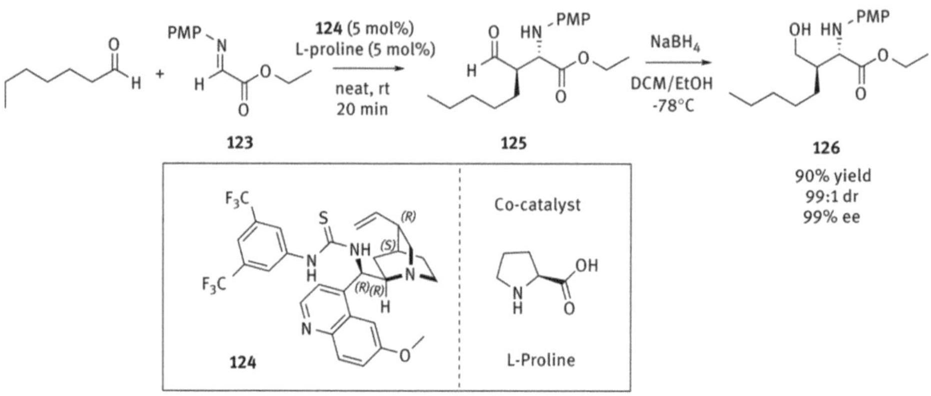

Figure 3.50: Example of a highly enantioselective *anti*-Mannich reaction catalysed by a modularly designed organocatalyst comprised of cinchona thiourea **124** and L-proline as described by Zhao and Konda.

3.7.3 Aza-Henry reactions

Lin, Duan and co-workers very recently described the first catalytic enantioselective aza-Henry reaction employing aryl ketimines and novel cinchona based bifunctional thiourea ammonium salt catalysts (PTC type catalysts) [54]. When *N*-tosyl-protected ketimine **127** was reacted with nitromethane in *m*-xylene at –10 °C in the presence of lithium hydroxide monohydrate and 10% thiourea phase-transfer catalyst **128** the chiral aza-Henry product **129** was obtained in 99% yield and 94% ee (Figure 3.51). X-ray crystal structure analysis confirmed the absolute stereochemistry of the newly formed chiral centre. Addition product **129** could be further derivatised in to α,β-diamino ester **130** in 81% yield without loss of enantiopurity.

It has to be noted that ordinary thiourea cinchona catalysts delivered lower yields and poor enantioselectivities in comparison to the quaternized PTC that was employed here. The method was also found to be applicable to a range of different aryl ketimine substrates.

3.7.4 Aldol reactions

Kayal and Mukherjee used a quinine based cinchona thiourea catalyst for cascade Aldol-cyclization reactions between 3-isothiocyanato oxindole **131** and α-ketophosphonate **132** [55]. With 2 mol% of catalyst **133** in toluene at –78 °C spiro-oxindole **134**

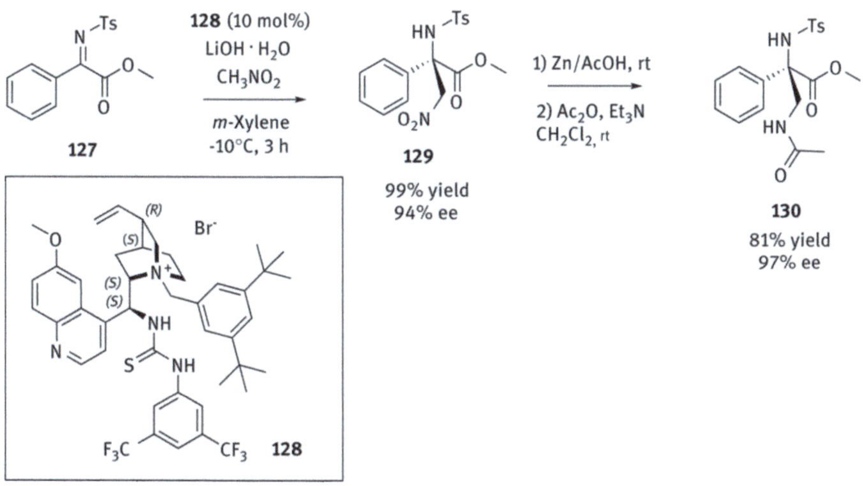

Figure 3.51: Cinchona alkaloid phase-transfer catalysed aza-Henry reaction of ketimines as described by Lin and Duan.

bearing a β-amino-α-hydroxyphosphonate motif was isolated in 80% yield, 10:1 dr and 98:2 er (Figure 3.52). These types of spiro-compounds bearing adjacent quaternary stereocenters are very rare and had not been previously reported. As a side note, these workers used the analogous cinchonidine based thiourea catalyst for carrying out a conjugate addition/halogenation reaction to produce dihydro-1,2-oxazines as described in Section 3.7.1 (Figure 3.49).

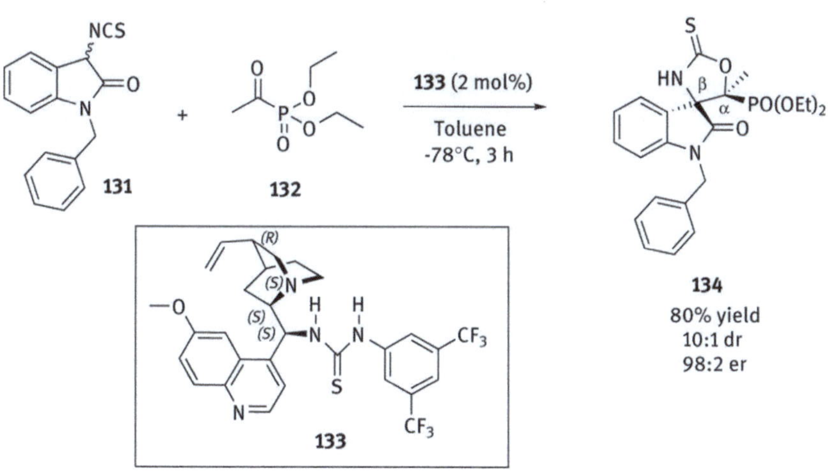

Figure 3.52: Example of the asymmetric synthesis of spiro-oxindole **134** bearing a β-amino-α-hydroxyphosphonate motif, catalysed by amino-cinchona thiourea organocatalyst **133** as developed by Kayal and Mukherjee.

3.7.5 Diels-Alder Reactions

Wang and co-workers reported in 2017 a catalytic asymmetric Diels–Alder reactions between 3-hydroxy-2-pyrones and cyclopentene-1,3-diones utilising a cinchona thiourea organocatalyst [56]. When 3-hydroxy-2-pyrone **136** was reacted with dione **137** in dichloromethane at −20 °C in the presence of catalyst **138**, Diels–Alder product **139** was obtained in 85% yield and 94% ee (Figure 3.53).

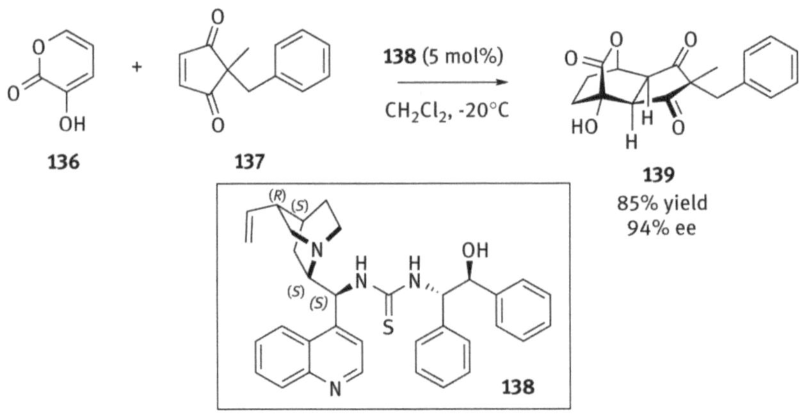

Figure 3.53: An example of a cinchona thiourea catalysed Diels–Alder reaction for the synthesis of multifunctional bridged tricyclic lactone **139** as described by Wang and co-workers.

Melchiorre's group reported the use of cinchona thiourea organocatalysts in connection with poorly explored enantioselective photoenolization/Diels–Alder (PEDA) sequences [57]. When *N-tert*-butylmaleimide **140** and 2-methylbenzophenone **141** were irradiated in toluene with a single black light emitting diode (black LED, $\lambda_{max} = 365$ nm) in the presence of 20 mol% bifunctional cinchona thiourea catalyst **142** the Diels–Alder product **143** was obtained in 76% yield and 90% ee. The reaction was conducted in a cyclohexane/toluene (3:1) mixture at −5 °C for 24 h (Figure 3.54).

The reaction follows a photoenolisation mechanism, whereby the formation of photoenol **144** does not require any catalyst. The utilised catalyst is actually a dichotomous system; the thiourea moiety with its bulky chiral isopentyl substituent functions as a catalytic centre to activate the maldimide diene acceptor **140** for stereoselective interception of photoenol **144**. The catalyst's quinuclidine moiety is an important inhibitory centre, which attenuates the photoenolization background process by reducing the amount of available photoenol and reverting it back into its benzophenone state.

The method could be applied to a range of differently substituted benzophenone derivatives and the authors propose that this enantioselective catalytic process could function as a catalytic blueprint for other applications, which they actually don't

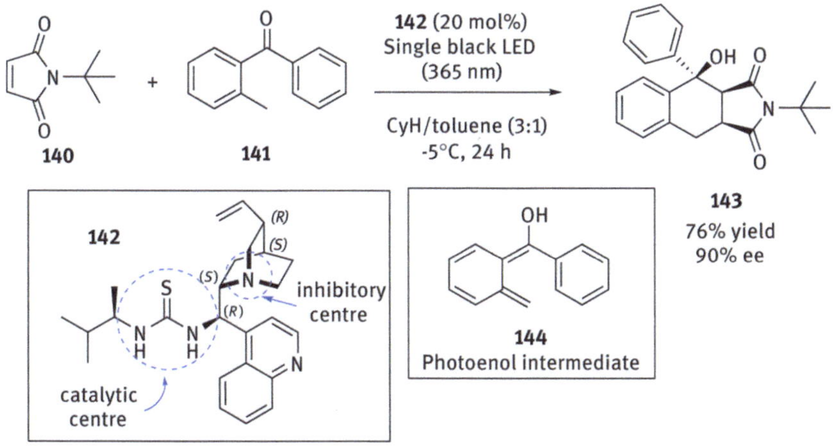

Figure 3.54: An example of an enantioselective photoenolization/Diels–Alder (PEDA) reaction catalysed by cinchona thiourea organocatalyst **142** as published by Melchiorre and co-workers.

further specify. A recent micro-review by Melchiore and Cuadros provides an insight into the historical background and more detailed mechanistic aspects of PEDA and related reactions [58].

3.7.6 Oxidation reactions

Yuan *et al.* reported a series of novel cinchona alkaloid derived thiourea catalysts and their application for asymmetric syntheses of oxaziridines [59]. The developed catalysts gave moderate to high yields and good selectivities for a range of aldimine substrates in different solvent systems. Figure 3.55 shows an example of an oxaziridination of compound **145** employing 12 mol% thiourea organocatalyst **146** and the oxidising agent *m*CPBA in xylene at −40 °C; the chiral oxaziridine **147** was obtained under mild conditions in 89% yield and with a remarkable 99% ee.

Lattanzi and co-workers disclosed the first asymmetric epoxidation reaction of alkylidenemalonitrile **148** employing a multifunctional thiourea cinchona catalyst and cumyl hydroperoxide (CHP) as an oxidant [60]. With 10 mol% organocatalyst **149** in toluene at −20 °C epoxide **150** was obtained in 78% yield and 85:15 dr (Figure 3.56). The reaction conditions were suitable for a number of different substrates and the oxidation product **150** could be successfully further converted into piperazin-2-one **151** in 60% yield and 86:14 er; piperazin-2-ones of this type represent attractive intermediates for the synthesis of biologically active compounds.

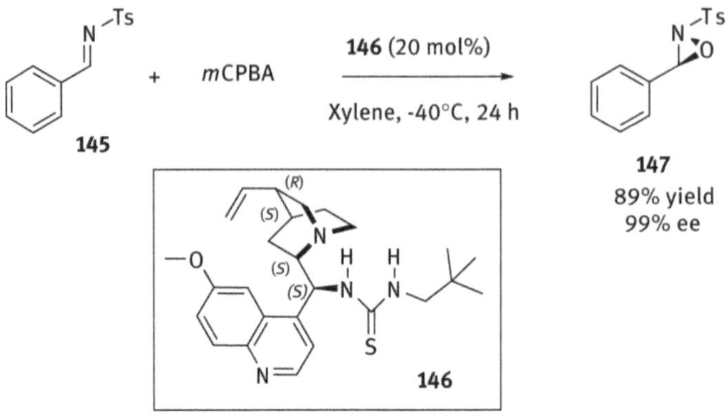

Figure 3.55: An example of an asymmetric oxaziridination of aromatic aldimine **145** catalysed by cinchona thiourea catalyst **146** as described by Yuan *et al.*

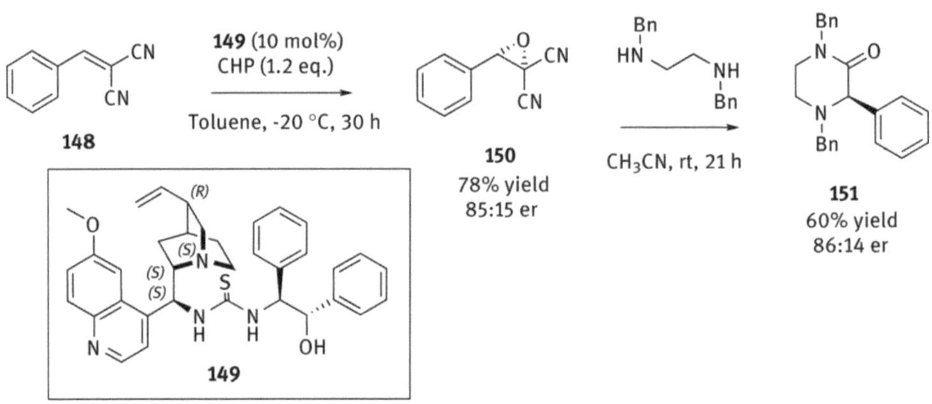

Figure 3.56: An example of an asymmetric epoxidation of alkylidenemalonitriles employing multifunctional thiourea cinchona catalyst **149** as developed by Lattanzi's group.

3.8 Cinchona-based diaminomethylenemalonitrile (DMM) organocatalysts

3.8.1 Conjugate addition reactions

3.8.1.1 Michael additions

Very recently Nakashima *et al.* established the first example of an asymmetric conjugate addition reaction between α-cyanoketones and vinyl ketones catalysed by cinchona-based diaminomethylene-malonitrile (DMM) organocatalysts [61]. When 2-oxocyclopentane-1-carbonitrile **152** was reacted with 1-phenyl-2-en-1-one **153** in toluene at −50 °C in the presence of 10 mol% catalyst **154** the Michael addition product **155a** was isolated in 94% yield and 92% ee (Figure 3.57). It should be noted that catalyst **154** performed very well in comparison with other cinchona derived catalysts, like thioureas, which delivered very poor ees.

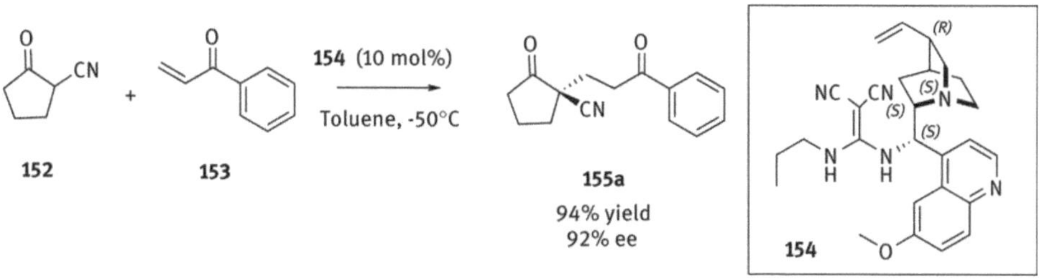

Figure 3.57: An example of an enantioselective conjugate addition reaction of 2-oxocyclopentane-1-carbonitrile **152** and 1-phenyl-2-en-1-one **153** with DMM organocatalyst **154** as described by Nakashima *et al.*

It was suggested by the authors that the two mildly acidic protons on the catalyst's DMM motif promote the activation the carbonyl group in **153** through hydrogen bonding. The catalyst's tertiary quinine group is believed to induce enolization on ketone **152**, which results in a highly selective orientation of the two carbonyl reaction partners to each other.

The method was applicable to a range of different α-cyanoketones and vinyl ketones. A small selection of reaction products (**155b**-**155g**) is shown in Figure 3.58.

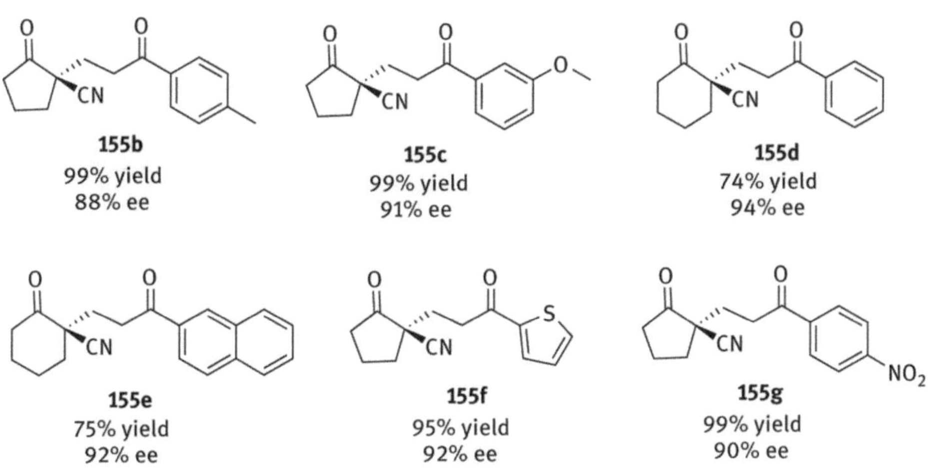

Figure 3.58: Selection of asymmetric Michael addition products (**155b**-**155g**) synthesised with cinchona-based diaminomethylenemalonitrile (DMM) catalyst **154**.

3.8.2 Hydrophosphonylation reactions

In 2018 Miura and co-workers also described the use of cinchona-based diaminomethylenemalonitrile organocatalysts to perform 1,2-hydrophosphorylation reactions of simple ketones and enones [62]. When 4'-nitro-acetophenone was reacted with diphenyl phosphonate and 20 mol% catalyst **156** in toluene at −60 °C the corresponding α-hydroxy-phosphonate product **157** was isolated in 99% yield and 92% ee (Figure 3.59). The employed catalyst achieved higher selectivities than related thiourea or squaramide catalysts. The absolute configuration of the newly introduced stereocentre was determined by X-ray crystallographic analysis.

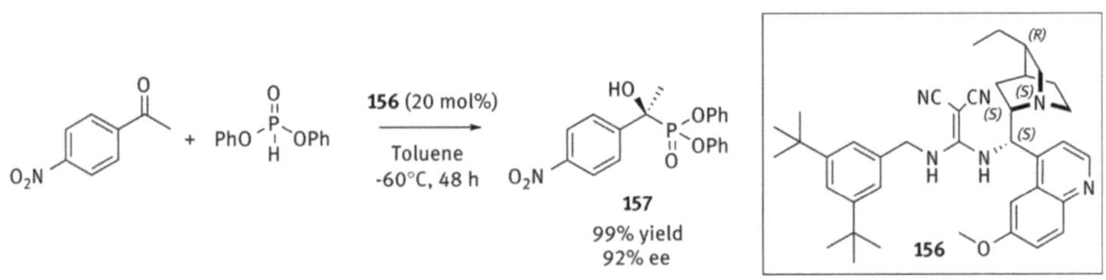

Figure 3.59: An example of a 1,2-hydrophosphorylation reaction of 4'-nitro-acetophenone catalysed by diamino-methylenemalonitrile (DMM) organocatalyst **156** as developed by Miura's group.

The method was applicable to a range of acetophenone derivatives and also α,β-unsaturated ketone substrates; a selection of synthesised products (**156b-156i**) is shown in Figure 3.60.

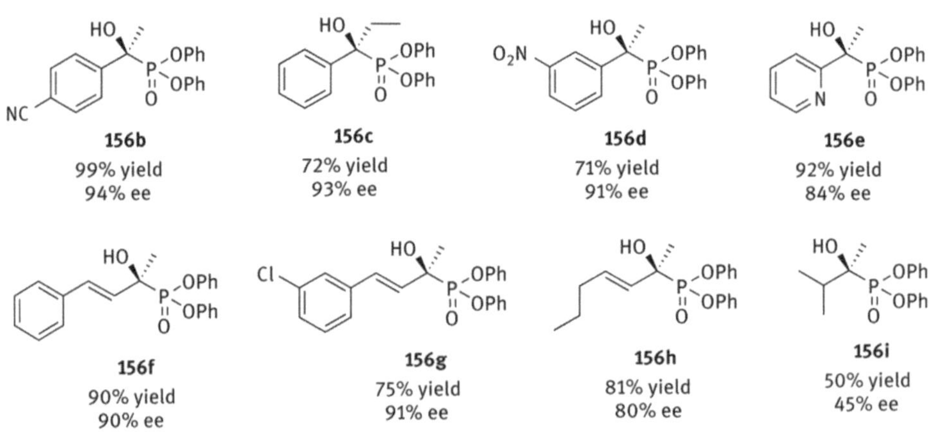

Figure 3.60: Selection of hydrophosphorylation products obtained from cinchona diaminomethylene-malonitrile **156** organocatalysed reactions.

3.9 Cinchona-squaramide catalysed asymmetric reactions

Cinchona-Squaramide organocatalysts were originally pioneered by Rawal's group in 2008, for their application in the conjugate addition of dicarbonyls to nitrostyrene electrophiles [63]. Compared[1] to thiourea (squaramides are recognized as bioisosteres of ureas), the properties of the squaramido functional group shows significant differences like: rigidity, H-bond spacing, H-bond angle, duality and pKa [64]. The distance between the two N–H groups is perhaps the most significant difference, which have been calculated by both the Rawal [63] and Takemoto groups [65] (Figure 3.61). The distance between the two N–H groups of the squaramide is longer than that for the thiourea, and in fact generally the squaramide moiety forms stronger hydrogen bonds with substrates bearing nitro, carbonyl, imino and nitrile functional groups, etc. In both thioureas and squaramides the lone pair on the nitrogen atom is delocalized, restricting the rotation of the C–N bond, however, only in squaramides can further delocalization occur through the cyclobutenedione unit, and as a consequence the NH acidity is greater and the squaramide unit is more rigid. The synthesis of cinchona-squaramide catalysts is very straight-forward [66].

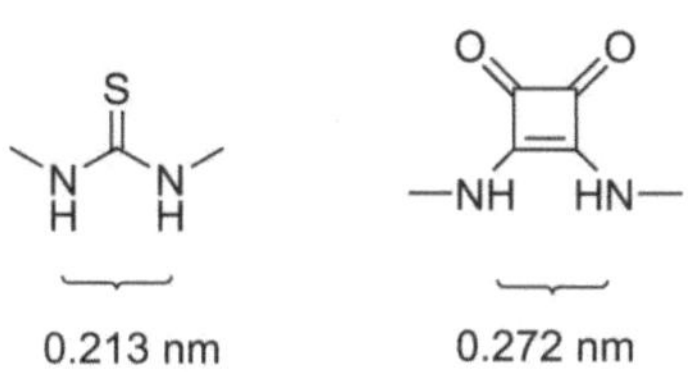

Figure 3.61: H-Bonding distances in *N,N'*-dimethylthiourea and *N,N'*-dimethylsquaramide.

In the case of the cinchona-squaramides, the overall structure allows for effective bifunctional hydrogen-bonding catalysis.

Several excellent recent reviews on squaramide organocatalysts are available and include the important cinchona-squaramides family [66].

In this section we review the principal cinchona-squaramides that have been used in a variety of transformations, including domino/cascade reactions (which are in fact very common), which generally include a conjugate addition step. It should also be noted that catalytic reactions involving these catalysts to form chiral 3-hydroxy and 3-aminooxindoles and derivatives have been recently extensively reviewed ([66]b). In this section we will encounter some organocatalysts that can give excellent results with loadings of 0.5–2 mol%, which is amazing for an organocatalyst!

3.9.1 Conjugate addition reactions

3.9.1.1 Michael additions

In 2008 Rawal and co-workers reported the first cinchona-squaramide bifunctional organocatalyst **157** (Figure 3.62) which was derived from cinchonine (**CN**) (Figure 3.1), it was first screened in the conjugate addition of 2,4-pentanedione to β-nitrostyrene, using only 2 mol% of **157** and afforded excellent results, with a highest yield of 97% for the addition product and an enantioselectivity of 96% [63]. Further screening of this organocatalyst in a number of other Michael addition reactions was conducted (Figure 3.63). A remarkable catalyst loading of 0.5 mol% could be used.

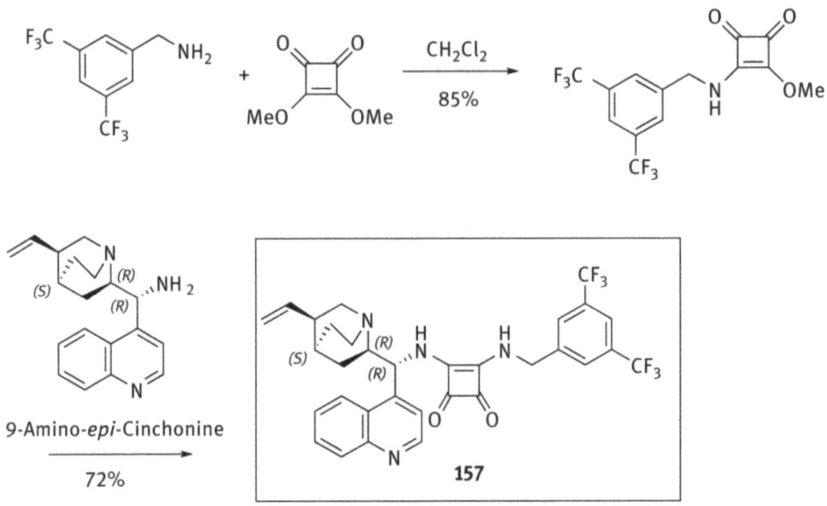

Figure 3.62: Synthesis of Cinchonine-squaramide organocatalyst **157**.

On analysing the results shown in Figure 3.63, one can see that excellent yields could be obtained in all cases (up to 99%) and enantioselectivities (up to 98% ee), but the diastereoselectivities were only moderate (1.4:1 to 4:1) with better results using the cyclic nucleophiles (18:1 in the case of **158p** and 50:1 in the case of **158q**). In these latter cases the significant rigidity of the nucleophile had a role to play in the diasterofacil selectivity. No significant electronic effects in either the nucleophile or the electrophile were apparent. No information on the proposed transition state was given in the paper, but on the basis of analogous work it should be akin to that shown in Figure 3.64 [67]. We presume there is *Re*-face attack of the nucleophile on the Michael acceptor, to give products with the (*R*)-configuration.

Since the report of Rawal there have been a number of other reactions reported in the literature [63]. In 2015, Du and Li reported on the stereoselective Michael addition of pyrazolin-5-ones to 3-nitro-2*H*-chromenes (Figure 3.65) [68]. The interest behind this

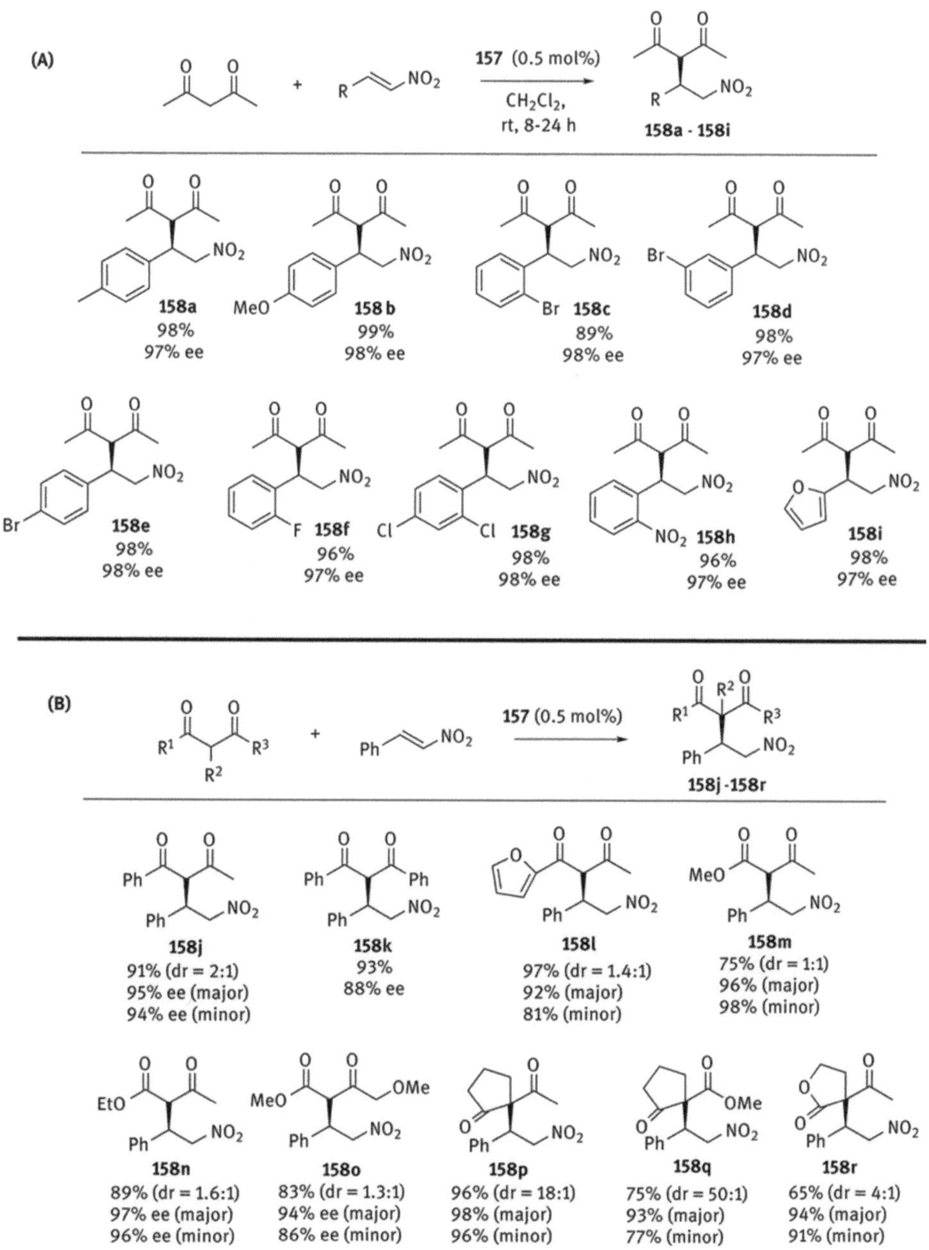

Figure 3.63: Conjugate additions with cinchonine-squaramide organocatalyst **157**, **(A)** variation of the electrophile and **(B)** variation of the nucleophile with β-phenyl styrene.

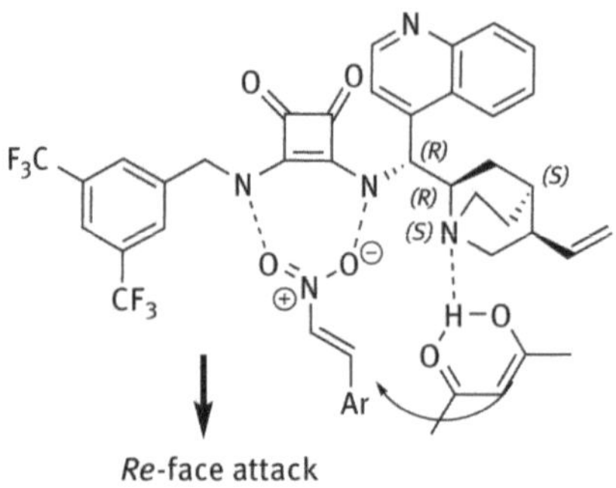

Figure 3.64: Proposed working model for the reactions described by Rawal.

work was the synthesis of chroman-pyrazolone hybrids with biological activity. Initial catalyst sceening using 3-methyl-1-phenyl-2-pyrazolin-5-one with 3-nitro-2*H*-chromene in the presence of organocatalyst **159** and **160** at a loading of 5 mol% furnished the chroman-pyrazolone hybrid **162** with excellent yields, diastereoselectivities and good enentioselectivities (Figure 3.65). However, it was organocatalyst **160** with only one CF_3 group in the aryl ring that gave a lower ee of 62%, which was probably due to weaker H-bonding with the electrophilic nitro-group. Infact other organocatalysts were screened, that gave equivalent results at a loading of only 0.2 mol%. It was necessary to trap the resulting enol as its acetate derivative to avoid unnecessary tautomerism.

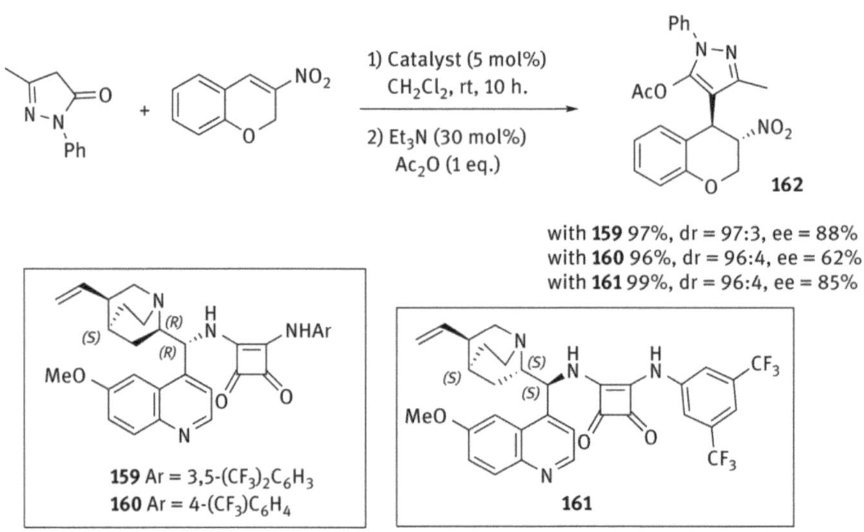

Figure 3.65: The asymmetric Michael addition to afford chromene-pyrazolinone hybrids by Li and Du.

In 2016 the same group also reported on a Michael addition of α-alkylidene succinimides to nitrostyrenes (Figure 3.66) [69]. In this study several types of cinchona-squaramide catalysts (**159–161, 163–165**) were used, including the ones described above. The reaction was run generally in CH_2Cl_2 at room temperature with 5 mol% organocatalyst present. The results were very good, with very good to excellent yields and almost complete diastereoselectivity right across the board. The enantioselectivities varied a bit, and again it was obvious that the presence of the 3,5-$(CF_3)_2$-C_6H_3 group was mandatory for the highest ees. In the case of organocatalyst **161** and **163** the opposite enantiomer was obtained. $CHCl_3$ was found to be a better solvent than CH_2Cl_2. The reaction with organocatalyst **164** could be run in $CHCl_3$ at −10 °C and when a catalyst loading of 2.5 mol% was used, the equivalent results were observed, except that the yield dropped to 91%!

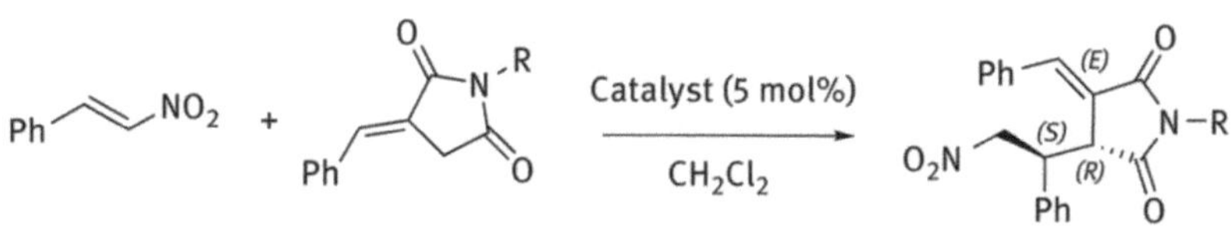

166

with **159** (R = Boc), 89%, dr = >99:1, ee = 95%
with **160** (R = Boc), 85%, dr = >99:1, ee = 82%
with **161** (R = Boc), 93%, dr = >99:1, ee = 91% (enantiopode)
with **163** (R = Boc), 82%, dr = >99:1, ee = 83% (enantiopode)
with **164** (R = Boc), 92%, dr = >99:1, ee = 96%
with **164** (R = Boc), 98%, dr = >99:1, ee = 98% ($CHCl_3$)
with **165** (R = Boc), 71%, dr = >99:1, ee = 85%

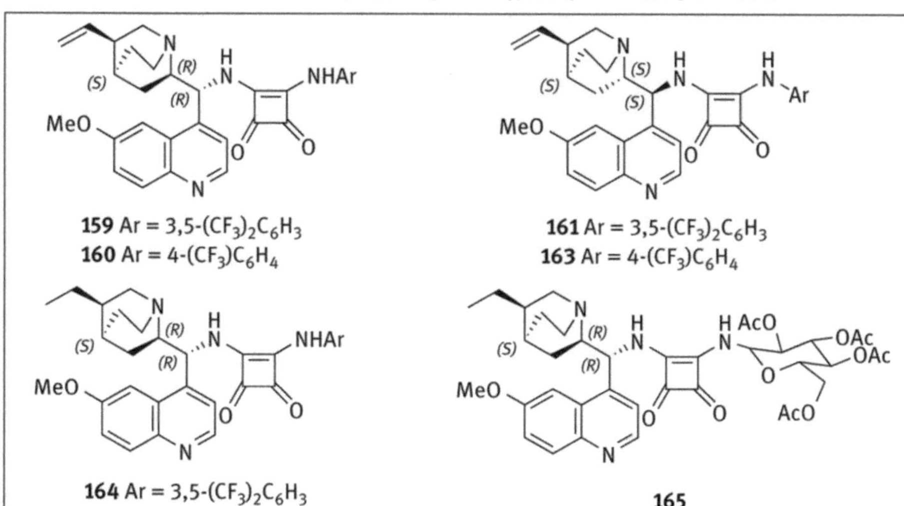

159 Ar = 3,5-$(CF_3)_2C_6H_3$
160 Ar = 4-$(CF_3)C_6H_4$

161 Ar = 3,5-$(CF_3)_2C_6H_3$
163 Ar = 4-$(CF_3)C_6H_4$

164 Ar = 3,5-$(CF_3)_2C_6H_3$

165

Figure 3.66: The asymmetric Michael addition of α-alkylidene succinimides to nitrostyrenes by Du and co-workers.

This method has been used with great effect to furnish several biologically active compounds. For instance, *Baclofen* (Lioresal® and Baclon®) is a GABA (γ-aminobutyric acid) antagonist that is used as a muscle relaxant to treat muscle spasms. In 2016 Šebesta's group harnessed the squaramide-catalysed Michael addition of malonate to β-nitrostyrene derivatives as the key step in its synthesis (Figure 3.67) [70]. Unfortunately, when both the cinchona-squaramides **167** (this was not obvious from the structure given in the paper but we presume it was this isomer) and **159** were used, the results were less than desired (in the case of **167**, despite a good ee of 90%, the conversion went to only 55% with a yield of 52%, and in the case of **159** a conversion of 26% and an isolated yield of 8% were observed). Other types of squaramide organocatalysts gave superior results in this study.

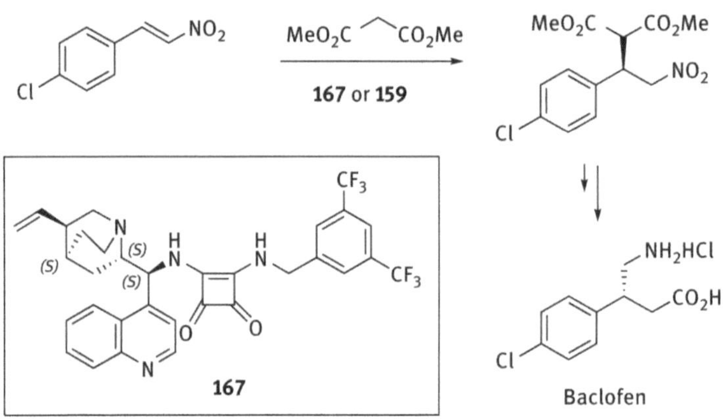

Figure 3.67: The asymmetric Michael addition to nitrostyrenes as an approach to *Baclofen* as reported by Šebesta and co-workers.

Aminophosphonic acids are transition state analogues of amino acids due to the ability of the phosphonate unit to mimic the tetrahedral transition state of peptide bond hydrolysis. In 2015, Bera and Namboothiri and co-workers used organocatalyst **161** for the Michael addition of α-nitrophosphonates to enones, giving quaternary α-nitrophosphonates (which can then be transformed to α-nitrophosphonates) in high yields and enantioselectivities. The full scope of this method is highlighted in Figure 3.68 [71]. These workers screened an array of cinchona-squaramides but it was the novel dihydroquinine-squaramide **164** (at a loading of 10 mol%) that gave marginally better selectivities, and it should be noted that the conditions were quite drastic with an optimum temperature of −65 °C. As can be seen in Figure 3.68, in many instances, both the yields and the enantioselectivities were high. However, in some cases the ees were low, as in the case of **168i** and **168k**, it seems that in the case of the former reaction, lack of π-π-stacking between the enone R group and the organocatalyst might be the reason, in the case of the latter reaction, poor π-π-stacking was probably the cause.

The authors furnished explicit transition state models which we will not go into here. The nitro group could easily be reduced to an amino group using Zn in HCl.

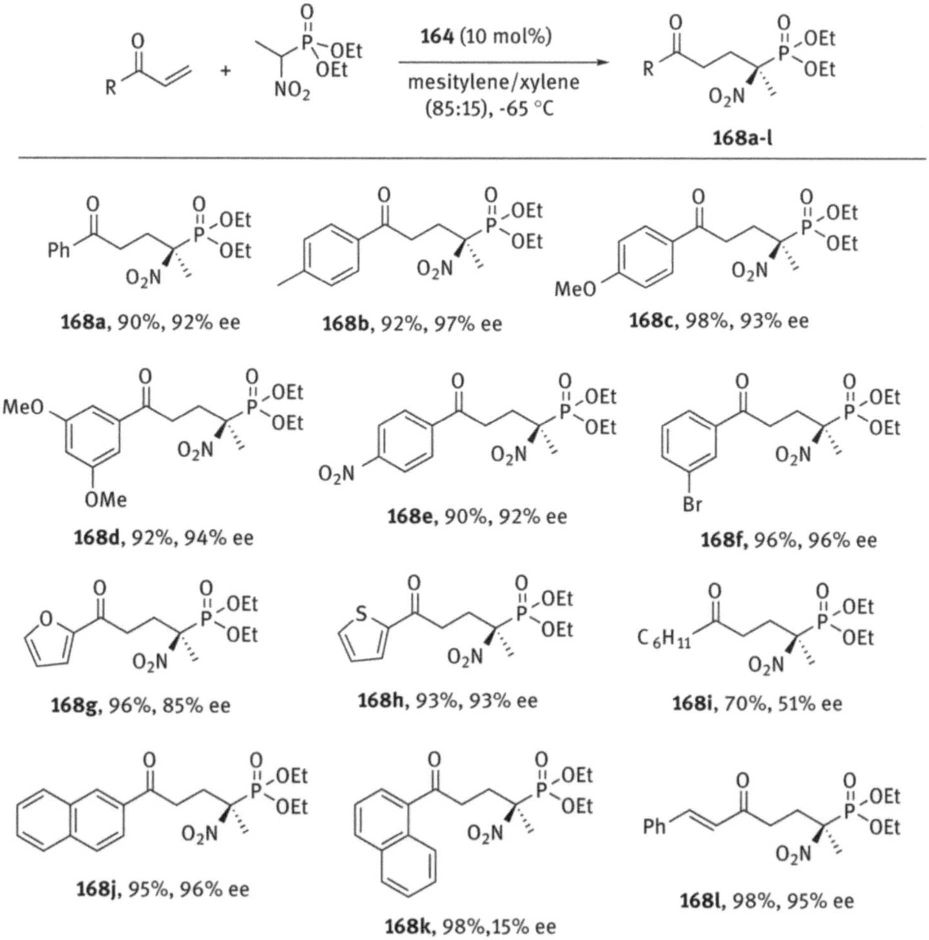

Figure 3.68: Selected examples for the asymmetric Michael addition to enones as an approach to quaternary α-nitrophosphonates as reported by Bera and Namboothiri.

Bae and Song reported in 2015 concise scalable pathways to the antidepressant (*S*)-*Rolipram* and the anticonvulsant (*S*)-*Pregabalin* using key cinchona-squaramide catalysed Michael additions [72]. In this study, water was used as the reaction medium for the noncovalent, hydrogen-bonding-promoted enantioselective Michael addition of the malonates to the nitroolefins (Figure 3.69). Due to the so-called hydrophobic hydration effect, a significant increase in the reaction rate was observed when the reaction was performed "on water" rather than in the conventional organic solvents. Moreover, due to the remarkable rate acceleration under "on water" conditions, the catalyst loading also significantly decreased to only 0.01 mol% (which is an amazing feat for an organocatalyst). This protocol gave excellent diastereoselectivity and enantioselectivity (up to >99:1 dr, up to 99% ee).

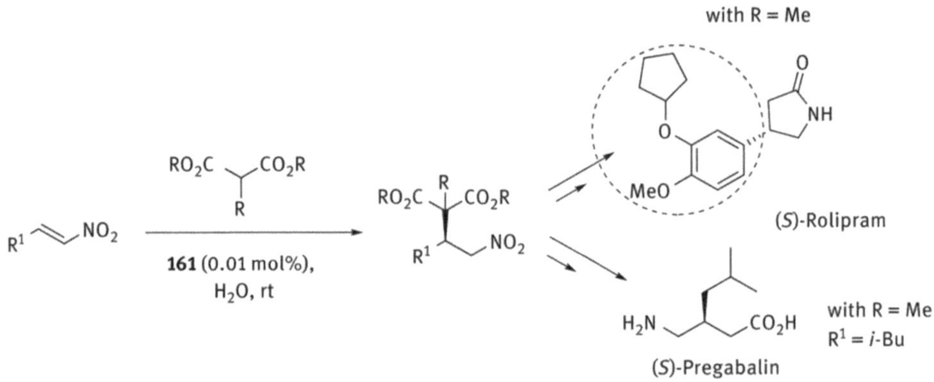

Figure 3.69: Application of the asymmetric Michael addition to nitrostyrene for the synthesis of (*S*)-*Rolipram* and the anticonvulsant (*S*)-*Pregabalin* as reported by Bae and Song.

3.9.1.1.1 Squaramide-catalysed sequential reactions initiated by a Michael addition

For some time, it has been common knowledge that sequential one-pot reaction procedures bring many advantages to the process of chemical synthesis [73]. Organocatalytic domino/cascade reactions are also known, and permit, through operationally simple methods the synthesis of hosts of enantiomerically pure complex molecular structures ([66]d). Some excellent reviews on this topic have been published by both Enders' and Rueping's groups in 2014 and 2015, respectively ([66]d, [66]h) and another more recent by Chanda and Zhao in 2018 ([66]e).

Some of the recent examples that caught our attention include the following (not exhaustive). As an additional note, we see here that most of the molecules produced via these pathways, are of relatively low molecular weight with very constrained architectures, many of which are spiro and not easy to access, although of major interest in medicinal chemistry.

In 2016 Zhao and Du reported an interesting organocatalytic cascade Michael/Michael reaction that afforded 5-membered spirooxindoles [74]. The spirooxindole structure is a privileged scaffold present in both medicinal compounds and natural products ([66]b). In their approach these workers treated α-alkylidene succinimides with 3-ylideneoxindoles in the presence of a variety of cinchona-squaramide catalysts. The hydroquinine organocatalyst **164** gave the best results. The conditions were fully optimized before screening the substrate scope to afford the spirooxindoles **169** (Figure 3.70).

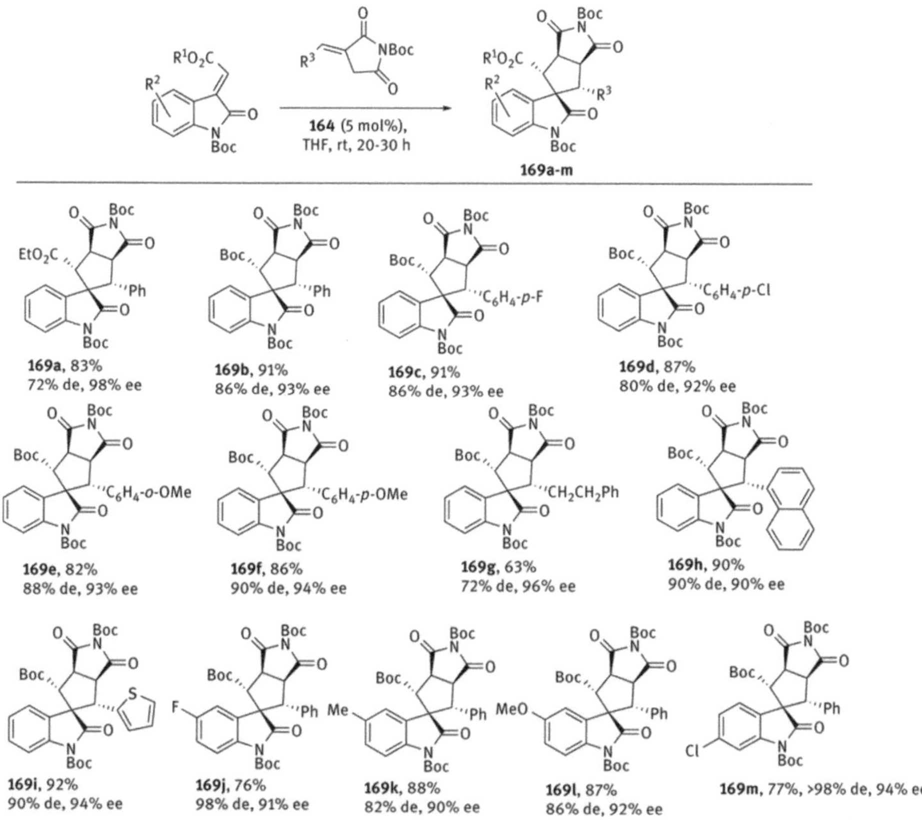

Figure 3.70: Application of the asymmetric catalysed cascade Michael/Michael addition for accessing 5-membered spirooxindoles **169** as reported by Zhao and Du.

The yields and the enantioselectivities were consistently high, and the electronic nature of the substituents did not influence both these results. The diastereoselectivity in general was relatively high, but however, in some cases, exceptional levels were obtained as was the case for both **169j** and **169m**. This might be attributed to some form of electronic effect.

In 2016 Sun *et al.* reported a sequential Michael/Michael/Aldol cascade that afforded fully-substituted cyclohexanes **170** bearing 6-contiguous stereocentres [75]. In this procedure the organocatalyst **167** was found to be the most efficient with a low loading of only 0.5 mol%. All that was required was to add acetylacetone, with nitrostyrene and the enal with 10 mol% pyrrolidine at room temperature in dichloromethane. The results for a study on the reaction scope are given in Figure 3.71.

In all cases the diastereomeric ratio was greater than 20:1 and the enantioselectivity above 99% ee. All the yields were basically the same (around 60%), only in the case of **170i** and **170j** they were lower (31% and 27%, respectively), the authors did not

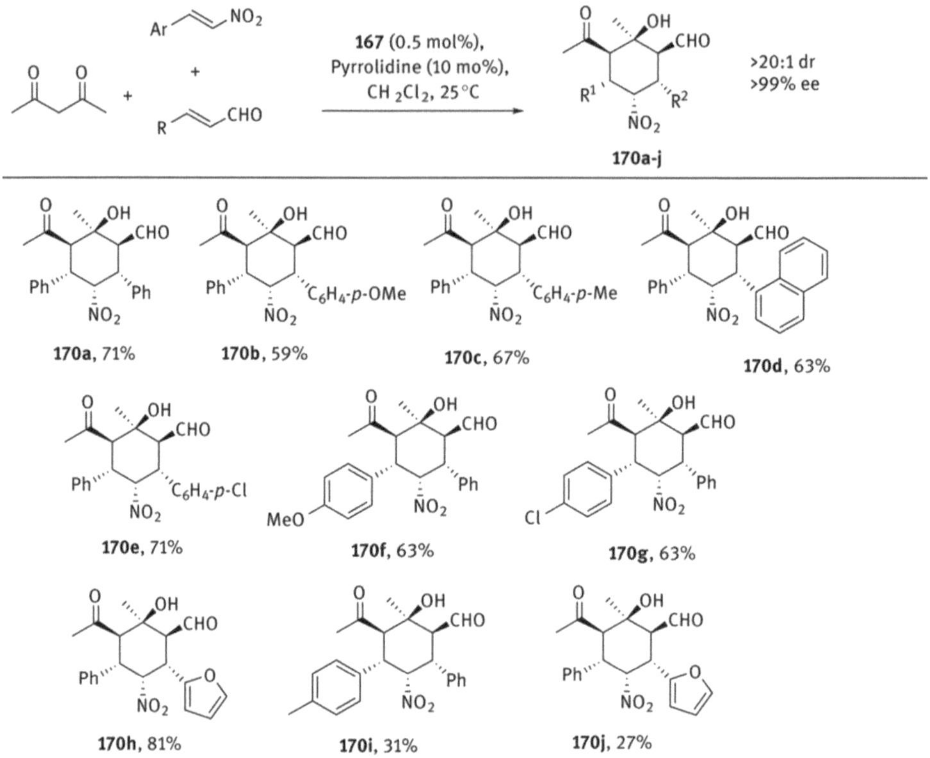

Figure 3.71: The cascade Michael/Michael/Aldol cascade for accessing fully substituted cyclohexanes as described by Sun *et al.*

provide an explanation for this. Overall this showed that most functional groups were tolerated. The authors provided a working model, and the Michael addition of the malonate to the electrophile was akin to the model described in Figure 3.64. The authors proposed that there was first a Michael addition of the acetylacetone enolate to the nitrostyrene to form the chiral substituted dicarbonyl, at the same time the pyrrolidine interacts with the enal to form the corresponding iminium ion, which is subsequently attacked by the substituted dicarbonyl intermediate in the second Michael addition. This is then followed by the ring closing step, which is an intramolecular Aldol condensation between one of the carbonyl electrophiles of the acetylacetone unit and the enamine that was formed after the second Michael addition.

In 2015 Amireddy and Chen reported a squaramide-catalysed Michael-Aldol cascade reaction of γ-nitro ketones and 2-arylideneindane-1,3-diones to afford spirocyclohexane indane-1,3-diones [76]. In fact, this structure is present in a plethora of naturally occurring and medicinal compounds, such as Coleophomone A, Coleophomone D and Fredericamycin. These workers screened a variety of cinchona-squaramide catalysts, and it was organocatalyst **151** (we assume that it is this catalyst from the drawing in the Scheme) which gave the best results. All the results can be observed in Figure 3.72, which clearly shows the scope of the reaction. Although in most cases the diastereoselectivity was very high (about 20:1) the best enantioselectivity of 86% ee was obtained in the case of **171q** (Figure 3.72). A remarkable loading of only 2 mol% of catalyst could be used. Some interesting results were observed. The reaction tolerated both electron-donating and electron-withdrawing substituents on the phenyl ring of the 2-arylidene-1,3-indanediones, but however, when the aryl substituent was a 2-thiophene or 2-furyl ring both the diastereoselectivities and enantioselectivities were only moderate (**171i** and **171j**, 34% and 44% ee, respectively). A low yield of 57% was obtained for **171k** which contained the strong electron-withdrawing group NO_2 in the *para* position. A very good yield of 90% was obtained for **171r** with the NO_2 in the *meta*-position, but the ee was only 30% as opposed to 70% ee in the case of **171k**. The absolute configuration of product **171j** was determined by X-ray crystallography.

The authors presented a simple mechanism to explain the reaction and the enantioselectivity (Figure 3.73). They claimed that the carbonyl of the arylideneindane-1,3-dione is activated by H-bonding with the squaramide, and that the Michael addition takes place between the nucleophilic γ-carbon of the nitroketone from the *re* face. A similar reaction has also been reported by Duan *et al.* [77].

In 2016, Soós' group reported the stereoselective synthesis of terpenoid decalin subunits bearing quaternary stereocenters through an organocatalysed Robinson Annulation of Nazarov's reagent [78]. Many important biological active or natural product compounds bearing a decalin core are known, these include: Bucidarasin A, Dysidiolide, Teucvin and Atisine etc. [73]. In their study, these workers used a variety of Nazarov reagents **172** with substituted 3-oxopropanoates **173** in the presence of the novel massive organocatalyst **174** (derived from 9-amino-epi-quinine, and obtained via an extensive screening study) at a loading of only 2 mol% to give the decalin products **175** (Figure 3.74). It was also discovered that running the reaction at room temperature and using 1,4-dioxane as solvent gave the best enantioselectivities. Yields of up to 78% and enantioselectivities of up to 91% ee were obtained, including in those specific cases which involved the creation of two stereogenic centres, diastereoselectivities of up to 27:1. The yields were lower in the case of **175i** and **175j**, which was probably due to the increase in stereochemical bulk of the Nazarov reagents. The reaction could be scaled-up to 60 mmol loading of substrate. The authors assigned the absolute configurations given in Figure 3.74 without discussing how this was established.

In 2018, Mondal and Pan reported an interesting domino Michael/Acyl transfer reaction between γ/δ-hydroxyenones and α-nitroketones [79]. In the case of the γ-hydroxyenones (Figure 3.75**(A)**) the reaction afforded γ-nitroketone target compounds **176** in high yields and with excellent enantioselectivities using the quinine derived organocatalyst **161** at a loading of 10 mol% in toluene at room temperature (Figure 3.75**(A)**). The reaction was tolerant of a wide functional group range, and it was observed that aromatic nitroketones were preferred over aliphatic nitroketones as the yields were lower

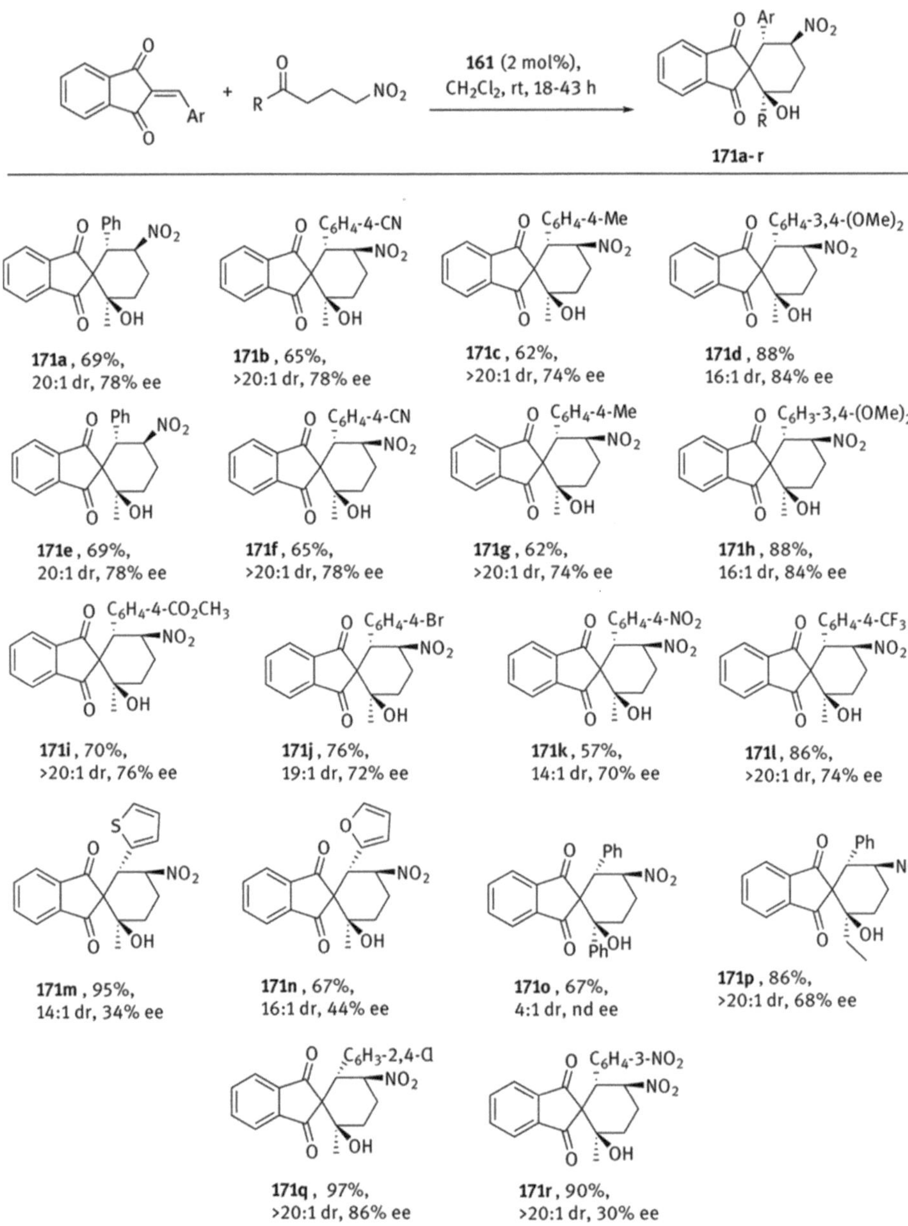

Figure 3.72: A cross-section of results for the Michael/Aldol cascade giving spirocyclohexane indane-1,3-diones as described by Amireddy and Chen.

(e. g. **176g** and **176h**). Upon changing from the γ-hydroxyenones to the δ-hydroxyenone substrates the following results were obtained (see Figure 3.75**(B)**).

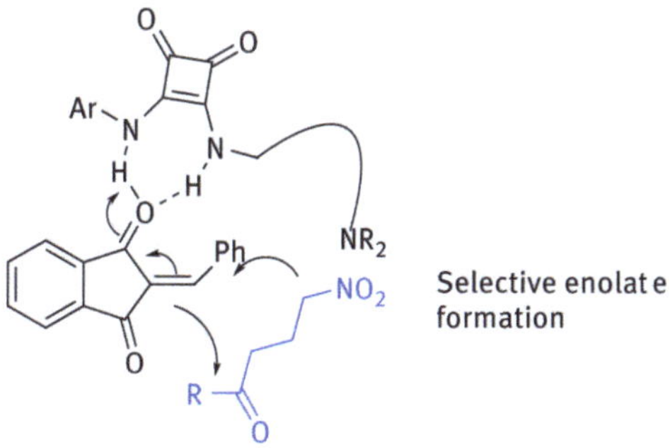

Figure 3.73: Proposed mechanism for the Michael/Aldol cascade as described by Amireddy and Chen.

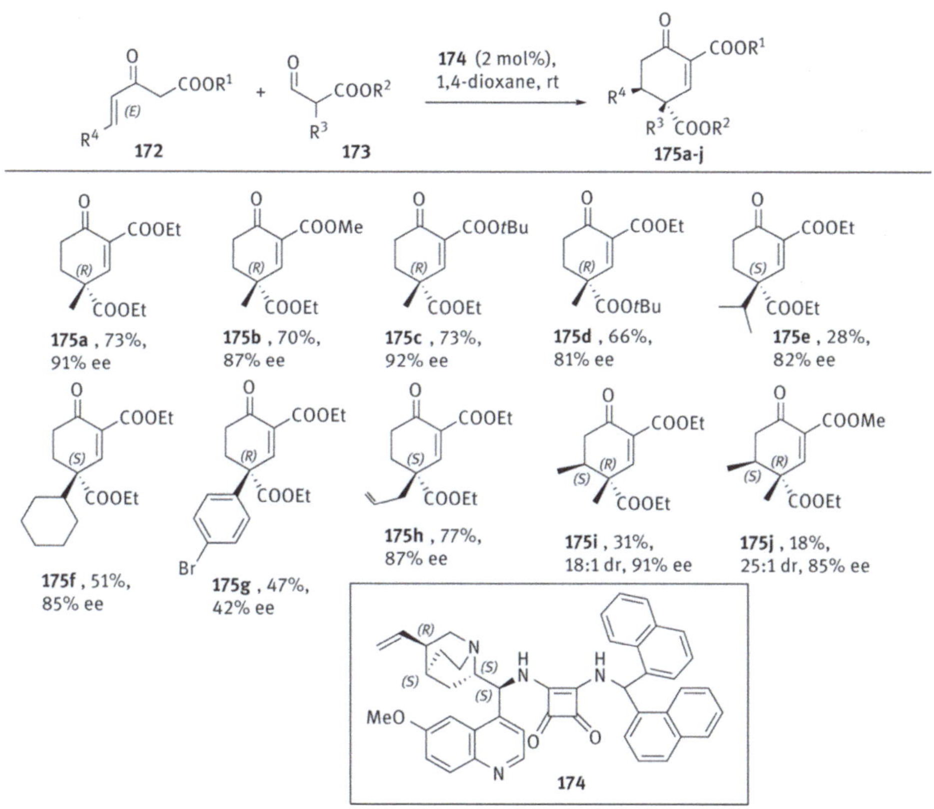

Figure 3.74: The organocatalysed Robinson Annulation of Nazarov reagents as reported by Soós' group.

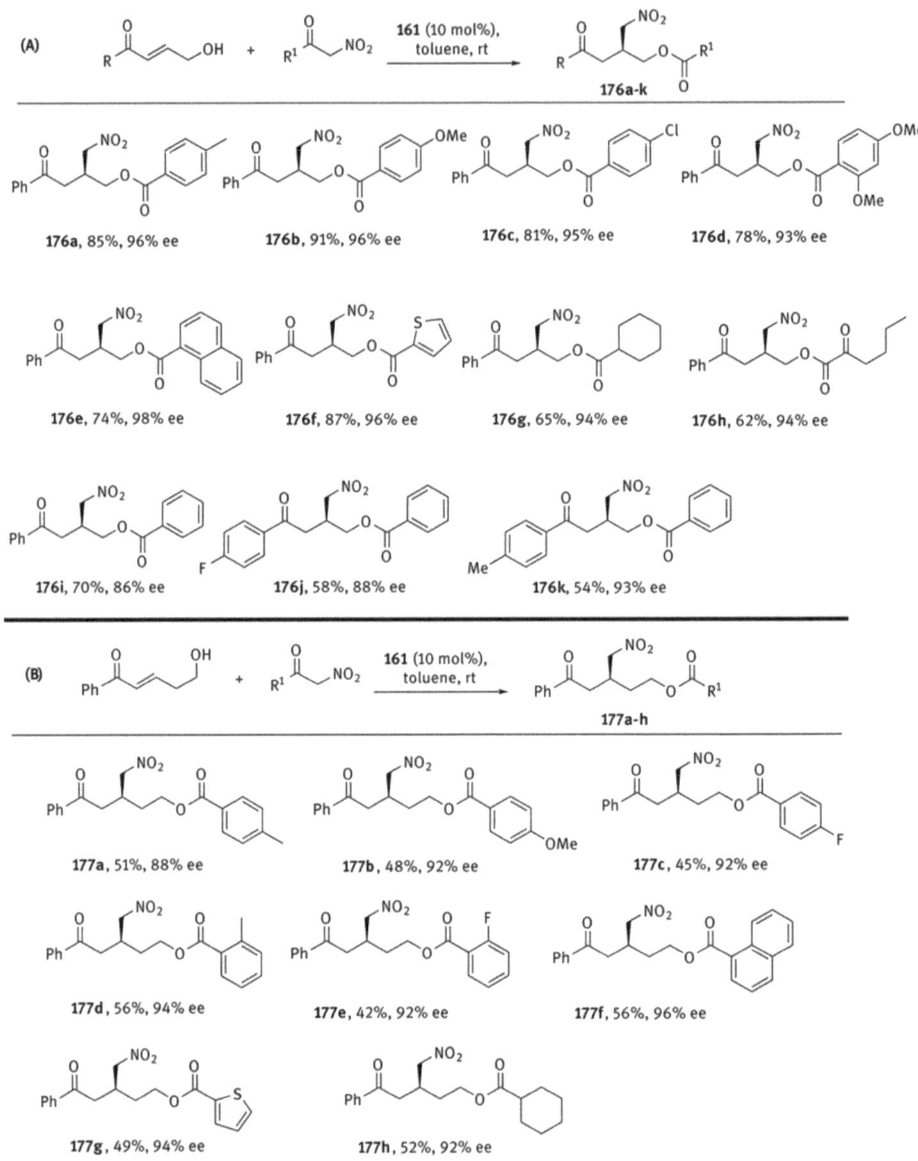

Figure 3.75: The organocatalysed Asymmetric Domino Michael/Acyl transfer reaction of Mondal and Pan.

The problem with this method was that the reaction required from 2 to 5 days to go to completion.

The absolute configuration of product **176c** was established to be of (*S*)-configuration using X-ray crystallography and by analogy all the other products were also expected to have this configuration.

The mechanism for this process proposed by the authors is described in Figure 3.76. This is very similar to other mechanistic proposals: once again the squaramide-cinchona catalyst will orient the enolate to the Michael acceptor, and after the formation of the conjugate addition product an intramolecular hemi-ketalization (the acyl transfer reaction) takes place, followed by the *retro*-Henry reaction to give the γ-nitroketone **176a.**

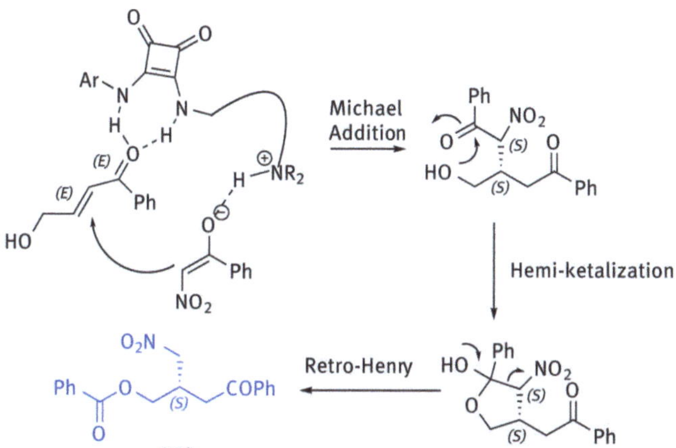

Figure 3.76: The mechanism of the organocatalysed asymmetric domino Michael/acyl transfer reaction proposed by Mondal and Pan.

Pyrano[2,3-*c*]pyrazoles are a hot topic in medicinal chemistry, due to their biological properties and moreover, there has been great progress in recent years with the use of organocatalysts for making these products. In 2015 Li and Du described an enantioselective synthesis of *N*-phenyl-dihydropyrano[2,3-*c*]pyrazoles **178** via cascade Michael addition/Thorpe-Ziegler type cyclization **[80]**. In the approach taken by Li and Du, unsaturated pyrazolones were reacted with malononitrile to form these targets using the quinine derived organocatalyst **163** at a loading of 5 mol% (Figure 3.77). The best conditions were encountered after various optimization studies. The reactions took place with excellent yields (all almost quantitatively) at room temperature between 40 and 120 min. A study of the scope of the reaction revealed that the reaction was tolerant of a variety of functionalities in the phenyl ring appendages, but in the case of the reaction enantioselectivity, they were only in the 41–79% ee range. No specific electronic effect trends were in evidence, although some steric effects were observed, for instance, when the dihydropyrano unit aryl group contained an *ortho* substituent, the ees were generally lower than for their *para*-substituted analogues (see, **178e, 178h** and **178i**). The absolute configuration of these products was determined by comparison with the literature. A proposed transition-state working model was furnished, similar to those presented by other authors for analogous processes (see Figure 3.77 for example).

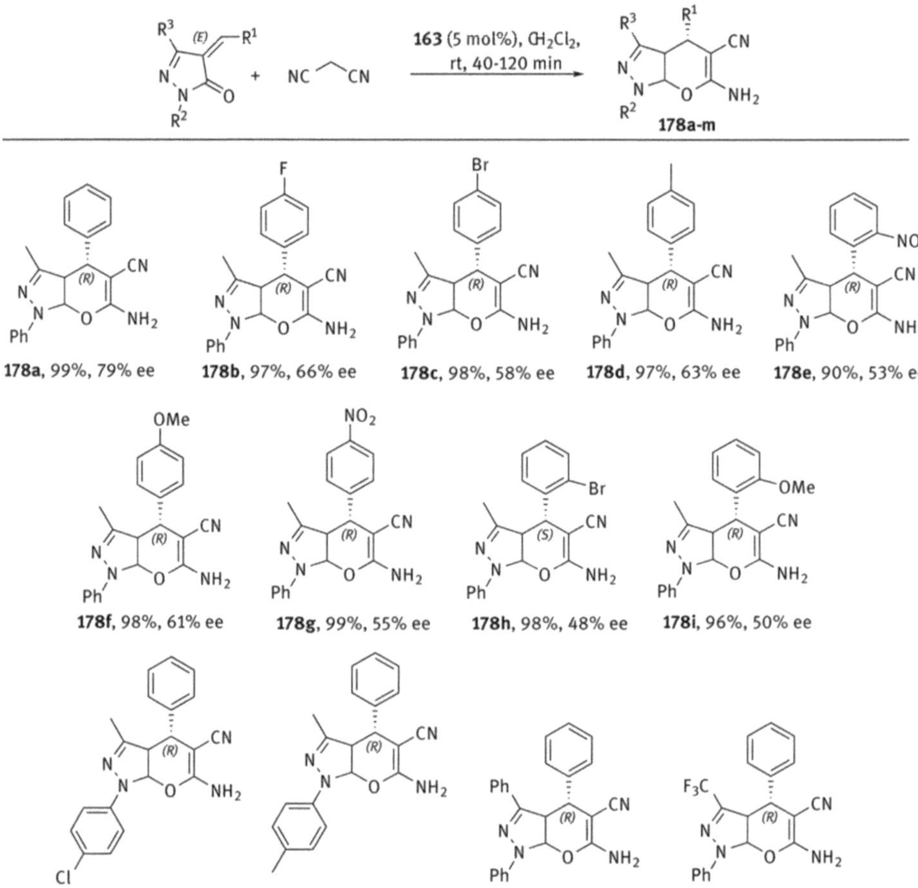

178a, 99%, 79% ee
178b, 97%, 66% ee
178c, 98%, 58% ee
178d, 97%, 63% ee
178e, 90%, 53% ee
178f, 98%, 61% ee
178g, 99%, 55% ee
178h, 98%, 48% ee
178i, 96%, 50% ee
178j, 99%, 50% ee
178k, 97%, 54% ee
178l, 98%, 41% ee
178m, 98%, 41% ee

Figure 3.77: The organocatalysed asymmetric domino Michael/Thorpe-Ziegler type reaction described by Li and Du.

Cascade Michael/alkylation reactions are also known, however due to space limitations this will not be discussed here (one can consult ref. ([66]c), for some nice examples).

3.9.1.2 Sulfa-Michael additions

Sulfur containing drugs are a very potent class of pharmaceuticals, and thus organocatalytic synthetic methods to such targets are very desirable. These reactions are well documented, and the reader is encouraged to consult key reviews like ref. ([66]c), for previous documented reactions (prior to 2015).

In 2015 Hiemstra's group reported on the cinchona catalysed sulfa-Michael addition of thiols including benzylthioalcohol to α,β-unsaturated *N*-acylated oxazolidine-2-ones and related α,β-unsaturated α-amino acid derivatives, however unfortunately the quinidine derived organocatalyst **161** failed to catalyse this reaction and in fact no reaction occurred [81].

In 2018, Hasilcioğullari and Tanyeli reported the enantioselective sulfa-Michael addition of methyl thioglycolate to chalcones [82]. A number of cinchona-squaramide organocatalysts were screened but it was the 2-adamentyl quinine-squaramide **179**, a congested catalyst, that gave the best results (Figure 3.78). The reactions were performed at −40 °C in toluene with 10 mol% of the catalyst and afforded the adduct **180** in excellent yields, having enantioselectivities in the range 68–99% ee. The reaction tolerated a wide range of functionality, and although the yields were excellent, the enantioselectivities were influenced by steric effects, for instance in the case of the *ortho*-substituted products – that were either electron-withdrawing or electron-donating groups – the ees were lower (as in the case of **180c, 180f** and **180g**). The absolute configuration of the product was determined by comparison with the literature. The reaction proved to be efficient at gram scale, too.

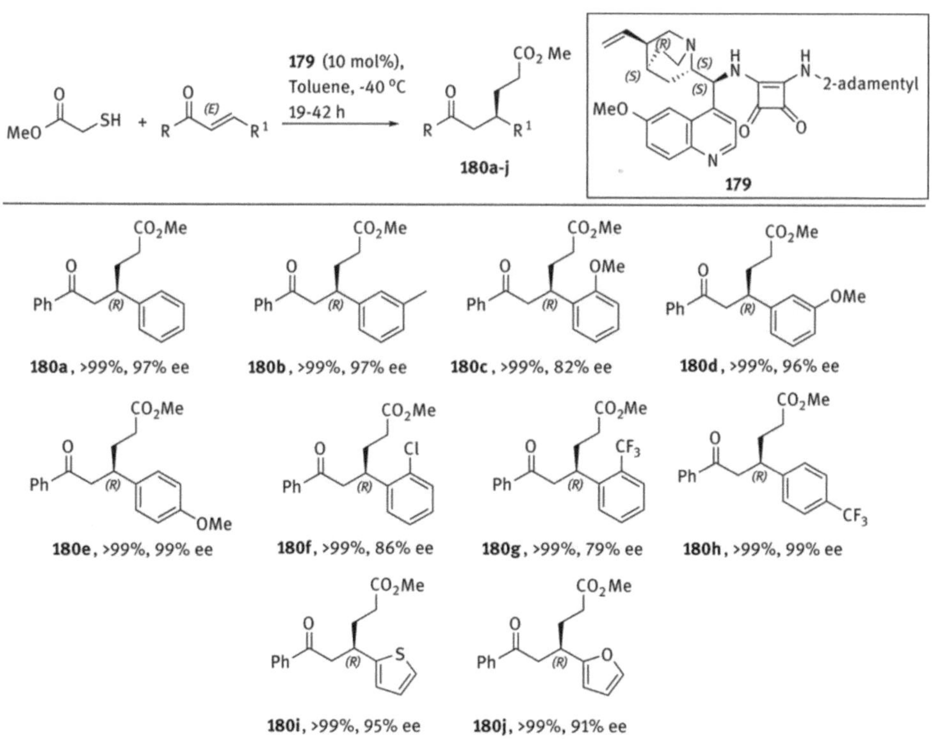

Figure 3.78: The organocatalysed asymmetric sulfa-Michael addition of methyl thioglycolate to chalcones described by Hasilcioğullari and Tanyeli.

Gu *et al.* recently reported the addition of tritylthiol to azadienes using a variety of cinchona-squaramide catalysts, but it was the dihydroquinine derived organocatalyst **181** that gave the best results (Figure 3.79) [83]. Excellent yields, with ample scope and enantioselectivities of up to 94% were achieved, using mild conditions. The absolute configuration of these products was established by carrying out an X-ray structural analysis on compound **182a.** The reaction showed reasonable scope – perhaps more diverse functional groups could have been included – and there didn´t seem to be any particular electronic or steric preferences within the azadiene substrate. It was also shown that a bulky nucleophile such as tritylthiol was necessary for high enantioselectivities, as was the case with the use of phenylthiol the reaction was racemic (for **182h**). The standard mechanism for cinchona-squaramide catalysis was invoked based on the Brønsted acid-hydrogen bonding transition state model.

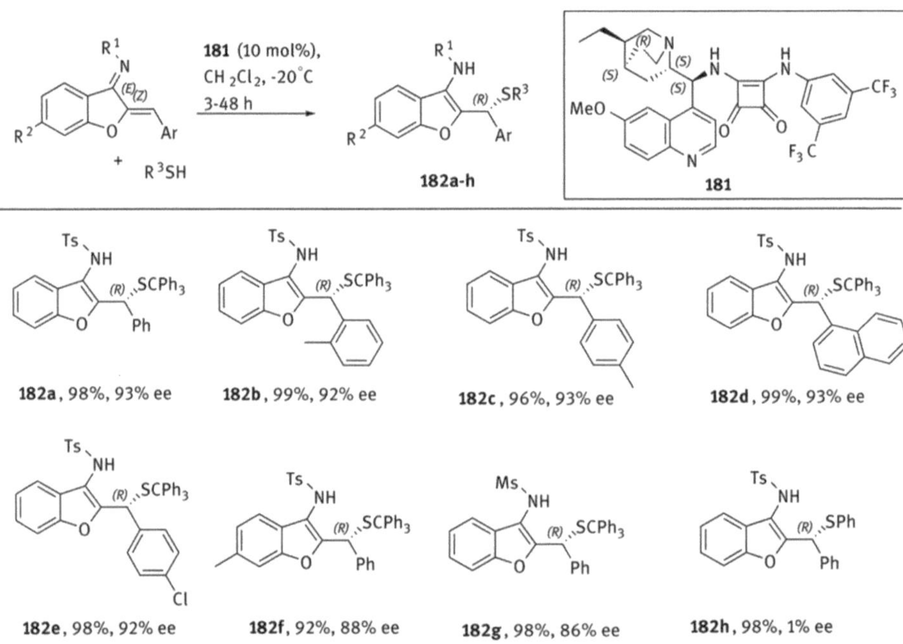

Figure 3.79: The organocatalysed asymmetric sulfa-Michael addition of tritylthiol to azadines described by Gu *et al.*

3.9.1.2.1 Squaramide-catalysed sequential reactions initiated by a sulfa-Michael addition

In 2016 Zhu *et al.* reported on the synthesis of CF_3- and indole-containing thiochromanes via a squaramide-catalysed Michael–Aldol reaction [84]. Thiochromanes are an interesting class of compounds for preparing various types of drugs, and indeed

2,2-di-substituted analogues are even more interesting [79]. After careful optimization studies, 2-mercaptobenzaldehydes were reacted with β-indole-β-trifluoromethyl enones in the presence of the cinchonine based organocatalyst **182** at a loading of 10 mol% in toluene at 0 °C (Figure 3.80). Excellent yields, and stereocontrol were observed for all reactions. The reaction showed very good functional group tolerance, and no specific electronic or stereo effects were in evidence. The absolute configuration of these products (2*S*,3*S*,4*R*) was determined by an X-ray crystallographic study that was performed on compound **183b.**

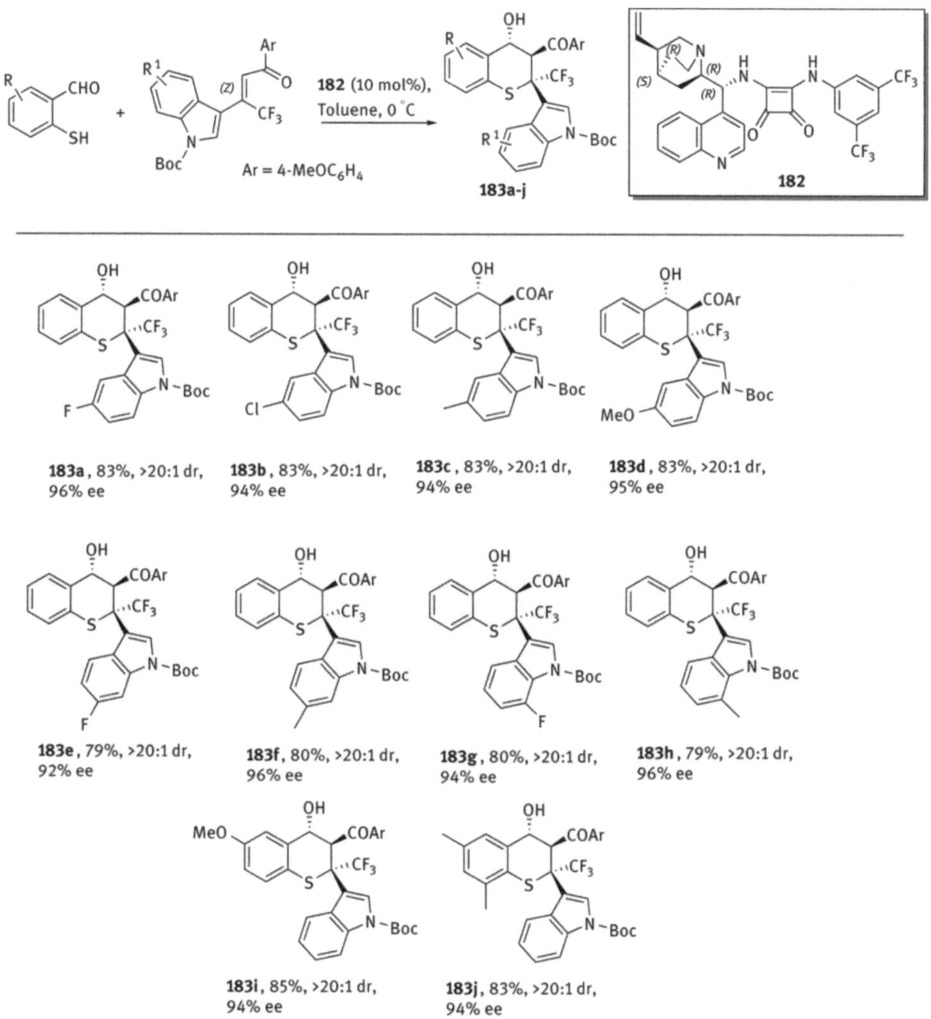

Figure 3.80: The squaramide-catalysed thio-Michael-Aldol reaction described by Zhu *et al.*

3.9.1.3 Aza-Michael additions

In this particular category, most of the examples given are those associated with a sequential process involving an initial aza-Michael reaction.

However, a very recent publication by Szekely and co-workers described the use of membrane grafted cinchona-squaramide organocatalyst for application in aza-Michael reactions [85]. The cinchona-squaramide catalyst – which was derived from hydroquinine – was immobilized to a polybenzimidazole-based nanofiltration membrane **184** (Figure 3.81) and used in a number of aza-Michael reactions. In the case of the reaction between pyrazole and β-nitrostyrene the best results were achieved in ethyl acetate (**185a**, 23% and 43% ee) and in the case of 1,2,3-triazole the best results were achieved in acetonitrile (**185b**, 23% and 39% ee).

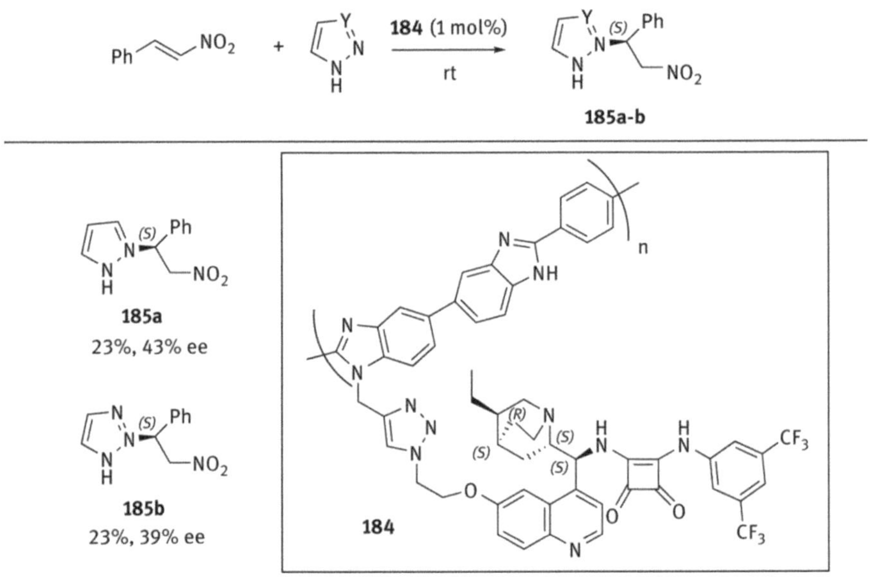

Figure 3.81: Membrane grafted cinchona-squaramide catalysts for the aza-Michael reaction as described by Szekely and co-workers.

3.9.1.3.1 Squaramide-catalysed sequential reactions initiated by an aza-Michael addition

In 2015, Da-Ming Du's group reported a cinchona-squaramide catalysed asymmetric sequential aza-Michael/Michael addition reaction for the synthesis of chiral trisubstituted pyrrolidines **186** (Figure 3.82) [86]. The best organocatalyst found was the hydroquinine **181** and was used at a loading of 10 mol% in CH_2Cl_2 at room temperature. The reaction showed good functional group tolerance, with yields in the range 47 to 99% and drs in the range 53:47 to 91:9 and enantioselectivities in the range 77 to > 99% ee. No discernible electronic or steric effects were in evidence, other than when switching to enoate substrates the diastereoselectivity dropped substantially (see **186l**, Figure 3.82). The absolute configuration of these products was assigned on the basis of comparison with literature data (by comparing specific rotations and NMR data.)

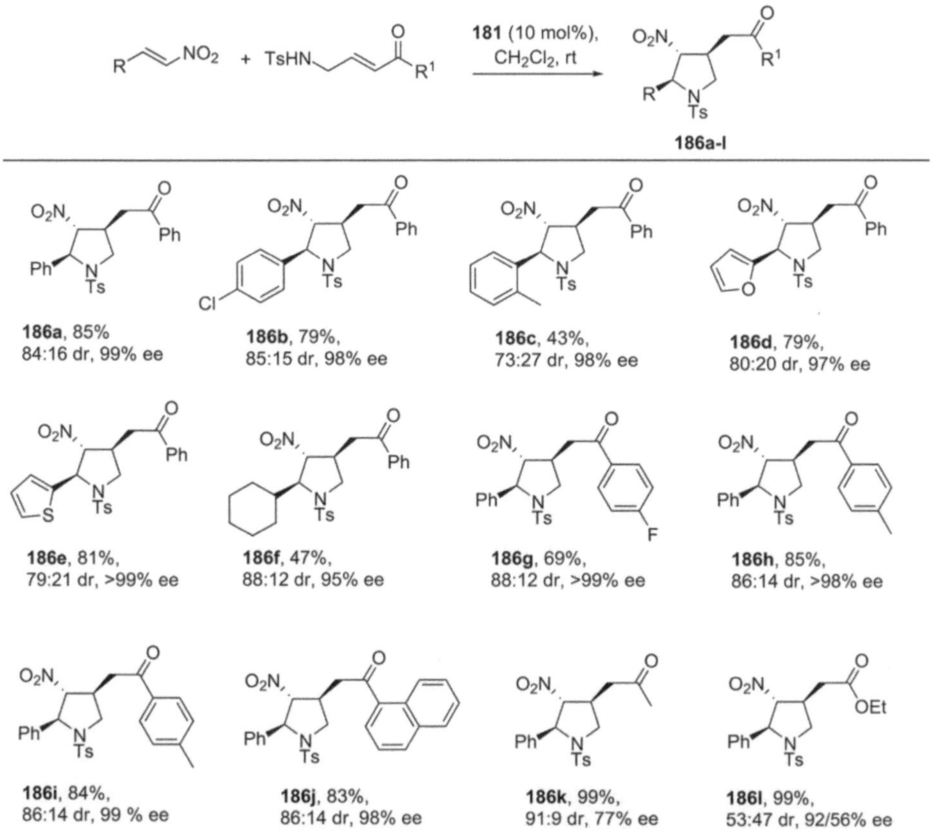

Figure 3.82: Cinchona-squaramide catalysis of a sequential aza-Michael/Michael reaction leading to chiral trisubstituted pyrrolidines as described by Du and co-workers.

Later in 2016, Du's group reported a diastereoselective and enantioselective synthesis of spiro-pyrrolidine-pyrazolones by cinchona-squaramide-catalysed cascade aza-Michael/Michael reactions (Figure 3.83) [87]. Spiro-compounds are a very important class of compounds in medicinal chemistry and have been extensively reviewed [88]. In this reaction 5 mol% of the organocatalysts **159, 160, 161** and **163** were used, and although the yields were good in the range 72–95% and with a dr of 96:4 to 98:2, the enantioselectivity was only moderate (52 to 73% ee). The authors observed that a non-cinchona squaramide gave better ees. The absolute configuration of these spiro products was unambiguously established by single-crystal X-ray crystallography.

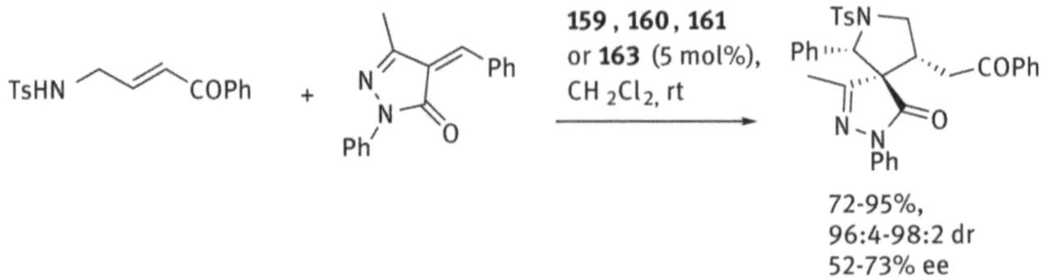

Figure 3.83: Cinchona-squaramide catalysis of a sequential aza-Michael/Michael reaction leading to a spiro-pyrrolidine-pyrazolone product as described by Du and co-workers.

3.9.1.3.2 Squaramide-catalysed sequential reactions involving an aza-Michael addition

In this section we consider some sequential processes that involve an aza-Michael step.

In 2018, Ghorai's group reported a sequential process involving a cinchona-squaramide catalysed aza-Michael reaction to form dihydroisoquinolines from ortho-homoformyl chalcone, which involved the formation of a key enamine as the first step, followed by the conjugate addition with concomitant ring closure (Figure 3.84) [89]. After careful optimization studies to encounter the best catalyst and solvent etc., the quinine derived organocatalyst **161** and $CHCl_3$ gave best results. The scope of the reaction was surveyed, a variety of functional groups were very well tolerated, and the reactions were observed to occur with generally good yields (up to 91%) and excellent enantioselectivities (up to 99% ee). It was noted that lower yields were obtained with those enones bearing furan (giving **187f**) and thiophene (giving **187g**) units (58 and 62% yields, respectively), including enoates (giving **187h-187j**) and the Weinreb amide containing substrate (giving **187e**). Lower enantioselectivities of 65% and 75% ee were encountered with the aliphatic enone (giving **187k**) and the Weinreb amide containing substrate (giving **187e**). The downside of this procedure was the requirement for using 25 mol% of the organocatalyst, but this was compensated by the ability of recycling the catalyst (the authors showed that it could be recycled up to 3 times, without changes in the yield nor the enantioselectivity). The absolute configuration was unambiguously determined by obtaining an X-ray crystal structure for one of the products.

Also in 2018, Pan and Mukhopadhyay reported something similar, they reported a route to 2,4-disubstituted imidazolidines that involved an imine addition reaction followed by an intramolecular aza-Michael addition (Figure 3.85) [90]. Upon screening a host of cinchona-squaramide catalysts, it was found that the quinine derived catalyst **188** that gave the best results. The reaction demonstrated excellent functional group tolerance, affording the products with excellent yields, diastereoselectivities and enantioselectivities. No electronic effect or steric effect trends were apparent. An X-ray crystal structure analysis of **189f** allowed the absolute stereochemical configuration of these products to be assigned. This analysis also allowed the authors to propose a

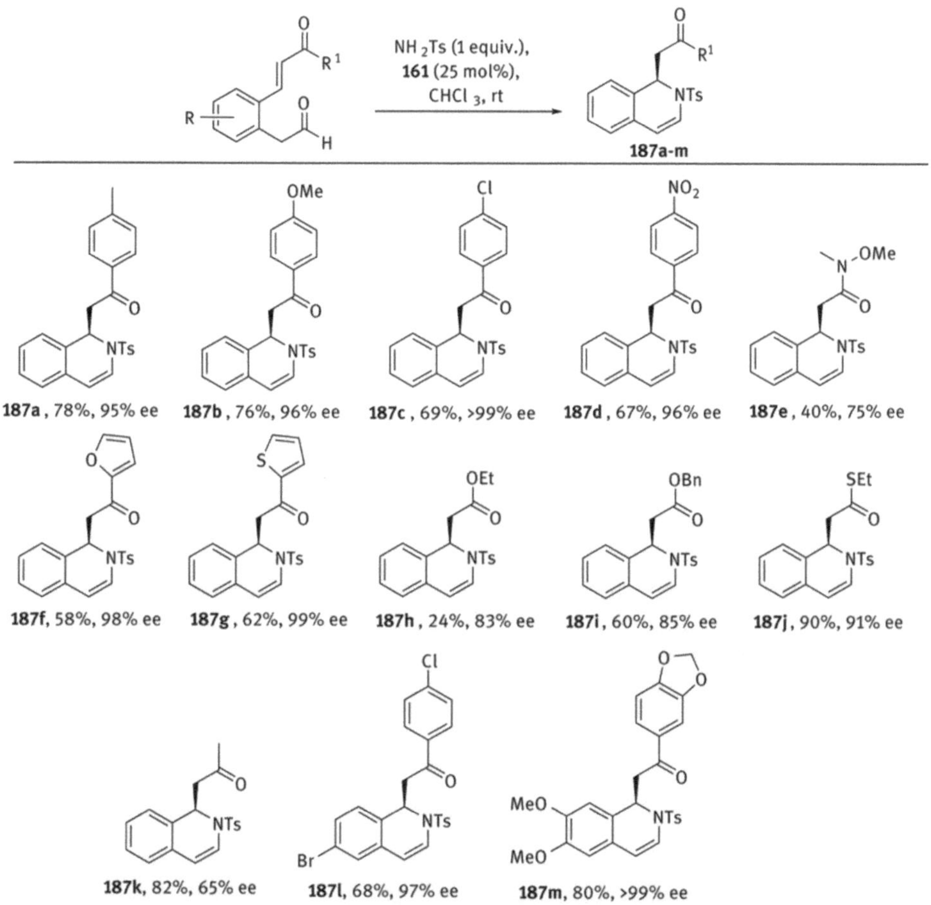

Figure 3.84: Cinchona-squaramide catalysis of a sequential enamintion/aza-Michael reaction leading to dihydroisoquinolines as described by Ghorai and co-workers.

reliable transition state model, that indicated stereofacial selectivity through preferential *Si* face attack on the imine, followed by preferential *Re* face conjugate addition to the enone.

3.9.1.4 Cinchona-squaramide catalysed reactions initiated by an oxa-Michael addition

Chromans considering their broad ranging biological activities are a very important target structure from the point of view of drug discovery and development. In 2018, Xu and co-workers reported a sequential oxa-Michael/nitro-Michael route to polysubstituted chiral chromans **190** using 2-hydroxynitrostyrenes and *trans*-β-nitroolefins catalysed by a variety of quinine-squaramide catalysts [91]. After testing a range of

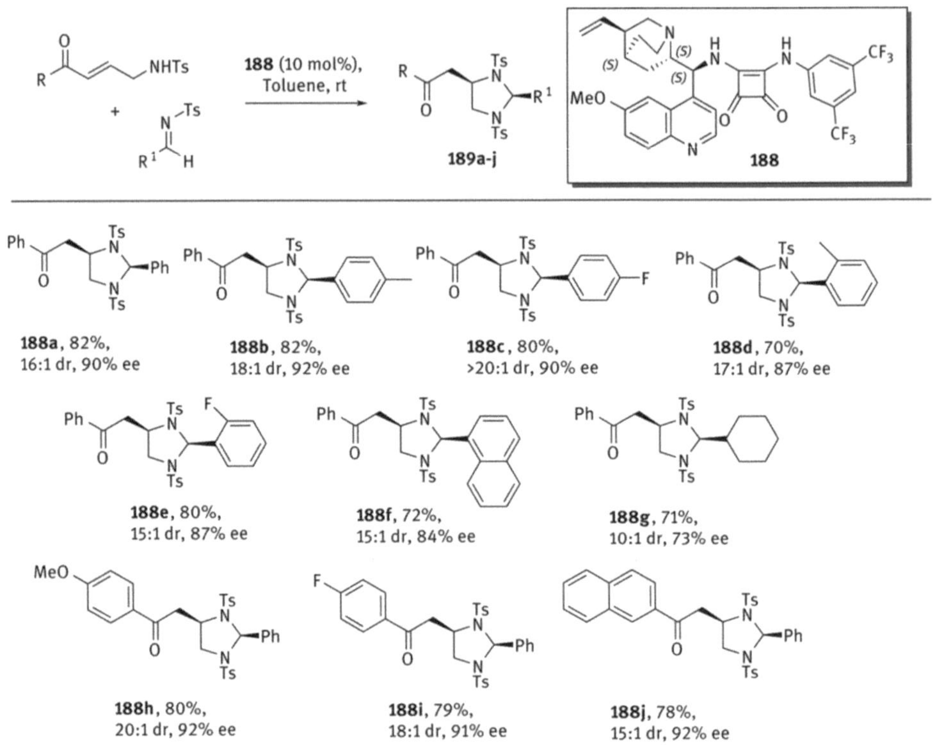

Figure 3.85: Cinchona-squaramide catalysis of a sequential imine addition/aza-Michael reaction leading to 2,4-disubstituted imidazolidines as described by Pan and Mukhopadhyay.

quinine-squaramide catalysts it was found that organocatalyst **161**, gave the best results, the reactions were run with a 5 mol% loading of the catalyst in CH_2Cl_2 at room temperature (Figure 3.86). The reactions gave good yields (60% on average), with excellent levels of both diastereoselectivity (mostly > 20:1 dr) and enantioselectivities (in many cases well over 90% ee). The absolute configuration was determined by recourse to an X-ray crystallographic study on compound **190k**.

In 2018 Mondal *et al.* reported the first diastereoselective and enantioselective synthesis of spiro-tetrahydrofuran-pyrazolones via a cascade oxa-Michael/Michael reaction between γ-hydroxyenones and unsaturated pyrazolones catalysed by cinchona-squaramide catalysts [92]. The best bifunctional squaramide catalyst found to be most effective for this reaction was the quinine-squaramide catalyst **161**, at a loading of 10 mol %, excellent results were attained for a variety of spiropyrazolones under mild reaction conditions (rt), however, the down-side of this reaction was the requirement to leave the reaction for 5 days (Figure 3.87). The stereoselectivities were generally high, with drs of > 20: 1 for many of the reactions (but were lower in the case of aliphatic γ-hydroxyenones (**191g**)), the thiophene containing enone (**191f**), the γ-hydroxyenoate (**191h**) and the pyrazolones without a phenyl or aryl substituent, and enantioselectivities of

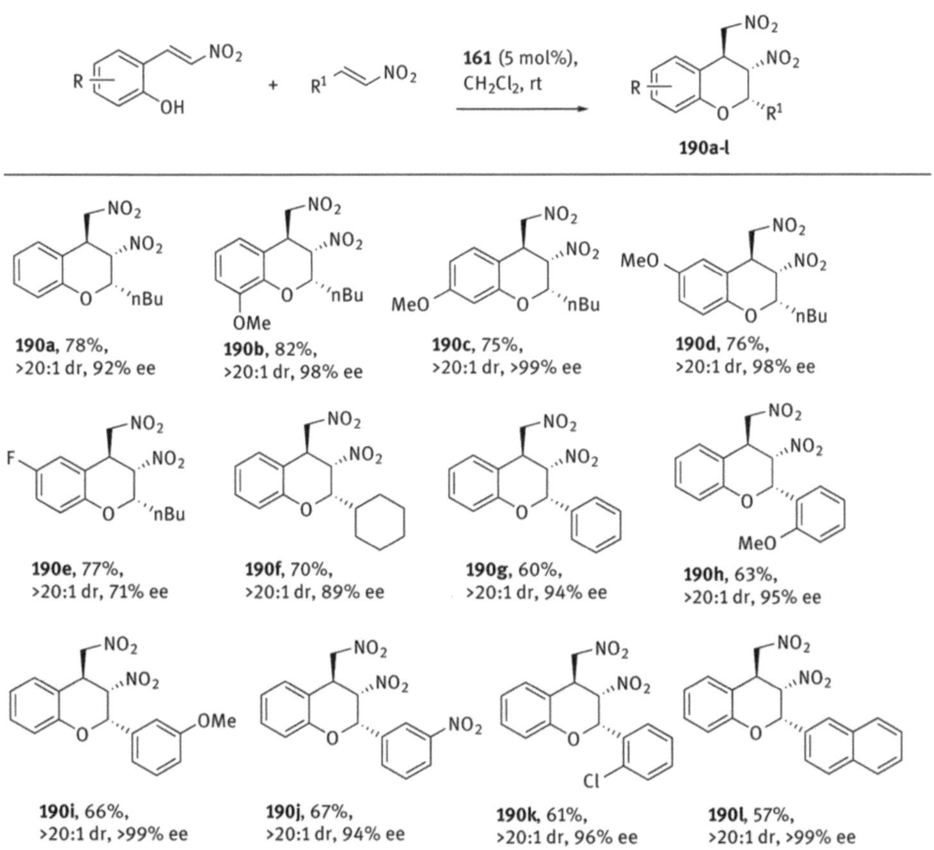

Figure 3.86: Cinchona-squaramide catalysis of a sequential oxa-Michael/nitro-Michael reaction leading to polysubstituted chiral chromans as described by Xu and co-workers.

over 90% ee. There were no apparent electronic or steric effects, and the reaction was tolerant of a variety of functional groups. The absolute configuration was determined once again by obtaining an X-ray crystal structure of one of the products.

3.9.1.5 Cinchona-squaramide catalysed reactions involving an oxa-Michael addition

1-Substituted isochromans are a very important family of compounds that constitute the core of many natural products and biologically active compounds. In 2015, Ghorai and co-workers reported a novel method to access this class of compound by way of a borane facilitated reduction followed by an oxo-Michael addition [93]. In the reaction a keto-aldehyde substrate was treated with pinacolborane and a tertiary amine in isopropanol in the presence of quinine-squaramide catalyst **161**, which allowed the formation of a critical boronate ester which according to the

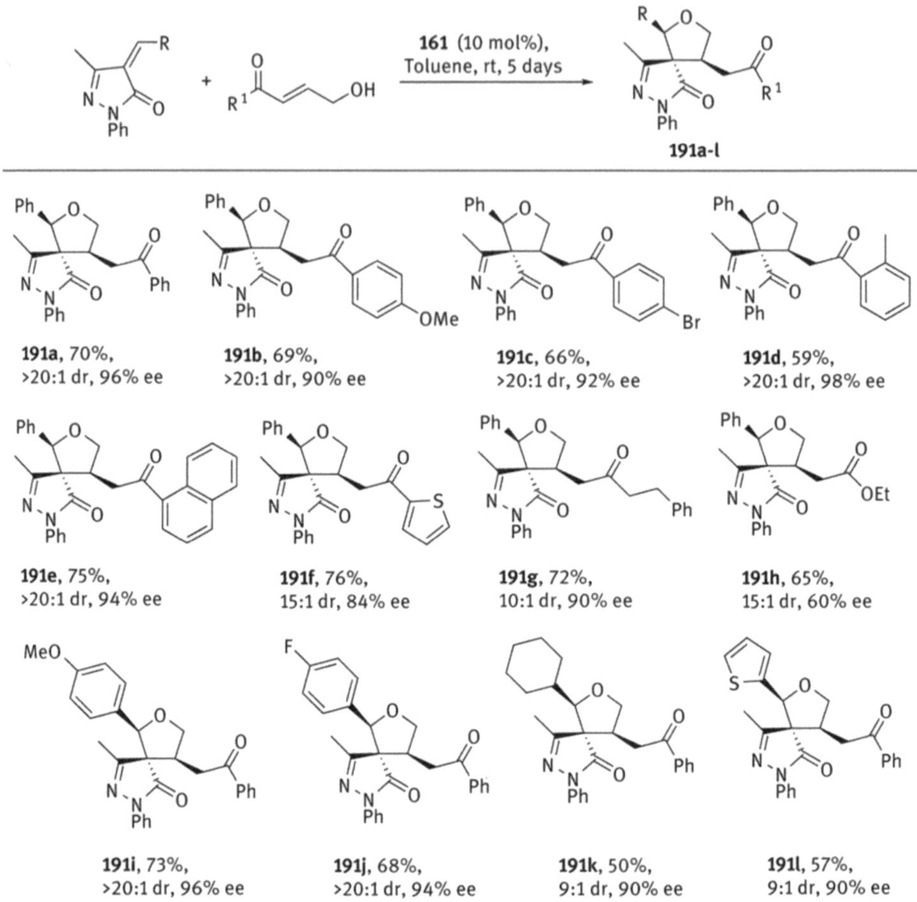

Figure 3.87: Cinchona-squaramide catalysis of a sequential oxa-Michael/Michael reaction leading to spiro-tetrahydrofuran-pyrazolones as described by Mondal and co-workers.

authors is hydrolysed by the alcohol to the free OH group that subsequently undergoes the oxo-Michael addition forming the 1-substituted isochroman products **192** (Figure 3.88(**A**)). This reaction required 10 mol% of the organocatalyst and was run in nitromethane at 45 °C. Very good yields and enantioselectivities were obtained, and the reaction showed generally good functional group tolerance, forming product irrespective of the position and electronic nature of the substituents. Nevertheless the reaction failed to work in the case of substrates with a methyl group in the α-position of the enone side-chain, substitution of the enone aryl group with a methyl group or replacement of the enone side-chain with a malonitrile derivative. The absolute configuration of these products was assigned based on an X-ray crystal structure of an analogous 3-substituted isochromane product (see Figure 3.88). It should be noted that **192h** is a precursor to the bioactive natural product PNU-142,633.

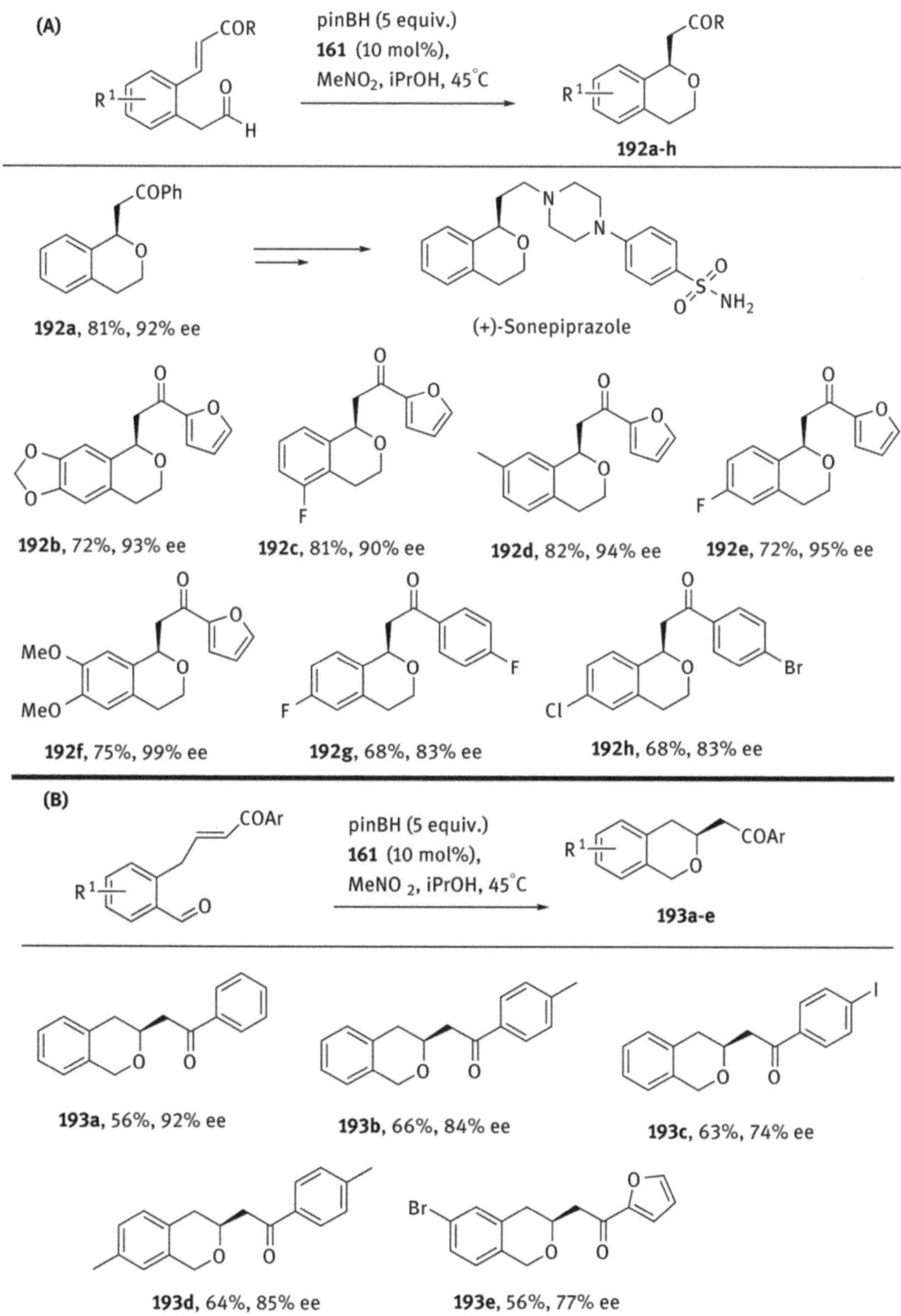

Figure 3.88: Cinchona-squaramide catalysis of a sequential borane reduction/oxo-Michael addition leading to 1- and 3-substituted isochromanes as described by Ghorai and co-workers.

These workers also investigated the synthesis of 3-substituted isochromanes, using ortho-formyl homochalcones (Figure 3.88(**B**)), however, although the reaction was tolerant of a variety of functional groups, in this case the yields and the enantioselectivities were lower than those obtained in the case of the 1-substituted analogues. An X-ray crystal structure determination of **193c** allowed the assignment of the absolute configuration of these products.

Benzoxaboroles are an important family of compounds, as many of which form the cores of key drug compounds, examples include AN2690 (*Tavaborole*, an antifungal drug), AN2728 (under clinical trials for psoriasis) and AN2718 (active against TeniaPedis) [94]. Ghorai's group have developed an intriguing sequential process forming 3-substituted benzoxaboroles **195** from 2-formyl arylboronic acids that employs first a Wittig reaction followed by an oxa-Michael reaction [94]. Once again a variety of cinchona-squaramide organocatalysts were screened, but it was the novel dimeric quinine derived catalyst **194** that gave the best results (Figure 3.89). The conditions included the use of an activated aroyl phosphonium ylide with 10 mol% **194** in chlorobenzene at room temperature. These products could be prepared also at the gram-scale.

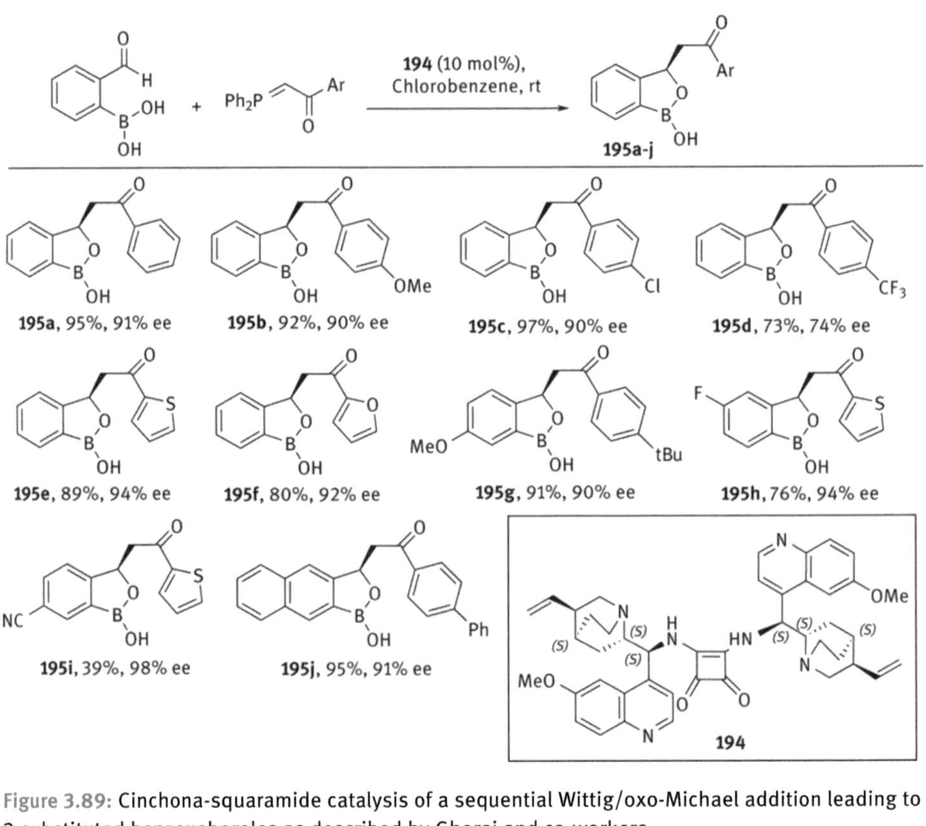

Figure 3.89: Cinchona-squaramide catalysis of a sequential Wittig/oxo-Michael addition leading to 3-substituted benzoxaboroles as described by Ghorai and co-workers.

Both the yields and the enantioselectivities were high. The reaction was tolerant of a variety of functional groups. The absolute configuration of these products was assigned by conducting an X-ray crystal structure study on **195c.**

The downside was that some of these reactions could take up to 69 h. These products could then be converted to the corresponding chiral β-hydroxy ketones without affecting the enantioselectivity.

This same group also developed an approach to *exo*-peroxy-benzannulated spiroketals which are another important group of structures as they are present in various natural products and pharmaceuticals [95]. This process is quite intriguing as it involves hydroperoxide addition to the aldehyde followed by a spiroketalization and then an oxa-Michael addition that involves a Dynamic Kinetic Resolution (DKR) to afford the peroxy-benzannulated spiroketal **197** (Figure 3.90). The oxa-Michael addition was catalysed by the novel quinine derived squaramide **196.**

The reaction took place with very good yields, diastereoselectivities and excellent enantioselectivities. The absolute configurations of the products were assigned again by recourse to an X-ray crystal structure for one of the compounds. There didn´t seem to be notable substituent effects in evidence, only in the case of the synthesis of **197c**, the diastereoselectivity was low, and the authors failed to furnish any explanation for this. Once again the down-side was the long reaction times needed (40–62 h).

3.9.1.6 Vinylogous Michael additions

In 2017, Feng and Li reported an enantioselective vinylogous Michael/Michael cascade reaction of 3-alkylidene oxindoles and nitroolefin enoates to give chroman-3-arylidene oxindoles (3-arylidene oxindoles are present in a wide array of pharmaceutically interesting compounds) (Figure 3.91) [96]. Organocatalyst **161** was again the catalyst of choice at a loading of 20 mol% in CH_2Cl_2 with molecular sieves at 35 ° C. The reactions occurred with excellent levels of enantioselectivity and generally high diastereoselectivity (In some cases the diastereo-selectivities were only moderate, 5:1 to 10:1, but it was difficult to determine the reason for this). The reactions were also tolerant of a plethora of functional groups. The absolute configuration of the products was assigned on the basis of an X-ray crystal structure for **198h**, and the author also proposed a transition-state working model to explain the stereofacial selectivity involved (Inset Figure 3.91). The reaction could be performed at the gram-scale. The downside (as usual) of this reaction was the requirement for extended reaction times (96 h) and a catalyst loading of 20 mol%.

3.9.2 Mannich reactions

The enantioselective Mannich reaction is a very useful method for obtaining β-amino carbonyl units, which are very useful in medicinal chemistry. A comprehensive review

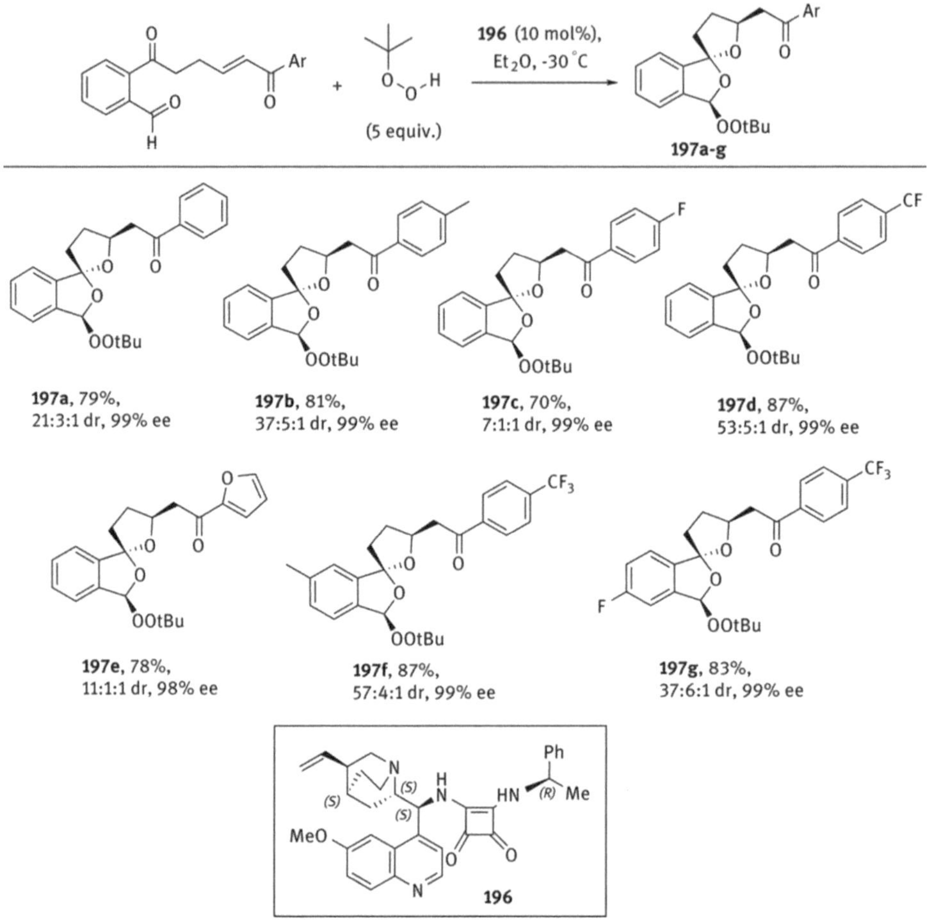

Figure 3.90: Cinchona-squaramide catalysis of a sequential hydroperoxide addition/oxo-Michael addition leading to peroxy-benzannulated spiroketals as described by Ghorai and co-workers.

of this reaction prior to 2015 has been published by Karahan and Tanyeli in 2018, see ref. ([66]i).

Based on previous work developed by Rao and co-workers, who used a quinine-squaramide catalyst in a Mannich reaction [97], Tanyeli and his team reported on the use of cinchona-squaramide catalysts at low loading for the synthesis of 3-amino-2-oxindoles **199** using *N*-carbamoyl isatin ketimine and acetylacetone (Figure 3.92) [98]. After testing some hindered quinine-squaramide catalysts, it was organocatalyst **199** which gave the best results. The reaction was conducted at room temperature in ether, with a loading of 1 mol% catalyst over 3–29 h.

The reaction demonstrated broad scope on the oxindole component but was limited to acetylacetone (Figure 3.92). Most reactions proceeded with high yields and excellent enantioselectivities, except for in the cases of **199c**, **199d** and **199e**,

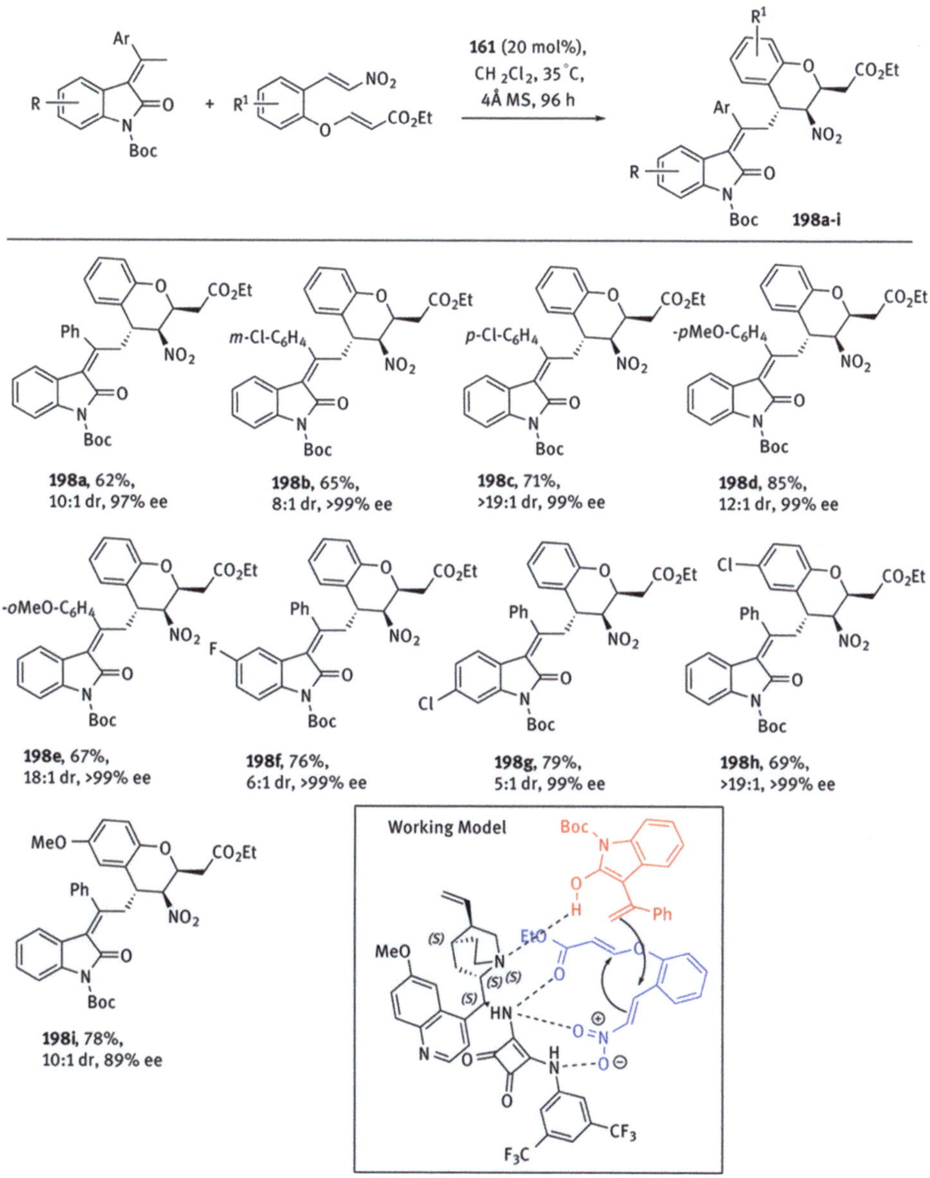

Figure 3.91: Cinchona-squaramide catalysis of a sequential vinylogous Michael/Michael addition leading to chroman-3-arylidene oxindoles described by Feng and Li.

and infact product **199f** did not form, which strongly indicated the necessitity of having an *N*-carbamoyl unit present. It was also noted that the presence of a bromo atom in the 5-position had a deleterious effect on the enantioselectivity as can be seen in the case of **199i**. The authors proposed that this was due to failure to form a

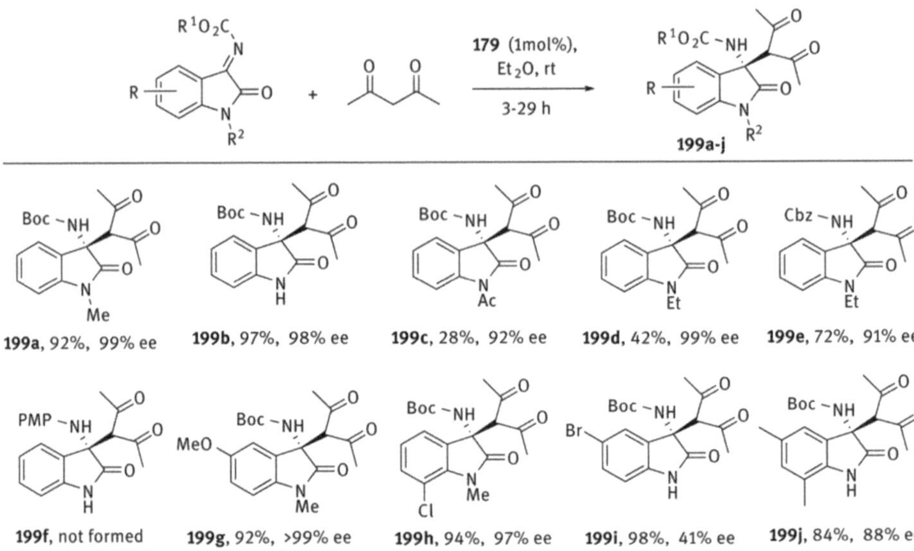

Figure 3.92: Cinchona-squaramide catalysis of a Mannich reaction leading to 3-amino-2-oxindoles **199** described by Tanyeli and co-workers.

rigid transition state, because of the bulky Br atom near the imine unit. The absolute stereochemistry was assigned on the basis of literature precedent.

In 2016 Song and co-workers reported an elegant asymmetric Mannich reaction with dithiomalonates using quinine- and hydroquinine-squaramide catalysts to give β-amino diesters (Mannich adducts) **201** (Figure 3.93) [99]. This method relied on the use of dithiomalonates, and was showcased in the synthesis of the antidiabetic drug, (-)-(*R*)-*Sitagliptin* (a selective DPP-4 inhibitor for the treatment of type 2 diabetes mellitus) – in fact it should be noted that this was the first organocatalytic route reported for this API (see Figure 3.94). In this study a number of organocatalysts were investigated, but it was the quinine derived organocatalyst **200** that gave best results. After optimization studies, the reaction could be performed at −50 °C in DCM over 24 h with a catalyst loading of only 0.5 mol%. Both high yields and excellent enantioselectivities could be obtained. The reaction showed good scope for the imine reagent, and both aryl and aliphatic imines could be used. The absolute configuration of these products was determined by comparison with literature values, this was particularly the case for the synthesis of (-)-(*R*)-*Sitagliptin*.

These workers took this a step forward and developed a one-pot method using bench stable alkyl-substituted α-amidosulfones as imine surrogates and gave the Mannich adducts with excellent yields and excellent levels of stereocontrol (Figure 3.94). In this particular case the loading of the organocatalyst **200** was increased to 1 mol%, it was necessary to add sodium carbonate and the reaction was run in a water/DCM biphasic mixture at 0 °C between 24–72 h.

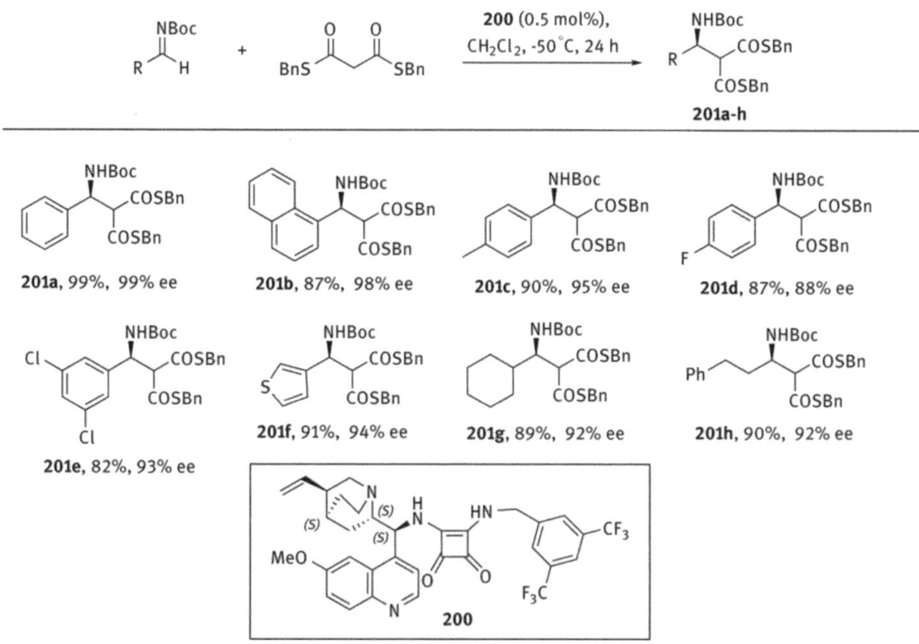

Figure 3.93: Cinchona-squaramide catalysis of a Mannich reaction leading to β-amino diesters **201** described by Song and co-workers.

3.9.2.1 Squaramide-catalysed sequential reactions involving a Mannich reaction

In 2018, Zhao and co-workers reported a stereoselective sequential Mannich reaction/ cyclization of isocyanoacetates with cyclic sulfamide ketimines to afford optically active 2,3,3a,4-tetrahydroimidazo[1,5-b][1,2,5]thiadiazole-1,1-dioxide derivatives **202** (Figure 3.95) [100]. The reactions were carried out using the quinine derived organocatalyst **161** (10 mol%), with AgOAc (3 or 5 mol%) in toluene at 0 °C for 2 mins, and furnished very high yields of the products **202**, with excellent diastereoselectivities and generally very good levels of enantiocontrol. The scope in both the isocyanoacetate and the ketimine was evaluated, and good functional group tolerance was demonstrated in either case. It was found that the enantiocontrol was sensitive to steric factors as can be seen in the case of **202f** and **202k**, where the ees were low.

3.9.2.2 The aza-Mannich reaction (aza-Henry reaction)

This is a well-known addition of nitronates to electrophilic imine species to form new C–C bonds (see Section 3.7.3). This reaction has been recently satisfactorily reviewed (see ref. 68c). In 2017, Susam and Tanyeli reported a number of Aza-Mannich reactions with the novel cinchona-squaramide catalyst **203**, the reactions could be performed at room temperature in dichloromethane (at a concentration of 0.1 M), with 10 mol% catalyst loading and gave enantioselectivities of up to 91% ee

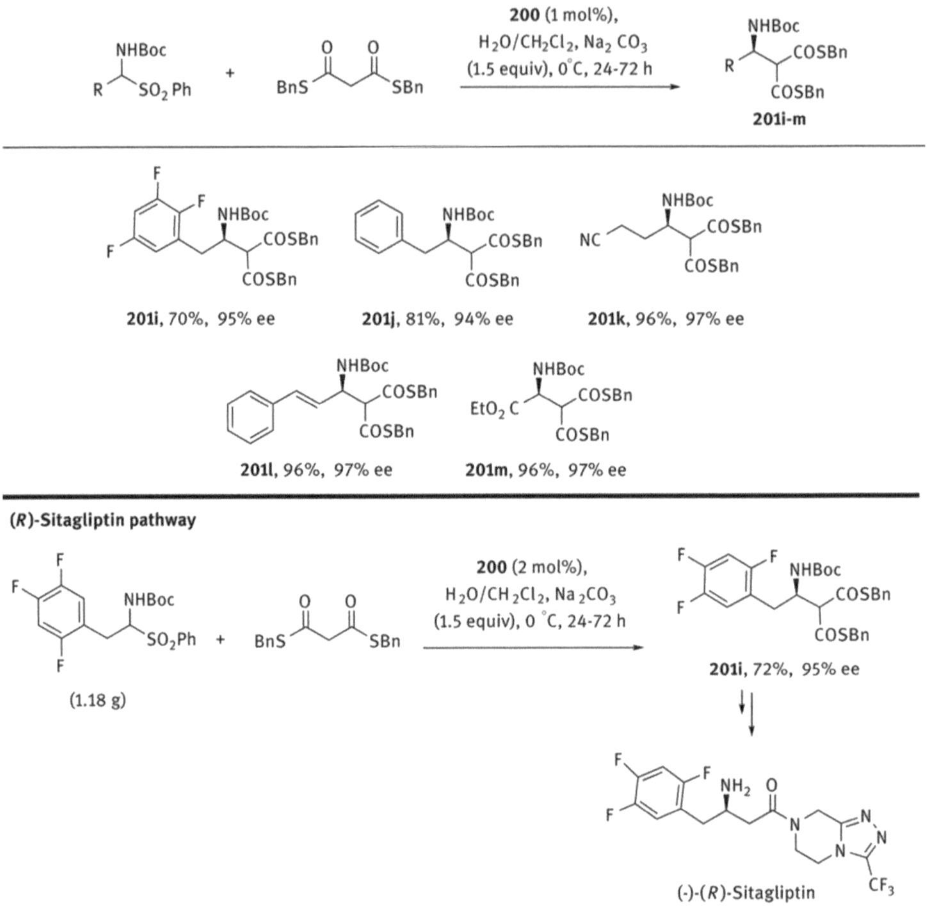

Figure 3.94: The Mannich reaction using an imine surrogate and the first organocatalytic route to (-)-(*R*)-*Sitagliptin* as described by Song and co-workers.

(Figure 3.96) [101]. The reaction demonstrated very good scope, and there was evidence of substituent effects, for instance it appeared that electron-donating groups on the phenyl ring increase the reaction time, whilst electron-withdrawing groups had the opposite effect. Apart from the fact that the 3-Cl group in **204d** gave an ee of only 64%, the direct effect of these groups on the reaction enantioselectivity was not in evidence. The stereochemical configurations were atributed on the basis of literature precedent. A transition state model was added to explain the origin of the enantioselectivity.

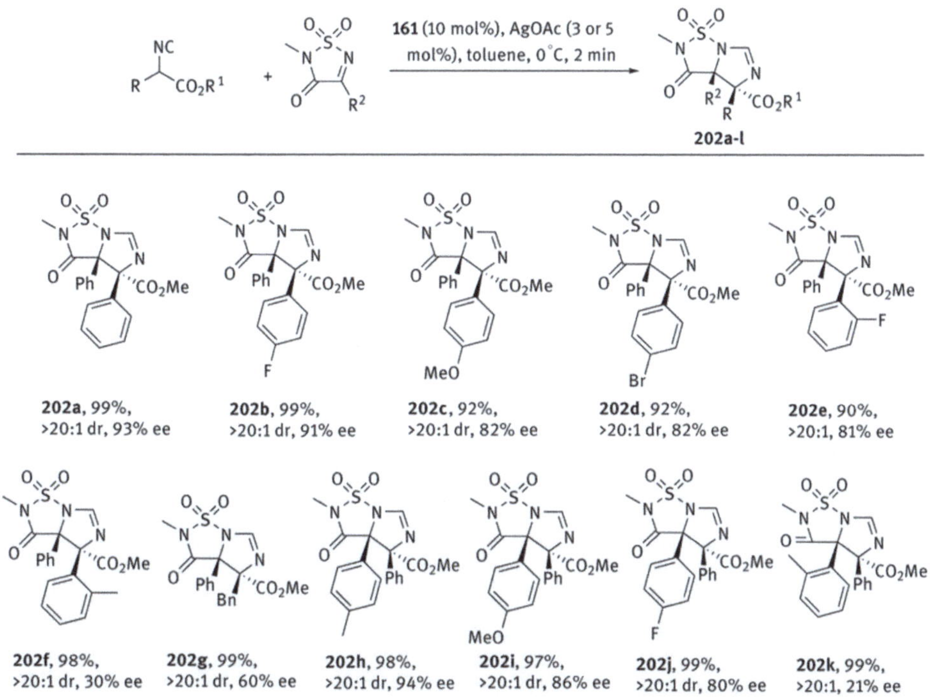

Figure 3.95: The stereoselective Mannich/cyclization cascade reaction as described by Zhao and co-workers.

3.9.3 The Rauhut–Currier reaction

The Rauhut–Currier (RC) reaction was discovered by Rauhut and Currier in 1963 [102] and is a very useful reaction for the construction of structurally complex natural products and bioactive compounds. Essentially it involves the addition of a nucleophile to an enone to form an enolate intermediate that undergoes inter- or intra-molecular attack (more interesting as it forms rings) at a carbonyl or other enone site to give a complex cyclic adduct [103]. The reaction remained relatively unexplored from the 60s to the late 90s due to selectivity issues, but then in 1999, and the early 00s some new developments were made with the intramolecular version, principally by the groups of Moore and Erguden, Roush and Krische [103].

Generally both amines and phosphines are used as catalysts for this reaction, but there have been only a few examples on the use of cinchona-squaramide catalysts (although in 2018, Ghorai and co-workers, reported the use of cyclohexenone phenylenone to give tricyclic dione products using a non-cinchona-squaramide catalyst [104]. However, the same year, He *et al.* reported a cross RC type reaction of tri-substituted alkenes contianing CF_3 groups where a cinchona-squaramide catalyst was successfully used (Figure 3.97) [105]. After verious optimization studies,

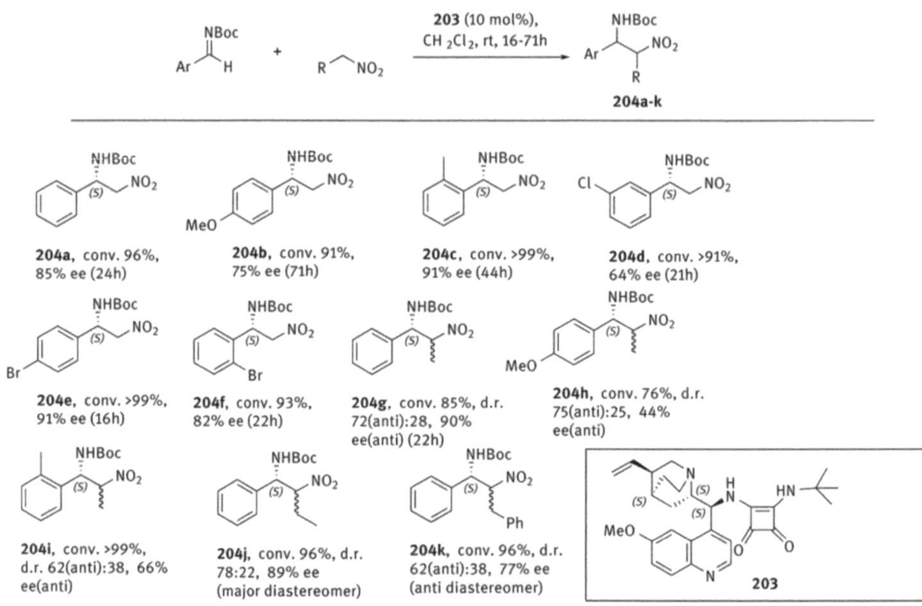

Figure 3.96: The aza-Mannich Reaction as described by Susam and Tanyeli.

these workers showed that it was possible to react ethyl (*E*)-4-cyano-3-(trifluoromethyl)but-3-enoate with a variety of nitro-olefins to give chiral γ-cyano-ε-nitro-β-trifluoromethyl-β,γ-unsaturated esters **205** with 20 mol% of the quinidine derived organocatalyst **159** (see Figure 3.97) in good yields, with excellent diastereocontrol and good enantiocontrol.

The scope of the reaction was only investigated on the nitro-olefin component and not on the ester substrate, and it was observed that the electronic properties of the olefin influenced the reaction efficency, for example, when electron donating groups were present the reaction was slow (12 h, in the case of **205b**-**205e**) but when electron withdrawing groups were present it was faster (6 h, **205f-205h**). The yields were consistently good right across the board, the diastereoselectivity was excellent in all cases, and the enantioselectivity consistent over most of the compounds, except in the case of the β-naphthyl derivative **205j**, where it was lower.

3.9.4 1,3-Dipolar cycloadditions

1,3-Dipolar cycloaddition reactions are a powerful means of accessing useful 5-membered rings, which are very common in many natural products and biologically active molecules [106]. (This reaction has been previously discussed in Section 3.3 above).

The first organocatalytic enantioselective [3+2] cycloaddition between ammonium salts and azomethine ylides was reported by Vicario *et al.* in 2007 [107].

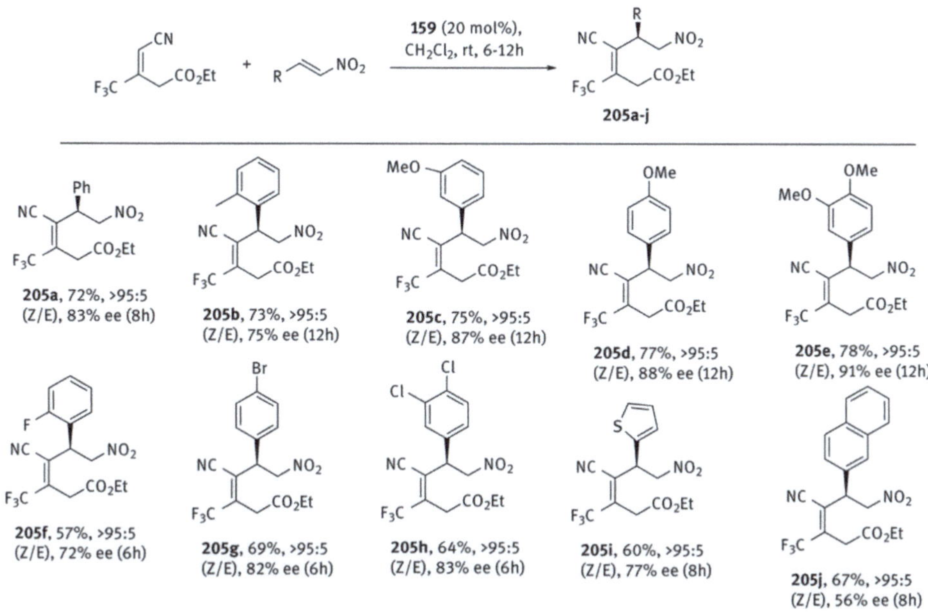

Figure 3.97: The asymmetric catalytic Rauhut-Currier (RC) reaction as described by He *et al.*

In 2017, Šebesta and co-workers reported an efficient 3-component cascade 1,3-dipolar cycloaddition reaction of an *in situ* generated azomethine ylide (from isatins and benzylamines) with α,β-unsatuared esters to afford 20 new highly substituted spirooxindole derivatives **206**, using a variety of organocatalysts, which included the quinidine derived organocatatalyst **159** that gave the target compounds in moderate to good yields (Figure 3.98) [108]. Althought most of the reactions afforded a single diastereomer the reactions gave only a recamic mixture. The reaction could also be performed under microwave conditions.

In 2017 Huang *et al.* reported the synthesis of trifluoromethyl-substituted 3,3´-pyrrolidinyl-dispirooxindoles **207** – which are very structurally congested and rigid architectures, with four contiguous stereocentres and two adjacent spiro quaternary stereocentres and useful for application in drug discovery – through organocatalytic 1,3-dipolar cycloaddition reactions using cinchona-squaramide catalysts, the best of which was the cinchonidine derivative **208** which contained an extra chiral centre on its side arm (Figure 3.99) [109]. The reactions took place under very mild conditions, invariably with very good to excellent yields, and excelent levels of stereocontrol (up to >20:1 dr and >99% ee). The reaction was tolerant of a wide-range of different substrate types, containing both electron-withdrawing and donating substituents, in either the ketimine or the azomethine ylide derived from the alkylidene substrate, however, it was noticed that the presence of a nitro group in either the

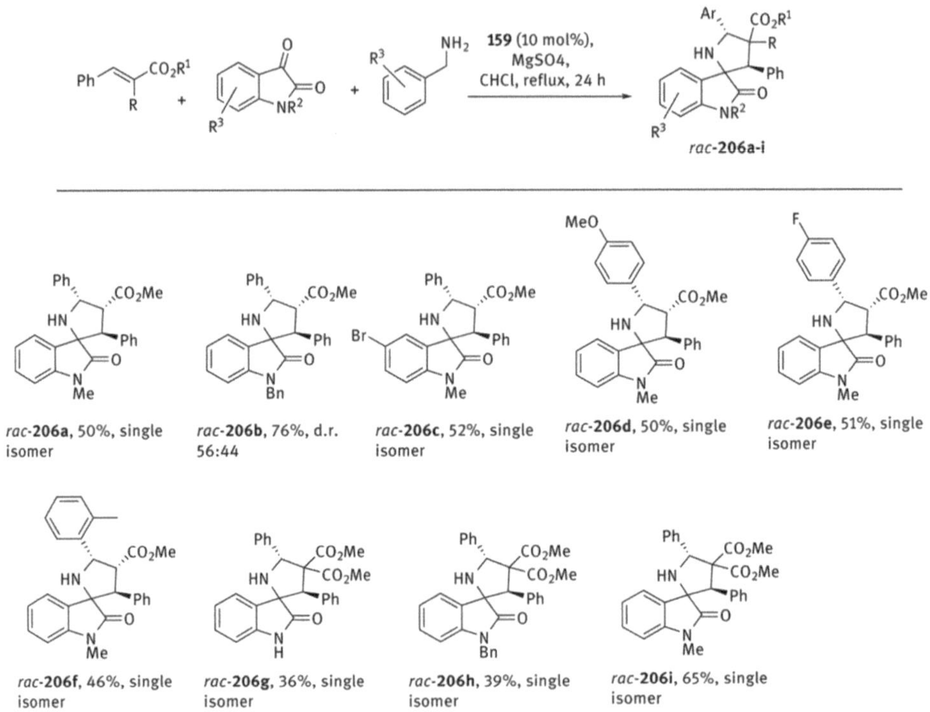

Figure 3.98: The asymmetric catalytic multi-component [3+2] reaction as described by Šebesta and co-workers.

ketimine or the alkylidene substrate reduced the yield, and increased the reaction time (see **207e** and **207j**).

The absolute configuration of products **207** was determined by obtaining an X-ray crystal structure for compound **207d.** Similar previous work has been done by Su *et al.* who used another cinchonidine-squaramide derivative to catalyse the reaction [110].

To finalize this section the cinchona-squaramide catalysed asymmetric [3+2] cycloaddition of isocyanoacetates with β-trifluoromethylated enones reported by Zhao and co-workers is discussed [111]. These workers demonstrated that the [3+2] cycloadditon of these reagents could be performed with novel cinchona-squaramides affording the chiral trifluoromethylated 2-pyrrolines **209** in excellent yields (up to 98%), with very good levels of stereocontrol (dr up to > 20:1, and enantioselectivities of up to > 99% ee) under mild conditions (Figure 3.100). The reaction was also found to be temperature-dependent.

A plethora of isocyanoacetates and β-trifluoromethylated enones with different electronic and steric properties are suitable in this [3+2] cycloaddition reaction leading to trifluoromethylated 2,3-dihydro-1*H*-pyrrole carboxylates with adjacent chiral tertiary-quaternary carbon centres. It was observed that the reaction was sensitive to steric effects, as can be seen by the fact that it was not possible to obtain either **209e**

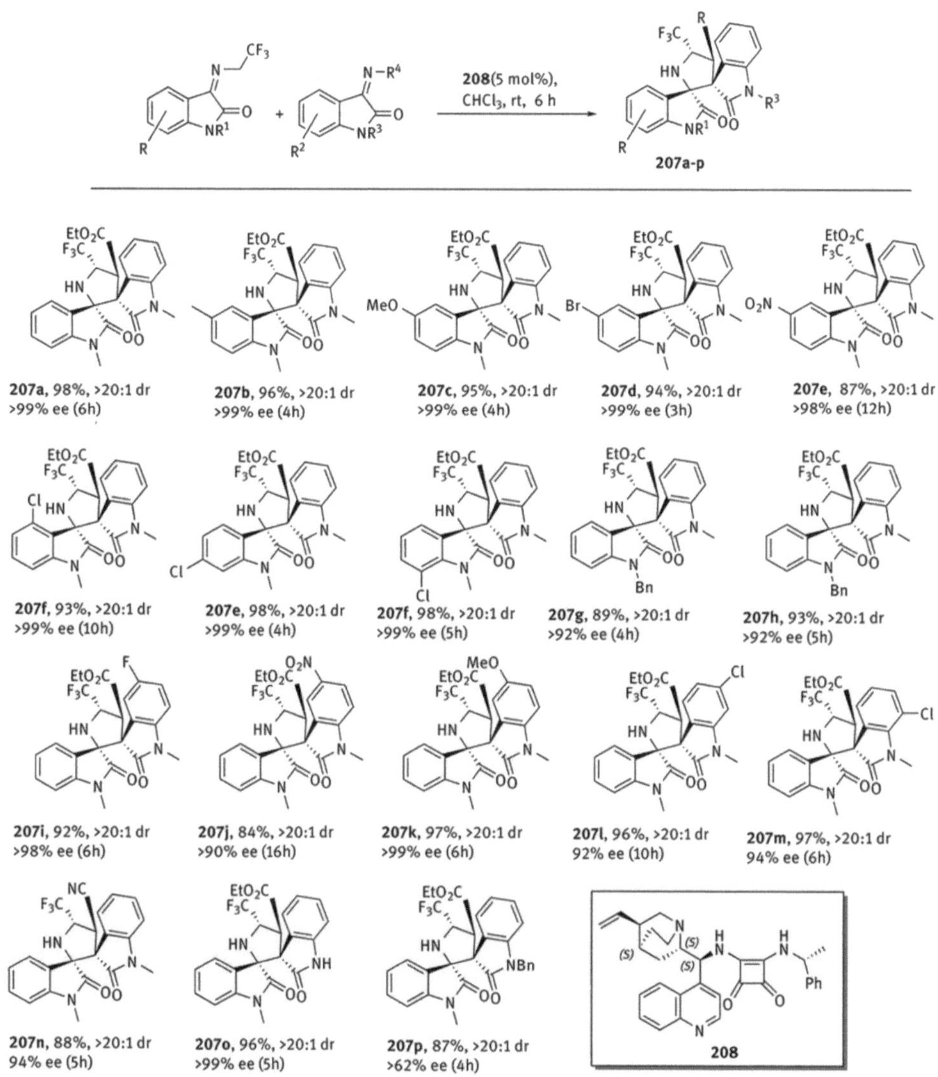

Figure 3.99: The asymmetric catalytic [3+2] reaction to give spiro oxindoles of type 207 as described by Huang *et al.*

nor **209f**, compounds with a substituent in the *ortho*-position and in the case of **209m** both the dr (5:1) and the enantioselectivity (61% ee) were lower than usual. It was also observed that weakly electron-donating groups in the *para* and *meta* positions afforded the cycloadducts in better yields than those containing strongly electron-donating or electron-withdrawing groups in these positions.

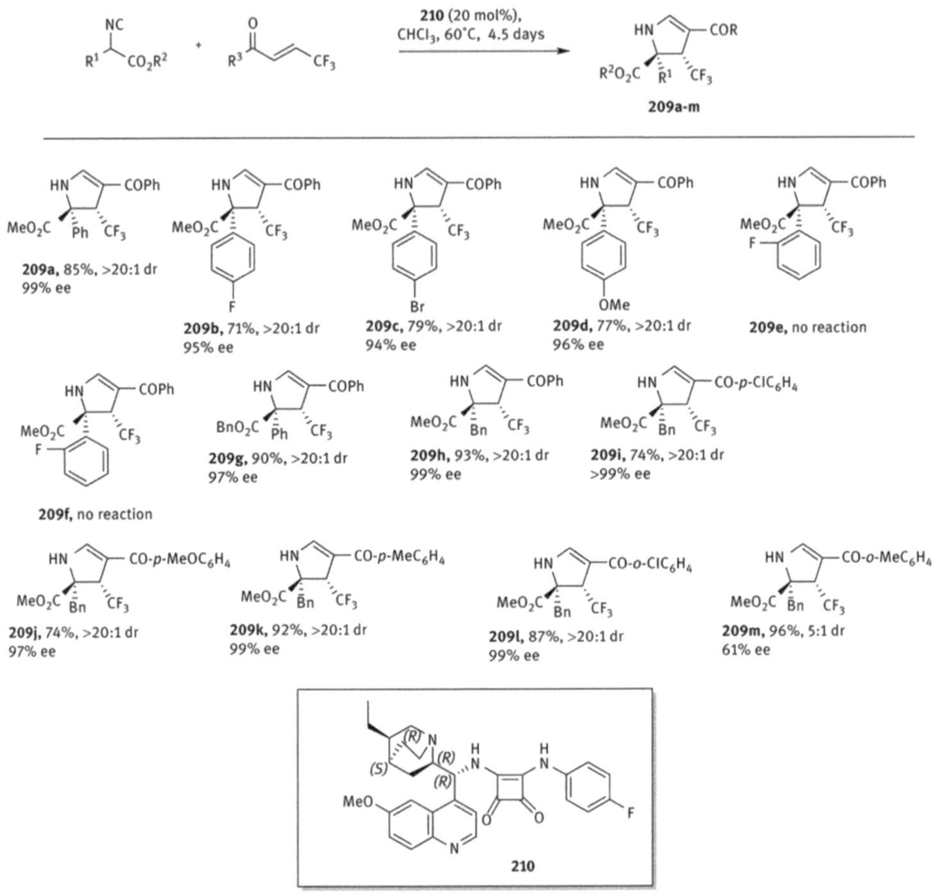

Figure 3.100: The asymmetric catalytic [3+2] cycloaddition of isocyanoacetates with β-trifluoromethylated enones affording trifluoromethylated 2-pyrrolines as described by Zhao *et al.*

The configuration of the products **209**, was determined by obtaining an X-ray crystal structure for product **209c**. The problem with this reaction was that it was very slow taking 4.5 days to reach completion!

3.10 Conclusions, outlook and perspectives

We have focused in this chapter on amino-cinchona or amino-cinchona derived organocatalysts, showing that amino-cinchona organocatalysts have gained increased interest and importance in the field of asymmetric catalysis over the last two decades. By reviewing the literature since 2015 it is apparent that simple amino-cinchona catalysts provide a solid foundation for asymmetric catalysis, those reactions that involve asymmetric conjugate additions, alkylations, oxidations or halogenations, to name a few. By derivatising the primary amine functionality found in these precursors an enormous toolbox of potent catalysts becomes available. These range from amides, thioamides, ureas, thioureas, sulphonamides, squaramides, DMMs to quaternized PTC analogues. This makes it possible to conduct other key transformations such as: Mannich reactions, Henry reactions, hydrosilylations, Diels–Alder reactions, dipolar cycloadditions and other useful transformations. Recently, it has also been successfully demonstrated that these catalysts can also be used in continuous flow set-ups. Generally, reaction conditions are mild with the majority of applications running at ambient temperature.

We have seen above that some very useful transformations, that afford very exciting molecules like the Rauhut–Currier reaction and the 1,3-dipolar cycloaddition, have not been fully developed, despite their potential for giving interesting cyclic and crowded structures (including relevant hard to access spiro-compounds which are very useful in medicinal chemistry) containing multiple contiguous stereocentres, etc.

Nontheless, despite the enormous potential these catalysts have for affording useful molecules, they are still being ignored (in genaral) by the chemical industry, and two of the main reasons for this include: the high catalyst loadings required, and the long reaction times required in many cases.

List of abbreviations

Ar	Aryl
ASD	Cinchona sulfonamide catalyst
Bn	Benzyl
Bu	Butyl
Boc	*tert*-Butylcarboxy
Cbz	Carboxybenzyl
CD	Cinchonidine
CHP	Cumyl hydroperoxide
CN	Cinchonine
Conv.	Conversion
CPME	Cyclopentyl methyl ether
d	Day
DES	Deep eutectic solvents
DFT	Density functional theory
DIAD	Diisopropyl azodicarboxylate
DKR	Dynamic kinetic resolution
DMF	Dimethylformamide
DMM	Diaminomethylenemalonitrile
DPPA	Diphenyl phosphoryl azide
dr	Diastereomeric ratio
ee	Enantiomeric ratio
eq.	Equivalents
er	Enantiomeric ratio

F-MAHT	Fluoromalonic acid half thioesters
F-MTM	Fluorinated monothiomalonates
GABA	γ-Butyric acid
h	Hour
IED	Inverse-electron demand
LED	Light emitting diode
MAHO	Malonic acid half oxyesters
MAHT	Malonic acid half thioesters
*m*CPBA	*m*-Chloroperbenzoic acid
MTM	Monothiomalonates
Ms	Methanesulfonyl group
MS	Molecular sieves
MDO	Molecularly designed organocatalyst
NBA	*N*-Bromoacetamide
NBS	*N*-Bromosuccinimide
NFSI	*N*-Bluorobenzenesulfonimide
NIS	*N*-Iodosuccinimide
NMP	*N*-Methylpyrrolidine
PEDA	Potoenolization Diels-Alder
Ph	Phenyl
PMP	*p*-Methoxyphenyl
PTC	Phase transfer catalyst
QD	Quinidine
QN	Quinine
rt	Room temperature
Ts	Tosyl

Note

1. This publication has already received a staggering 538 citations on ISI-Web of science (date 11-9-2018).

References

[1] Chen W, Du W, Yue L, Li R, Ding LS, Chen YC. Organocatalytic enantioselective indole alkylations of α,β-unsaturated ketones. Org Biomol Chem. 2007;5:816–21.

[2] Bartoli G, Bosco M, Carlone A, Pesciaioli F, Sambri L, Melchiorre P. Organocatalytic asymmetric friedel–crafts alkylation of indoles with simple α,β-unsaturated ketones. Org Lett. 2007;9:1403–5.

[3] Reviews and leading literature on amino-cinchona-based organocatalysts: (a) Melchiorre A. Cinchona-based primary amine catalysis in the asymmetric functionalization of carbonyl compounds. Angew Chem Int Ed 2012;51:9748–70, (b) Govender T, Arvidsson PI, Maguire GE, Kruger HG, Naicker T. Enantioselective organocatalysed transformations of β-Ketoesters. Chem Rev. 2016;116:9375–437, (c) Duan J, Li P. Asymmetric organocatalysis mediated by primary amines derived from cinchona alkaloids: recent advances. Catal Sci Technol. 2014;4:311–20, (d) Xu, L-W, Luo J, Lu Y. Asymmetric catalysis with chiral primary amine-

based organocatalysts. Chem Commun. 2009:1807–21, (e) Bartoli G, Melchiorre P. A novel organocatalytic tool for the iminium activation of α,β-Unsaturated ketones. Synlett. 2008;12:1759–72, (f) Jiang L, Chen Y-C. Recent advances in asymmetric catalysis with cinchona alkaloid-based primary amines. Catal Sci Technol. 2011;1:354–65, (g) Kaur J, Chauhan P, Singh S, Singh Chimni S. Journey heading towards enantioselective synthesis assisted by organocatalysis. Chem Rec. 2018;18:137–53, Chen Y-C. The development of asymmetric primary amine catalysts based on cinchona alkaloids. Synlett. 2008;13:1919–30.

[4] (a) Cassani C, Martín-Repún R, Arceo E, Bravo F, Melchiorre P. Synthesis of 9-amino(9-deoxy) epi cinchona alkaloids, general chiral organocatalysts for the stereoselective functionalization of carbonyl compounds. Nat Protoc. 2013;8:325–44, (b) He W, Liu P, Zhang BL, Sun XL, Zhang SY. Efficient iridium and rhodium-catalyzed asymmetric transfer hydrogenation using 9-amino (9-deoxy) cinchona alkaloids as chiral ligands. Appl Organomet Chem. 2006;20:328–34, (c) Brunner H, Bügler J, Nuber B. Tetrahedron asymmetry. 1995;6:1699–702, (d) Sundermeier U, Döbler C, Mehltretter GM, Baumann W, Beller M. Synthesis of 9-N-cinchona alkaloid peptide hybrid derivatives: preparation and conformational study of 9-N-acylamino(9-deoxy)cinchona alkaloids. Chirality. 2003;15:127–34.

[5] Tanvier G, Dedeoglu B, Catak S, Aviyente V. Computational studies on cinchona alkaloid-catalyzed asymmetric organic reactions. Acc Chem Res. 2016;49:1250–62.

[6] Li H, Hong X. Computational studies of cinchona alkaloid-catalyzed asymmetric Michael additions. Chin Chem Lett. 2018;29:1585–90.

[7] Lam YH, Houk KN. Origins of stereoselectivity in intramolecular aldol reactions catalyzed by cinchona amines. J Am Chem Soc. 2015;137:2116–27.

[8] Massolo E, Palnieri S, Benaglia M, Capriati V, Perna FM. Stereoselective organocatalysed reactions in deep eutectic solvents: highly tunable and biorenewable reaction media for sustainable organic synthesis. Green Chem. 2016;18:792–7.

[9] Porta R, Benaglia M, Coccia F, Cozzi F, Puglisi A. Solid supported 9-amino-9-deoxy-epi-quinine as efficient organocatalyst for stereoselective reactions in batch and under continuous flow conditions. Adv Synth Catal. 2015;357:377–83.

[10] Izquierdo J, Ayats C, Henseler AH, Pericàs MA. A polystyrene-supported 9-amino(9-deoxy)epi quinine derivative for continuous flow asymmetric Michael reactions. Org Biomol Chem. 2015;13:4204–9.

[11] Lee J, Ko MK, Kim SG. Asymmetric synthesis of 2,3-disubstituted indolines via an organocatalytic intramolecular Michael addition. RSC Adv. 2017;7:56457–62.

[12] Seitz T, Millán RE, Lentz D, Jiménez C, Rodruez J, Christmann M. Synthesis of thelepamide via catalyst-controlled 1,4-addition of cysteine derivatives and structure revision of thelepamide. Org Lett. 2018;20:594–7.

[13] Ricci P, Carlone A, Bartoli G, Bosco M, Sambri L, Melchiorre P. Organocatalytic asymmetric sulfa-Michael addition to α,β-unsaturated ketones. Adv Synth Catal. 2008;350:49–53.

[14] Shi ML, Zhan G, Zhou SL, Du W, Chen YC. Asymmetric inverse-electron-demand oxa-diels–alder reaction of allylic ketones through dienamine catalysis. Org Lett. 2016;18:6480–3.

[15] Arimitsu S, Yonamine T, Higashi M. Cinchona-based primary amine catalyzed a proximal functionalization of dienamines: asymmetric α-fluorination of α-branched enals. ACS Catal. 2017;7:4736–40.

[16] Lam YH, Houk KN. How cinchona alkaloid-derived primary amines control asymmetric electrophilic fluorination of cyclic ketones. J Am Chem Soc. 2014;136:9556–9.

[17] Hu L, Lu X, Deng L. Catalytic enantioselective peroxidation of α,β-unsaturated aldehydes for the asymmetric synthesis of biologically important chiral endoperoxides. J Am Chem Soc. 2015;137:8400–3.

[18] Peng W, Wan J, Xie B, Ma X. 9-amino-(9-deoxy)cinchona alkaloid-derived new chiral phase-transfer catalysts. Org Biomol Chem. 2014;12:8336–45.

[19] Lygo B, Wainright PG. A new class of asymmetric phase-transfer catalysts derived from cinchona alkaloids — Application in the enantioselective synthesis of α-amino acids. Tetrahedron Lett. 1997;38:8595–9598; Corey FJ, Xu F, Noe MC. A rational approach to catalytic enantioselective enolate alkylation using a structurally rigidified and defined chiral quaternary ammonium salt under phase transfer conditions. J Am Chem Soc. 1997;119:12414-5; see also Catalytic Processes for API Synthesis: From Laboratory to Industry, Burke AJ, Marques CS, Turner NJ, Hermann GJ. Ch. 11, Wiley-VCH, 2018:359-86.

[20] Maddox SM, Dawson GA, Rochester NC, Ayonon AB, Moore CE, Rheingold AL, et al. Enantioselective synthesis of biaryl atropisomers via the addition of thiophenols into aryl-naphthoquinones. ACS Catal. 2018;8:5443–7.

[21] Ray B, Mukherjee S. Direct catalytic enantioselective vinylogous aldol reaction of allyl ketones to pyrazole-4,5-diones. J Org Chem. 2018;83:10871–80.

[22] George J, Kim HY, Oh K. Silver-catalyzed asymmetric desymmetrization of cyclopentenediones via [3+2] cycloaddition with α-substituted isocyanoacetates. Org Lett. 2018;20:2249–52.

[23] Shiomi N, Yamamoto K, Nagasaki K, Hatanaka T, Funahashi Y, Nakamura S. Enantioselective oxidative ring-opening reaction of aziridines with α-nitroesters using cinchona alkaloid amide/nickel(II) catalysts. Org Lett. 2017;19:74–7.

[24] Barrulas PC, Genoni A, Benaglia M, Burke AJ. Cinchona-derived picolinamides: effective organocatalysts for stereoselective imine hydrosilylation. Eur J Org Chem. 2014:7339–442; Genoni A, Benaglia M, Mattiolo E, Rossi S, Raimondi L, Barrulas PC, Burke AJ. Synthesis of an advanced precursor of Rivastigmine: Cinchona-derived quaternary ammonium salts as organocatalysts for stereoselective imine reductions. Tetrahedron Lett. 2015;56:5752-6.

[25] Singjunla Y, Pigeaux M, Laporte R, Baudoux J, Rouden J. Thioamide-substituted cinchona alkaloids as efficient organocatalysts for asymmetric decarboxylative reactions of MAHOs. Eur J Org Chem. 2017;4319–23.

[26] Nakamura S, Hayama D, Miura M, Hatanaka T, Funahashi Y. Catalytic enantioselective reaction of 2H-azirines with thiols using cinchona alkaloid sulfonamide catalysts. Org Lett. 2018;20:856–9.

[27] Endo Y, Takata S, Kumpuga BT, Itsuno S. Synthesis of cinchona alkaloid sulfonamide polymers as enantioselective catalysts for the michael addition reaction of β-ketoester and nitrostyrene. Chemistry Select. 2017;2:10107–11.

[28] March T, Murata A, Kobayashi Y, Takemoto Y. Enantioselective synthesis of anti-β-hydroxy-α-amino esters via an organocatalyzed decarboxylative aldol reaction. Synlett. 2017;28:1295–9.

[29] Nakamura S, Takahashi S. Organocatalytic enantioselective peroxidation of ketimines derived from isatins. Org Lett. 2015;17:2590–2.

[30] Takata S, Endo Y, Ullah S, Isuno S. Synthesis of cinchona alkaloid sulfonamide polymers as sustainable catalysts for the enantioselective desymmetrization of cyclic anhydrides. RSC Adv. 2016;6:72300–5.

[31] Blise K, Cvitkovic MW, Gibbs NJ, Roberts SF, Whitaker RM, Hofmeister GE, et al. A theoretical mechanistic study of the asymmetric desymmetrization of a cyclic meso-anhydride by a bifunctional quinine sulfonamide organocatalyst. J Org Chem. 2017;82:1347–55.

[32] Li XT, Gu QS, Dong XY, Meng X, Liu XY. A copper catalyst with a cinchona-alkaloid-based sulfonamide ligand for asymmetric radical oxytrifluoromethylation of alkenyl oximes. Angew Chem Int Ed. 2018;57:7668–72.

[33] Cosimi E, Saadi J, Wennemers H. Stereoselective synthesis of α-fluoro-γ-nitro thioesters under organocatalytic conditions. Org Lett. 2016;18:6014–17.

[34] Phelan JP, Ellman JA. Catalytic enantioselective addition of pyrazol-5-ones to trisubstituted nitroalkenes with an N-sulfinylurea organocatalyst. Adv Synth Catal. 2016;358:1713–18.
[35] Jiang X, Liu S, Yang S, Jing M, Xu L, Yu P, et al. Enantioselective bromolactonization of deactivated olefinic acids. Org Lett. 2018;20:3259–62.
[36] Saadi J, Wennemers H. Enantioselective aldol reactions with masked fluoroacetates. Nature Chem. 2016;8:276–80.
[37] Liu Y, Liu Y, Wang J, Wei Z, Cao J, Liang D, et al. Asymmetric phase-transfer catalysts bearing multiple hydrogen-bonding donors: synthesis and application in nitro-Mannich reaction of isatin-derived N-Boc ketimines. Tetrahedon Lett. 2017;58:2400–3.
[38] Liu X, Wang J, Wei Z, Cao J, Liang D, Lin Y, et al. Diastereo- and enantioselective nitro-Mannich reaction of isatin-derived N-Boc ketimines catalyzed by chiral phase-transfer catalysts. New J Chem. 2018;42:1608–11.
[39] Lu N, Li R, Wei Z, Cao J, Liang D, Lin Y, et al. Enantio- and diastereoselective nitro-Mannich reaction of α-aryl nitromethanes with amidosulfones catalyzed by phase-transfer catalysts. J Org Chem. 2017;82:4668–76.
[40] Wang B, Liu Y, Sun C, Cao J, Liang D, Lin Y, et al. Asymmetric phase-transfer catalysts bearing multiple hydrogen-bonding donors: highly efficient catalysts for enantio- and diastereoselective nitro-Mannich reaction of amidosulfones. Org Lett. 2014;16:6432–5.
[41] Miyaji R, Asano K, Matsubara S. Induction of axial chirality in 8-arylquinolines through halogenation reactions using bifunctional organocatalysts. Chem Eur J. 2017;23:9996–10000.
[42] Zilate B, Castrogiovanni A, Sparr C. Catalyst-controlled stereoselective synthesis of atropisomers. ACS Catal. 2018;8:2981–8.
[43] Craig R, Sorrentino E, Connon SJ. Enantioselective alkylation of 2-oxindoles catalyzed by a bifunctional phase-transfer catalyst: synthesis of (–)-debromoflustramine B. Chem A Eur Journal. 2018;24:4528–31.
[44] Vakulya B, Varga S, Csámpai A, Soós T. Highly enantioselective conjugate addition of nitromethane to chalcones using bifunctional cinchona organocatalysts. Org Lett. 2005;7:1967–9.
[45] Wang Y, Milkiewicz KL, Kaufman ML, He L, Landmesser NG, Levy DV, et al. Plant process for the preparation of cinchona alkaloid-based thiourea catalysts. Org Process Res Dev. 2017;21:408–13.
[46] Liu H, Yuan J, Tian Q, Ji N, He W. An efficient synthesis of enantiomerically pure γ-aminobutyric acid (GABA) derivatives. J Mater Sci Chem Eng. 2017;5:25–32.
[47] Medina S, Harper MJ, Balmond EI, Miranda S, Crisenza GE, Coe DM, et al. Stereoselective glycosylation of 2-nitrogalactals catalyzed by a bifunctional organocatalyst. Org Lett. 2016;18:4222–5.
[48] Bhaskararao B, Sunoj RB. Two chiral catalysts in action: insights into cooperativity and stereoselectivity in proline and cinchona-thiourea dual organocatalysis. Chem Sci. 2018;9:8738–47.
[49] Zhou MJ, Li Y, Dang L. Theoretical studies on the mechanism of the Michael addition reaction catalyzed by a thiourea-cinchona-amine: triple activation. Asian J Org Chem. 2015;4:904–11.
[50] Ramachary DB, Shruthi KS, Madhavachary R. A modularly designed supramolecular organocatalyst as an effective stimulant for the highly asymmetric Michael addition of ketones to nitro olefins. Eur J Org Chem. 2015;29:6413–18.
[51] Yarlagadda S, Sankaram GS, Balasubramanian S, Reddy BV. Asymmetric Robinson annulation of 3-indolinone-2-carboxylates with cyclohexenone: access to chiral bridged tricyclic hydrocarbazoles. Org Lett. 2018;20:4195–9.
[52] Suresh R, Simlandy AK, Mukherjee S. A catalytic enantioselective iodocyclization route to dihydrooxazines. Org Lett. 2018;20:1300–3.

[53] Konda S, Zhao JC. Enantioselective anti-Mannich reaction catalyzed by modularly designed organocatalysts. Tetrahedron. 2018;74:6166–72.
[54] Lu N, Fang Y, Gao Y, Wei Z, Cao J, Liang D, et al. Bifunctional thiourea–ammonium salt catalysts derived from cinchona alkaloids: cooperative phase-transfer catalysts in the enantioselective Aza-Henry reaction of ketimines. J Org Chem. 2018;83:1486–92.
[55] Kayal S, Mukherjee S. Catalytic aldol–cyclization cascade of 3-isothiocyanato oxindoles with α-ketophosphonates for the enantioselective synthesis of β-amino-α-hydroxyphosphonates. Org Lett. 2015;17:5508–11.
[56] Shi LM, Dong WW, Tao HY, Dong XQ, Wang CL. Catalytic asymmetric desymmetrization of cyclopentendiones via diels–alder reaction of 3-hydroxy-2-pyrones: construction of multifunctional bridged tricyclic lactones. Org Lett. 2017;19:4532–5.
[57] Dell'Amico L, Vega-Peñaloza A, Cuadros S, Melchiorre P. Enantioselective Organocatalytic diels–alder trapping of photochemically generated hydroxy-o-quinodimethanes. Angew Chem Int Ed. 2016;55:3313–17.
[58] Cuadros S, Melchiorre P. Organocatalytic strategies to stereoselectively trap photochemically generated hydroxy-o-quinodimethanes. Eur J Org Chem. 2018;23:2884–91.
[59] Ji N, Yuan J, Xue S, Zhang J, He W. Novel chiral thiourea organocatalysts for the catalytic asymmetric oxaziridination. Tetrahedron. 2016;72:512–17.
[60] Meninno S, Vidal-Albalat A, Lattanzi A. Asymmetric epoxidation of alkylidenemalononitriles: key step for one-pot approach to enantioenriched 3-substituted piperazin-2-ones. Org Lett. 2015;17:4348–51.
[61] Nakashima K, Noda Y, Hirashima SI, Koseki Y, Miura T. Asymmetric conjugate addition of α-cyanoketones to enones using diaminomethylenemalononitrile organocatalyst. J Org Chem. 2018;83:2402–8.
[62] Arai R, Hirashima S-I, Kondo J, Nakashima K, Koseki Y, Miura T. Cinchona–diaminomethylenemalononitrile organocatalyst for the highly enantioselective hydrophosphonylation of ketones and enones. Org Lett. 2018;20:5569–72.
[63] Malerich JP, Hagihara K, Rawal VH. Chiral squaramide derivatives are excellent hydrogen bond donor catalysts. J Am Chem Soc. 2008;130:14416–17.
[64] Zhao BL, Li JH, Du DM. Squaramide-catalyzed asymmetric reactions. Chem Rec. 2017;17:994–1018.
[65] Okino T, Hoashi Y, Furukawa T, Xu XN, Takemoto Y. Enantio- and diastereoselective Michael reaction of 1,3-dicarbonyl compounds to nitroolefins catalyzed by a bifunctional thiourea. J Am Chem Soc. 2005;127:119–25.
[66] (a) Han X, Zhou HB, Dong C. Applications of chiral squaramides: from asymmetric organocatalysis to biologically active compounds, Chem Rec. 2016;16:897–906. (b) Brandão P, Burke AJ. Recent advances in the asymmetric synthesis of chiral 3-Hydroxy and 3-Aminooxindoles and derivatives: Medicinally Relevant Compounds. Tetrahedron. 2018;74:4927–57. (c) Zhao B-L, Li J-H, Du D-M. Squaramide-catalyzed asymmetric reactions. Chem Rec. 2017;17:994–1018. (d) Chauhan P, Mahajan S, Kaya U, Hack D, Enders D. Bifunctional amine-squaramides: powerful hydrogen-bonding organocatalysts for asymmetric domino/cascade reactions. Adv Synth Catal. 2015;357:253–81. (e) Govender T, Arvidsson PI, Maguire GE, Kruger HG, Naicker T. Enantioselective organocatalyzed transformations of β-Ketoesters. Chem Rev. 2016;116:9375–9437. (e) Chanda T, Zhao JC. Recent progress in organocatalytic asymmetric domino transformations. Adv Synth Catal. 2018;360:2–79. (f) Alonso DA, Baeza A, Chinchilla R, Gómez C, Guillena G, Pastor IM, Ramón DJ. Recent advances in asymmetric organocatalyzed conjugate additions to nitroalkenes. Molecules. 2017;22:895 (1-51). (g) Li P, Hu X, Dong X-Q, Zhang X. Recent advances in dynamic kinetic resolution by chiral bifunctional (Thio)urea- and squaramide-based organocatalysts.

Molecules. 2016;21:1327 (1-14). (h) Volla CN, Atodiresei I, Rueping M. Catalytic C-C bond-forming multi-component cascade or domino reactions: Pushing the boundaries in asymmetric organocatalysis. Chem Rev. 114:2390–2431. (i) Karahan S, Tanyeli C. Squaramide catalyzed α-chiral amine synthesis. Tetrahedron Letts. 2018;59:3725–37.

[67] Dong Z, Qiu G, Zhou HB, Dong C. Chiral squaramide as multiple H-bond donor organocatalysts for the asymmetric Michael addition of 1,3-dicarbonyl compounds to nitroolefins. Tetrahedron: Asymmetry. 2012;23:1550–56.

[68] Li JH, Du DM. Enantioselective synthesis of chiral heterocycles containing both chroman and pyrazolone derivatives catalysed by a chiral squaramide. Org Biomol Chem. 2015;13:5636–45.

[69] Zhao BL, Zhang D, Liu L, Du DM. Organocatalytic asymmetric Michael addition of α-alyylidene succinimides to nitrostyrenes. Org Biomol Chem. 2016;14:6337–45.

[70] Veverková E, Bilka S, Baran R, Šebesta R. Squaramide-catalyzed Michal addition as a key step for the direct synthesis of GABAergic drugs. Synthesis. 2016;48:1474–82.

[71] Bera K, Namboothiri IN. Quinine-derived thiourea and squaramide catalyzed conjugate addition of α-nitrophosphonates to enones: asymmetric synthesis of quaternary α-aminophosphonates. J Org Chem. 2015;80:1402–13.

[72] Bae HY, Song CE. Unprecedented hydrophobic amplification in noncovalent organocatalysis "on Water": hydrophobic chiral squaramide catalyzed Michael addition of malonates to nitroalkenes. ACS Catal. 2015;5:3613–19.

[73] Burke AJ, Furtado OM, Marinho VR. Recent multiple transition metal catalysed single-pot reactions. Curr Org Synth. 2010;7:94–119.

[74] Zhao BL, Du DM. Organocatalytic cascade Michael/Michael reaction for the asymmetric synthesis of spirooxindoles containing five contiguous stereocenters. Chem Comm. 2016;52:6162–5.

[75] Sun QS, Zhu H, Lin H, Tan Y, Yang XD, Sun XW, et al. Dual-organocatalytic Michael/Michael/aldol cascade reaction for the asymmetric construction of fully-substituted cyclohexane. Tetrahedron Letts. 2016;57:5768–70.

[76] Amireddy M, Chen K. Organocatalytic synthesis of spirocyclohexane indane-1,3-diones via a chiral squaramide-catalyzed Michael/aldol cascade reaction of γ-nitro ketones and 2-arylideneindane-1,3-diones. Tetrahedron. 2015;71:8003–8.

[77] Duan J, Cheng J, Li P. Enantioselective construction of spiro-1,3-indandiones with three stereocenters *via* organocatalytic Michael-aldol reaction of 2-arylideneindane-1,3-diones and nitro aldehydes. Org Chem Front. 2015;2:1048–52.

[78] Berkes B, Ozsváth K, Molnár L, Gáti T, Holczbauer T, Kardos G, et al. Expedient and diastereodivergent assembly of terpenoid decalin subunits having quaternary stereocenters through organocatalytic Robinson annulation of nazarov reagent. Chem Eur J. 2016;22:18101–6.

[79] Mondal K, Pan SC. Organocatalytic asymmetric domino Michael/Acyl transfer reaction between γ/δ-hydroxyenones and α-nitroketones. J Org Chem. 2018;83:5301–12.

[80] Li J, Du D. Enantioselective synthesis of N-phenyl-dihydropyrano[2,3-c]pyrazoles via cascade Michael addition/thorpe-ziegler type cyclization catalyzed by a chiral squaramide. Chin J Chem. 2015;33:418–24.

[81] Breman AC, Telderman SE, van Santen RP, Scott JI, Maarseveen JH, Ingemann S, et al. Cinchona alkaloid catalyzed sulfa-Michael addition reactions leading to enantiopure β-functionalized cysteines. J Org Chem. 2015;80:10561–74.

[82] Hasilcioğullari D, Tanyeli C. Enantioselective sulfa-Michael addition reaction of methyl thioglycolate to chalcones derivatives with sterically encumbered quinine squaramide organocatalyst. Tetrahedron Lett. 2018;59:1414–6.

[83] Gu Z, Xie JJ, Jiang GF, Zhou YG. Catalytic asymmetric conjugate addition of tritylthiol to azadienes with a bifunctional organocatalyst. Asian J Org Chem. 2018;7:1561–4.

[84] Zhu Y, Dong Z, Cheng X, Zhong X, Liu X, Lin L, et al. Asymmetric synthesis of CF_3- and indole-containing thiochromanes via a squaramide-catalyzed Michael-aldol reaction. Org Lett. 2016;18:3546–9.
[85] Didaskalou C, Kupai J, Cseri L, Barabas J, Vass E, Holtzl T, et al. Membrane-grafted asymmetric organocatalyst for an integrated synthesis-separation platform. ACS Catal. 2018;8:7430–8.
[86] Zhao BL, Lin Y, Yan HH, Du DM. Squaramide-catalysed asymmetric cascade aza-Michael/Michael addition reaction for the synthesis of chiral trisubstituted pyrrolidines. Org Biomol Chem. 2015;13:11351.
[87] Li JH, Wen H, Liu L, Du DM. Diastereo- and enantioselective synthesis of spiro-pyrrolidine-pyrazolones by squaramide-catalyzed cascade aza-Michael/Michael reactions. Eur J Org Chem. 2016;2492–9.
[88] For a very recent review on this topic see: Mei, GJ, Shi, F. Catalytic asymmetric synthesis of spirooxindoles: recent developments. Chem Commun. 2018, 54, 6607–21.
[89] Roy TK, Parhi B, Ghorai P. Cinchonamine squaramide catalyzed asymmetric aza-Michael reaction: dihydroisoquinolines and tetrahydropyridines. Angew Chem Int Ed. 2018;57:9397–401.
[90] Mukhopadhyay S, Pan SC. Organocatalytic asymmetric synthesis of 2,4-disubstituted imidazolidines via domino addition-aza-Michael reaction. Chem Commun. 2018;54:964–7.
[91] Tang CK, Feng KX, Xia AB, Li C, Zheng YY, Xu ZY, et al. Asymmetric synthesis of polysubstituted chiral chromans via an organocatalytic oxa-Michael-nitro-Michael domino reaction. RSC Adv. 2018;8:3095–8.
[92] Mondal B, Maity R, Pan SC. Highly diastereo- and enantioselective synthesis of spiro-tetrahydrofuran-pyrazolones via organocatalytic cascade reaction between γ-hydroxyenones and unsaturated pyrazolones. J Org Chem. 2018;83:8645–54.
[93] Ravindra B, Maity S, Das BG, Ghorai P. Organocatalytic, enantioselective synthesis of 1- and 3-substituted isochromans via intramolecular oxa-Michael reaction of alkoxyboronate: synthesis of (+)-sonepiprazole. J Org Chem. 2015;80:7008–18.
[94] Hazra G, Maity S, Bhowmick S, Ghorai P. Organocatalytic, enantioselective synthesis of benzooxaboroles via wittig/oxa-Michael reaction cascade of α-formyl boronic acids. Chem Sci. 2017;8:3026–30.
[95] Midya A, Maity S, Ghorai P. Dynamic kinetic spiroketalization/oxa-Michael addition cascade of alkoxyboronates and peroxyacetals: enantio- and diastereoselective synthesis of benzannulated spiroketals. Chem Eur J. 2017;23:11216–20.
[96] Feng J, Li X. Enantioselective vinylogous Michael/Michael cascade reactions of 3-alkylidene oxindoles and nitroolefin enoates. J Org Chem. 2017;82:7317–23.
[97] Rao KS, Ramesh P, Chowhan LR, Trivedi R. Asymmetric Mannich reaction: highly enantioselective synthesis of 3-amino-oxindoles *via* chiral squaramide based H-bond donor catalysis. RSC Adv. 2016;6:84242–7.
[98] Işibol D, Karahan S, Tanyeli C. Asymmetric organocatalytic direct Mannich reaction of acetylacetone and isatin derived ketimines: low catalyst loading in chiral cinchona-squaramides. Tetrahedron Letts. 2018;59:541–5.
[99] Bae HY, Kim MJ, Sim JH, Song CE. Direct catalytic asymmetric Mannich reaction with dithiomalonates as excellent Mannich donors: organocatalytic synthesis of (*R*)-sitagliptin. Angew Chem Int Ed. 2016;55:10825–9.
[100] Zhao MX, Dong ZW, Zhu GY, Zhao XL, Shi M. Diastereo- and enantioselective Mannich/cyclization cascade reaction of isocyanoacetates with cyclic sulfamide ketimines by cinchona alkaloid squaramide/AgOAc cooperative catalysis. Org Biomol Chem. 2018;16:4641–9.
[101] Susam D, Tanyeli C. Enantioselective aza-Henry reaction of t-Boc protected imines and nitroalkanes with bifunctional squaramide organocatalysts. New J Chem. 2017;41:3555–61.
[102] Rauhut MM, Currier H. 1963-01-22 Preparation of dialkyl 2-methyleneglutarates, US3074999.

[103] For a comprehensive review see: Aroyan CE, Dermenci A, Miller SJ. The Rauhut-Currier reaction: a history and its synthetic application. Tetrahedron. 2009;65:4069–84.
[104] Maity S, Sar S, Ghorai P. Primary aminothiourea-catalyzed enantioselective synthesis of Rauhut-Currier adducts of 3-arylcyclohexenone with a tethered enone on the aryl moiety at the ortho-position. Org Lett. 2018;20:1707–11.
[105] He XH, Yang L, Ji YL, Zhao Q, Yang MC, Huang W, et al. Chemo- and stereoselective cross Rauhut-Currier-type reaction of tri-substituted alkenes containing trifluoromethyl groups. Chem Eur J. 2018;24:1947–55.
[106] Burke AJ, Marques CS, Turner NJ, Hermann GJ. Active pharmaceutical ingredients in synthesis, ch. 8, Wiley-VCH, 2018:291–319. Colham, I, Hufton, H. Intramolecular dipolar cycloaddition reactions of azomethine ylides. Chem Rev. 2005;105:2765–810.
[107] Vicario JL, Reboredo S, Badia D, Carrillo L. Organocatalytic enantioselective [3+2] cycloaddition of azomethine ylides and α, β-unsaturated aldehydes. Angew Chem Int Ed. 2007;46:5168–70.
[108] Peňaška T, Ormandyová K, Meĉiarová M, Filo J, Šebesta R. Organocatalytic diastereoselective synthesis of spirooxindoles via [3+2] cycloadditions of azomethine ylides with α, β-unsaturated esters. New J Chem. 2017;41:5506–12.
[109] Hang WJ, Chen Q, Lin N, Long XW, Pan WG, Xiong YS, et al. Asymmetric synthesis of trifluoromethyl-substituted 3,3´-pyrrolidinyl-dispirooxindoles through organocatalytic 1,3-dipolar cycloaddition reactions. Org Chem Front. 2017;4:472–82.
[110] Su J, Ma Z, Li X, Lin L, Shen Z, Yang P, et al. Asymmetric synthesis of 2'-trifluoromethylated spiro-pyrrolidine-3,3'-oxindoles via squaramide-caatalyzed umpolung and 1,3-dipolar cycloaddition. Adv Synth Catal. 2016;358:3777–85.
[111] Zhao MX, Zhu GY, Zhu HK, Zhao XL, Ji M, Shi M. Temperature-dependent cinchona alkaloid squaramide-catalyzed asymmetric formal [3+2] cycloaddition of isocyanoacetates with β-trifluoromethylated enones. Eur J Org Chem. 2018:3997–4005.

Laura Raimondi, Chiara Faverio and Monica Fiorenza Boselli

4 Chiral imidazolidinones: A class of priviliged organocatalysts in stereoselective organic synthesis

Abstract: Chiral molecules hold a mail position in Organic and Biological Chemistry, so pharmaceutical industry needs suitable strategies for drug synthesis. Moreover, Green Chemistry procedures are increasingly required in order to avoid environment deterioration. Catalytic synthesis, in particular organocatalysis, in thus a continuously expanding field. A survey of more recent researches involving chiral imidazolidinones is here presented, with a particular focus on immobilized catalytic systems.

Keywords: organocatalysis, aminocatalysis, iminium ion, imidazolidinones, supported catalyst

4.1 Introduction

Chiral molecules do hold a main position as numerous biological functions rely on the recognition between an unsymmetrical system and the correct stereoisomer. On these basis stereoisomers activities in pharmacologically active compounds can be completely different, one providing the highest therapeutic effect while the others being less active, inert or toxic. Despite a substantial quantity of chiral drugs is still sold in the racemic form – i. e. a mixture of enantiomers – the pharmaceutical industry needs to switch to the commercialization of single enantiomers; in fact, in 1992 FDA and in 1994 EU issued guidelines concerning the development of new chiral drugs, which mandate the development of enantiopure compounds [1].

Development of strategies for preparing enantiomerically enriched chiral molecules represents therefore a field of very active research. Synthetic methods based on the use of chiral auxiliaries (chiral enantiomerically pure molecules covalently linked to a reaction partner to control the stereochemical outcome of a reaction and removed once the desired product is obtained) are gradually being abandoned in favor of *chiral catalysts*, regarded as a fundamental tool in the context of *Green Chemistry*. Recognizing the progressive deterioration of environment and natural resources, a special commission of the United Nations General Assembly (the World

This article has previously been published in the journal Physical Sciences Reviews. Please cite as: Raimondi, L., Faverio, C., Boselli, M. F. Chiral imidazolidinones: A class of priviliged organocatalysts in stereoselective organic synthesis *Physical Sciences Reviews* [Online] 2020, 5. DOI: 10.1515/psr-2018-0087

https://doi.org/10.1515/9783110590050-004

Commission on Environment and Development) released an official report entitled "Our Common Future" stating the concept of *sustainable development* [2].

Catalytic strategies are included in the twelve Principles of Green Chemistry and are classified according to the kind of application or to the structure of the catalyst, i. e. the molecule that "accelerates a chemical reaction without affecting the position of the equilibrium" on the basis of their aggregation state or their nature and composition [3]. Catalysts, promoting and controlling reaction selectivity, can be metal complexes, enzymes or organocatalysts. Specifically, in 2000 the term *organocatalysis* was coined by McMillan to identify reactions where catalysts are "organic compounds which do not contain a metal atom" [4]. Perhaps better described as a process where a metal is not directly involved in the reaction mechanism, so to include also ferrocene-based organocatalysts, organocatalysis is a continuously expanding field in organic chemistry. Relying on the use of relatively small molecules, typically cheap and readily accessible, easy to handle and non-toxic, organocatalysis finds economic- and eco-friendly applications.

While organocatalysts' biocompatibility was previously mainly hypothesized, a recent study on different cell lines demonstrated these molecules are actually harmless over a wide range of concentrations, the only exception being a thiourea derivative [5]. Despite the high catalyst loading is often considered a requirement limiting organocatalysis viability in industry, the low cost of organocatalysts as well as bypassing the problem of metals removal from final products make this approach nevertheless appealing [6] and highly selective organocatalytic synthesis of pharmaceutical compounds have already been reported and developed in large scale [7].

Stereoselective organocatalysis relies on different modes of substrate activation. This can be classified as *covalent* or *non-covalent* and further distinguished considering the molecular orbital involved (Figure 4.1) [8]. It has to be mentioned that the same catalyst can operate through distinct modalities depending on the interacting substrate. Besides, organocatalysts featuring more than one active site, which are simultaneously activating two reaction partners or different positions of the same compound, have been developed and dubbed as *multifunctional catalysts.*

When the energy level of the lowest unoccupied molecular orbital of the substrate is lowered, nucleophilic attack is favored. Within covalent catalysis, *iminium ion*, generated upon condensation of a (chiral) primary or secondary amine, is one of the most common intermediate exploited to make the LUMO of α,β–unsaturated compounds more accessible to electrophiles. To achieve carbonyl carbon activation of carboxylic acid derivatives, *acylammonium* catalysis is instead applied: initiated by the nucleophilicity of an acyltransfer organocatalyst, reactions including transesterifications, kinetic resolutions, desymmetrizations, and Steglich rearrangements have been reported. Activation of the LUMO orbital is also achieved by non-covalent *Brønsted acid* or *hydrogen bonding* catalysis, both based on substrate electronic depletion.

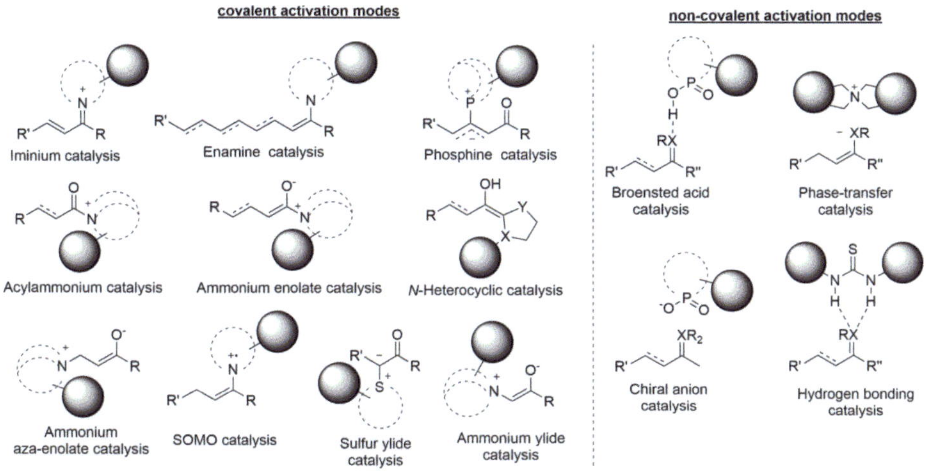

Figure 4.1: covalent and non-covalent activation modes of substrate activation

On the contrary, increasing of the energy level of the highest occupied molecular orbital enhances the nucleophilicity of the substrate. The most traditional approach is represented by *enamine* – subsequently extended to di- and trienamine catalysis: parallel to iminium ion mode, the first step of the catalytic cycle consists in the reaction between an amine catalyst and a carbonyl compound, followed by α–deprotonation and generation of an intermediate highly reactive towards electrophilic carbons or heteroatoms. Alternatively, a nucleophilic enolate equivalent (*ammonium enolate*) can be obtained either by addition of a chiral tertiary amine catalyst to a ketene or through α-deprotonation of an acylammonium intermediate. A different kind of activated nucleophile is the *Breslow intermediate*; in this case, the α-carbon of an aldehyde acquires an electron rich character upon reaction with carbenes. As for non-covalent catalysis, HOMO activation is achieved *via phase-transfer* principle.

In aminocatalysis [9], three main catalytic mechanisms are operating, based on HOMO (via enamine), LUMO (via iminium-ion) and SOMO activation (Figure 4.2). Stereoselectivity is given either by steric-shielding, in which one of the reactive intermediate faces is blocked by a bulky group, or by a hydrogen-bonding [10] directed approach, in which both substrate and reagent are simultaneously coordinated. In particular, in enamine-based catalytic processes a carbonyl compound is transformed into a more nucleophilic intermediate. Upon condensation between the catalyst and the substrate, an iminium ion is formed, that tautomerizes to the favored nucleophilic enamine form, in contrast to what happens in the keto-enol equilibrium, mainly lying in the keto-form. The stronger nucleophilic character achieved thanks to the amine activation is not only due to the abundance of the enamine species, but also to the enamine HOMO energy that is higher than that of the corresponding enol, the less electronegative nitrogen lone-pair electrons being higher in energy than that of the more electronegative oxygen. Iminium ion catalysis acts lowering the substrate LUMO energy: when an α,β-carbonyl compound condenses with a chiral secondary amine, it forms a positively charged intermediate prone to electrophilic attack on the β-position. SOMO activation, instead, is based on the *in situ* generation of single electron species, generated through single electron oxidation of an enamine intermediated or by attack of a radical species.

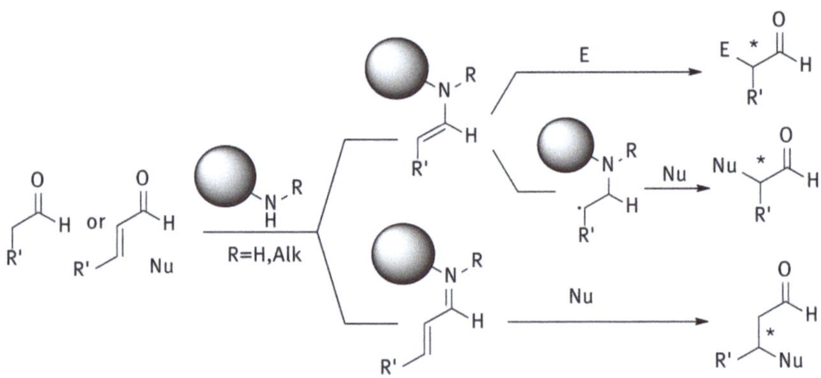

Figure 4.2: Operating catalytic mechanisms in aminocatalysis

The field evolved allowing stereoselective functionalization of position that are farther from the carbonyl group, thanks to the vinylogous activation of poly-unsaturated substrates.

4.2 Imidazolidinones: Iminium ion activation

The fundamental work of MacMillan and coworkers with chiral imidazolidinones represents the first example of the iminium ion activation mode and is now considered a landmark in the field of organocatalysis [11]. While usually inactive for enamine-mediated transformations, imidazolidinones have proven to be very versatile catalysts for the activation of α,β-unsaturated aldehydes through formation of a transient iminium ion.

In 2000, the MacMillan research group reported the use of a chiral imidazolidinone derived from natural amino acid *L*-phenylalanine as organocatalyst to promote Diels-Alder cycloadditions [4]. The catalyst was developed in the context of a general strategy for stereoselective organocatalytic reactions promoted by chiral amines alternatives to Lewis acid catalyzed ones. The reversible formation of iminium ions from α, β-unsaturated aldehydes and amines emulates the equilibrium dynamics and π-orbital electronics that are inherent to Lewis acid catalysis resulting in a LUMO-lowering activation (Figure 4.3). The iminium ion formation is enhanced by the presence of acid additives so the catalyst operates in its protonated form.

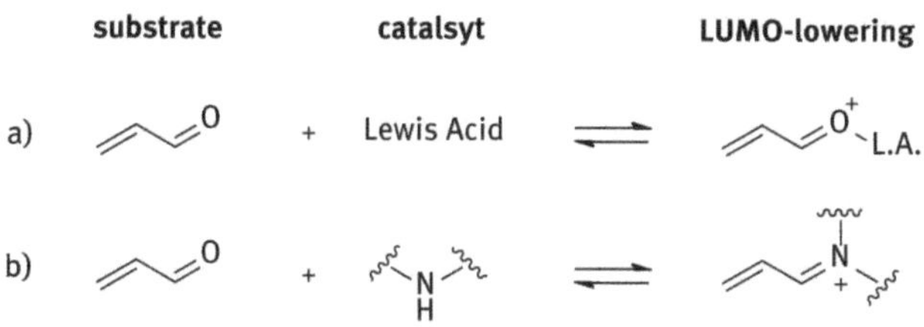

Figure 4.3: Modes of activation of α,β-unsaturated aldehydes.

The catalytic cycle for the stereoselective Diels–Alder reaction promoted by MacMillan catalyst is outlined in Figure 4.4: the condensation of the α,β-unsaturated aldehyde with the enantiopure secondary amine leads to the formation of an iminium ion that is sufficiently activated to engage a diene reaction partner. Accordingly, Diels–Alder cycloaddition would lead to an adduct which upon hydrolysis provides the enantioenriched product, while releasing the chiral amine catalyst.

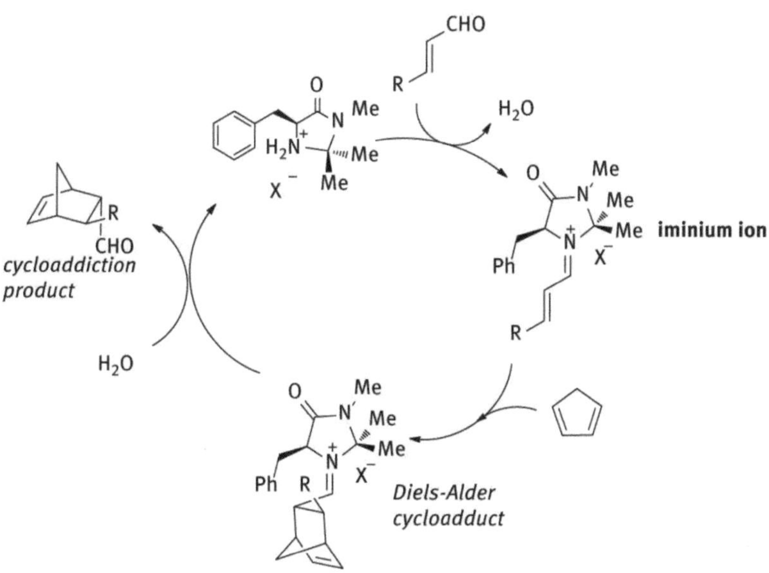

Figure 4.4: Reaction mechanism for stereoselective Diels Alder cycloaddition promoted by chiral imidazolidinones.

MacMillan group reported many other asymmetric reactions such as the 1,3-dipolar cycloadditions, Friedel-Crafts alkylations, α-chlorinations, α-fluorinations, and intramolecular Michael reactions using chiral imidazolidinones. It was also shown that by combining the organocatalyst and Hantzsch ester it was possible to promote the first enantioselective organocatalytic hydride reduction of α,β-unsaturated aldehydes in what can be defined as a biomimetic approach, somehow analogous to nature's stereoselective enzymatic transfer hydrogenation with NADH cofactor [12].

A second generation of imidazolidinones-based organocatalysts was developed soon after, to further improve the stereochemical efficiency of the chiral scaffold by exploiting the presence not only of the stereocenter on the aminoacid moiety, but also of an additional stereogenic center on the imidazolidinone scaffold, that offers further possibilities to control the behavior of the catalyst, employed either as *trans* or *cis* diastereoisomers, depending on the reactive substrates and the type of transformation (Figure 4.5). Starting from an amino acid the reaction with an aldehyde or a ketone allows to prepare a high number of compounds, where it is possible to fine tuning the performance of the catalyst and its ability to direct the formation of a well defined conformer in the iminium ion intermediate, key structure for all the stereoselective transformations promoted by this class of catalysts. The reaction of a nucleophile with such intermediate brings to the formation of β–substituted aldehydes, often with very high enantioselectivity.

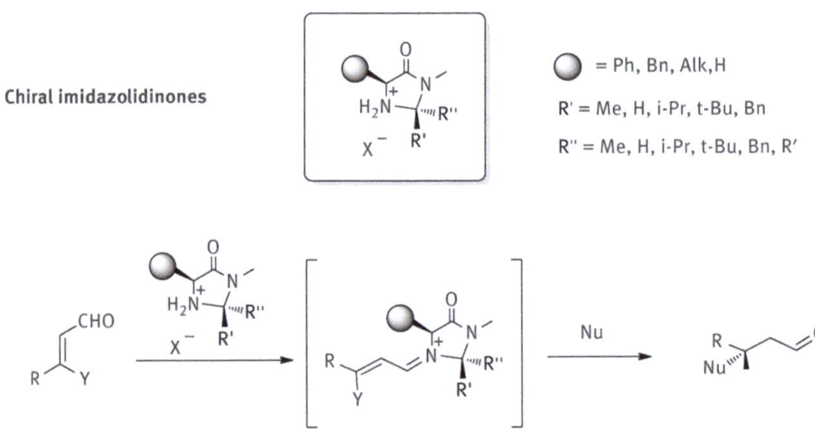

Figure 4.5: Stereoselective nucleophile addition to unsaturated aldehydes.

Iminium-ion based catalytic cycle has been the subject of mechanicistic investigation, confirming the importance of a protic additive in assisting a proton transfer that facilitates the formation of the hemiacetal intermediate and of water as proton shuttle for the enamine resulting after C-C bond formation. Moreover, kinetic measurements by Mayr group established an electrophilicity scale for eleven iminium ions obtained from differently structured secondary amines, which revealed dominated by imidazolidinone-based ones (Figure 4.6) [13]. Rationalization on the stereochemical outcome of the reaction has also been addressed through NMR and theoretical studies. These assessed the conformation of the catalysts, showing that steric shielding has to be ascribed to the benzyl substituent of oxazolidinones. Additionally, the configuration of the iminium ion intermediate was elucidated being in an E/Z ratio > 9:1, the higher level of enantioselection justified by a kinetic resolution process operated by the nucleophile itself [14].

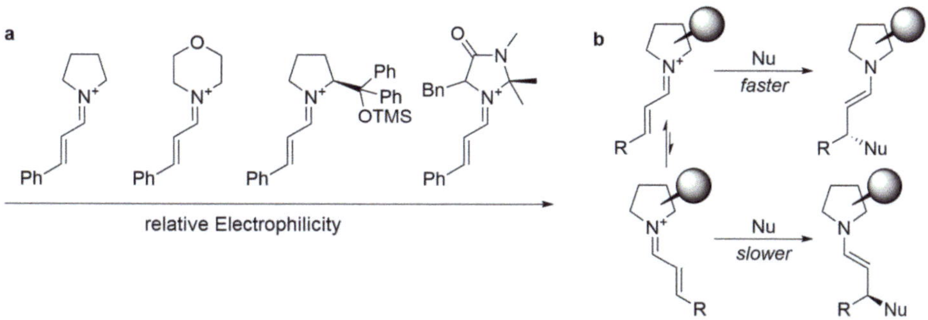

Figure 4.6: electrophilicity scale for iminium ions.

Such organocatalysts have found later wide application in the so called SOMO catalysis (Single Occupied Molecular Orbital), where the chiral imidazolidinone is employed in combination with a photoredox catalyst to promote radical stereoselective reactions, and the organocatalytic cycle is interlocked with the photoredox catalytic cycle. However, since a whole chapter of the book is devoted to organo-photoredox catalysis, the topic will not be discussed in the present chapter.

4.3 Chiral imidazolidinones: Other activation modes

4.3.1 Stereoselective alkylation of aldehydes

Metal-free synthetic α-alkylation methodologies of aldehydes have been developed by S_N1-type reactions, in which carbocations of sufficient stability generated in situ from alcohols, or stable carbenium ions, are employed to perform enantioselective α-alkylation of aldehydes (Figure 4.7) catalyzed by MacMillan-type catalysts, according to the mechanism reported in Figure 4.8 [15]. Differently from the previous cases, the organocatalyst is here employed to activate the alfa position of the carbonyl derivative as enamine that reacts with the electrophilic species to afford α-substituted enantiomerically enriched aldehydes.

High yields and remarkable enantioselectivities, often higher than 90% e.e. were obtained with linear and not branched aldehydes. Enantiomerically pure α–alkyl-

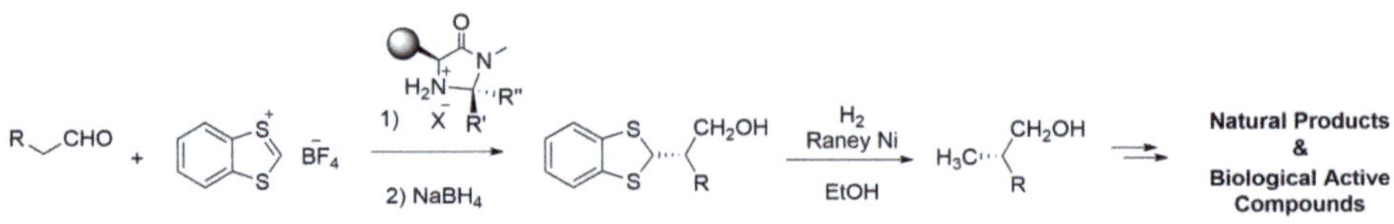

Figure 4.7: Stereoselective alkylation of aldehydes.

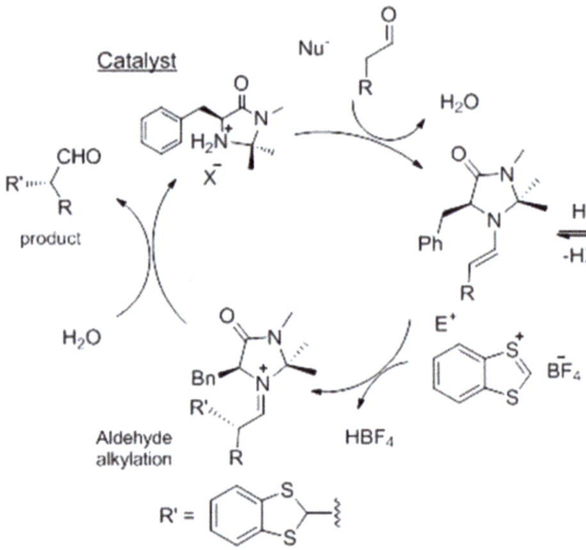

Figure 4.8: Reaction mechanism for the stereoselective alkylation of aldehydes promoted by chiral imidazolidinone.

substituted aldehydes are remarkably important key substrates for the synthesis of more complex molecules [16].

4.3.2 Stereoselective imine reduction

Despite all the recent achievements in the field of enantioselective organocatalytic reduction of C=N double bonds, the combination of inexpensive, non-toxic, easily disposable reagents with very low catalyst loadings, comparable to those of the metal-based chiral catalysts, is an unmet challenge in organocatalysis today [17].

Trichlorosilane-based reduction methodology may offer this opportunity. Recently the design, the synthesis, the optimization, and the application of a new class of easy accessible chiral Lewis bases, prepared with few steps starting from cheap and commercially available aminoacids, have been reported (Figure 4.9) [18]. Picolinamides, that could be easily obtained by condensation between picolinic acid/chloride/

anhydride with different chiral scaffolds, have been extensively studied as activators of trichlorosilane [19]. The inspiration, for the design of this new class of LB*, was taken by the well-established and efficient scaffold of imidazolidinones, able to successfully control the stereochemical outcome of the reaction of carbonyl compounds. The picolinamide residue is responsible of $HSiCl_3$ coordination and chemical activation, while the imidazolidinone scaffold offers many opportunities to tune and modify the steric hindrance and the stereoelectronic properties of the catalytic system.

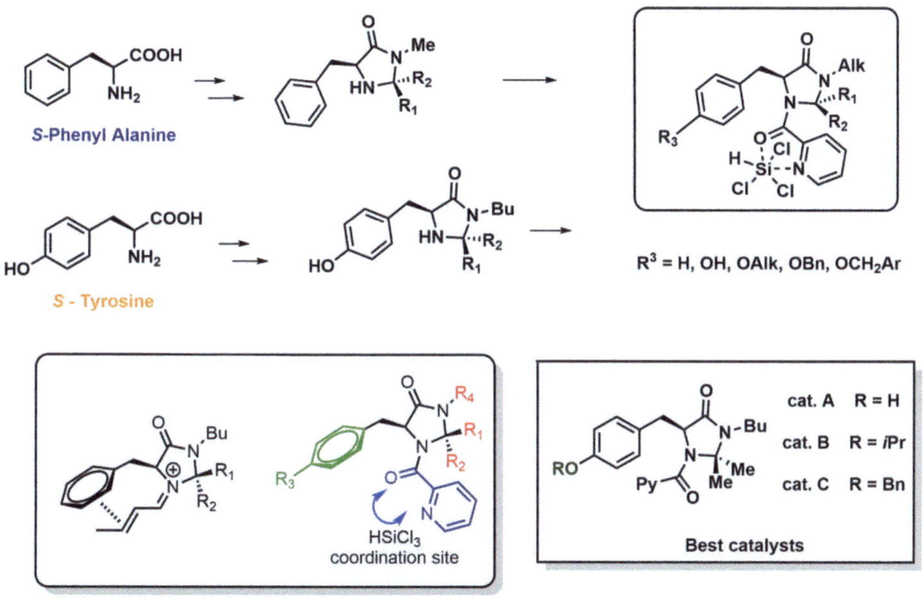

Figure 4.9: Chiral Lewis bases for enantioselective catalytic hydrosilylation of imines.

A series of new chemical entities, featuring different substituents on the imidazolidinone ring and on the aromatic ring of the amino acid moiety, were synthesized. Surprisingly the introduction of a para substituent on the aromatic ring leads to the identification of very efficient catalysts. **A-C** showed very high activity and incredible stereocontrol, *ee* up to 98% in the enantioselective reduction of ketoimines. In the model reduction of *N*-PMP imine of acetophenone, with catalyst **C** the product was obtained with 80% of yield and 97% of *ee*, using only 0.1% mol of catalyst. Working under such conditions, the ACE (Asymmetric Catalyst Efficiency) [20] was evaluated to be 375–400, a comparable value with organometallic systems and better then most organocatalysts [21].

To test the general applicability of the new catalysts the reduction of a wide variety of imines was investigated. The reduction of 3-alkoxy-substituted acetophenone imines, either *N*-benzyl protected (as precursor of primary amine derivatives), or *N*-(3-phenylpropyl) substituted, represents a practical and efficient entry to the synthesis of enantiomerically pure valuable pharmaceutical compounds, used in the treatment of Alzheimer and Parkinson diseases, hyperparathyroidism, neuropathic pains and neurological disorders (Figure 4.10).

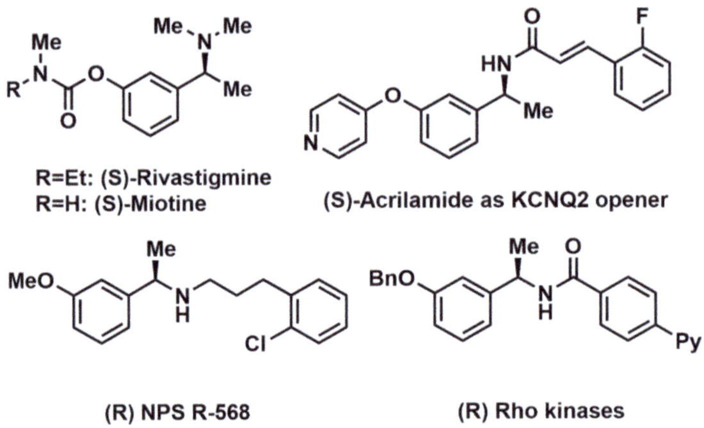

Figure 4.10: Chiral amines as valuable pharmaceutically active compounds.

4.4 Immobilized imidazolidinones

Given its popularity, due to wide applicability and easy availability, it is not surprisingly that the catalyst has been covalently anchored onto different supports [22]. The immobilization of a catalytic species on a solid support may represent a solution to some of the problems related to the use of chiral catalysts in organic synthesis [23]. Not only the recovery and the possible recycle of a catalyst may be investigated and successfully realized through its immobilization, but also the studies of other issues like the stability, the structural characterization and the catalytic behavior may be better conducted on a supported version of the enantioselective catalyst.

These general considerations are true also for organic catalysts. Even if the transformation of a stoichiometric into a catalytic process can be regarded as a significant step toward the development of a truly green chemistry, catalytic reactions are amenable to a variety of improvements that can make them greener and greener. Among these, the replacement of metal-based catalysts with equally efficient metal-free counterparts, the so-called "organic catalysts", can be extremely important.

The immobilization of the catalyst on a support with the aim of facilitating the separation of the product from the catalyst, and thus the recovery and recycling of the latter, can also be regarded as an important improvement for a catalytic process [24]. In this line, the immobilization of organic catalysts seems particularly attractive, because the metal-free nature of these compounds avoids from the outset the problem of the leaching of the metal that often negatively affects and practically prevents the efficient recycling of a supported organometallic catalyst [25].

Despite the great number of proposed solutions, the development of a supported chiral imidazolidinone for the application in catalytic flow reactor was not reported until 2013, when the use of chiral imidazolidinones supported on mesoporous silica nanoparticles to build catalytic reactors was investigated (Figure 4.11) [26].

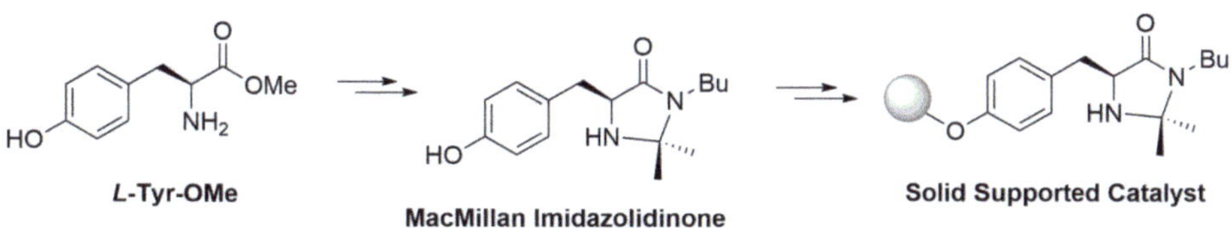

Figure 4.11: Immobilization of chiral imidazolidinones.

Later, the use of two supporting materials with different properties, inorganic silica and organic polystyrene was compared [27]. The silica supported catalysts were prepared by grafting on silica particles of different size. In order to evaluate the influence of the linker on catalyst performances, a different anchoring strategy was evaluated: the supported catalyst was synthesized through a click reaction between azido-functionalized silica and imidazolidinone in the presence of CuI as catalysts and Hunig's base in chloroform as solvent. Catalyst with the triazole ring as spacer was obtained (Figure 4.12).

Silica-grafted catalysts promoted the model Diels Alder cycloaddition between cyclopentadiene and cinnamic aldehyde in good yields and *ee* up to 82%, while the material which features a triazole linker on its structure, promoted the reaction in good yields and very good *ee*, up to 93%. This suggests that a longer linker between the solid support and the catalytically active site positively influences the enantioselectivity of the process.

The solid supported catalysts could be easily recovered after reaction time by simple filtration of the crude mixture. However, attempts to recycle the catalysts were unsuccessful, thus suggesting that the organic catalyst supported onto silica is deactivated under reaction conditions (probably the presence of the acid co-catalyst accelerates the degradation of the imidazolidinone ring).

The use of packed-bed flow reactors was then investigated. As catalytic reactor, an empty HPLC column (*l*: 12.5 cm, *id*: 0.4 cm, V: 1.57 mL) containing 1 g of silica-supported catalyst was employed. The void volume was determined experimentally for each reactor and it was necessary to calculate the residence time. The model cycloaddition between cyclopentadiene and cinnamaldehyde was studied. The reagents were pumped in the reactor by a syringe pump and the products were collected at the way out of the reactor. The catalytic reactor containing showed a good activity in terms of chemical yields and enantioselectivity (up to 85%). By using both TFA and

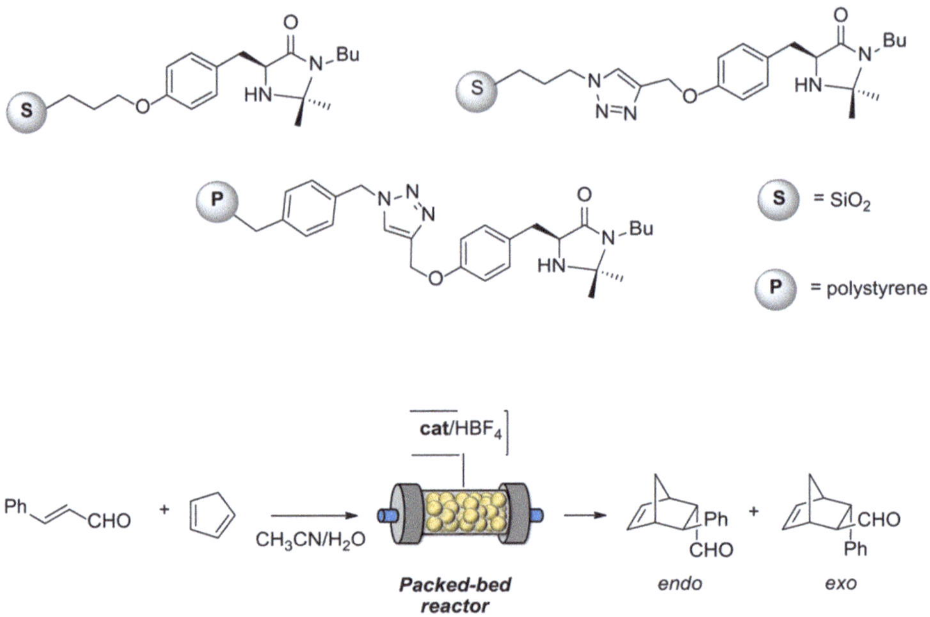

Figure 4.12: Synthesis and use of silica and polystyrene-supported catalysts.

HBF_4 as co-catalysts the flow system produced cycloadducts for more than 150 h, demonstrating that the flow process could extend catalyst lifetime and reduce the imidazolidinone deactivation observed in batch [28]. Trifluoroacetate salt of the supported imidazolidinones promoted a faster reaction than its tetrafluoroborate counterpart.

The synthesis of polymer-supported catalysts involved a click reaction between imidazolidinone and 1-(azidomethyl)-4-vinylbenzene in the presence of CuI and Hunig's base. Resulting intermediate was subjected to radical copolymerization with divinylbenzene as co-monomer in the presence of toluene and 1-dodecanol as porogens and AIBN as initiator. Polystyrene-supported catalyst was obtained with a loading of 0.57 mmol/g that was determined by the stoichiometry of the polymerization reaction. In the benchmark cycloaddition under the same experimental conditions employed for silica-supported catalysts, the cycloadducts were obtained in very good yields and excellent *ee*, up to 90%. Notably, when the catalyst was recycled, it maintained a good level of chemical activity and a very good enantioselectivity.

The results indicate that polystyrene as supporting material is better than silica. This is probably due to the a polar nature of the organic polymer compared to silica that presents Si-OH groups on his surface that can interfere with the activity of the supported catalyst.

A packed bed reactor containing polystyrene-anchored imidazolidinone was then prepared. As flow reactor an HPLC column (*l*: 6 cm, *id*: 0.4 cm, V: 0.75 mL) filled with 0.3 g of supported catalyst was employed. The system was used for the enantioselective cycloadditions of cyclopentadiene and three different aldehydes. For the first 24 h product was formed in 75% yield and 91% and 92% *ee* for *endo* and *exo* isomers, respectively. The catalytic reactor operated for 150 h with no appreciable loss in chemical or stereochemical activity.

Although the flow system proved to be stable for a long period of time, a major limitation was related to the low flow rate employed (2 μL/min). The flow reactor was not suitable for operation at high flow rates, as these prevented the supported catalyst to have sufficient contact time with the reagents.

To overcome this problem the use of monolithic reactors was explored [29]. As already reported above, these reactors feature a large surface area and can be employed at high flow rates. Monolithic reactor was prepared inside a HPLC column (*l*: 6 cm, *id*: 0.4 cm, V: 0.75 mL) by radical copolymerization of the chiral monomer featuring the imidazolidinone moiety with divinylbenzene as co-monomer, in the presence of toluene and 1-dodecanol as porogens and AIBN as initiator (Figure 4.13).

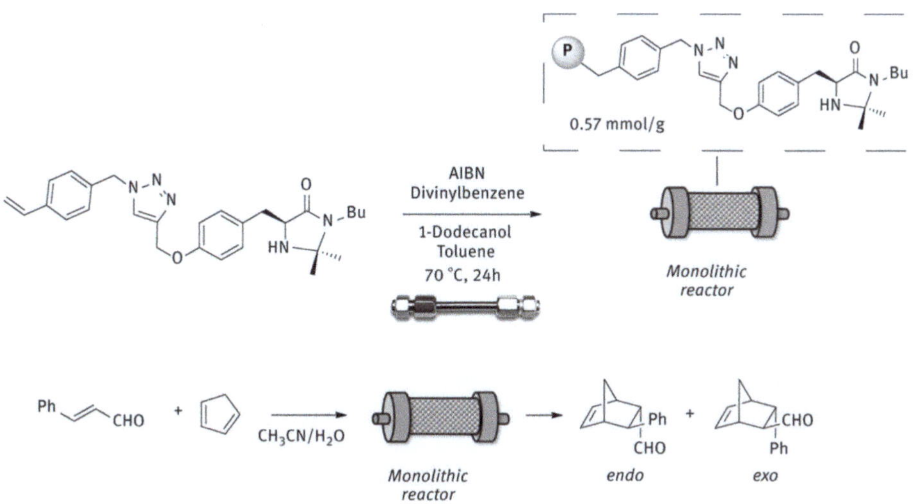

Figure 4.13: Synthesis of monolithic reactor.

The reactor was sealed and heated at 70 °C for 24 h. The reactor was then washed with THF in order to remove excess of porogens. The void volume was determined experimentally and it was necessary to calculate the residence time. Monolithic reactor was tested in the model cycloaddition between cyclopentadiene and cinnamaldehyde in the presence of HBF_4 as co-catalyst. Differently from packed bed reactors, the monolithic reactor could work at high flow rates with no appreciable loss in chemical or stereochemical activity. After 70 h of continuous operation the desired cycloadducts were produced in 73% yield and 88% *ee*.

Comparing the Turn Over Numbers (TON) and productivities (calculated as 1000* mmol product*mmol catalyst^{-1} * time^{-1}) for the three different flow systems (silica packed bed, polystyrene packed bed and polystyrene monolith reactors), the monolithic reactor proved to be much more efficient in terms of TONs and productivity than packed bed reactors, especially for longer reaction times. Differently from packed bed reactors, the monolithic reactor could work at high flow rates with no appreciable loss in chemical or stereochemical activity (after 24 h, 75% yield and 92% *ee*, after 44 h 68% yield and 90% *ee*). After 70 h of continuous operation the desired cycloadducts were produced in 73% yield and 88% *ee* with a residence time of 1.2 hours (entry 5).

To further demonstrate the advantages related to the use of monolithic reactors with respect to packed bed-ones a comparison of TON and productivities (calculated as 1000* mmol product*mmol catalyst^{-1} * time^{-1}) obtained with three different flow systems was made. The productivity was determined to be 64 h^{-1} for the silica-based packed bed reactor, 120 h^{-1} for the polystyrene-based packed bed reactor and 339 h^{-1} for the monolithic reactor.

The greater performances are ensured by the possibility to work at higher flow rates with no erosion in the chemical or stereochemical activity. Thus in the same unit of time, larger quantities of product can be obtained.

In order to develop a continuous flow process with the catalytic reactors previously described, the alkylation of propanal with three different cationic electrophiles (**I**, **II** and **III**) [15], was first studied in batch, in the presence of 30 mol% of supported imidazolidinones, HBF_4 as co-catalysts, NaH_2PO_4 as scavenger (of HBF_4 released from the electrophile) and CH_3CN/H_2O mixture as solvent (Figure 4.14) on two different supports, silica and polystyrene [30].

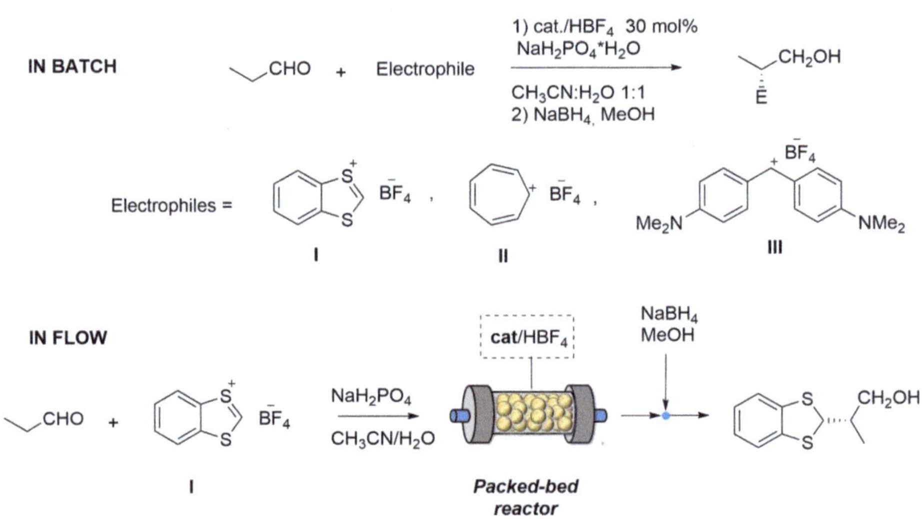

Figure 4.14: Alkylation of propanal with different cationic electrophiles.

Products were obtained in good yields and fair to very good enantioselectivity. For example, reaction of propionic aldehyde with electrophile **I** afforded the product in 75% yield and 90% *ee* with silica-supported catalyst, and 67% yield and 86% *ee* with PS-supported system. However, with electrophiles **II** and **III** polymer supported imidazolidinone promoted the reactions with higher enantioselectivity compared to silica supported, affording the products in 90% *ee* with cation **II** (compared to 79% *ee* with silica-immobilized species) and in 95% *ee* with cation **III** (compared to 67% *ee* with silica-immobilized catalyst).

Once the compatibility of supported catalysts in the batch reaction was demonstrated the use of packed-bed reactors for the continuous flow alkylation of aldehydes was studied. Two stainless-steel HPLC columns (*l*: 6 cm, *id*: 0.4 cm, V: 0.75 mL) were filled with polymer (0.3 g) and silica (0.3 g) supported catalysts, respectively. The alkylation of propanal with **I** in the packed bed reactor containing silica-support imidazolidinone afforded the product in 65% yield and 80% *ee*. Packed-bed reactor containing the polystyrene-immobilized catalyst promoted the reaction in moderate yield and very good *ee* (up to 95%).

The flow systems prepared showed a good activity in the stereoselective alkylation of aldehydes, affording the desired products in very good *ee* (up to 95%) and high productivities. However when high flow rates were employed quite low yields were obtained. Even if some improvements are still required, this procedure represent the first continuous flow stereoselective alkylation of aldehydes using solid supported imidazolidinones.

The stereoselective reduction of imines with $HSiCl_3$ using picolinamides with a chiral imidazolidinone as scaffold has been recently reported [18]. Starting from (*L*)-Tyrosine, a chiral picolinamide was prepared and the phenolic residue exploited as an handle to anchor the catalyst onto a solid support (Figure 4.15) [31].

Silica supported picolinamides were prepared by grafting commercially available SiO_2 (Apex Prepsil Silica Media 8 μm); the loading was determined by weight difference. Another silica supported picolinamide which features a longer linker longer between the active site and the solid support was synthesized by a click reaction.

The synthetic strategy for the preparation of polymer supported picolinamide involved the preparation of the chiral monomer that was then polymerized with divinylbenzene as co-monomer, in the presence of toluene and 1-dodecanol as porogens and AIBN as initiator. The loading was determined by the stoichiometry of the polymerization reaction.

The reduction of imines derived from acetophenone with trichlorosilane was chosen as model reaction to test the solid supported picolinamides (Figure 4.16).

Silica-supported catalyst promoted the reaction in excellent yield and very good *ee*. The presence of a longer linker in the structure of catalyst did not improve neither the yield nor the stereoselectivity of the process. However, polystyrene supported picolinamide proved to be the best catalytic system, affording the desired products in very good yields and excellent *ee* (97%).

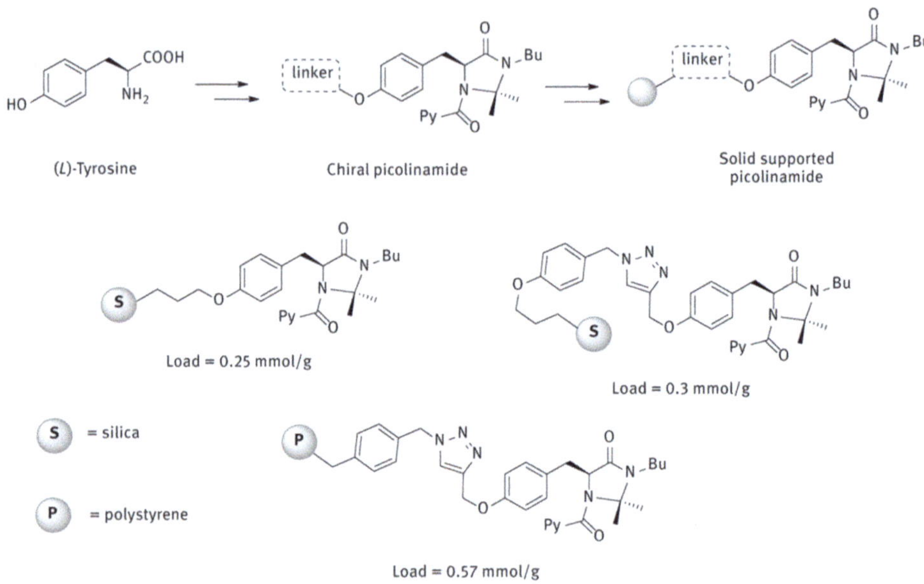

Figure 4.15: Synthesis of immobilized imidazolidinones-based chiral picolinamides for enantioselective reduction of imines.

On the basis of these preliminary results, the amount of the polymer-supported catalyst could be reduced to 1 mol% without any loss in the chemical or stereochemical activity. When the reaction time was reduced to 4 h, a loading of 5 mol% of immobilized catalyst gave the chiral amine in 82% yield and 94% *ee*. The methodology showed a wide reaction scope.

In order to demonstrate the practical applicability of the supported catalyst, the recycle was studied. When recycling experiments were conducted for 18 h for each cycle, amine was obtained as racemic mixture already at the second run. It was necessary to reduce reaction time to 2 h to maintain a high level of stereoselectivity. Under these conditions, it was possible to run four reaction cycles without any loss in the chemical or stereochemical activity of the process (ee up to 95%). In the fifth cycle, the product was obtained in very low yield and reduced *ee*. At the sixth cycle the catalysts was completely deactivated.

Polystyrene supported catalyst was then employed for the preparation of packed bed reactors for the continuous flow reduction of imines with $HSiCl_3$. A 0.05 M solution of imine and $HSiCl_3$ in CH_2Cl_2 was pumped into the reactor through a syringe pump at 0.4 mL/h (res time: 1 h) at room temperature. The outcome of the reactor was collected into a flask containing NaOH 10% solution. After phase separation and concentration the product was isolated. Every hour the product was collected and analyzed in order to determine the conversion and the *ee* as a function of time.

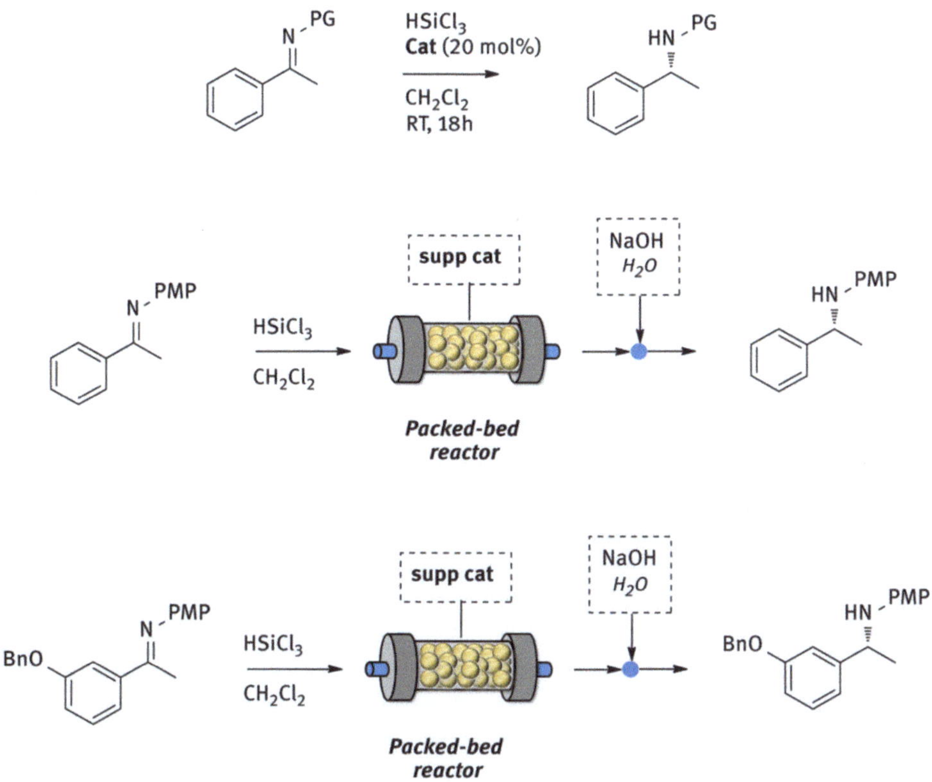

Figure 4.16: In batch and in flow enantioselective organocatalytic reductions of imines.

The flow system was stable for 7 h, producing the expected chiral amine in excellent yield and very good *ee* (up to 91% *ee*). The stereoselectivity of the process slightly decreased with time, probably because of a partial decomposition of the catalyst.

Finally the continuous flow synthesis of chiral precursor of Rivastigamine was performed. The continuous flow process, afforded the desired product in 82% yield and 83% *ee* for the first 2.5 h. Then the product was continuously produced for additional 3 h in 79% yield and 74% *ee*.

4.5 Conclusion and perspectives

Since their introduction in asymmetric catalysis, chiral imidazolidinones have shown to be one of the most successful class of organocatalysts. Their popularity is due to many positive features:

- they are highly tunable in different positions of the scaffold, thus modifying the behavior of the catalyst playing either with electronic or steric effects
- they are easy to synthesize by a few steps short reaction sequence
- they are reasonable cheap catalysts, since they are derived by readily available, enantiopure starting materials such alfa-aminoacids
- they are quite stable and can be stored as salts for long time
- in aminocatalysis they can be used to activate carbonyl derivatives both through iminium ion catalysis (preferred) and by enamines activation (less frequently)
- chiral imidazolidinones have found widespread use also in stereoselective organophoto-redox catalysis, opening the way to the development of unprecedented reactions
- imidazolidinone structure has been employed also as chiral platform to build novel metal-free catalysts to be used in fields other than aminocatalysis.

All these advantages well explain the great popularity of this family of organocatalysts, which have been also immobilized on different solid supports in order to develop a recyclable version of the chiral catalyst. The use of solid supported imidazolidinones in packed or monolithic reactors opened the avenue to the development of in-flow reactions, for the synthesis also of active pharmaceutical ingredients (APIs) or advanced intermediates [32].

Chiral imidazolidinones can be considered without doubt "privileged ligands" [33] and even more, novel applications in asymmetric catalysis can be envisaged in the future.

References

[1] Kasprzyk-Hordern B. Pharmacologically active compounds in the environment and their chirality. Chem Soc Rev. 2010;39:4466–503 and references therein.

[2] Our Common Future: Report of the World Commission on Environment and Development, United Nations (UN) Commission on Environment and Development (Brundtland Commission), 1987, Published as Annex to General Assembly document A/42/427, Development and International Cooperation: Environment, August 1987.

[3] Shaikh IR. Organocatalysis: key trends in green synthetic chemistry, challenges, scope towards heterogenization, and importance from research and industrial point of view. J Catal. 2014:2014. Article ID 402860. http://dx.doi.org/10.1155/2014/402860.

[4] Ahrendt KA, Borths CJ, MacMillan DW. New strategies for organic catalysis: the first highly enantioselective organocatalytic Diels–Alder reaction. J Am Chem Soc. 2000;122:4243–4.

[5] Nachtergael A, Coulembier O, Dubois P, Helvenstein M, Duez P, Blankert B, et al. Organocatalysis paradigm revisited: are metal-free catalysts really harmless? Biomacromolecules. 2015;16:507–14.

[6] MacMillan DW. The advent and development of organocatalysis. Nature. 2008;455:304–8.

[7] Howell GP. Asymmetric and diastereoselective conjugate addition reactions: C–C bond formation at large scale. Org Process Res Dev. 2012;16:1258–72.

[8] Abbasov ME, Romo D. The ever-expanding role of asymmetric covalent organocatalysis in scalable, natural product synthesis. Nat Prod Rep. 2014;31:1318–27.

[9] a) Denmark SE, Beutner GL Lewis base catalysis in organic synthesis. Angew Chem Int Ed, 2008, 47, 1560–638; b) Melchiorre P, Marigo M, Carlone A, Bartoli C. Asymmetric aminocatalysis – gold rush in organic chemistry. Angew Chem Int Ed. 2008;47:6138–71; c) Nielsen M, Worgull D, Zweifel T, Gschwend B, Bertelsen S, Jørgensen KA. Mechanisms in aminocatalysis. Chem M Commun. 2011:632–49.

[10] For a recent summary on recent progress in the field of hydrogen-bonding aminocatalysis XE "aminocatalysis" using proline-derived systems see Albrecht L, Jiang H, Jørgensen KA. Hydrogen-bonding in aminocatalysis: from proline and beyond. Chem Eur J. 2014;20:358–68.

[11] Lelais G, MacMillan DW. Modern strategies in organic catalysis: the advent and development of iminium activation. Aldrichimica Acta. 2006;39:79–87.

[12] Selected papers: a) Paras NA, MacMillan DW New strategies in organocatalysis: the first enantioselective organocatalytic Friedel-Crafts alkylation, J Am Chem Soc 2001, 123, 4370–1; b) Brochu MP, Brown AP, MacMillan DW. Direct and enantioselective organocatalytic α-chlorination of aldehydes. J Am Chem Soc. 2004;126:4108–409; c) Lee S, MacMillan DW. Organocatalytic vinyl and Friedel-Crafts alkylations with trifluoroborate salts. J Am Chem Soc. 2007;129:15438–39; d) Fonseca MT, List B. Catalytic asymmetric intramolecular Michael reaction of aldehydes. Angew Chem Int Ed. 2004;43:3958–60; e) Ouellet SG, Tuttle JB, MacMillan DW. Enantioselective organocatalytic hydride reduction. J Am Chem Soc. 2005;127:32–3.

[13] Lakhdar S, Tokuyasu T, Mayr H. Electrophilic reactivities of alpha,beta-unsaturated iminium ions. Angew Chem Int Ed. 2008;47:8723.

[14] See ref. 6 and also: Seebach D, Gilmour R, Groselj U, Deniau G, Sparr C, Mo E, et al. Stereochemical models for discussing additions to α,β-unsaturated aldehydes organocatalyzed by diarylprolinol or imidazolidinone derivatives – Is there an (E)/(Z)-dilemma? Helv Chim Acta. 2010;93:603–34.

[15] Cozzi PG, Benfatti F, Zoli L. Organocatalytic asymmetric alkylation of aldehydes by S(N)1-type reaction of alcohols. Angew Chem Int Ed. 2009;48:1313.

[16] a) Wakabayashi T, Mori K, Kobayashi S Total synthesis and structural elucidation of Khafrefungin. J Am Chem Soc 2001, 123, 1372; b) Fürstner A, Bonnekessel M, Blank JT, Radkowski K, Seidel G, Lacombe F, Gabor B, Mynott R. Total Synthesis of Myxovirescin A_1. Chem Eur J. 2007;13:8762.

[17] For a review on low-loading chiral organocatalysts see: Giacalone F, Gruttadauria M, Agrigento P, Noto R. Low loading asymmetric organocatalysis. Chem Soc Rev. 2012;41:2406–47.

[18] Brenna D, Porta R, Massolo E, Raimondi L, Benaglia M. A new class of low-loading catalysts for a highly enantioselective, metal-free imine reduction of wide general applicability. ChemCatChem. 2017;9:941–5.

[19] Review: Rossi S, Benaglia M, Massolo E, Raimondi L. Organocatalytic strategies for enantioselective metal-free reduction. Catal Sci Technol. 2014;9:2708–23.

[20] The definition of ACE was recently proposed in the attempt to compare and evaluate the efficiency of different catalysts, considering the level of enantioselectivity and the yield guaranteed by the catalyst, the molecular weight of the product and of the catalysts itself. See: El-Fayyoumy S, Todd MH, Richards CJ. Can we measure catalyst efficiency in asymmetric chemical reactions? A theoretical approach. Beilstein J Org Chem. 2009;5:67.

[21] Jones S, Warner CJ. Trichlorosilane mediated asymmetric reductions of the CN bond. Org Biomol Chem. 2012;10:2189–200.

[22] a) Benaglia M, Celentano G, Cinquini M, Puglisi A, Cozzi F Poly(ethylene glycol)-supported chiral imidazolidin-4-one: an efficient organic catalyst for the enantioselective diels-alder

cycloaddition. Adv Synth Catal 2002, 344, 149–52; b) Selkälä SA,Tois J, Pihko P-M, Koskinen AM. Asymmetric organocatalytic Diels-Alder reactions on solid support. Adv Synth Catal. 2002;344:941; c) Zhang Y, Zhao L, Lee SS, Ying JY. Enantioselective catalysis over chiral imidazolidin-4-one immobilized on siliceous and polymer-coated mesocellular foams. Adv Synth Catal. 2006;348:2027; d) Hagiwara H, Kuroda T, Hoshi T, Suzuki T. Immobilization of MacMillan imidazolidinone as Mac-SILC and its catalytic performance on sustainable enantioselective Diels–Alder cycloaddition. Adv Synth Catal. 2010;352:909; e) Shi JY, Wang CA, Li ZJ, Wang Q, Zhang Y, Wang W. Heterogeneous organocatalysis at work: functionalization of hollow periodic mesoporous organosilica spheres with MacMillan catalyst. Chem Eur J. 2011;17:6206; f) Guizzetti S, Benaglia M, Siegel JS. Poly(methylhydrosiloxane)-supported chiral imidazolinones: new versatile, highly efficient and recyclable organocatalysts for stereoselective Diels-Alder cycloaddition reactions. Chem Commun. 2012;48:3188–90; g) Riente P, Yadav J, Pericas MA. click strategy for the immobilization of Macmillan organocatalysts onto polymers and magnetic nanoparticles. Org Lett. 2012;14:3668.

[23] a) Trindade F, Gois PM, Afonso CA Recyclable stereoselective catalysts. Chem Rev 2009, 109, 418–514; b) Benaglia M. Recoverable and recyclable catalysts. Weinheim: Wiley-VCH, 2009; c) Jimeno C, Sayalero S, Pericàs MA. In: Barbaro, P, Liguori, F, editors. Catalysis by metal complexes, vol. 33: heterogenized homogeneous catalysis for fine chemicals production. Berlin: Springer, 2010:123–70.

[24] Ding KJ, Uozomi FJ. Handbook of asymmetric heterogeneous catalysts. Weinheim: Wiley-VCH, 2008.

[25] For a perspective on chiral supported metal-free catalysts see: Benaglia M. Recoverable and recyclable chiral organic catalysts. New J Chem. 2006;30:1525–33.

[26] Chiroli V, Benaglia M, Cozzi F, Puglisi A, Annunziata R, Celentano G. Continuous-flow stereoselective organocatalyzed Diels Alder reactions in a chiral catalytic "homemade" HPLC column. Org Lett. 2013;15:3590.

[27] Puglisi A, Benaglia M, Annunziata R, Chiroli V, Porta R, Gervasini A. Chiral hybrid inorganic-organic materials: synthesis, characterization and application in stereoselective organocatalytic cycloadditions. J Org Chem. 2013;78:11326–33.

[28] For recent reviews on stereoselective catalytic reactions in flow see: a) Tsubogo T, Ishiwata T, Kobayashi S Asymmetric carbon-carbon bond formation under continuous-flow conditions with chiral heterogeneous catalysts. Angew Chem Int Ed 2013, 52, 6590; b) Puglisi A, Benaglia M, Chiroli V. Stereoselective organic reactions promoted by immobilized chiral catalysts in continuous flow systems. Green Chem. 2013;15:1790; c) Atodiresei I, Vila C, Rueping M, Asymmetric organocatalysis in continuous flow: opportunities for impacting industrial catalysis. ACS Catal 2015;5:1972.

[29] Chiroli V, Benaglia M, Puglisi A, Porta R, Jumde RP, Mandoli A. A chiral organocatalytic polymer-based monolithic reactor. Green Chem. 2014;16:2798–806.

[30] Porta R, Benaglia M, Puglisi A, Mandoli A, Gualandi A, Cozzi PG. A catalytic reactor for organocatalyzed enantioselective continuous flow alkylation of aldehyde. ChemSusChem. 2014;7:3534–40.

[31] Porta R, Benaglia M, Annunziata R, Puglisi A, Celentano G. Solid supported chiral *N*-picolyl-imidazolidinones: recyclable catalysts for the enantioselective,metal- and H_2-free reduction of imines in batch and in flow mode. Adv Synth Catal. 2017;359:2375–82.

[32] Porta R, Benaglia M, Puglisi A. Flow chemistry: recent developments in the synthesis of pharmaceutical products. Org Process Res Dev. 2016;20:2–25.

[33] a) Zhou QL, editor. Privileged chiral ligands and catalysts, 2011, NJ, USA: Wiley; b) Yoon, TP, Jacobsen, EN. Privileged chiral catalysts. Science. 2003;299:1691.

Florenci V. González Adelantado

5 Phase-transfer catalysis and the ion pair concept

Abstract: This review outlines the recent advances in the field of asymmetric phase-transfer catalysis and the ion-pair concept including alkylation of amino acids and peptides, oxyindoles and other substrates, conjugate additions, fluorinations, photo-induced phase-transfer catalysis, Nitro-Mannich reactions, heterocyclizations and cycloadditions for the preparation of heterocycles, derivatization of isoxazoles, umpolung conjugate addition of imines and other three asymmetric reactions.

Keywords: phase-transfer catalysis, ion-pair concept, chiral ammonium salts

5.1 Introduction

Phase-transfer catalysis (PTC) was firstly introduced by Starks for displacement reactions of alkyl halides in a biphasic system using quaternary ammonium and phosphonium salts as catalysts [1]. General scheme of the process is given in Figure 5.1. As an example, a quaternary ammonium halide (Q^+X^-), dissolved in the aqueous or in the organic phase, in dependence of its lipophilicity, exchanges its anion at the interphase to form a new quaternary ammonium halide (Q^+Y^-) which can diffuse into the organic phase (phase-transfer step), where the reaction takes place and the original catalyst form Q^+X^- is restored.

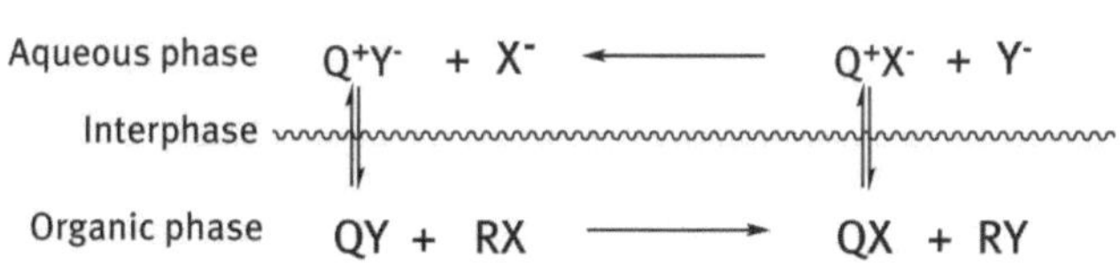

Figure 5.1:

PTC has become an important method in organic synthesis and has made an important impact on industry because of advantages associated to these catalysts such as low cost, metal-free and solvent amount reduction [2]. The efficiency of the phase-transfer catalysts (PTCs) has been also demonstrated operating in organic media without the use of an aqueous phase (ion-pair concept). Interestingly, chiral PTCs facilitate asymmetric processes. Lygo [3] and Corey [4] reported enantioselective

This article has previously been published in the journal Physical Sciences Reviews. Please cite as: Gonzalez Adelantado, F. V. Phase-transfer catalysis and the ion pair concept *Physical Sciences Reviews* [Online] 2020, 12. DOI: 10.1515/psr-2018-0094

https://doi.org/10.1515/9783110590050-005

alkylations by using quaternary ammonium salts derived from cinchona alkaloids (Figure 5.2) as chiral PTCs. These catalysts are easily prepared via N-alkylation of pseudoenantiomeric cinchonine/cinchonidine or quinine/quinidine alkaloids.

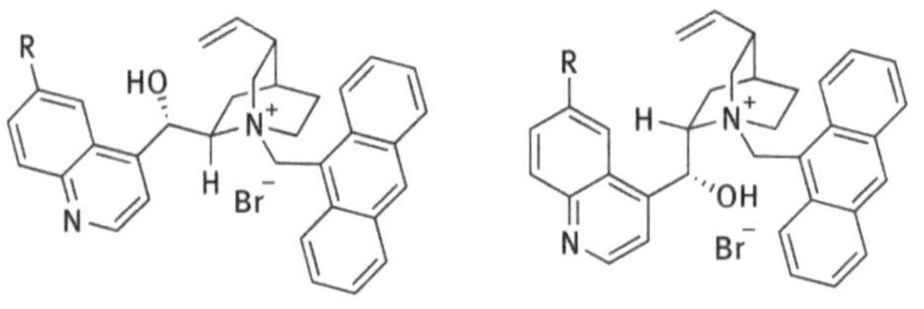

R = H, cinchoninium
R = OMe, quinidinum

R = H, cinchonidinium
R = OMe, quininium

Figure 5.2:

Maruoka reported the preparation of C2-symmetric chiral PTCs (Figure 5.3) and their use for the enantioselective synthesis of α-amino acids [5]. A number of asymmetric processes with these ammonium quaternary salts have been reported [6].

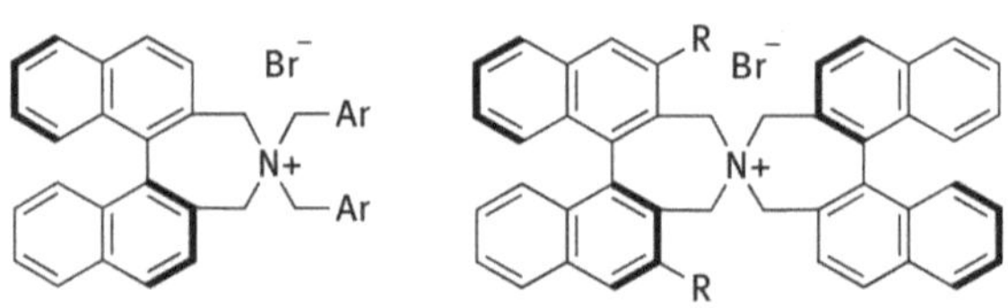

Ar = Ph, α-Np

R = H, Ph, α-Np

Figure 5.3:

This paper focuses on recent (2017–2018) progresses made in new PTC chemical transformations, the use of new PTCs and of novel methods, such as the combination of PTCs with light.

5.2 Alkylation

5.2.1 Preparation of amino acids and peptides

α-Methyl-p-boronophenylalanine, used in boron neutron capture therapy (BNCT), was prepared through a Maruoka catalyst promoted enantioselective alkylation of racemic N-protected alanine tert-butyl ester with a Maruoka catalyst (Figure 5.4) [7].

Highly enantioselective alkylation of a glycinate ester was accomplished using a benzophenone-bridged dimeric cinchonium salts as catalysts. A dual function of these dimeric catalysts was proposed for their high efficiency (Figure 5.5) [8].

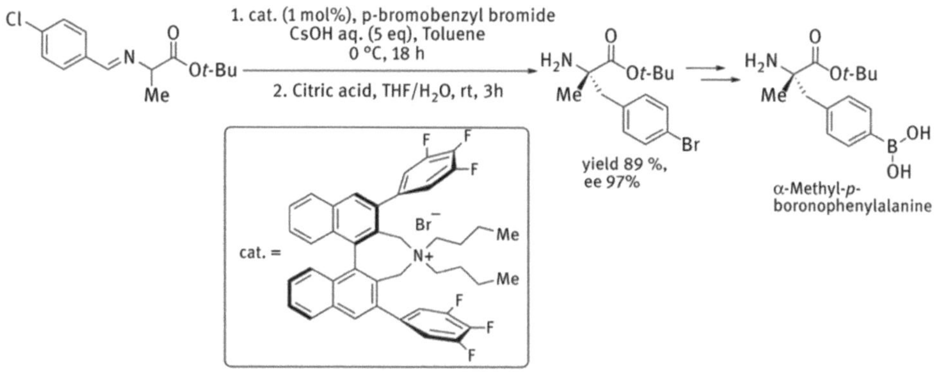

Figure 5.4:

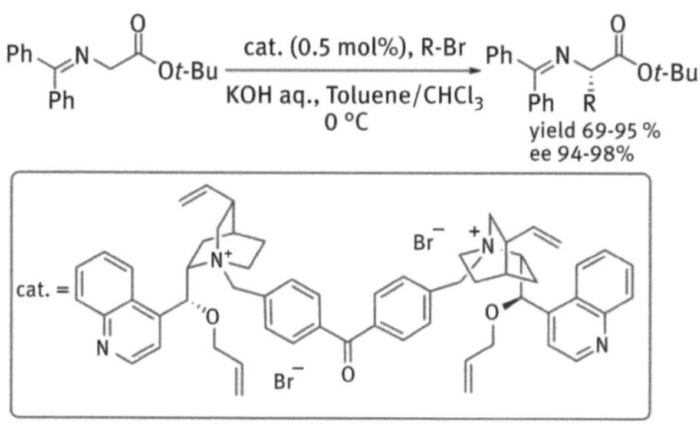

Figure 5.5:

Radiolabeled peptides have become indispensable tools for the *in vivo* localization of tumors and positron emission tomography (PET). Two examples about the preparation of 11C-labeled peptides through alkylation of peptide Schiff base under PTC conditions have been recently reported by the same group: a cinchoninium salt-afforded high diastereomeric excess and radiochemical conversion for the preparation of a 11C-labeled tetrapeptide [9] (Figure 5.6a), whereas a Maruoka catalyst was used for the synthesis of ^{11}C-labeled Ala-Leu or Phe-Leu (Figure 5.6b) [10].

Both R and S enantiomers of unnatural phenylalanine derivatives were prepared through asymmetric alkylation of glycine Schiff base with substituted benzyl bromides, by using the pseudoenantiomeric O-allyl-N-(9-anthracenylmethyl) cinchoninium bromide (cat. A) and O-allyl-N-(9-anthracenylmethyl) cinchonidinium bromide (cat. B), as catalysts (Figure 5.7) [11].

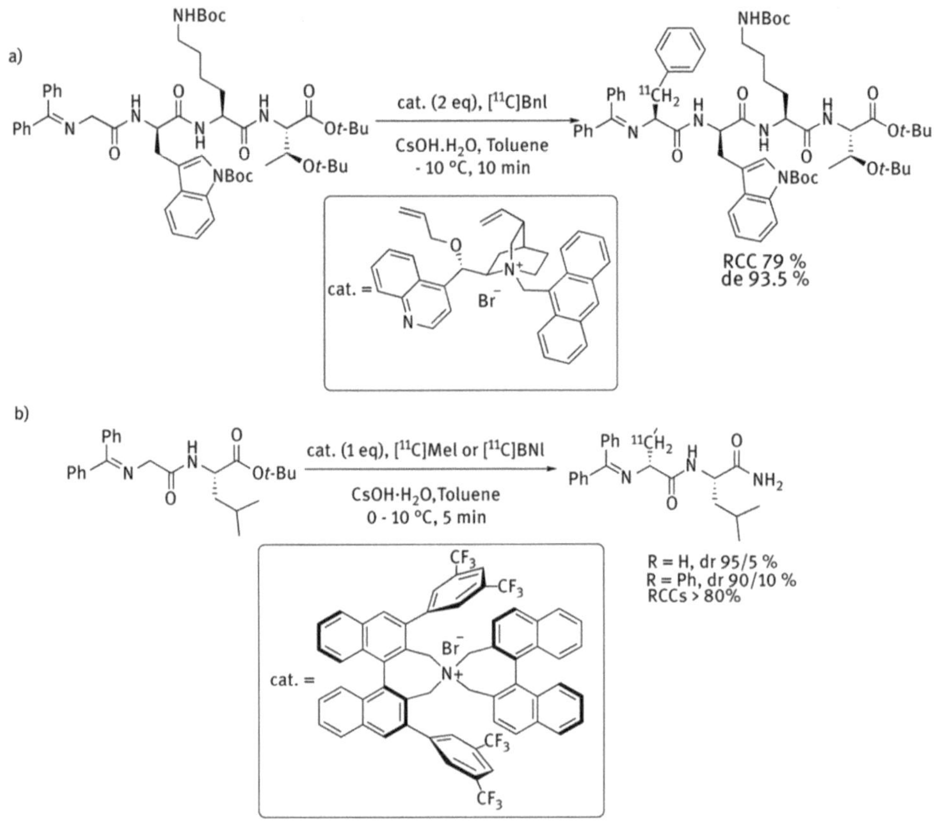

Figure 5.6:

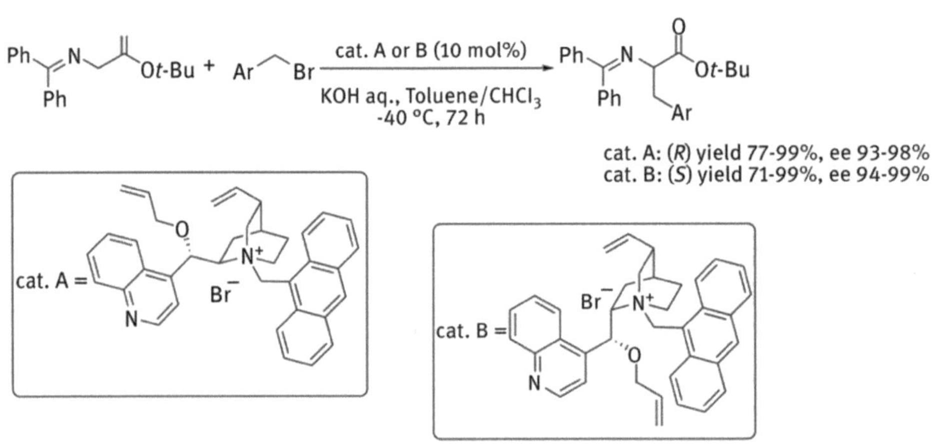

Figure 5.7:

5.2.2 Alkylation of oxindoles

A novel chiral spirocyclic amide-derived triazolium catalyst was used for asymmetric homo- and heterodialkylations of various bisoxindoles. As an application of this methodology, the first asymmetric total synthesis of (−)-chimonanthidine was achieved (Figure 5.8) [12].

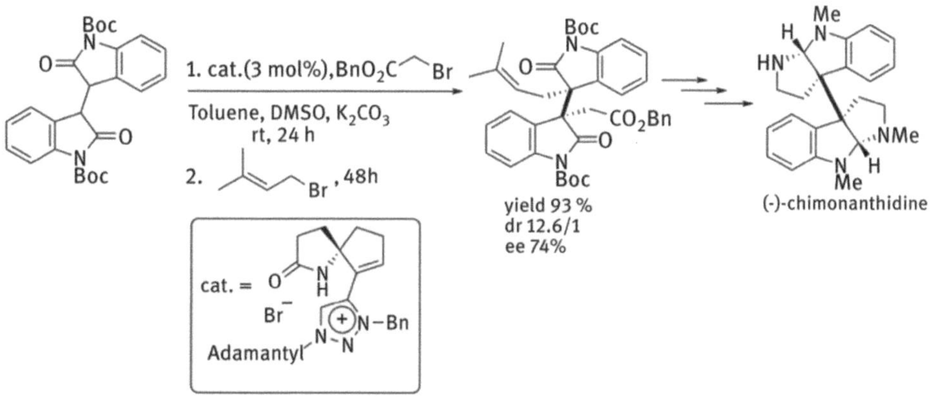

Figure 5.8:

A highly enantioselective SN2 alkylation of a malleable 3-substituted oxindole was recently reported using a new bifunctional PTC displaying a urea moiety as a hydrogen bonding element. The utility of the methodology was demonstrated by the synthesis of the natural product (-)-debromoflustramine B. (Figure 5.9) [13].

3-Substituted oxindoles have also been alkylated in high diastereo and enantioselective fashion using chiral triazolium ions as PTCs (Figure 5.10) [14].

5.2.3 Other alkylations

An enantioselective synthesis of (*R*)-(+)-1-(5-bromopentyl)-1-methyl-7-methoxy-2-tetralone, a key intermediate of opioid analgesic dezocine, was accomplished through PT alkylation using a p-trifluorobenzyl cinchonidinium catalyst (Figure 5.11) [15].

An improved enantioselective synthesis of potent anti-inflammatory agent PH46A has been reported through an elimination-alkylation process of the corresponding indane under PTC conditions (Figure 5.12) [16].

Maruoka et al. have recently reported the asymmetric synthesis of bioactive 2,2-disubstituted 1,4-benzoxazin-3-ones by alkylation of 2-aryl substituted 1,4-benzoxazin-3-ones with benzylic, allylic and propargylic bromides under PTC (Figure 5.13) [17].

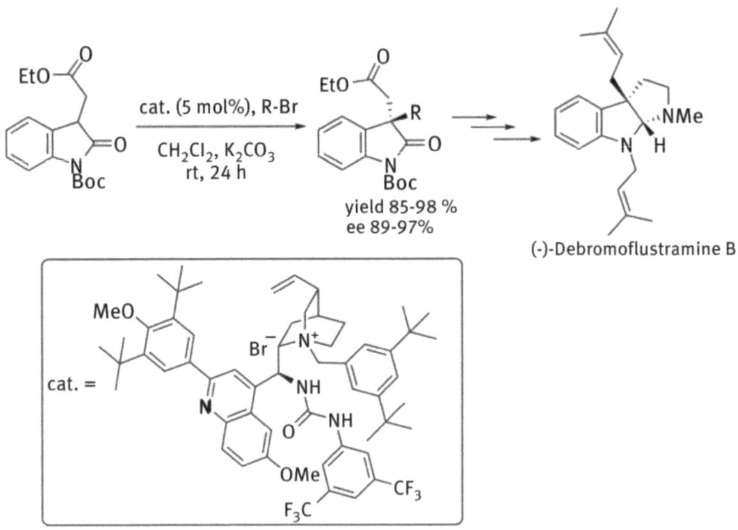

Figure 5.9:

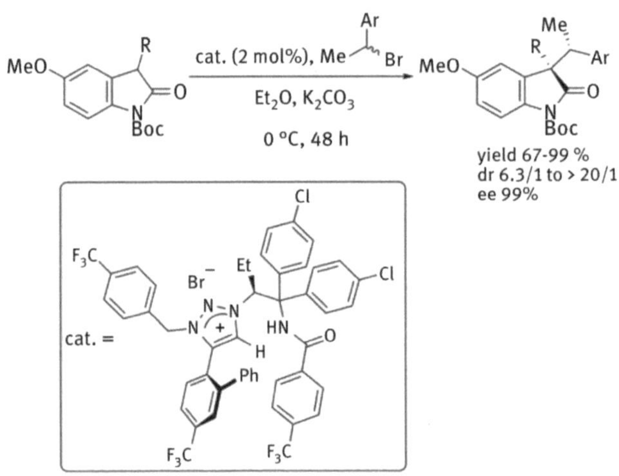

Figure 5.10:

Maruoka et al. have also described the asymmetric alkylation of N-arylhydrazones derived from α-keto-esters and isatin derivatives using PTC. The chemical transformation produces enantioenriched azo compounds bearing a tetra-substituted carbon stereocenter. The resulting compounds can be further converted into amino esters, hydrazines and aza-β-lactams, without loss of enantiopurity (Figure 5.14) [18].

Enantioselective alkylation of diphenylmethyl *tert*-butyl-α-bromomalonate under PTC gave the corresponding α-bromo-α-alkylmalonates. The resulting compounds

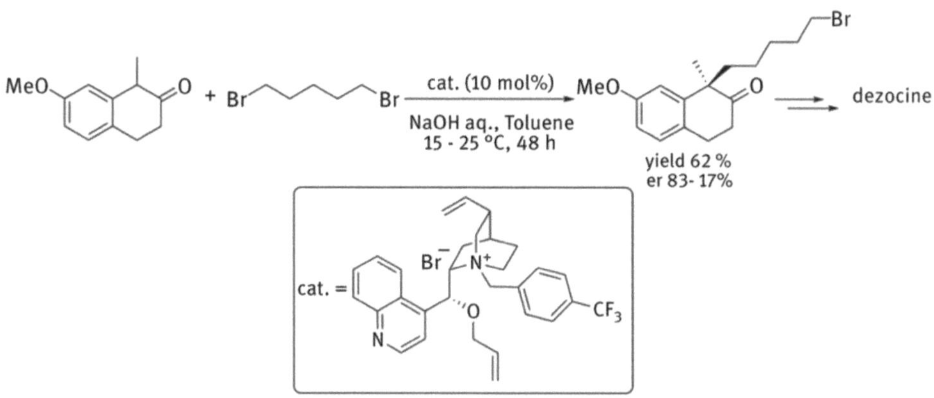

Figure 5.11:

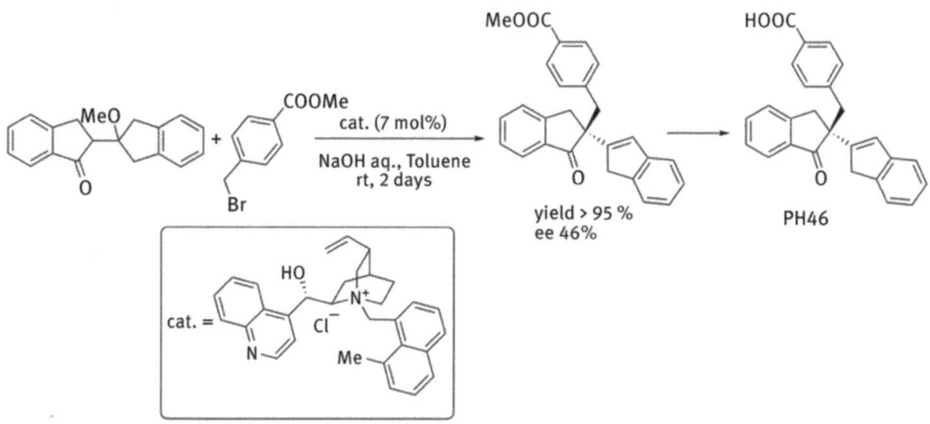

Figure 5.12:

can be further derivatized into α-azido-α-alkylmalonates and α-aryloxy-α-alkylmalonates in high enantiomeric excess by S_N2 substitution with sodium azide and aryloxides (Figure 5.15) [19]. A similar chemical transformation has been reported by same authors but starting from diphenylmethyl-*tert*-butyl α-thioacetyl malonate which provides a method to prepare chiral molecules containing α-sulfur quaternary stereogenic centers (Figure 5.15) [20].

Xiang et al. reported Cinchona alkaloid derivatives quaternized at both quinuclidine nitrogen and quinoline nitrogen as efficient catalysts for intramolecular alkylations to afford spirocycles (Figure 5.16) [21]. Recently, Houk et al. have proposed a model, using DFT calculations, that explains how doubly quaternized cinchona alkaloid derivatives function as PTCs. Like the Dolling and Pliego models, the interaction of the C9-hydroxyl group with the enolate oxygen is important. The cinchona alkaloid

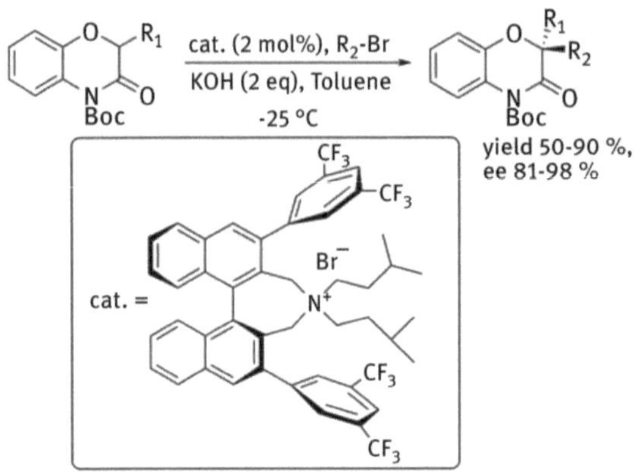

Figure 5.13:

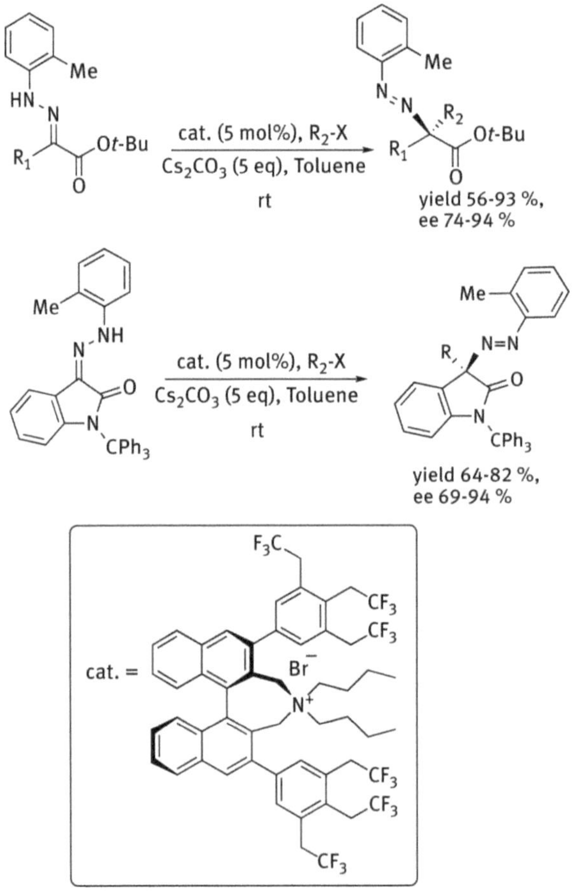

Figure 5.14:

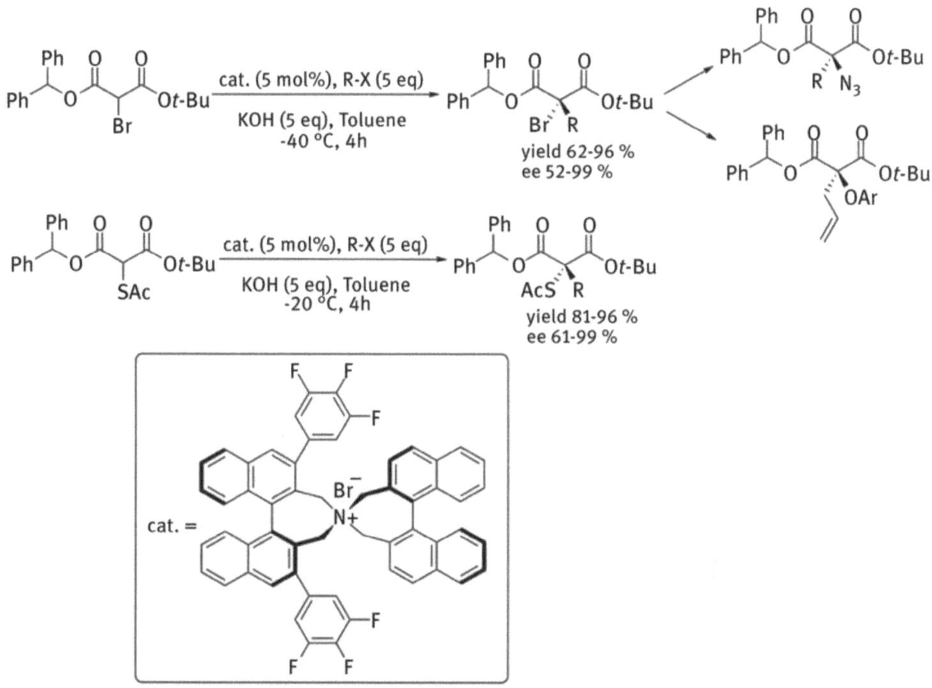

Figure 5.15:

stabilizes the transition structure by electrostatic interactions with the leaving chloride ion, and two more interactions are required for enantioselectivity: a chloride–CH interaction for the leaving group, and a π–π stacking interaction between the pyridine of the substrate and the chiral catalyst (Figure 5.16) [22].

5.3 Conjugate additions

Conjugate addition of a cyanide ion to β-trifluoromethyl-β-disubstituted alkylidenemalonates was reported under PTC and acetone cyanohydrin as a cyanide source. Among all assayed catalysts, dihydroquininium anthracenylmethyl chloride gave the best results (Figure 5.17) [23].

Vinylogous Michael addition of γ-lactams to enones has been recently reported by Maruoka et al. Regiocontrol was a challenge associated with this reaction since α-adducts are usually more favored than desired γ-adducts. Authors circumvented this issue by choice of a proper catalyst and Michael acceptor (Figure 5.18) [24].

Asymmetric Michael reactions between 3-carboxylate substituted isoindolinones and enones to access chiral derivatives containing a tetrasubstituted carbon stereocenter were performed using cinchona ammonium salts as catalysts. The highest

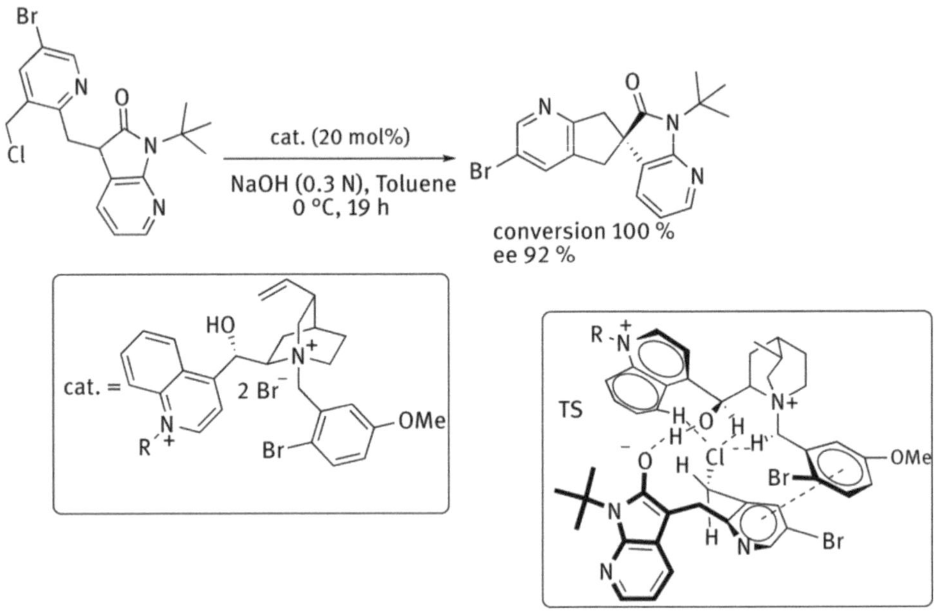

Figure 5.16:

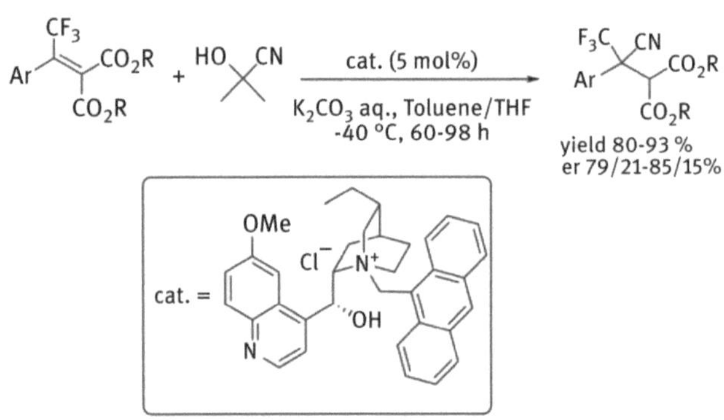

Figure 5.17:

enantioselectivity was obtained with a cinchonidinium salt bearing a benzotriazole group (Figure 5.19) [25].

Zhao et al. reported enantioselective conjugate additions of 3-substituted oxindoles to β-haloalkene ketones/esters using a quaternary ammonium PTC containing an amide as a hydrogen-bonding group (Figure 5.20) [26]. These additions afford chiral oxindoles derivatives with a quaternary stereogenic carboncenter in high yields, enantioselectivity and *E*/*Z* selectivity.

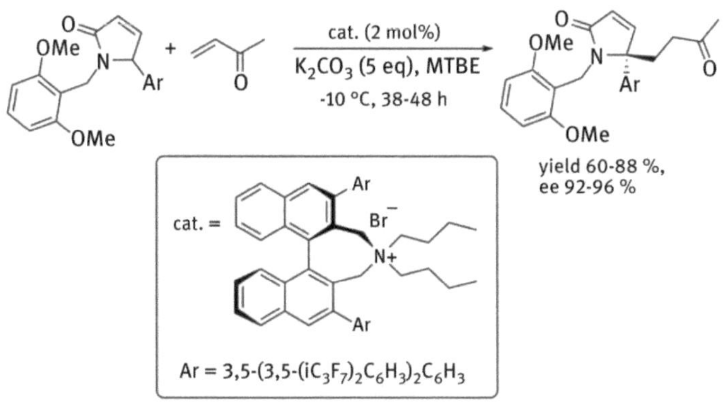

Figure 5.18:

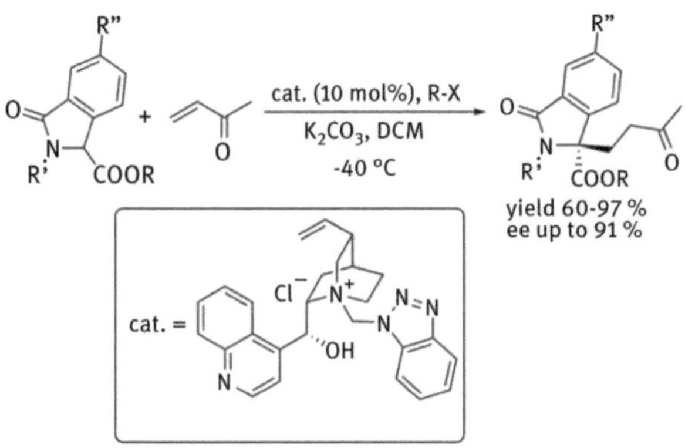

Figure 5.19:

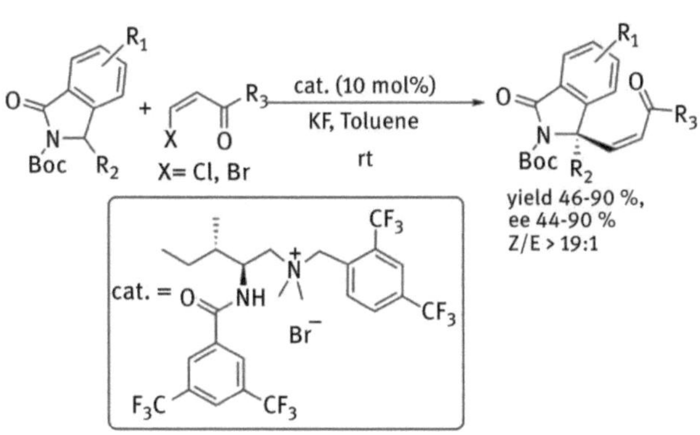

Figure 5.20:

Cyclic β-keto esters react with *O*-derivatized Morita-Baylis-Hillman adducts through an enantioselective Michael reaction using Jorgensen PTC. Resulting compounds display an acrylate moiety with a quaternary stereocenter (Figure 5.21) [27].

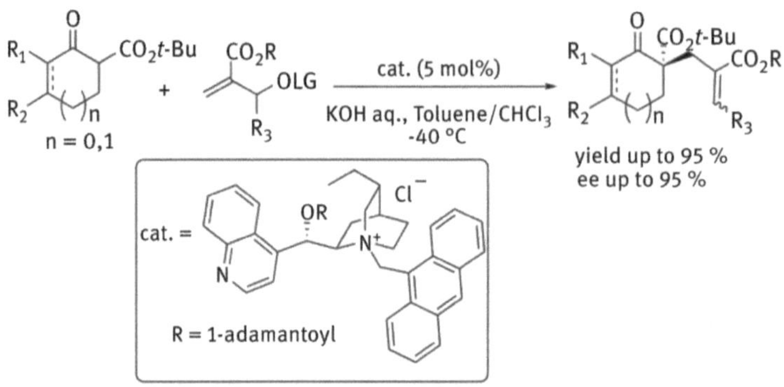

Figure 5.21:

Kobayashi et al. have performed asymmetric Michael additions of glycine-derived imines to enones using a novel chiral heterogeneous catalyst in which 2-oxopyrimidinium was immobilized in a polystyrene-based polymer (Figure 5.22). Interestingly, the catalyst can be recovered and reused without significant reduction of activity [28].

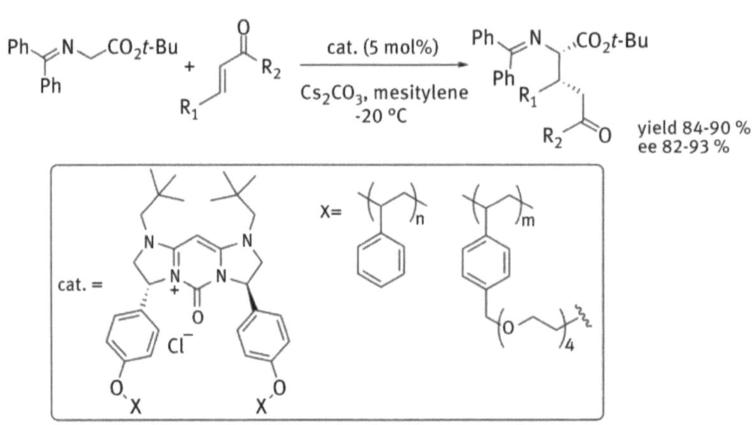

Figure 5.22:

An asymmetric 1,6-conjugate addition of *para*-quinone methides, generated *in situ* from 4-hydroxybenzyl *p*-tolyl sulfones with tritylthiol, was developed by Li et al. affording a range of optically active α-substituted benzyl thioethers (Figure 5.23) [29].

A new bifunctional catalyst derived from squaramide was found to be effective for the asymmetric amination of nitroalkenes (Figure 5.24). Noteworthy, the reactions

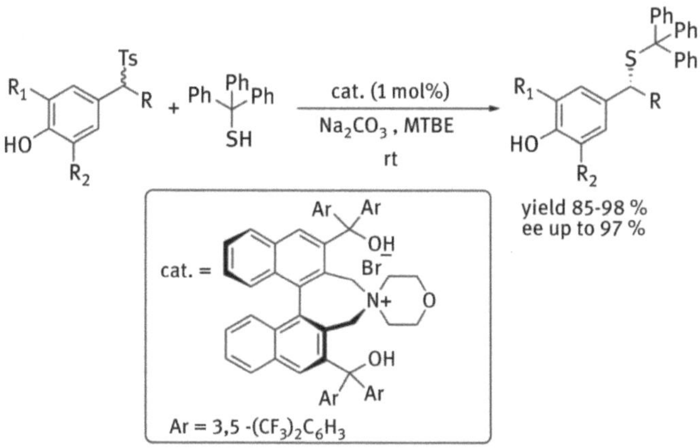

Figure 5.23:

have been performed by using these chiral bifunctional PTCs under base-free and water-rich conditions [30].

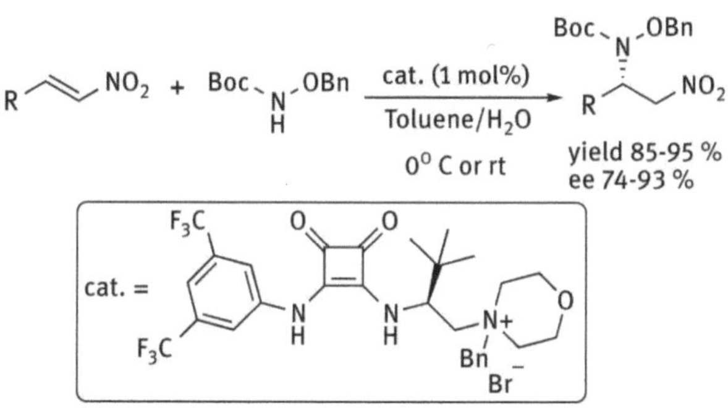

Figure 5.24:

5.4 Fluorinations

Although catalytic asymmetric dearomatization (CADA) reactions through fluorination are very interesting a few works were reported. An asymmetric fluorinative dearomatization reaction of tryptamine derivatives was developed by using a BINOL-derived phosphate anion as a PTC and Selectfluor as F^+ donor (Figure 5.25) [31].

A very similar process but starting from tryptophol was reported by You et al. Evidence for tryptophol boronic ester as the key intermediate was given (Figure 5.26) [32].

An enantioselective synthesis of γ-fluoroalkenols starting from the corresponding homoallylic alcohols using chiral anion PTC has been recently described by Toste et al.

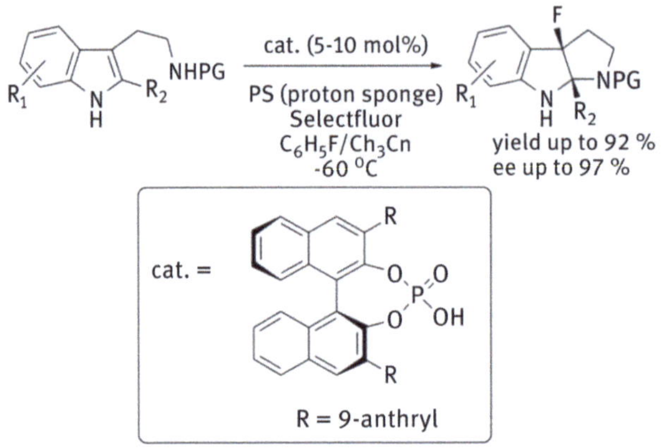

Figure 5.25:

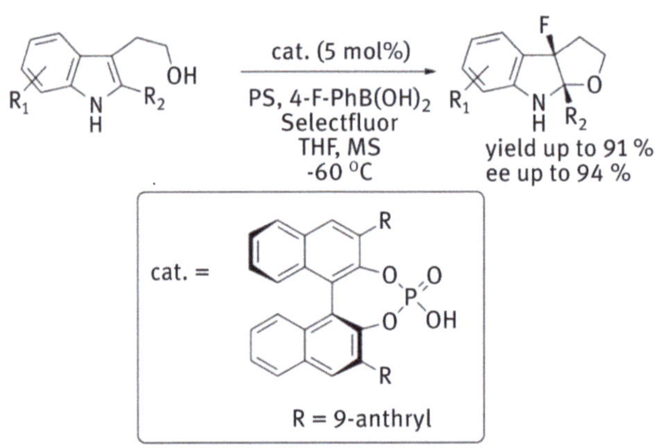

Figure 5.26:

Enantioselectivities were high, but chemical yields were modest. Multivariate correlation analysis were done to explain key structural features affecting the selectivity of the process (Figure 5.27) [33].

Hamashima et al. have reported the fluorination of tetrasubstituted alkenes with an amide at the homoallylic position.

Unlike previous work of the same authors with disubstituted alkenes, in the presence of a dianionic catalyst and Selectfluor only deprotonative fluorination takes place, without any 6-endo fluorocyclization traces (Figure 5.28) [34].

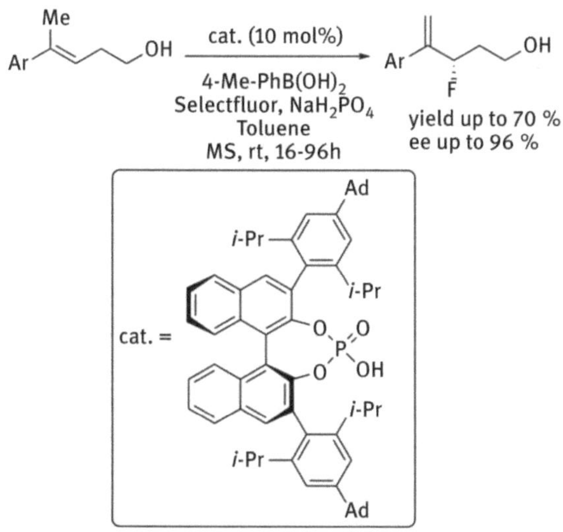

Figure 5.27:

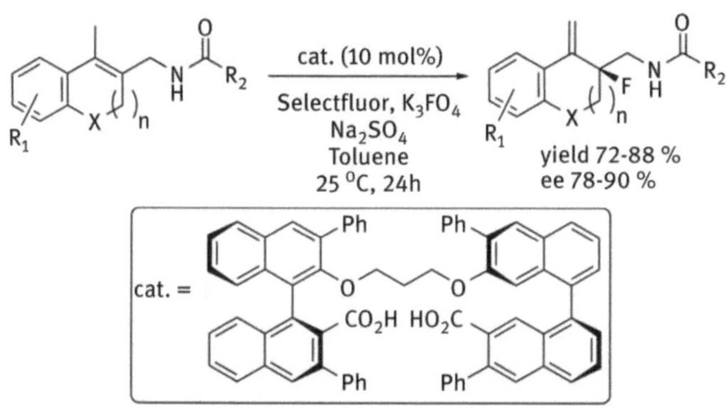

Figure 5.28:

5.5 Photoinduced PTC

During the last decade, the combination of organocatalysis in general, and PTC in particular, with light afforded new synthetic tools to stereoselective synthesis [35]. Interesting examples using β-dicarbonyl compounds have been recently reported on this field.

Asymmetric photoinduced PT catalytic perfluoroalkylations of β-ketoesters have been reported by Melciorre et al. [36]. A theoretical study of these reactions was reported in 2017. The reaction proceeds through the initial formation of a free radical via a chiral EDA complex, followed by combination with the β-ketoester enolate. TD-DFT calculations indicate that asymmetric induction is controlled by hydrogen bonding rather than π–π stacking interactions (Figure 5.29) [37].

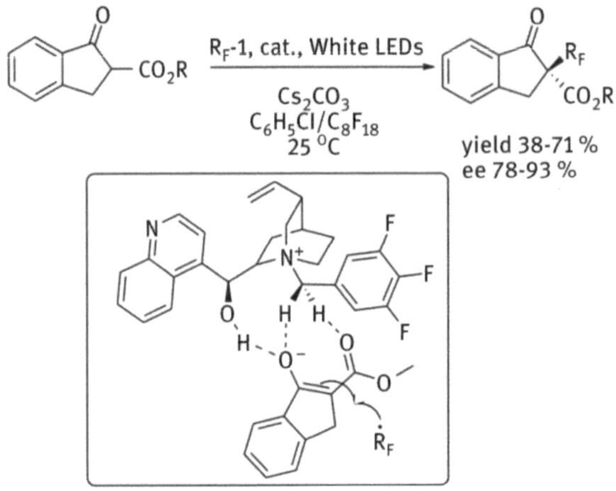

Figure 5.29:

β-Ketoesters and β-ketoamides have been also enantioselectively hydroxylated in a photo-organocatalytic process using a cinchona-derived N-oxide asymmetric PTC and tetraphenylporphyrin (TPP) as a photosensitizer (Figure 5.30) [38].

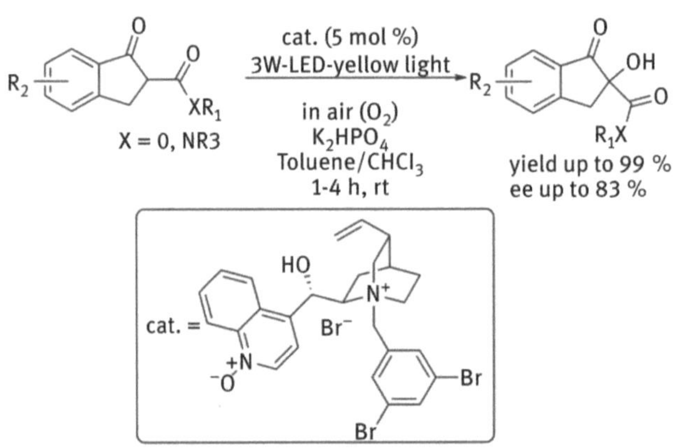

Figure 5.30:

The above chemical transformation has been reported later by the same authors, using a PTC linked to a photosensitizer TPP unit (Figure 5.31) [39].

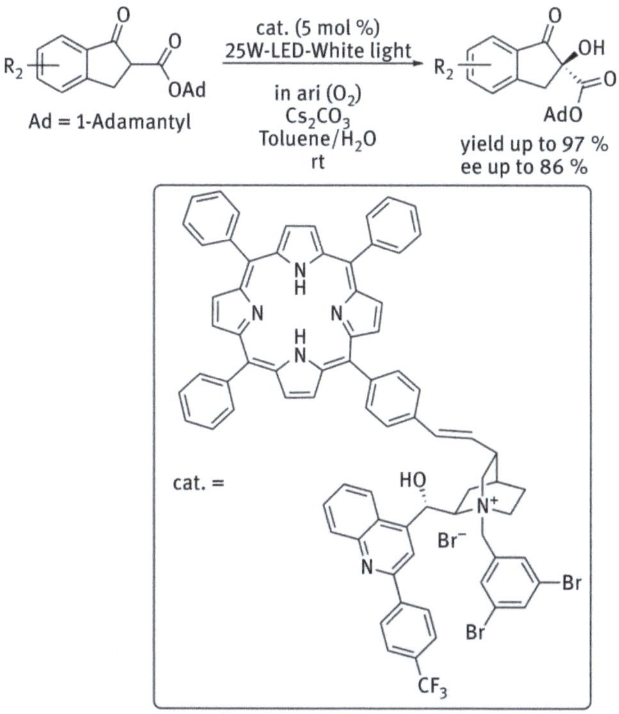

Figure 5.31:

5.6 Nitro-Mannich reactions

Lin et al. reported the use of new bifunctional PTCs bearing multiple H-bonding donors in asymmetric nitro-Mannich over the last few years. For example, they have studied the nitro-Mannich reaction of isatin-derived N-Boc ketimines (Figure 5.32) [40] and of amidosulfones (Figure 5.32) [41] using a quaternary ammonium Cinchona derivative containing a urea moiety and an alcohol group. Authors demonstrated the importance for the catalytic process of hydrogen-bonding by the free hydroxyl group: when this group, loss of enantioselectivity was observed.

These authors have also reported new bifunctional asymmetric PTCs, bearing multiple hydrogen-bonding donors derived from α-amino acids. These catalysts have been applied to asymmetric nitro-Mannich reactions of amidosulfones (Figure 5.33) [42].

They have also reported the nitro-Mannich reaction of ketimines *N*-activated with 6-methyl-2-pyridylsulfonyl group, catalyzed by a quaternary ammonium salt derived from quinine, bearing a thiourea group. DFT calculations support the bifunctional catalytic mechanism (Figure 5.34) [43].

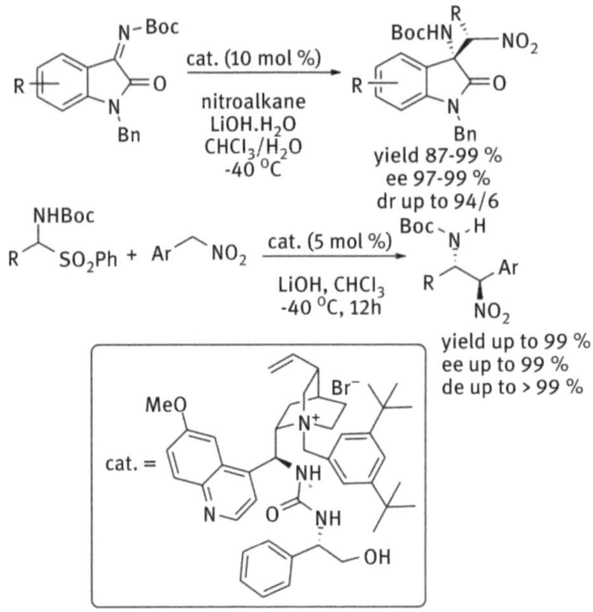

Figure 5.32:

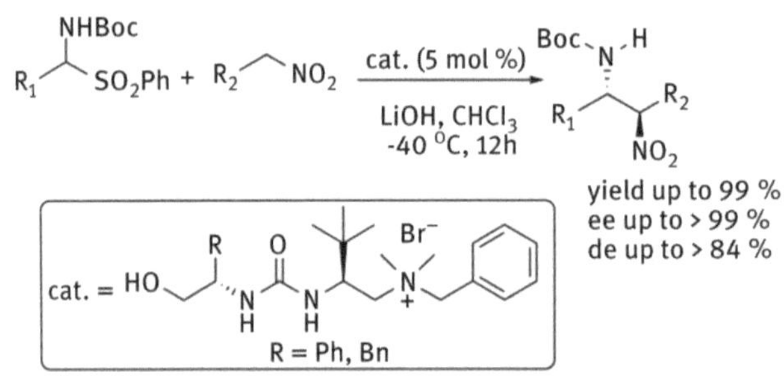

Figure 5.33:

5.7 Preparation of heterocycles

5.7.1 Heterocyclizations

Tantillo and Smith have reported an enantioselective synthesis of indolines with two asymmetric stereocenters one of which is all-carbon and quaternary. A quinine-derived ammonium salt with unprotected hydroxyl group was used as a catalyst. If the hydroxyl group of the catalyst is protected then reaction is slow and nonselective, denoting that the reaction is facilitated by a protonation event. Interestingly,

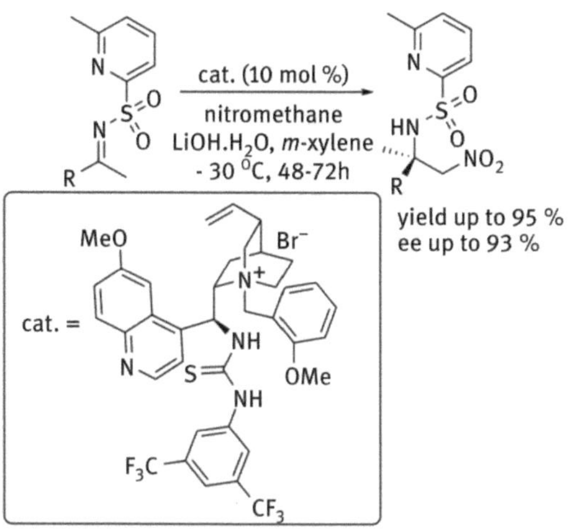

Figure 5.34:

authors suggest that the reaction may be better described as a phase-transfer initiated rather than a PT catalyzed process (Figure 5.35) [44].

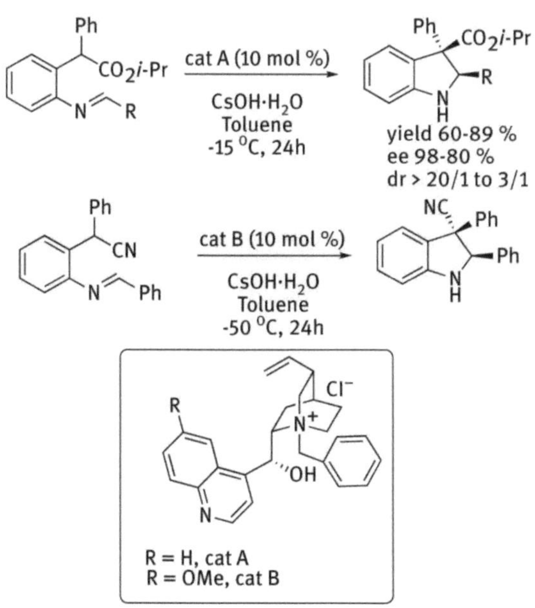

Figure 5.35:

A related study for the preparation of aza-indolines, but using *N*-2-naphthylmethyl cinchonidinium bromide as a catalyst has been recently carried out by Smith et al. (Figure 5.36) [45].

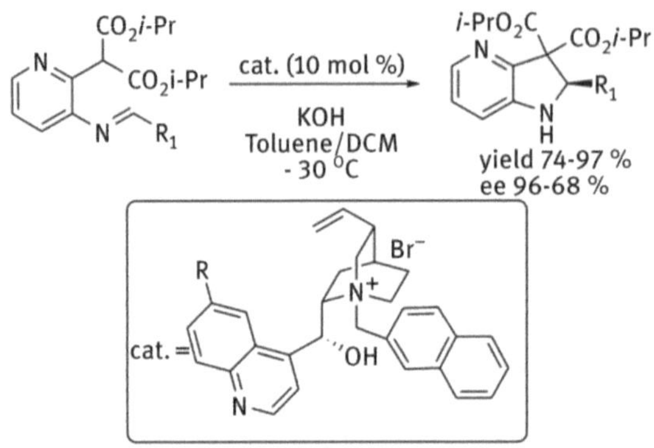

Figure 5.36:

An asymmetric synthesis of isoindolinones was realized through an intramolecular aza-Michael reaction by using both a cinchoninium salt as a catalyst and a chiral auxiliary (Figure 5.37) [46]. As an application of this methodology, an analog of anxiolytic drug pazinaclone was prepared.

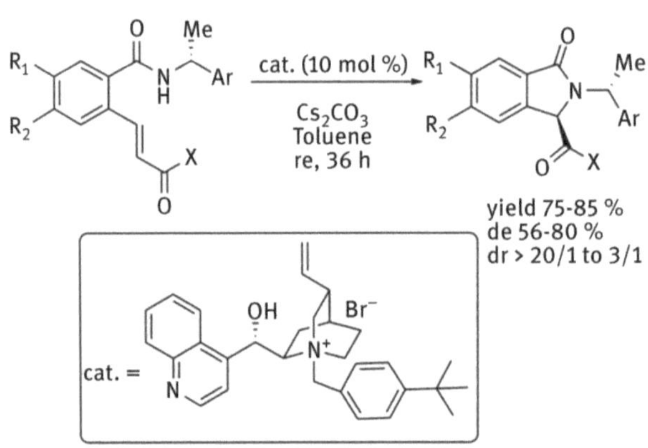

Figure 5.37:

Benzoxazinones were prepared through an asymmetric esterification/aza-Michael cascade between 2-hydroxyphenyl-substituted enones and isocyanates with a bifunctional bisguanidinium salt (Figure 5.38) [47].

Carboxy-substituted 2-isoxazolines have been prepared from β-carboxy-substituted α,β-unsaturated ketones through an enantioselective cascade oxa-Michael-cyclization

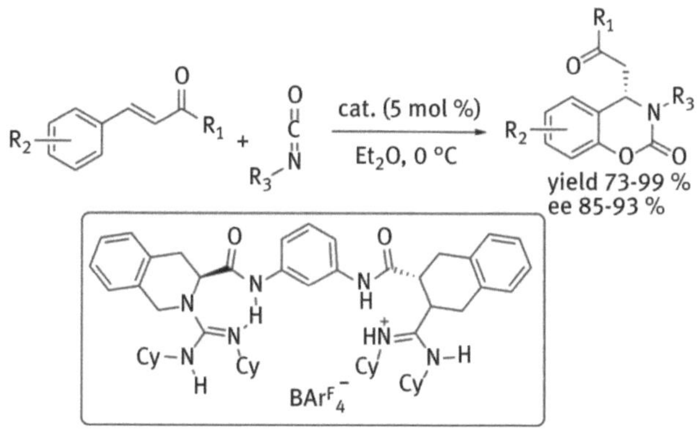

Figure 5.38:

reaction using hydroxylamine and a chiral quininium salt as a catalyst. Herbicide (S)-methiozolin was prepared as an application of this methodology (Figure 5.39) [48].

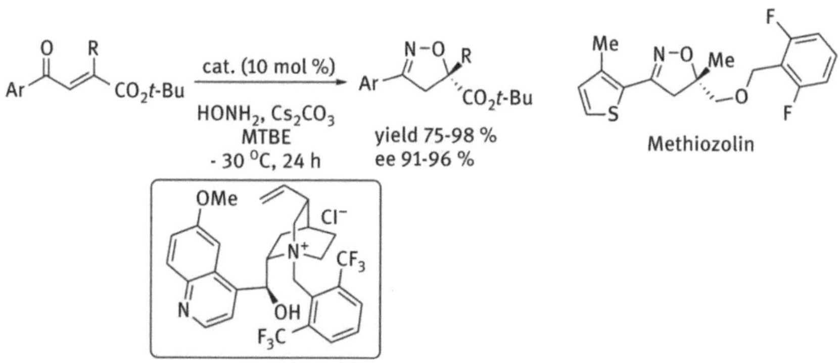

Figure 5.39:

Asymmetric synthesis of uncommon thio-isoindolinimine heterocycles was reported. This transformation was performed by cascade reaction of thiols and 2-cyano-N-tosyl-benzylidenimine in the presence of a PTC or an organocatalyst. NMR experiments determined that, after addition of thiol to the imine, a dynamic kinetic resolution (DKR) takes place assisted by the catalyst (Figure 5.40) [49].

5.7.2 Cycloadditions

An asymmetric synthesis of dihydrobenzofurans through a catalyzed [4 + 1] cycloaddition reaction of ortho-quinone methides and bromomalonates was reported

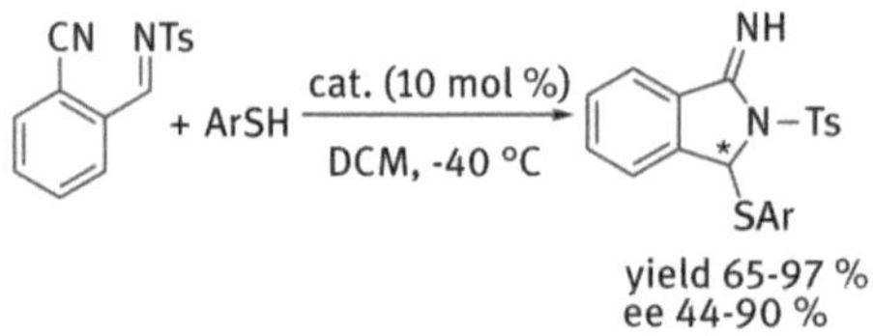

Figure 5.40:

(Figure 5.41) [50]. The best catalyst was a previously reported quinine and (*R*)-BINOL-derived chiral ammonium salt [51].

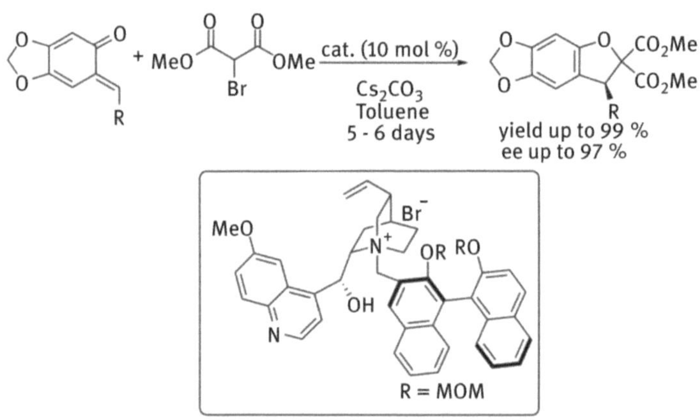

Figure 5.41:

An asymmetric formal [3 + 2] cycloaddition of *N,N*-cyclic azomethine imines with azalactones was developed using a new bisguanidinium hemisalt as a catalyst. The reaction affords tetrahydropyrazolo[1,2-a]pyrazole-1,7-dione derivatives with vicinal aza-quaternary and tertiary carbon centers that represent scaffolds with potential bioactivity (Figure 5.42) [52].

5.8 Derivatizations of isoxazoles

Jiang et al. have reported two works about chemical derivatizations of 5-alkyl-4-nitro isoxazoles using a novel chiral dipeptide based urea-amide-guanidinium as a PTC and sodium acetate as a base. An enantioselective vinylogous amination was realized when azodicarboxylates were employed (Figure 5.43) [53]. DFT calculations demonstrated that HOMO–LUMO gap is reduced by activation of nucleophile by catalyst with acetate anion through non-covalent interactions. Also an asymmetric aldol

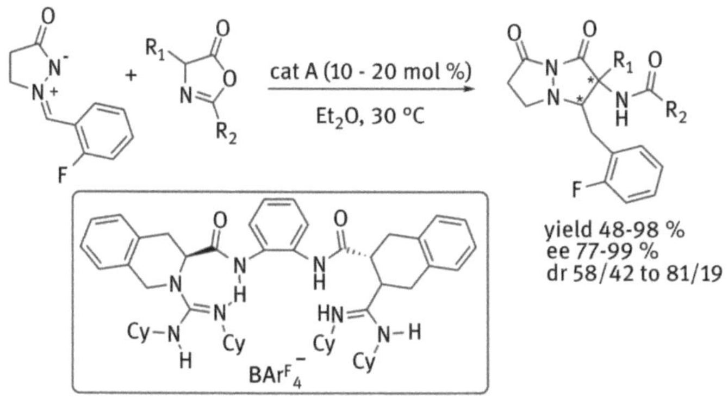

Figure 5.42:

reaction of 5-alkyl-4-nitroisoxazoles with paraformaldehyde was studied under similar conditions (Figure 5.43) [54].

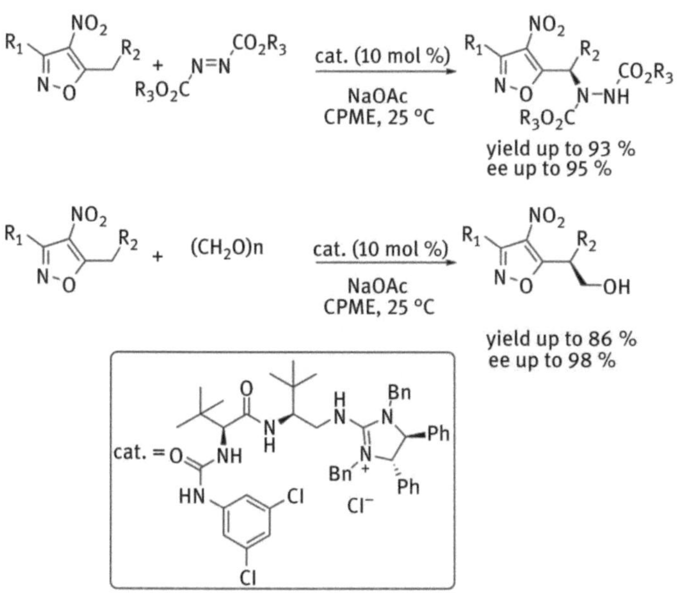

Figure 5.43:

5.9 Umpolung conjugate additions of imines

Deng et al. reported enantioselective umpolung additions of imines using PTCs. They also reported the preparation of chiral amines with non-adjacent stereocenters through an asymmetric tandem conjugate addition–protonation reaction between trifluoromethyl imines and α-alkyl acroleins. A phenolic proton donor catalyst, in addition to a cinchonium-derived PTC, was used to favor the desired asymmetric tandem conjugate addition–protonation pathway over a number of side-reactions (Figure 5.44) [55].

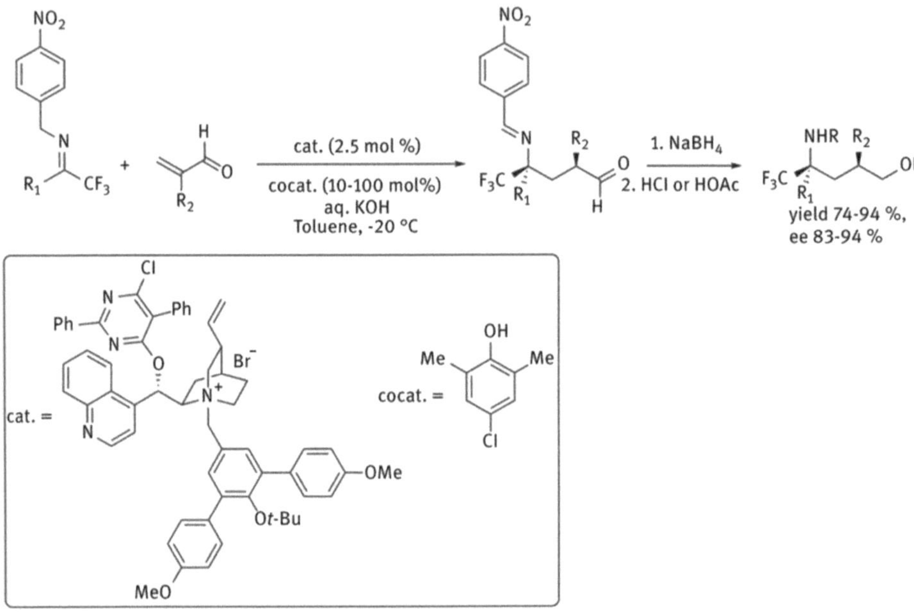

Figure 5.44:

Hu and Deng reported an enantioselective umpolung addition of trifluoromethyl imines to α,β-unsaturated N-acyl pyrroles, under similar conditions, but using different catalysts (Figure 5.45) [56].

A related study by Yoshida et al. showed that α-imino esters react with acrolein to produce amino acid derivatives with a tetrasubstituted carbon center which are useful as pharmaceuticals and chiral building blocks (Figure 5.46) [57].

5.10 Other reactions

An enantioselective decarboxylative protonation process of Meldrum's acid derivatives, using Lygo's catalyst, affords medically interesting 2-aryl propionic ester derivatives. The mechanism of the chemical process proceeds through four steps: addition of phenoxide followed by fragmentation to give a carboxylate anion, which suffers decarboxylation to give an enolate that affords the final chiral product by enantioselective protonation (Figure 5.47) [58].

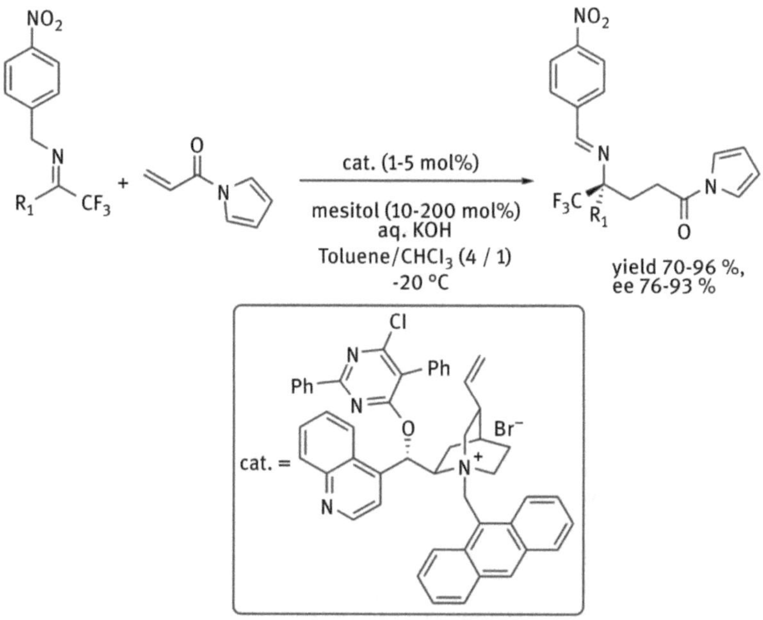

Figure 5.45:

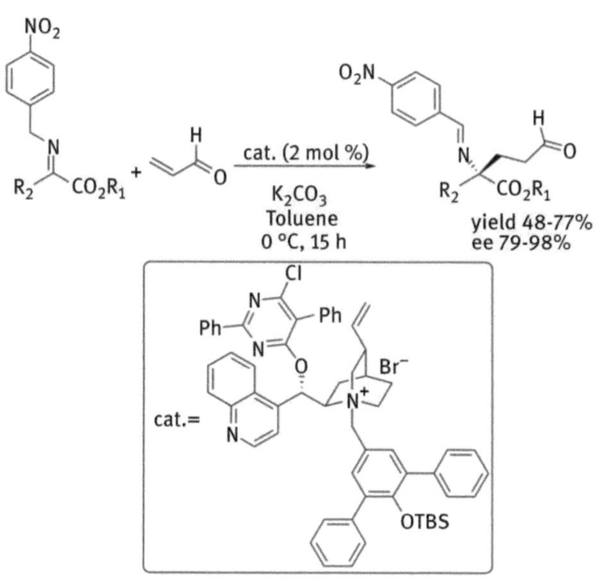

Figure 5.46:

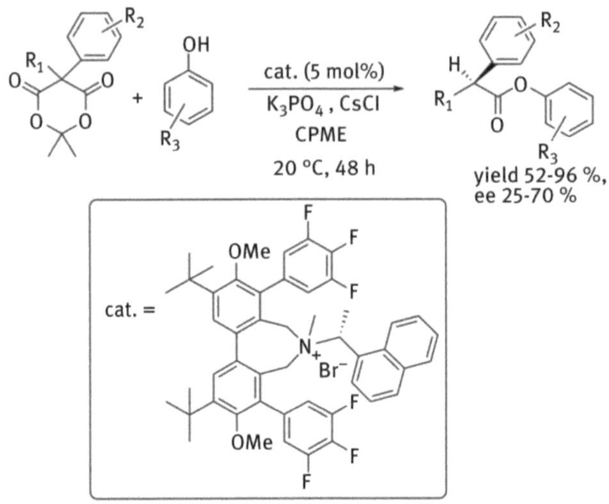

Figure 5.47:

A desymmetrisation of *meso*-anhydrides in the presence of methanol, potassium fluoride, and a PTC was reported. The catalyst is a cinchona derived ammonium salt with a hydrogen-bond donating urea moiety which forms a chiral ammonium fluoride *in situ*. Synthetic and NMR experiments unambiguously demonstrated the formation of the corresponding acyl fluoride as an intermediate (Figure 5.48) [59].

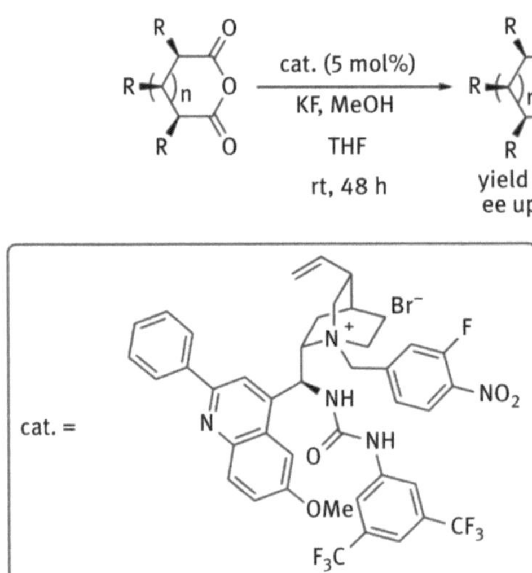

Figure 5.48:

An asymmetric base hydrolysis of N-protected amino acid hexafluoroisopropyl esters on PTC was reported. The reaction proceeds via dynamic kinetic resolution, affording the hydrolyzed products (Figure 5.49) [60].

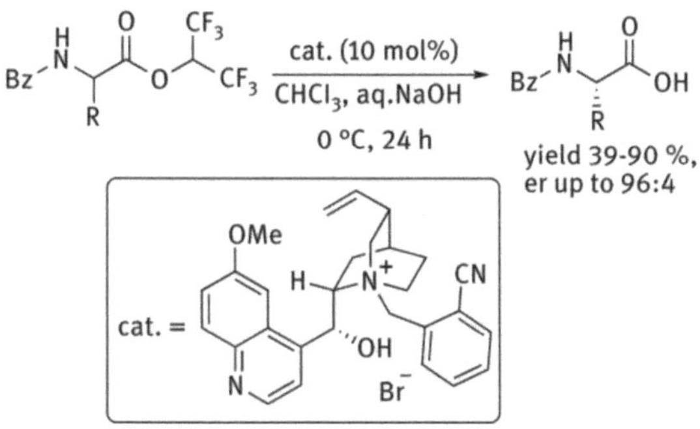

Figure 5.49:

5.11 Outlook and perspectives

Asymmetric reactions using binaphtyl-derived or cinchona alkaloid quaternary ammonium salts as catalysts have been reported, such as alkylations of glycine derivatives including peptides, conjugate additions using nitroalkenes, amines or thiols as nucleophiles, and umpolung conjugate additions of imines. Other catalysts such as bisguanidinium salts have been used in heterocyclizations, and phosphate anion catalysts were employed to perform enantioselective fluorinations and related processes. Furthermore, bi-functional ammonium cations having a hydrogen-bonding donor group are efficient catalysts to perform derivatizations of isoxazoles and nitro-Mannich reactions. Interestingly, photoinduced PTC is a recent field which opens up new possibilities by combining organocatalysis with light, for example alkylations and hydroxylations shown in this review. It is expected new catalysts will be developed in the next future affording useful tools to investigate unexplored chemical processes.

References

[1] Starks CM. Phase-transfer catalysis. I. Heterogeneous reactions involving anion transfer by quaternary ammonium and phosphonium salts. J Am Chem Soc. 1971;93:195–9.

[2] Freedman HH. Industrial applications of phase transfer catalysis (PTC): past, present and future. Pure Appl Chem. 1986;58:857–68.

[3] (a) Lygo B, Wainwright PG. A new class of asymmetric phase-transfer catalysts derived from cinchona alkaloids - application in the enantioselective synthesis of α-amino acids. Tetrahedron Lett. 1997;38:8595–8. (b) Lygo B, Wainwright PG. Asymmetric phase-transfer mediated epoxidation of α,β-unsaturated ketones using catalysts derived from Cinchona alkaloids. Tetrahedron Lett. 1998;39:1599–602. (c) Lygo B, Crosby J, Peterson JA.

Enantioselective synthesis of bis-α-amino acid esters via asymmetric phase-transfer catalysis. Tetrahedron Lett. 1999;40:1385–8.

[4] (a) Corey EJ, Xu F, Noe MC. A rational approach to catalytic enantioselective enolate alkylation using a structurally rigidified and defined chiral quaternary ammonium salt under phase transfer conditions. J Am Chem Soc. 1997;119:12414–5. (b) Corey EJ, Noe MC, Xu F. Highly enantioselective synthesis of cyclic and functionalized α-amino acids by means of a chiral phase transfer catalyst. Tetrahedron Lett. 1998;39:5347–50. (c) Corey EJ, Bo Y, Busch-Petersen J. Highly enantioselective phase transfer catalyzed alkylation of a 3-oxygenated propionic ester equivalent; applications and mechanism. J Am Chem Soc. 1998;120:13000–1.

[5] Ooi T, Kameda M, Maruoka K. Molecular design of a C_2-symmetric chiral phase-transfer catalyst for practical asymmetric synthesis of α-amino acids. J Am Chem Soc. 1999;121:6519–20.

[6] (a) Shioiri T, Handbook of phase-transfer catalysis, Chap. 14. (Eds.: Y. Sasson, R. Neumann), London: Blackie Academic & Professional, 1997:462. (b) O'Donnell MJ, Catalytic asymmetric synthesis, Chap. 10, 2nd ed. (Ed.: I. Ojima), New York: Wiley-VCH, 2000:727. (c) O´Donnell MJ. Aldrichimica acta. 2001;34:3–15. (d) Maruoka K, Ooi T. Enantioselective amino acid synthesis by chiral phase-transfer catalysis. Chem Rev. 2003;103:3013–28. (e) O'Donnell MJ. The enantioselective synthesis of α-amino acids by phase-transfer catalysis with achiral schiff base esters. Acc Chem Res. 2004;37:506–17. (f) Lygo B, Andrews BI. Asymmetric phase-transfer catalysis utilizing chiral quaternary ammonium salts: asymmetric alkylation of glycine imines. Acc Chem Res. 2004;37:518–25. (g) Hashimoto T, Maruoka K. Recent development and application of chiral phase-transfer catalysts. Chem Rev. 2007;107: 5656–82. (h) Shirakawa S, Maruoka K. Recent developments in asymmetric phase-transfer reactions. angew. Chem Int Ed. 2013;52:4312–48.

[7] Harada S, Kajihara R, Muramoto R, Jutabha P, Anzai N, Nemoto T. Catalytic asymmetric synthesis of α-methyl-p-boronophenylalanine. Bioorg Med Chem Lett. 2018;28:1915–8.

[8] Woo S, Kim Y, Lim B, Oh J, Lee Y, Gwon H, et al. Dimeric cinchona ammonium salts with benzophenone linkers: enantioselective phase transfer catalysts for the synthesis of α-amino acids. RSC Adv. 2018;8:2157–60.

[9] Pekošak A, Rotstein BH, Collier TL, Windhorst AD, Vasdev N, Poot AJ. Stereoselective ^{11}C labeling of a "native" tetrapeptide by using asymmetric phase-transfer catalyzed alkylation reactions. Eur J Org Chem. 2017;5:1019–24.

[10] Filp U, Pekošak A, Poot AJ, Windhorst AD. Stereocontrolled [^{11}C] alkylation of N-terminal glycine schiff bases to obtain dipeptides. Eur J Org Chem. 2017;37:5592–6.

[11] Jin L, Shuai Zhao S, Chen X. Synthesis of both enantiomers of chiral phenylalanine derivatives catalyzed by cinchona alkaloid quaternary ammonium salts as asymmetric phase transfer catalysts. Molecules. 2018;23:1421–37.

[12] Chen SZ, Ma W, Yan Z, Zhang F, Wang S, Tu Y, et al. Organo-cation catalyzed asymmetric homo/heterodialkylation of bisoxindoles: construction of vicinal all-carbon quaternary stereocenters and total synthesis of (–)-chimonanthidine. J Am Chem Soc. 2018;140:10099–103.

[13] Craig R, Sorrentino E, Connon SJ. Enantioselective alkylation of 2-oxindoles catalyzed by a bifunctional phase-transfer catalyst: synthesis of (-)-debromoflustramine B. Chem Eur J. 2018;24:4528–31.

[14] Ohmatsu K, Furukawa Y, Kiyowaka M, Ooi T. Diastereo- and enantioselective phase-transfer alkylation of 3-substituted oxindoles with racemic secondary alkyl halides. Chem Commun. 2017;53:13113–6.

[15] Li R, Liu Z, Chen L, Pan J, Zhou W. Enantioselective phase-transfer catalyzed alkylation of 1-methyl-7-methoxy-2-tetralone: an effective route to dezocine. Beilstein J Org Chem. 2018;14:1421–7.

[16] Zhang T, Scalabrino G, Frankish N, Sheridan H. Bioactive indanes: proof of concept study for enantioselective synthetic routes to PH46A, a new potential anti-inflammatory agent. Molecules. 2018;23:1503–21.
[17] Pawliczek M, Shimazaki Y, Kimura H, Oberg KM, Zakpur S, Hashimoto T, et al. Phase-transfer-catalysed asymmetric synthesis of 2,2-disubstituted 1,4-benzoxazin-3-ones. Chem Comm. 2018;54:7078–80.
[18] Kang Q, Selvakumar S, Arumugam N, Almansour AI, Kumar RS, Maruoka K. Enantioselective alkylation of N-arylhydrazones derived from α-keto esters and isatin derivatives through asymmetric phase-transfer catalysis. Chem Asian J. 2018;13:1780–3.
[19] Kim D, Ha MW, Hong S, Park C, Kim B, Yang J, et al. Enantioselective synthesis of chiral α-azido and α-aryloxy quaternary stereogenic centers via the phase-transfer-catalyzed α-alkylation of α-bromomalonates, followed by SN2 substitution. J Org Chem. 2017;82:4936–43.
[20] Ha MW, Lee JY, Kim D, Lee G, Lee JK, Hong S, et al. Enantioselective synthesis of chiral α-thio-quaternary stereogenic centers via phase-transfer-catalyzed α-alkylation of α-acylthiomalonates. J Org Chem. 2018;83:1011–8.
[21] Xiang B, Belyk KM, Reamer RA, Yasuda N. Discovery and application of doubly quaternized cinchona-alkaloid-based phase-transfer catalysts. Angew Chem Int Ed. 2014;53:8375–8.
[22] He CQ, Simon A, Lam Y, Brunskill AP, Yasuda N, Tan J, et al. Model for the enantioselective of asymmetric intramolecular alkylations by bis-quaternized cinchona alkaloid-derived catalysts. J Org Chem. 2017;82:8645–50.
[23] Jakhar A, Nazish M, Gupta N, Khan NH, Kureshy RI. Enantioselective addition of cyanide to CF_3-substituted alkylidenemalonates: construction of trifluoromethylated all-carbon quaternary stereocenters. ChemistrySelect. 2018;3:4838–43.
[24] Arlt A, Toyama H, Takada K, Hashimoto T, Maruoka K. Phase-transfer catalyzed asymmetric synthesis of α, β-unsaturated γ, γ-disubstituted γ-lactams. Chem Commun. 2017;53:4779–82.
[25] Scorzelli S, Di Mola A, De Piano F, Consiglia Tedesco C, Palombi L, Filosa R, et al. A systematic study on the use of different organocatalytic activation modes for asymmetric conjugated addition reactions of isoindolinones. Tetrahedron. 2017;73:819–28.
[26] Jin Q, Zheng C, Zhao G, Zou G. Bifunctional quaternary ammonium salts catalyzed stereoselective conjugate addition of oxindoles to electron-deficient β-haloalkenes. J Org Chem. 2017;82:4840–50.
[27] Takagi R, Fujii E, Kondo H. Enantioselective Michael reaction of cyclic β-ketoesters with Morita–Baylis–Hillman derivatives using a phase-transfer catalyst. J Org Chem. 2018;83:11191–203.
[28] Miguélez J, Miyamura H, Kobayashi S. a polystyrene-supported phase-transfer catalyst for asymmetric Michael addition of glycine-derived imines to α, β-unsaturated ketones. Adv Synth Catal. 2017;359:2897–900.
[29] Fan Y, Zhou L, Li S. Catalytic asymmetric 1,6-conjugate addition of in situ generated para-quinone methides with tritylthiol. Org Chem Front. 2018;5:1820–4.
[30] Zhu J, Cui D, Li Y, He J, Chen W, Wang P. Enantioselective amination of nitroolefins under base-free and water-rich conditions using chiral bifunctional phase-transfer catalysts. Org Biomol Chem. 2018;16:3012–7.
[31] Liang X, Liu C, Zhang W, You S. Asymmetric fluorinative dearomatization of tryptamine derivatives. Chem Commun. 2017;53:5531–4.
[32] Liang X, Cai Y, You S. Asymmetric fluorinative dearomatization of tryptophol derivatives by chiral anion phase-transfer catalysis. Chin J Chem. 2018;36:925–8.
[33] J A S C, Matsumoto A, Orlandi M, M J H, Sigman MS, Toste FD. Enantioselective fluorination of homoallylic alcohols enabled by the tuning of non-covalent interactions. Chem Sci. 2018;9:7153–8.

[34] Niwa T, Ujiie K, Sato H, Egami H, Hamashima Y. Asymmetric fluorination of cyclic tetrasubstituted alkenes with a pendant amide groups under dianionic phase-transfer catalysis. Chem Pharm Bull. 2018;66:920–2.
[35] Silvi M, Melchiorre P. Enhancing the potential of enantioselective organocatalysis with light. Nature. 2018;554:41–9.
[36] Woźniak Ł, Murphy JJ, Melchiorre P. Photo-organocatalytic enantioselective perfluoroalkylation of β-ketoesters. J Am Chem Soc. 2015;137:5678–81.
[37] Yang C, Zhang W, Li Y, Xue X, Li X, Cheng J. Origin of stereoselectivity of the photoinduced asymmetric phase-transfer-catalyzed perfluoroalkylation of β-ketoesters. J Org Chem. 2017;82:9321–7.
[38] Wang Y, Yin H, Tang X, Wu Y, Meng Q, Gao Z. A series of cinchona-derived N-oxide phase-transfer catalysts: application to the photo-organocatalytic enantioselective α-hydroxylation of β-dicarbonyl compounds. J Org Chem. 2016;81:7042–50.
[39] Tang X, Feng S, Wang Y, Yang F, Zheng Z, Zhao J, et al. Bifunctional metal-free photo-organocatalysts for enantioselective aerobic oxidation of b-dicarbonyl compounds. Tetrahedron. 2018;74:3624–33.
[40] Liu Y, Liu Y, Wang J, Wei Z, Cao J, Liang D, et al. Asymmetric phase-transfer catalysts bearing multiple hydrogen bonding donors: synthesis and application in nitro-Mannich reaction of isatin-derived N-Boc ketimines. Tetrahedron. 2017;73:2400–3.
[41] (a) Lu N, Li R, Wei Z, Cao J, Liang D, Lin Y, et al. Enantio- and diastereoselective nitro-mannich reaction of α-aryl nitromethanes with amidosulfones catalyzed by phase-transfer catalysts. J Org Chem. 2017;82:4668–76. (b) Liu Y, Wang J, Wei Z, Cao J, Liang D, Lin Y, Duan H. Diastereo- and enantioselective nitro-mannich reaction of α-aryl nitromethanes with isatin-derived n-boc ketimines catalyzed by chiral phase-transfer catalysts bearing multiple hydrogen-bonding donors. New J Chem. 2018;42:1608–11.
[42] Liu Y, Wei Z, Liu Y, Cao J, Liang D, Lin Y, et al. Novel α-amino acid-derived phase-transfer-catalyst application to a highly enantio- and diastereoselective nitro-Mannich reaction. Org Biomol Chem. 2017;15:9234–42.
[43] Wang B, Xu T, Zhu L, Lan Y, Wang J, Lu N, et al. Highly enantioselective nitro-Mannich reaction of ketimines under phase-transfer catalysis. Org Chem Front. 2017;4:1266–71.
[44] Sharma K, Wolstenhulme JR, Painter PP, Yeo D, Grande-Carmona F, Johnston CP, et al. Cation-controlled enantioselective and diastereoselective synthesis of indolines: an autoinductive phase-transfer initiated 5-endo-trig process. J Am Chem Soc. 2015;137:13414–24.
[45] Lamb AD, Davey PD, Driver RW, Thompson AL, Smith MD. Enantioselective synthesis of 4- and 6-azaindolines by a cation-directed cyclization. Org Lett. 2016;18:5372–5.
[46] Sallio R, Lebrun S, Capet F, Agbossou-Niedercorn F, Michon C, Deniau E. Diastereoselective auxiliary- and catalyst-controlled intramolecular aza-Michael reaction for the elaboration of enantioenriched 3-substituted isoindolinones. Application to the synthesis of a new pazinaclone analogue. Beilstein J Org Chem. 2018;14:593–602.
[47] Guo S, Liu X, Shen B, Lin L, Feng X. Organocatalytic asymmetric cascade reaction of 2-hydroxyphenyl-substituted enones and isocyanates to construct 1,3-benzoxazin-2-ones. Org Lett. 2016;18:5070–3.
[48] Lee H, Eun B, Sung E, G T H, Ko YK, Cho C. Catalytic enantioselective synthesis of carboxy-substituted 2-isoxazolines by cascade oxa-Michael-cyclization. Org Biomol Chem. 2018;16:657–64.
[49] Capaccio V, Capobianco A, Stanzione A, Pierri G, Tedesco C, Di Mola A, et al. Organocatalytic heterocyclization driven by dynamic kinetic resolution: enantioselective access to multi-heteroatomic cyclic structures mediated by Cinchona alkaloid-based catalysts. Adv Synth Catal. 2017;359:2874–80.

[50] Lian X, Adili A, Liu B, Tao Z, Han Z. Enantioselective [4 + 1] cycloaddition of ortho-quinone methides and bromomalonates under phase-transfer catalysis. Org Biomol Chem. 2017;15:3670–3.

[51] Tao ZL, Adili A, Wu X, Gong LZ. A highly enantioselective Mannich-type reaction of glycine schiff base catalyzed by a cinchoninium salt. Chin J Chem. 2014;32:969–73.

[52] Zhang Q, Guo S, Yang J, Yu K, Feng X, Lin L, et al. Asymmetric formal [3 + 2]-cycloaddition of azomethine imines with azlactones to synthesize bicyclic pyrazolidinones. Org Lett. 2017;19:5826–9.

[53] Zhu B, Lee R, Yin Y, Li F, Coote ML, Jiang Z. Enantioselective vinylogous amination of 5-alkyl-4-nitroisoxazoles with a dipeptide-based guanidinium phase-transfer catalyst. Org Lett. 2018;20:429–32.

[54] Bo Zhu B, Fuyuan Li F, Lu B, Chang J, Jiang Z. Organocatalytic enantioselective vinylogous aldol reaction of 5-alkyl-4-nitroisoxazoles to paraformaldehyde. J Org Chem. 2018;83:11350–8.

[55] Li Z, Hu B, Wu Y, Fei C, Deng L. Control of chemoselectivity in asymmetric tandem reactions: direct synthesis of chiral amines bearing nonadjacent stereocenters. Pnas. 2018;115:1730–5.

[56] Hu B, Deng L. Catalytic asymmetric synthesis of trifluoromethylated γ-amino acids through the umpolung addition of trifluoromethyl imines to carboxylic acid derivatives. Angew Chem Int Ed. 2018;57:2233–7.

[57] Yoshida Y, Mino T, Sakamoto M. Organocatalytic highly regio- and enantioselective umpolung michael addition reaction of a-imino esters. Chem Eur J. 2017;23:12749–53.

[58] Legros F, Martzel T, Brière J, Oudeyer S, Levacher V. Organocatalytic enantioselective decarboxylative protonation reaction of Meldrum's acid derivatives under PTC conditions. Eur J Org Chem. 2018;17:1975–83.

[59] Craig R, Litvajova M, Sarah A, Cronin SA, Connon SJ. Enantioselective acyl-transfer catalysis by fluoride ions. Chem Commun. 2018;54:10108–11.

[60] Yamamoto E, Wakafuji K, Furutachi Y, Kobayashi K, Kamachi T, Tokunaga M. Dynamic kinetic resolution of N-Protected amino acid esters via phase-transfer catalytic base hydrolysis. ACS Catal. 2018;8:5708–13.

Elisabetta Massolo and Maurizio Benaglia

6 Stereoselective organocascades: from fundamentals to recent developments

Abstract: Reaction sequences where more bonds are sequentially formed (cascade reactions) may be started either by a stoichiometric or by a catalytic reagent, and proceed in an enantio- diastereo- or non-stereo- selective manner. A wide variety of such strategies has been developed, including both stoichiometric and catalytic ones.

Within the widely developed cascade reactions field, this chapter is not meant to be omni-comprehensive, but to offer an as much as possible complete overview on organocatalytic stereoselective methods. We embrace the more general definitions by Tietze and Denmark, considering as *cascade reactions* all those one-pot processes that involve two or more bond formations, where each subsequent step is enabled by a structural change caused by the previous one. We will include both two- and multi-component reactions where one or more organocatalysts may be responsible either for all or just some of the occurring transformations. Organocascades will be reported according to the number of involved catalytic cycles.

In the following paragraphs, only cascade reactions that are stereoselective by means of a chiral catalyst will be considered. It will be shown that multiple possibilities, relying on different catalysis modes, are available to achieve the same reaction sequence.

Keywords: domino reactions, organocascade, stereoselectivity, organocatalysis, synthetic methodologies

Notes: 1. TS are drawn to clearly show the activation mechanism involved in the transformation, but conformational details related to the stereochemical outcome are neglected.
2. In the reaction mechanisms schemes, only those parts of the catalysts' structures that are involved in the catalytic steps are drawn

This article has previously been published in the journal Physical Sciences Reviews. Please cite as: Massolo, E., Benaglia, M. Stereoselective organocascades: from fundamentals to recent developments *Physical Sciences Reviews* [Online] 2021, 2. DOI: 10.1515/psr-2018-0096

https://doi.org/10.1515/9783110590050-006

6.1 Introduction

Over 150 years of research made Chemistry progress to a point where no synthetic wish can stay unfulfilled. Strategies are available to reach high level of structural and stereochemical complexity, so that every desired target, including natural isolates, would finally be achieved. The ultimate challenge of synthesis, now, is "to be able to provide large quantities of complex natural products with a minimum amount of labor and material expense" [1] and *sustainability* has progressively become a central topic. Efforts are now devoted to the development of methods to be environmentally conscious and resource-effective, aiming to save raw materials, energy and time and to minimize waste.

It clearly appears that the number of steps is a preeminent factor in defining the feasibility of a synthesis route, as it influences the (man)power and material input, the cost and the environmental impact. Thus, *step* [2] and *pot economy* [3] represent significant leading principles to be considered when elaborating a new synthetic strategy. In this context, stop-and-go sequences are progressively outcompeted by methods where multiple transformations are combined into one synthetic operation. For their effective implementation the key issue to be addressed is the compatibility of substrates and conditions across the different occurring steps. Inspired by Nature, although not so efficient and sophisticated yet, these approaches enable indeed a rapid increase in molecular complexity, improving operational simplicity and making processes more feasible from both an economic and ecological standpoint [4].

Reaction sequences where more bonds are sequentially formed may get started either by a stoichiometric or by a catalytic reagent and proceed in an enatio- diastereo- or nonstereo-selective manner. A wide variety of such strategies has been developed, including both stoichiometric and catalytic ones and, simultaneously, a plethora of definitions aiming to classify those emerged.

6.1.1 Taxonomy

In Tietze's work, *domino catalysis* is defined as "a process involving two or more bond-forming transformations which take place under the same reaction conditions without adding additional reagents and catalysts and in which subsequent reactions result as a consequence of the functionality in the previous step" [5] and Faber added that "each individual reaction belong tightly together and are rather difficult to perform in a stepwise fashion" [6]. Additionally, Fogg stated that "the transformations must be effected by a single catalytic mechanism" [7] and Chapman and Frost completed the definition expressing that "a catalyst must be integral to both of the bond-forming transformations" [8]. Previous to these specification, Denmark integrated Tietze's work defining *tandem cascade reactions*, wherein each subsequent stage can occur by virtue of the structural change brought about by the previous step under the same reaction conditions, *tandem consecutive reactions*, wherein the first reaction is necessary but not sufficient for the tandem process, so external reagents or changes in reaction conditions are also required, and *tandem sequential reactions*, wherein the second stage requires the addition of another reagent [9].

The term domino was thus placed side by side to the term *cascade*. When employed alone, this term is meant to indicate a transformation where a single catalyst is responsible for subsequently building new bonds; however, additional features may accompany *cascade*, making the term suitable to different situations including even those where more than one catalytic species is employed. In fact, MacMillan and Walji introduced three other denominations: *iterative cascade catalysis*, i.e. cascade catalysis involving one catalyst and one iterative reaction type, *cascade catalysis based on multiple reaction types*, i.e. cascade catalysis involving one catalyst but multiple reaction types and *cycle-specific cascade catalysis*, i.e. cascade catalysis involving multiple catalysts and multiple reaction types [10].

These partially overlap to Fogg's definition of *tandem catalysis*, involving more than one catalytic cycle and including the cases of *orthogonal catalysis*, where more catalysts are required, and of *assisted tandem* and *auto-tandem catalysis*, where a chemical trigger is or not needed to transform the catalyst or to cause a change in mechanism, respectively [6].

In 2005 Bazan group specified the concept of *concurrent tandem catalysis* (CTC), which involves the cooperative action of two or more catalytic cycles in a single reactor, and classified generic CTC cycles according to the number of unique catalytic cycles and the manner in which the products from each cycle are distributed in subsequent reactions [11]. When a set of starting materials A reacts with catalyst I to produce intermediate B and then, upon the addition of C but still under the action of I, the final product P forms, the cycle below is designated $(A_IB)(BC_IP)$. When, instead, a second catalyst (II) is added together with C to react with B toward P, the cycle is indicated ad $(A_IB)(BC_{II}P)$. The generic term $(A_IB)(B_{II}C)..(S_{nIn}P)$ refers to a situation where Sn is the nth substrate and In is the nth catalyst.

While being highly specific, these classification approaches lack in giving the direct macroscopic distinction between *organomulticatalysis* and *organocascade*, the difference lying in the number of catalysts, more or one, respectively, employed. This was adopted in 2014 by Volla et al. who, reviewing this field, pointed out double, triple and even quadruple cascades [12].

A further specification within *organomulticatalysis* was given by Wende and Schreiner. The situation represented by the case $(A_IB)..(S_{In}P)$, in which the intermediate generated from one catalytic cycle is the substrate for the subsequent one promoted by a different catalyst or by an independent catalytic moiety on the same catalyst, is precisely defined as *multicatalysis*. This strategy was also translated in the design of *multicatalysts*, i.e. a catalyst equipped with an appropriate spacers [13]. Going back to the origin of this rather articulated taxonomy system, it is possible to notice that multicatalysis can be considered a kind of subgroup of domino reactions as defined by Tieze.

In 2012, Allen and McMillan made clear the difference between *double activation catalysis*, where two different catalysts activate the same starting material, and *synergistic catalysis*, where two different catalyst act on two different reaction partners [14].

Shifting the attention from the number of catalytic species to the number of substrates, it is possible to identify single-component, two-component, and multicomponent transformations. *Multicomponent* reactions are defined as domino reactions involving at least three substrates and, according to Ramon and Yus, "these should be clearly differentiated from other one-pot processes [..] that involve the reaction between two reagents to yield an intermediate which is captured by the successive addition of a new reagent" [15].

A tetra-coordinated [a,b,c,d] system was very recently introduced by Indu and Kaliappan, applicable to multistep reactions independently from the number of components and even transferable to sequences occurring in different reaction vessels. In this nomenclature (a) indicates the number of pots, (b) the number of reactions taking place in one-pot, (c) the number of rings formed in one-pot and (d) the number of bonds formed in the same one-pot sequence [16].

A different systematic description and classification of one-pot reactions was introduced by Jorgensen group in 2011 [17]. Considering that this kind of transformations have the ultimate goal of reducing time demand and waste production and have a evident link to industrial processes, the author proposed not to focus on the involved activation modes, but on the number of "manual operations", which is easily counted and may provide an indication of the complexity of the overall reaction and the required manual effort. This system relies on three parameters: type, indicating the position of the enantiodifferentiating manual operation; order, indicating total number of manual operations that are defined as "interruptions of the cascade by the addition of reagents or the removal of the solvent"; fingerprint, indicating the number of C-C (m) and C-X (n) bonds formed and abbreviate as mCnX. In particular, the authors proposed three types underlying the different "chemical purposes" on the basis of which the position of the stereoselective step is chosen. Specifically, in Type A reactions (with asymmetric catalysis as the first manual operation) rapid assembly of structurally diverse chiral frameworks, subsequently modified by in situ modification may lead to highly complex target molecules; in this case, the main concerns are racemization and decomposition of the assembled chiral framework. Instead, Type C strategies (with asymmetric catalysis at the end of the sequence), are performed to avoid handling of delicate starting materials, reaching the stereoselective step as last one; in this case, success is threatened by contaminants somehow hampering the late-stage reaction. In the paper, a flowchart is offered to make the assignment easier.

$$Z = \sqrt[b]{\frac{Y}{100}} \times 100 \qquad Y_{PMO} = \sqrt[nmo]{\frac{Y}{100}} \times 100 \qquad P_f = nmo - 1 - n(INI) + x$$

The remarkable analysis work carried out in the last twenty years offered a deep insight in to the area of one-pot transformation, even though sometimes directness and univocity have been sacrificed. Herein, we will use domino, cascade and tandem as synonymous terms without aiming to a strict categorization but to illustrate the main concepts and logics on which the field is based and develops.

6.1.2 Reaction sequences

Within the widely developed cascade reactions field [12], this Chapter is not meant to be omni-comprehensive but to offer an as much as possible complete overview on organocatalytic stereoselective methods. We embrace the more general definitions by Tietze and Denmark, considering as *cascade reactions* all those one-pot processes that involve two or more bond formations where each subsequent step is enabled by a structural change caused by the previous one. We will include both two- and multi-component reactions where one or more organocatalysts, which may be mono- or multifunctional, may be responsible either for all or just some of the occurring transformations.

Organocascades will be reported according to the number of involved catalytic cycles. Here below, a synopsis illustrating the main classes of organocatalysts and the activation modes through which they operate. The employed catalysts often feature more than one functionality participating in the activation mechanisms.

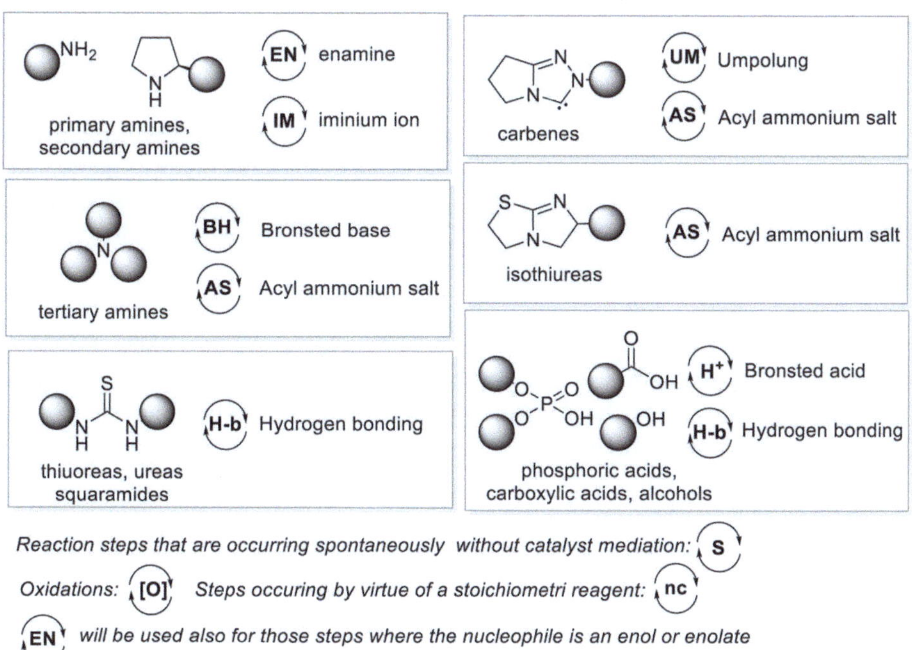

Figure 6.1: Main classes of organocatalysts and their activation modes.

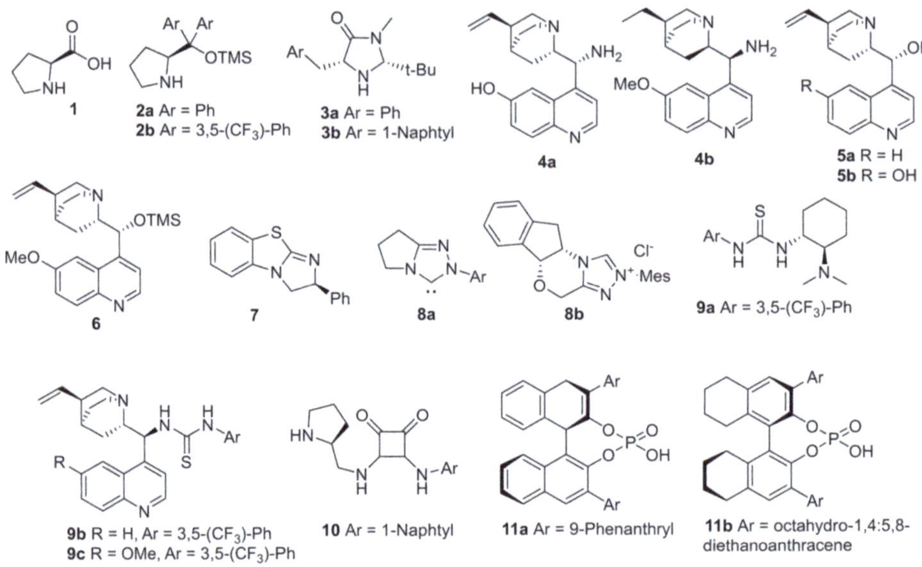

1; **2a** Ar = Ph, **2b** Ar = 3,5-(CF$_3$)-Ph; **3a** Ar = Ph, **3b** Ar = 1-Naphtyl; **4a**; **4b**; **5a** R = H, **5b** R = OH; **6**; **7**; **8a**; **8b**; **9a** Ar = 3,5-(CF$_3$)-Ph; **9b** R = H, Ar = 3,5-(CF$_3$)-Ph, **9c** R = OMe, Ar = 3,5-(CF$_3$)-Ph; **10** Ar = 1-Naphtyl; **11a** Ar = 9-Phenanthryl; **11b** Ar = octahydro-1,4:5,8-diethanoanthracene

As for the substrates of domino reactions, they typically need to fulfil structural requirements that become stricter as the number of sequential steps increases, as the proper functional groups are to be in key positions to allow the desired bond forming events.

In the following paragraphs, only cascade reactions that are stereoselective by means of a chiral catalyst will be reported. Organocascades will be classified according to the reaction types they are constituted by a grouped based on the reaction type that starts the sequence. It will be shown that multiple possibilities, relying on different catalysis mode, are available to achieve the same reaction sequence.

6.2 Double cascade sequences

For the concept of cascade to be realized, at least two consecutive transformations need to take place. The simplest sequences that can be implemented are thus double cascades, arising from the combination of different activation modes. A wide variety of these two-steps tandem processes have been developed, relying either on one or more mono- or bifunctional catalysts.

6.2.1 Michael-type reaction initiated sequences

The most of cascade sequences gets started by a 1,4-addition event, generating a nucleophilic intermediate that may undergo different fates. Selected cases will be illustrated to highlight the main synthetic concepts.

The field of stereoselective organocatalyzed domino reactions [18] was opened by Barbas et al. The group demonstrated the ability of Ab38C2 to accomplish the Michael addition/aldol condensation sequence proceeding via amine catalysis exerted by a lysine residue [19]. Shortly after the disclosure of this antibody promoted Robinson annulation, the authors reported secondary amines as catalysts for the Michael/Aldol sequence. In particular, with methyl vinyl ketone (MVK) and 2-methylcyclohexane-1,3-dione, proline afforded the cyclized product in 49% yield and 76% enantiomeric excess (ee) [20]. According to the proposed mechanism, both reaction partners are simultaneously activated: the LUMO lowering of the electrophile is achieved by condensation with the amine function leading to a chiral iminium ion; the presence of a carboxylate group on the catalyst – and, thus, on this active intermediate – allows coordination of the nucleophilic diketone in its enolate form. After the conjugate addition takes place, the resulting enamine intermediate acts as the nucleophile in the intramolecular aldol via a Zimmerman – Traxler-type transition state, proposed by Houk et al. [21].

Michael/Aldol sequence

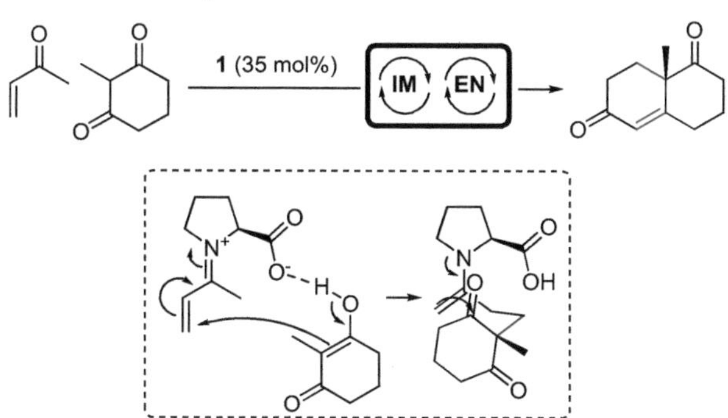

Over the years, several contributions were published where domino sequences led to Wieland – Miescher ketone analogues and to a wide variety of mono-, bi- and spirocyclic derivatives. Few examples are reported here below to illustrate the variety of structures achieved by this method.

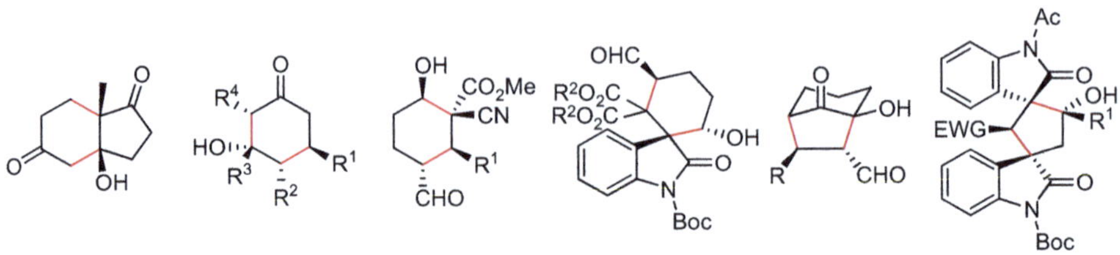

Vinylgous versions of this cascade have been reported as well, setting remote stereocenters. Key to the success of this strategy is the identification of well suited substrates, characterized by the needed substitution pattern [22]. The first example was published by Tian and Melchiorre employing dienones and 3-substituted oxindoles. After a primary amine promoted 1,6-addition, the resulting dienamine attacks the ketone substituent on the indole core. Interestingly, the whole sequence is dragged by the final aldol step, which provides the thermodynamic advantage needed to overcome the intrinsic low efficiency of the γ-addition. In fact, no reaction occurs employing oxindoles lacking the carbonyl group that allows the intramolecular aldol. Spirocyclopentanes bearing four contiguous stereocenters and the unaltered α,β-unsaturated carbonyl system form in good yield an ee >96% [23].

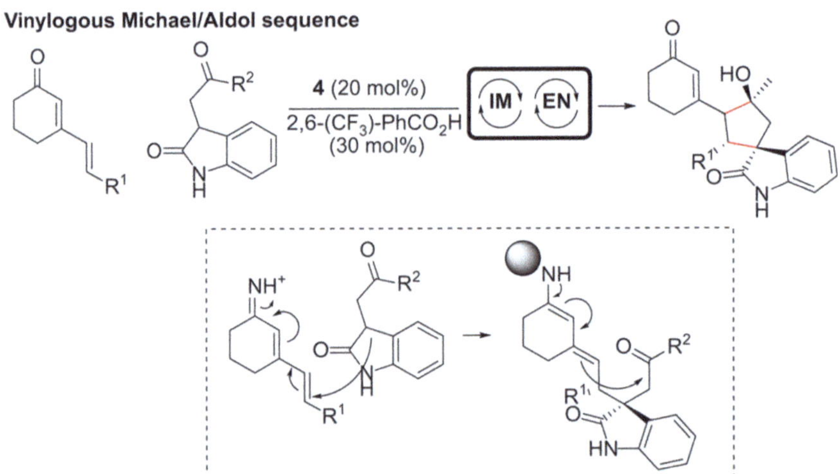

Domino reactions proceeding through the same sequence where, instead, the aminocatalyst is involved in the activation of the nucleophile, are also known. These consist in dienamine conjugate additions followed by a spontaneous cyclization event. The Jørgensen group reported the preparation of highly functionalized tricyclic cores from 2-(cyclohepta-1,3,5-trien-1-yl)acetaldehyde and 3-olefinic oxindoles. Condensation between the aldehyde and the TMS-protected prolinol catalyst leads to a trienamine species; this reacts at its δ-position in a conjugate addition affording an anionic intermediate. No such 1,4-addition occurs when 2-cycloheptylideneacetaldehyde, that features one less unsaturation and thus can be activated only as the corresponding dienamine: on the other hand, when a stoichiometric amount of prolinol is used, it ends up trapped in the final product. This evidence attests the need for a trienamine system for the catalyst turnover. The intramolecular aldol reaction takes place in a high diastereostereoselective way after the catalyst is released via iminium ion hydrolysis. Indeed, products are typically obtained in >95:5 diastereomeric ratio (d) and ee higher than 90% [24].

Vinylogous Michael/Aldol sequence

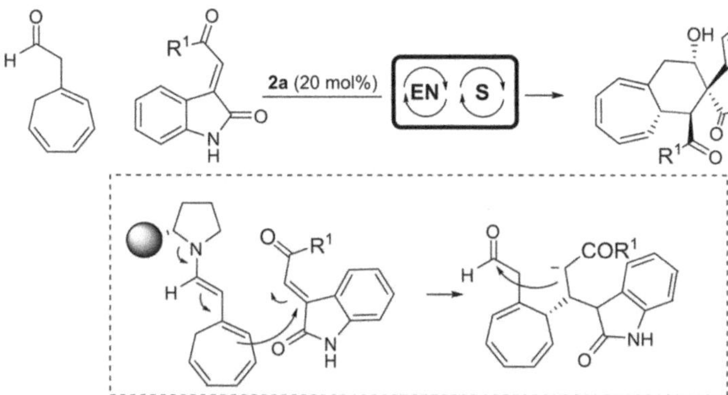

Robinson-type annulations have also been afforded by exploiting activation modes other than those provided by aminocatalysis. In 1998, Terashima et al. employed (-)-cinchonidine as both a Brønsted base and a hydrogen bonding donor for the preparation of an intermediate toward (–)-huperzine A, a natural product that features biological activity. Although results were modest in terms of enantioselectivity and yield, this represent the first synthetically practical way to access this tricyclic core [25].

Michael/Aldol sequence

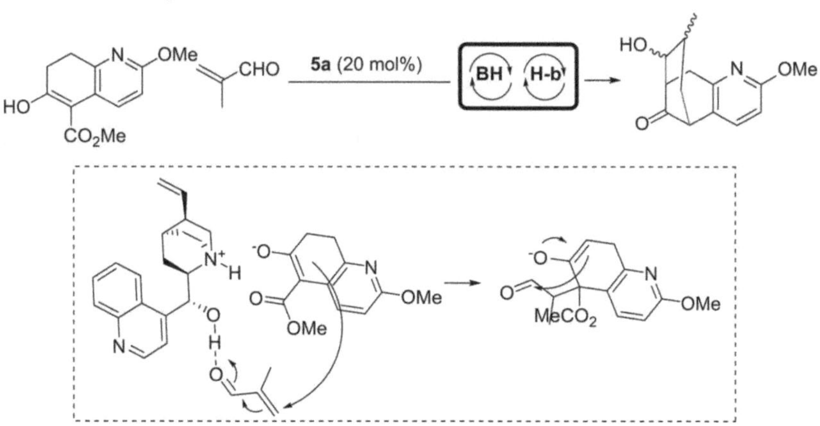

As a variant of the Michael/Aldol sequence, the Jørgensen group showed that the ring-closing event could occur via a Morita – Baylis – Hillman reaction. Here, after generating the iminium ion to promote the nucleophilic attack by a Nazarov reagent, the prolinol catalyst is hydrolysed. At this stage, the released prolinol acts as a nucleophilic catalyst via conjugate addition on the intermediate and thus starts the second catalytic cycle. This was the first example of an enantioselective Morita–Baylis–Hillman reaction catalyzed by a secondary amine. The proposed mechanism was supported by the isolation of the Michael addition product and control experiments revealed the stereoselectivity of the cyclization step to be substrate controlled. Products were obtained in good yields and ees generally higher than 90% and they were subjected to subsequent synthetic transformations [26].

Attack on a carbonyl function, following a 1,4-addition, may also be performed by a Breslow intermediate in a Michael/benzoin sequence. Lathrop and Rovis combined a prolinol ether and a triazolium salt in the presence of an additional base to prepare α-hydroxy-cyclopentanones from α,β-unsaturated aldehydes and either diketones or β-keto-esters. Within the fairly wide reaction scope, only two out of the four possible product diastereomers were observed. In this domino, the enal was first activated as an iminium ion undergoing a Michael reaction; the aldehyde intermediate is then attacked by the carbene, thus causing its unpolung: the originally carbonyl carbon thus becomes nucleophilic and responsible for the ring closing 1,2-adddition. group The authors demonstrated that trapping the aldehyde in the Breslow intermediate avoids the prolinol-mediated retro-Michael to occur, thus ensuring the high ee of the final product (80%< ee <97%) [27].

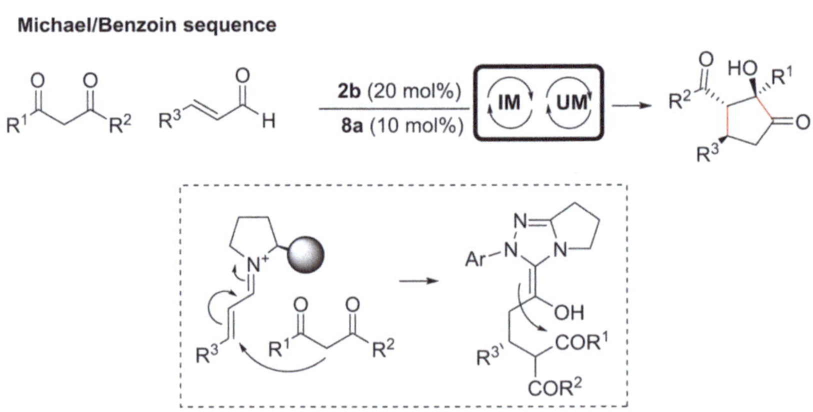

Also for Michael/Henry sequences, most methods rely on amine catalysts that may start the cascade activating either the nucleophile or the electrophile. In an example by the Hayashi group, the enamine of pentane-1,5-dial and a prolinol derivative react with nitrostyrene and the so generated nitronate intermediate attacks the remaining carbonyl group to afford cyclic structures with four contiguous stereogenic centers in ees mainly higher than 98% [28]. On the other hand, in the approach by Jørgensen et al., the same catalyst is employed for enal activation as iminium ion, while the nitroalkane partner deprotonation is performed by an achiral base [29].

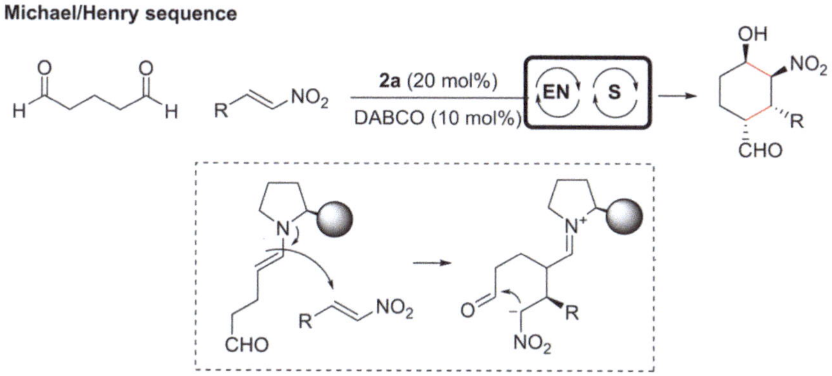

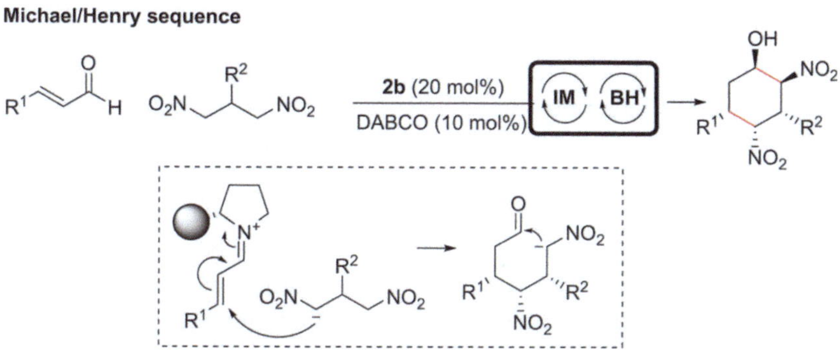

Dicarbonyl compounds and nitroalkenes have often been combined in the presence of a bifunctional catalyst; two different activation mechanisms are possible. Various bicyclic products have been obtained with very high selectivity [30–32].

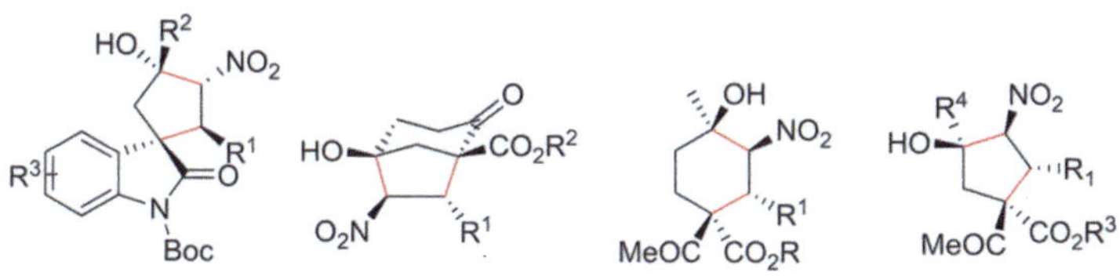

Again, vinilogous versions are available. In a work by Jørgensen, a bifunctional secondary amine-squaramide catalyst turns the enal into the active dienamine via condensation while increases the nitroolefin electrophilicity via hydrogen bonding [33]. In an analogous transformation, Vicario et al. used instead two distinct organocatalysts [34].

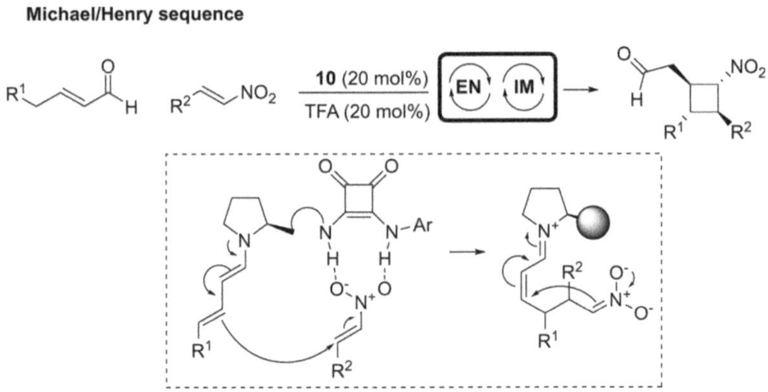

Michael/lactamization cascades have also been developed. The Romo group reported unsaturated acyl ammonium salts as the active γ-electrophiles generated from an acyl chloride and a chiral tertiary amine catalyst in the presence of a Brønsted base. N-protected benzylamines functionalized with an *ortho*-malonate pendant were chosen as substrates to achieve benzazepinones, bezo-fused lactams of pharmaceutical interest. The malonate function was used as nucleophile for the conjugate addition, followed by intramolecular amide bond formation that lead to the targeted eight-membered lactams in good yields and er higher than 97:3 [35]. In a contribution by Jin and co-authors, homobenzotetramisole was employed as nucleophilic catalyst for a lactamization leading to dihydrothiopyranone derivatives from indoline-2-thiones and α,β-unsaturated anhydrides [36].

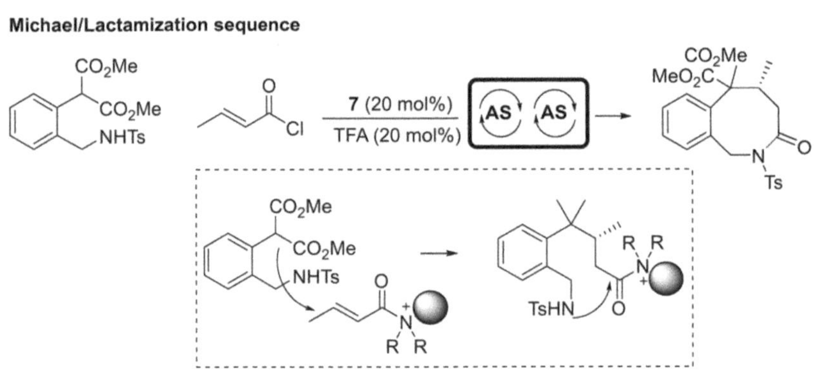

Extending the idea to the use of the hydride anion as nucleophile, the transfer hydrogenation of pyridines and quinolones promoted by Brønsted acids may be seen as a 1,4/1,2-addition sequence. Reported by Rueping et al. [37], it allows the preparation of the corresponding tetrahydroetherocycles with higher chemoselectivity with respect to metal catalysed transfer hydrogenation and a broad scope. According to the author's hypothesis, the cascade is initiated by formation of the quinoline substrate to the corresponding iminium ion. This protonated intermediate will hydride transfer; the generated enamine intermediate will be subjected to a further reduction via the same protonation/hydride attack sequence. This strategy allow accessing tetrahydrquinolines typically in higher than 90% yield and ee often close to 99% and was applied to the synthesis of biologically active alkaloids [38].

Michael/Mannich sequence

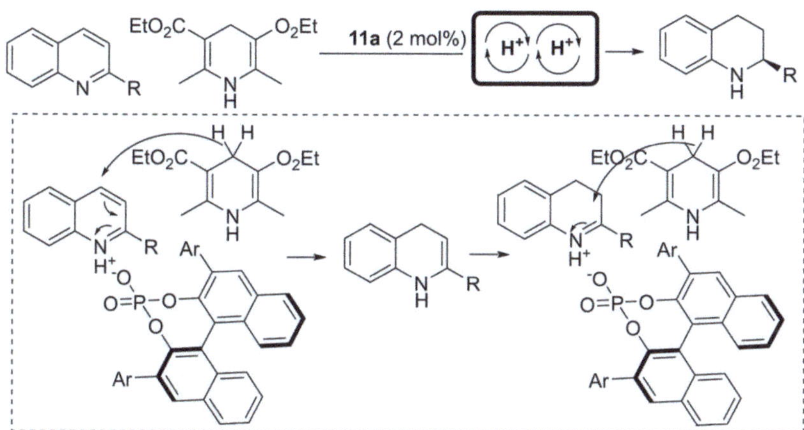

Michael/Michael sequences are also well established, implemented by means of different activation modes [39].

Using primary amine catalysts, dienamines derived from enones proved effective nucleophiles toward different unsaturated acceptors. A representative example was published by the Melchiorre group: a primary amine promoted the formation of cyclohexanones with up to four contiguous stereogenic center in good yield, complete diastereoselectivity and enantioselection higher than 94%, starting from enones and nitrostyrenes, *trans*-α-cyanocinnamate and maleimides. The transformation is proposed as a stepwise sequence that, oppositely to what observed in the analogous secondary amine promoted Diels-Alder reaction, gives access to the *endo* product. Reactivity suppression by polar reaction media and isolation of the first conjugate addition intermediate strongly support the double-Michael mechanicistic hypothesis [40].

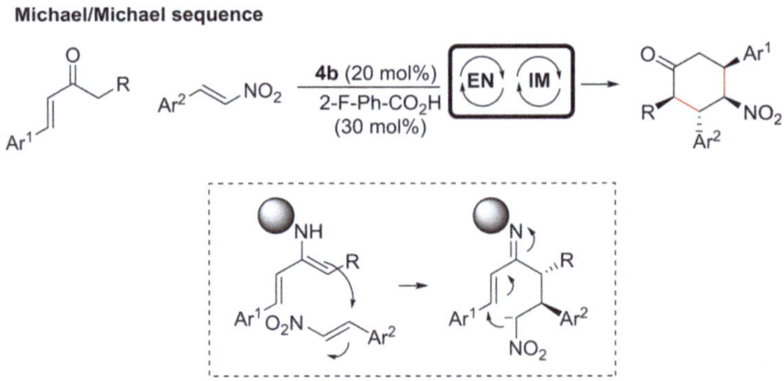

The same cascade, catalysed either by primary or secondary amines, successfully lead to a variety of cyclic structures with typically excellent stereocontrol [41– 44].

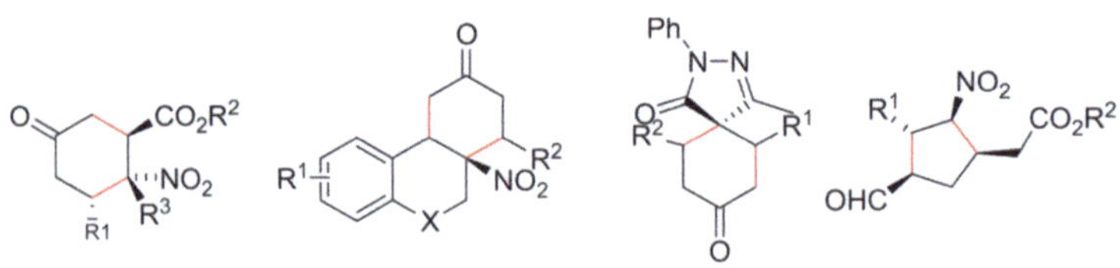

The opposite activation sequence, namely the iminium/enamine cascade, has been even more widely applied. One of the first report was published by the Wang group [45], consisting in the synthesis of densely functionalized cyclopentanes from enals and unsaturated esters bearing a malonate moiety at the γ-position. This activate methylene was the nucleophile attacking the iminium ion generated by condensation between a prolinol derivative catalyst and the aldehyde; this conjugate addition resulted in an intermediate chiral enamine responsible for the subsequent intramolecular Michael addition, affording the five membered products in 85–95% yield, 9:1–20:1 dr a and 94–99% ee.

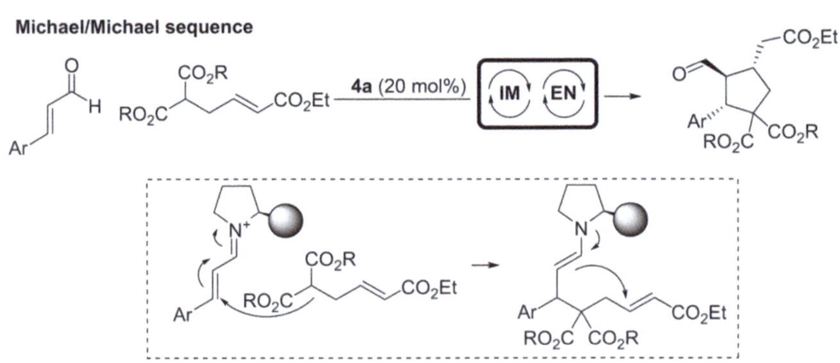

Application of this strategy, based on substrate activation by different primary and secondary amines successfully afforded several cyclic compounds, including heterocycles and spiro-derivatives.

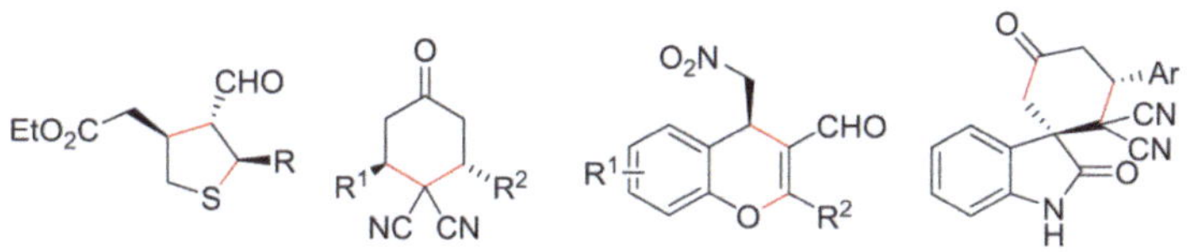

A different approach relies on noncovalent catalysis, as showcased by a work from the Takemoto group and pursued by several other authors. The use of bifunctional catalysts allows in fact deprotonation of methylene nucleophiles and simultaneous activation of the electrophilic counterpart. While the simple diaminocyclohexane derived thiourea allowed only modest ees [46], more hindered catalysts offering more hydrogen bonding sites ensured higher enantioselectivity [47].

Michael/Michael sequence

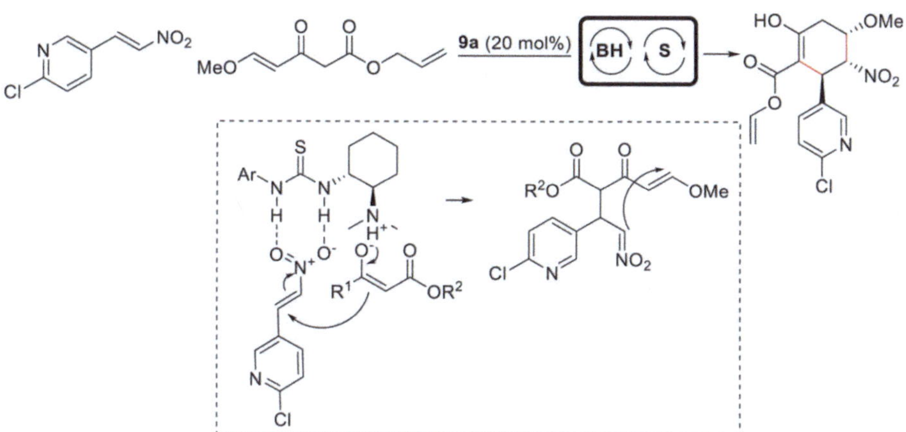

6.2.2 1,2-Additions initiated sequences

Cascades initiated by 1,2-additions have also been developed. When the domino is started by an aldol or a Mannich reaction, the first nucleophilic event is accomplished by α,β-unsaturated carbonyl species, which may be activated as enamines or dienamines. The intermediate iminium ion will then undergo conjugate addition with consequent carbon-carbon or carbon-heteratom bond formation. This strategy was applied by the Itoh group to access in 85% yield and almost complete stereocontrol a key intermediate en route to *ent*-dihydrocorynantheol, an archetypal indole alkaloid exhibiting antiparasitic, antiviral or analgesic activities. As for the proposed mechanism in this specific case, the enone substrate is catalytically turned into the corresponding dienamine, nucleophilic at its C2 position to attack the endocyclic imine. The resulting iminium ion is instead attacked by the amine function.

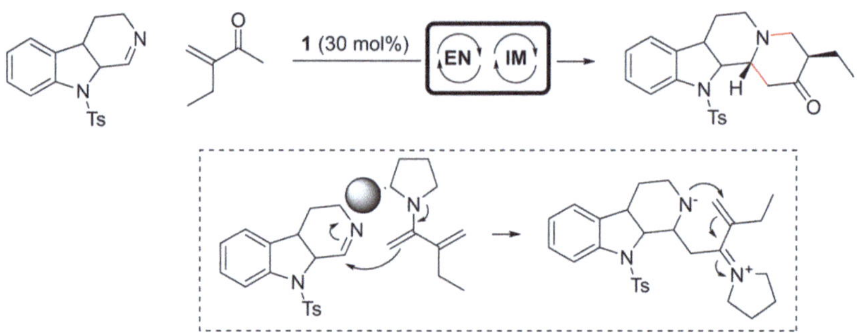

A convergent method was developed by Ramachary et al., who looked at it as at an asymmetric three-component Diels–Alder (ATCDA). Here, the amino acid catalyst, namely proline, is responsible for the formation of both reaction partners of the subsequent Diels-Alder step. In fact, the dienophile is represented by the alkylidene Meldrum's acid, generated by proline-promoted Knowenagel reaction, and the diene by the dienamine deriving from condensation between proline and the enal. Different pyrrolidine-like secondary amines were tested and the best results in terms of yield (up to 85%) and ee (up to 91%) was obtained with amino acids [48].

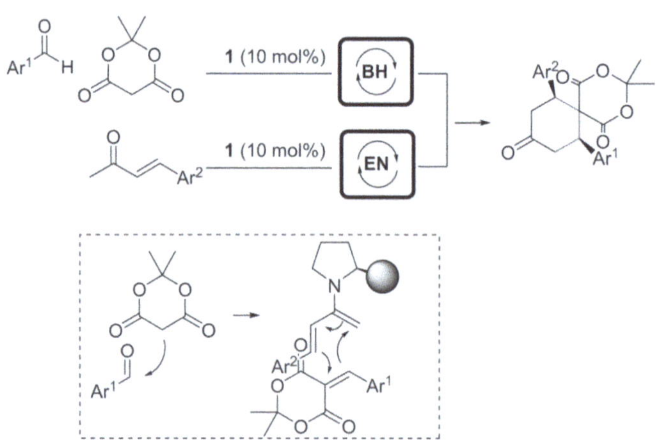

Exploiting Brønsted acid catalysis by chiral phosphoric acids, List et al. developed a stereoselective synthesis of β-hydroxythiols in yields generally higher than 80%. As supported by NMR analysis, the reaction proceeds through the formation of a dimeric complex between the phosphoric acid and thiobenzoic acid. In analogy to what has been found for self-assembled complexes between phosphoric acid catalysts and carboxylic acids, the observed dimer shows an increase in both the nucleophilicity of the thiocarboxylic acid and in the acidity of the complex. Indeed, epoxide ring opening occurs by the thiol group with enantioselectivity affording higher that 95:5 er. Mildly pushing reaction conditions allows the rate determining acyl transfer, thus realizing a tandem process [49].

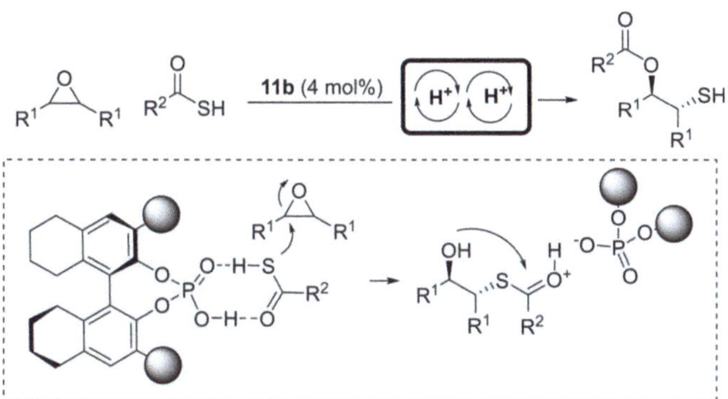

6.2.3 Diels-Alder-type reaction initiated sequences

Domino sequences can also be started by Diels-Alder reactions, known to be intrinsically highly efficient transformations especially in their stereoselective versions that allow to rapidly generate enantioenriched heterocycles and carbocycles featuring complex architectures and dense functionalization.

The McMillan group developed a Diels-Alder/Mannich reaction cascade between an ad-hoc prepared indole derivative and propynal, promoted by an imidazolidinone catalyst. According to the working hypothesis, the unsaturated aldehyde was activated as an iminium ion and underwent a [4 + 2] cycloaddition. In the so formed tricyclic intermediate, an enamine-iminium ion tautomerization involving the PMB protected heterocyclic nitrogen, allowed then to unmask the electrophilic function, which is thus intramolecularly attacked by the *N*-Boc amine. Optimization of the catalysts/co-catalyst system allowed obtaining, in no less than 80% yield, ees higher than 94% for this tetracyclic core that is only five steps away from Minfiensine, an isolation product from *Strychnos minfiensis* [50].

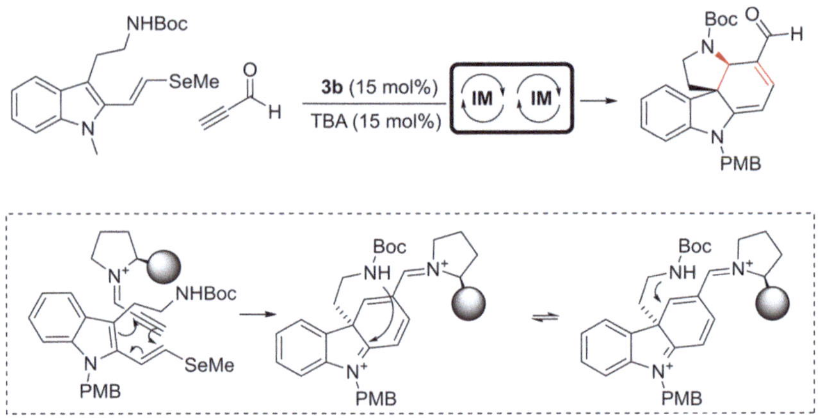

Shortly after, the same group developed a closely related Diels-Alder/Michael addition sequence exploiting the same catalytic system and starting from the same aldehyde and an analogous indole derivative, where the sulfide pendant is replaced by a selenide one. After the cycloaddition step, elimination occurs giving access to a conjugated intermediate that could potentially undergo both a 1,2- and a 1,4-addition step. Likely, a series of equilibria take place and, differently from what is observed in the previous case, the non-bridged tetracyclic intermediate is the one that evolves toward the product [51].

As alternative dienophiles, α,β–unsaturated acylammonium salts have been employed by the Romo group in a Diels-Alder/lactonization sequence. Employing chiral isothioureas catalysts, the authors developed the first enantioselective organocatalytic Diels-Alder reaction involving in situ activated α,β-unsaturated acid chlorides. With a Danishefsky diene derivative featuring a hydroxyl pendant, the reaction is postulated to proceed via the formation of a chiral ammonium salt upon condensation between a tertiary amine and the acyl chloride, [4 + 2]-cyclization and final lactonization. To suppress the direct participation of acyl chlorides in the cyclization step, their slow addition to the reaction mixture proved effective: avoiding a large excess of this reaction partner, the relative concentration between the acyl chloride itself and its catalytically generated chiral derivative was such to enable the stereoselective pathway to compete with the background one. An additional Brønsted base is needed and revealed influencing the *endo/exo* selectivity, even though no rationalization was yet proposed. Working under optimized conditions, bicyclic products were obtained in good yields (between 46 and 99%) and excellent ee (between 91 and 99%). Using a racemic diene, stereodivergent resolution was observed leading to separable tricyclic γ-lactones. The method was extended to in situ activated carboxylic acids, and applied on gram scale as well [52].

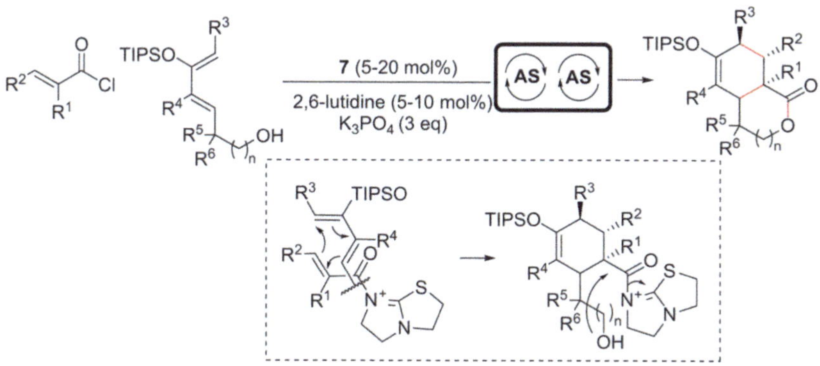

6.3 Triple cascade sequences

Evolution within the field lead to methods where more than two events were concatenated.

6.3.1 Michael-type reaction initiated sequences

In the contribution by the Enders group, three components were combined in the presence of a single organocatalyst leading to a cyclohexene with four stereogenic centers with high selectivity [53]. The secondary amine catalyst gave aldehyde activation via enamine formation; difference in reactivity between the two Michael acceptors allowed the 1,4-addition to be selective for the nitroalkene over the enal. The enal underwent condensation with the prolinol catalyst performing the typical iminium ion/enamine sequence. Only two out of the eight possible diastereoisomers were observed, the major one being enantiopure. The same kind of reactivity was exploited by Melchiorre et al. to obtain analogous cyclic derivatives featuring a quaternary stereocenter [54].

Michael/Michael/Aldol sequence

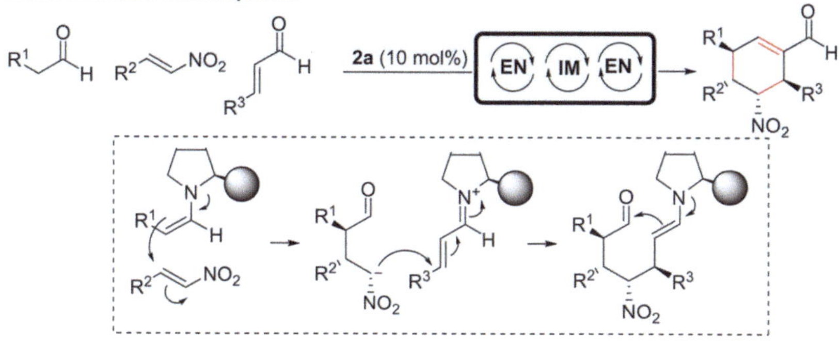

Aminocatalysis has been also applied to develop diverse iminium ion/iminium ion/enamine tandem reactions [55], giving access to highly functionalized cyclic products in good yield and enantioselectivity typically higher than 95%.

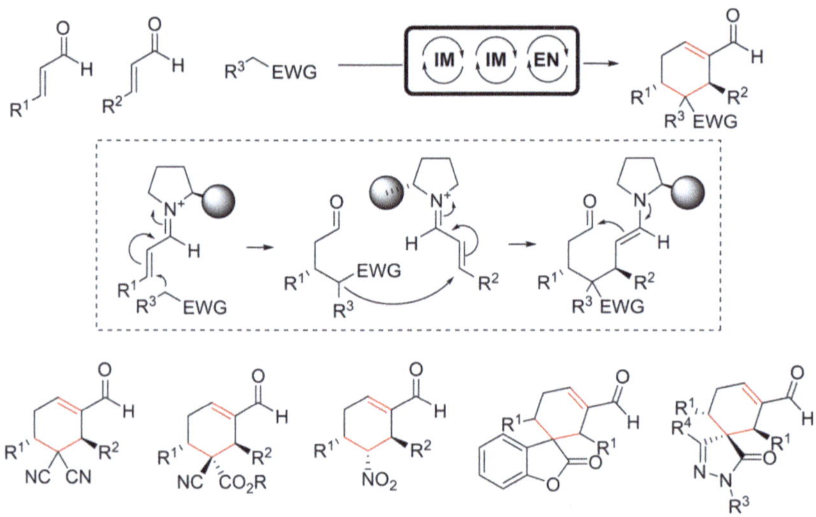

The Michael/Michael/Aldol sequence was also achieved relying on Brønsted base catalysis by a bifunctional thiourea for the first step, followed by iminium ion activation. In this example by Dixon et al., diesters are used as activated methylene nucleophiles attacking the nitrostyrene derivative in a stereoselective way thanks to simultaneous coordination of the two reaction partners to the Brønsted base/hydrogen bonding catalyst. The α,β-unsaturated aldehyde condenses with the prolinol to undergo the second 1,4-addition and subsequent Aldol condensation [56]. This procedure led to a mixture of two diastereoisomers, the major of which in >98% ees.

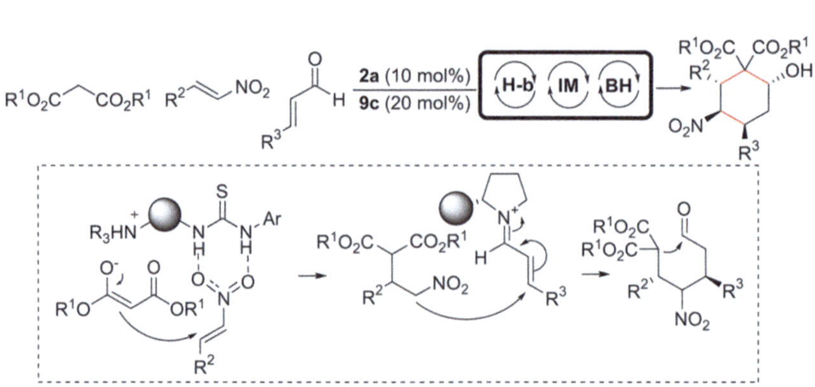

Replacing the ring-closing Aldol step with a Henry reaction, the group of Wang prepared cyclohexanols featuring six stereogenic centers with good diastereoselection and excellent enantiocontrol (ees >99%). Two catalysts, namely a thiourea bifunctional one and a secondary amine, were employed and proved to act independently. Despite relying on the same catalysts combination, this reaction proceeds though a different activation sequence with respect to that reported by the Dixon group. Indeed, the protected prolinol provides the formation of the chiral nucleophile, spontaneously undergoing addition to the first equivalent of nitroalkene; this hypothesis was supported by the sole formation of this first addition intermediate when carrying out the reaction without the thiourea catalyst. For the second Michael reaction, instead, hydrogen bonding activation of the electrophile is needed. The final base- promoted Henry step proceeded immediately, as in fact no acyclic product was ever isolated [57].

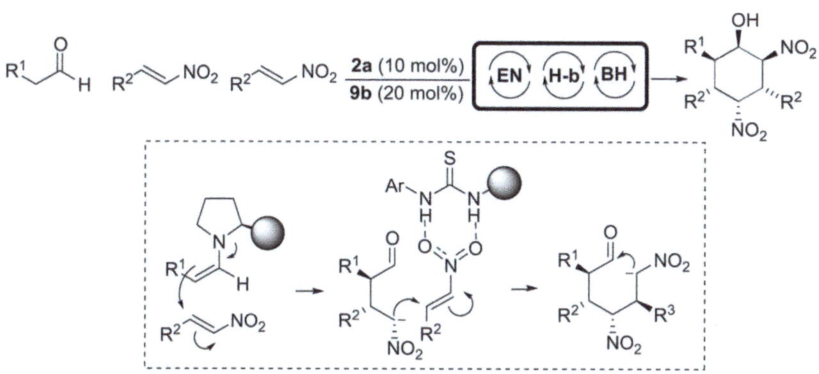

6.3.2 1,2-Additions initiated sequences

One of the first examples was reported by Còrdova et al., in continuation with their previous work on the proline-catalyzed assembly of three acetaldehyde molecules providing (+)-5-hydroxy-(2E)-hexenal [58]. A proline-promoted Aldol/Aldol/acetalization sequence led to trimeric lactols with four stereogenic centers. These pyranoses were isolated as 1:2 mixtures of α/β anomers upon slow addition of two equivantes of propionaldehyde to acceptor aldehydes in up to 53% yield and 33% ee [59].

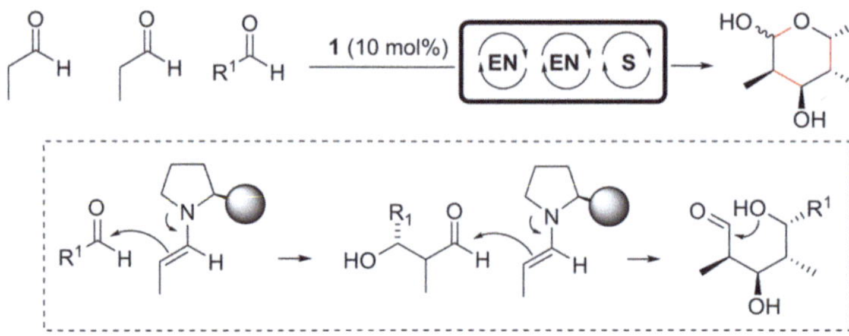

Exploiting the enamine/iminium ion activation cascade, the Woggon group got access to the lactol deriving from the condensation between phytenal and an *ortho*-hydroxybenzaldehyde in 58% yield and 97% ee. This intermediate, generated upon an Aldol/Michael/acetalization domino process, is a key intermediate in the total synthesis of α-tocopherol [60].

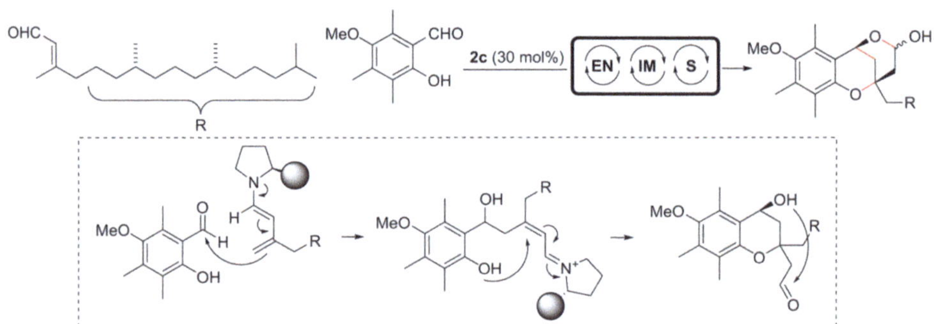

To build a longer-than-two-steps cascades, Knoevenagel reactions can be exploited as first transformation to allow a subsequent Michael addition, which is known to turn a β-acceptor into an α-donor intermediate able to react on its turn. An application of this strategy was reported by the Yuan group and consisted in a three component reaction between differently *N*-protected isatins, malonitrile and diketones or β-ketoesters mediated by cupreine, a cinchona alkaloid [61]. According to the proposed mechanism, the reaction proceeds via condensation between the 2,3-dioxoindoline and malonitrile leading an α,β-unsaturated intermediate. The subsequent conjugate addition is promoted by the bifunctional catalyst: its tertiary amine function deprotonates the 1,3-dione and stays associated with it being its counteranion while engaging in hydrogen bonding with the electrophile reaction partner. The final cyclization affords the desired spiro[4 H-pyran-3,3′-oxindoles]. Under optimized conditions, this Knoevenagel/Michael/cyclization sequence proceeds with yields typically higher than 90% and ee up to 97% [62–64].

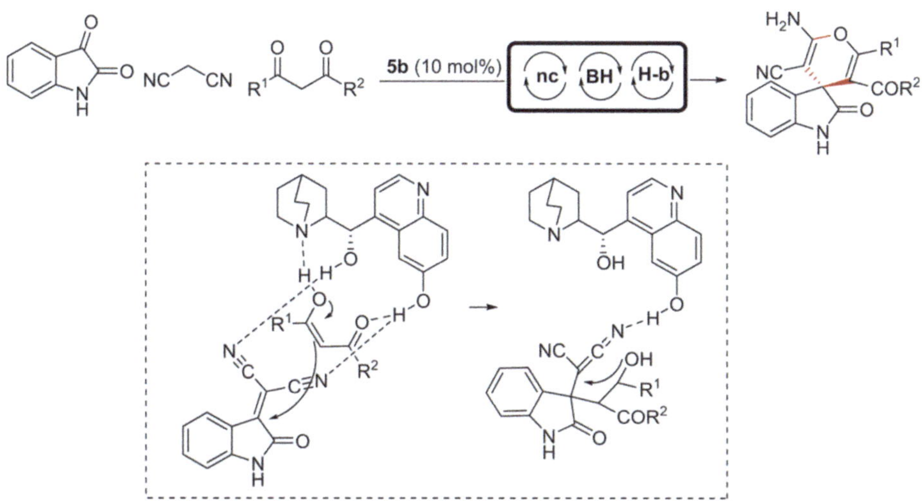

6.3.3 Miscellaneous

Rueping et al. reported a convergent catalysis approach, where two oxidative cycles lead to the in situ generation of an aldehyde and an enal derivative starting from an allylic alcohol and a 2-amino benzyl alcohol. A prolinol catalyst allowed their subsequent activation for an aza-Michael initiated iminium ion-enamine sequence leading to the desired 1,2-dihydroquinolines in 40–80% yield and ee often up to 99%. The relevance of this strategy does not lie only in the originality of the cascade design, but also in its formally enabling aminocatalysis on primary alcohol, more desirable substrates than the corresponding carbonyl compounds [65].

oxidation/aza-Michael/Aldol sequences

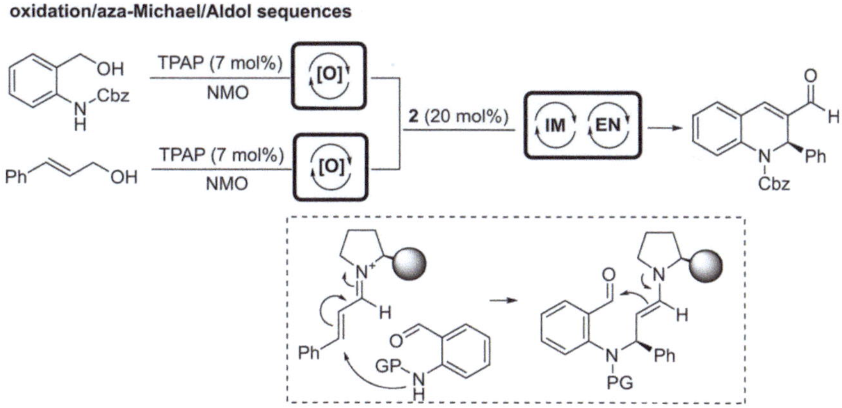

6.4 Quadruple cascade sequences

The need for methods giving access to complex skeletons in a single step, avoiding lengthy and time-consuming paths, prompt to reach a further level of sophistication in designing domino events. Indeed, transformations involving four sequential reactions have been implemented. These elaborate dominos acquire a higher synthetic impact when heading to articulated molecular structures. The employed starting materials need to feature proper functionalities in the key positions, and this translates in different levels of substrates customization, ranging from the simple introduction of aromatic substituents to the preparation of tailor made derivatives.

6.4.1 Michael-type reaction initiated sequences

Coherently with what observed in shorter sequences, also in the context of quadruple ones the most start with a Michael-type addition, where the nucleohpìile is either a carbon, and oxygen or a nitrogen center.

Again, exploiting the iminium/enamine activation sequence allowed preparing cyclic structure with multiple stereocenters combining simple enals, nitroalkenes and aldehydes. The Hong and Gong groups simultaneously reported an oxa-Michael initiated quadruple reaction. In both cases, the electrophile is represented by an enal activated as chiral iminium ion. Selectivity for the iminium ion over the nitroalkene as an electrophile is explained by the hard nature of the oxygen nucleophile. In a three component reaction, Hong demonstrated that two different enals can be simultaneously employed and still have the formation of a single product thanks to a steric hindrance-based selectivity [66]. Gong developed instead a four component reaction where the same unsaturated aldehyde is incorporated twice in the final product [67]. In both cases, pretty good yields (about 50%) where accompanied by complete enantioselection.

Oxa-Michael/Michael/Michael/Aldol Condensation Sequence

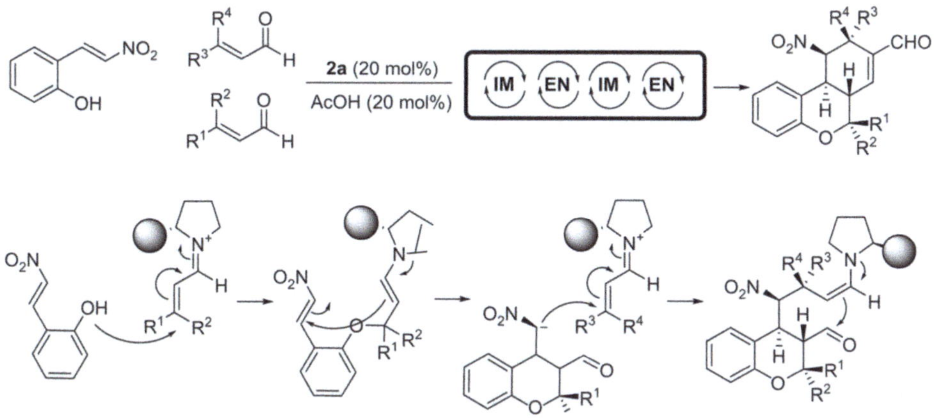

An aza-Michael initiated sequence as well as all carbon-carbon bond forming sequences were reported by the Enders group [68–70].

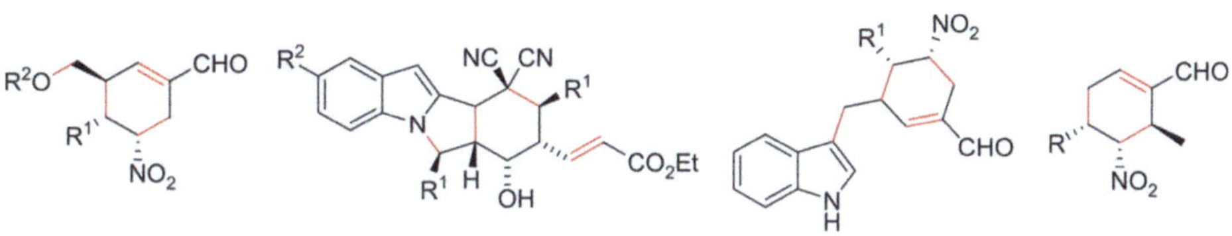

Besides aminocatalysis, involving enamine and iminium ion activation by primary and secondary amines, Lewis base catalysis by tertiary amines has also been exploited within quadruple cascades, although not being responsible for all the steps. In fact, the Romo group developed a Michael/Michael/aldol/β-lactonization to form highly functionalized bi- and tricyclic β-lactones where the only the last three steps are indeed born by the chiral isothiourea catalysts, while the first conjugate addition occurs by means of a stoichiometric base. The initial racemic Michael adduct undergoes kinetic resolution in the subsequent 1,4-addition on the chiral acyl ammonium salt, generated in situ by condensation between the tertiary amine and the acyl chloride. The proposed mechanism involves intermediate species existing in rig conformations due to noncovalent interactions, including a mo – s*s donation that keeps the acyl ammonium in an s-cis geometry and Lithium bidentate chelation of the amide enolate. As the resolution process was not made effectively dynamic, yields rarely exceeded 50%; however, almost complete diastereoselection and er between 82:18 and 97.5:2.5 were obtained [71].

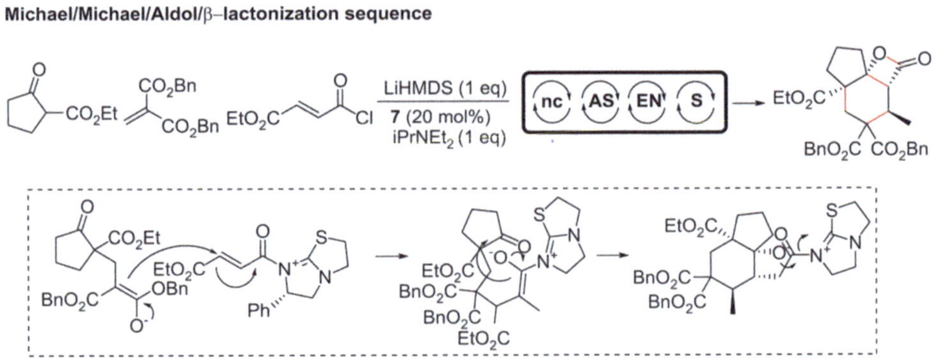

6.4.2 1,2-Additions initiated sequences

A Knoevenagel reaction initiated sequence was reported by Chang et al. Starting from simple commercially available compounds, this three component quadruple reaction gave access to spirocyclic products in good yield (about 70%) and excellent dr (often higher than 19:1) and ee (between 85 and 98%). After the addition/elimination of the dione on benzaldehye. the quinine-derived thiourea catalyst acts both as a hydrogen bond donor and Brønsted base to promote and stereochemically control the subsequent steps via noncovalent interactions with both the reaction partners [72].

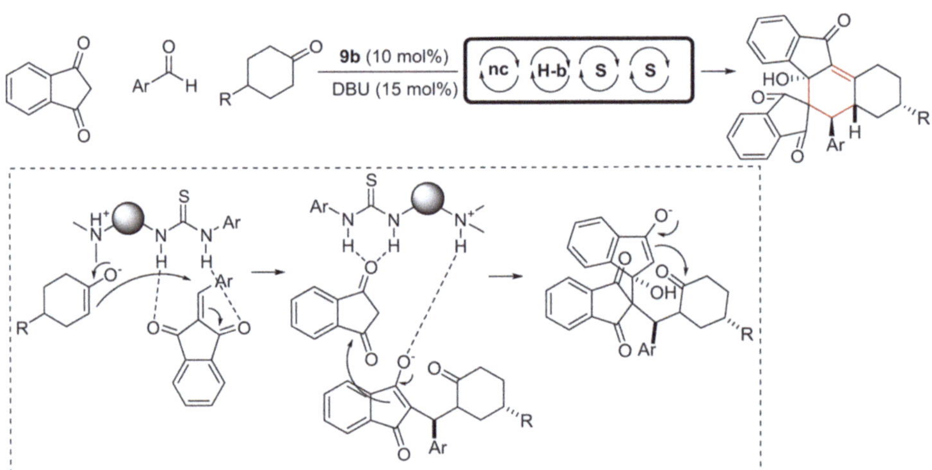

Maybe less exploited but still well established as a powerful enantioselective transformation, the benzoin reaction has also been applied as first step in cascades. The Bode group has been active in studying N-heterocyclic carbene as chiral catalysts and in this context developed a highly enantioselective *cis*-cyclopentene-forming annulation via intermolecular benzoin condensation-oxy-Cope rearrangement followed by an intramolecular Aldol reaction. In particular, the authors found that the use of a triazolium salt in the presence of a strong base allowed a cross-benzoin between enals and 4-oxoenones, giving rise to an intermediate that spontaneously undergoes a sigmatropic rearrangement. These couples reactions led to a species featuring both an enolate and an enol: as the latter tautomerizes, its attacked in a ring-closing 1,2-addition followed by lactonization and decarboxylaton. The products were obtained in good yields (ranging from 50 t 93%) and almost complete diastereo- and enantioselection. The stereochemical outcome was justified considering the electrophile was mainly living in an *s-cis* conformation and the oxy-Cope step proceeded via a boat TS.

Cross-benzoin/oxy-Cope/Aldol/decarboxylation sequence

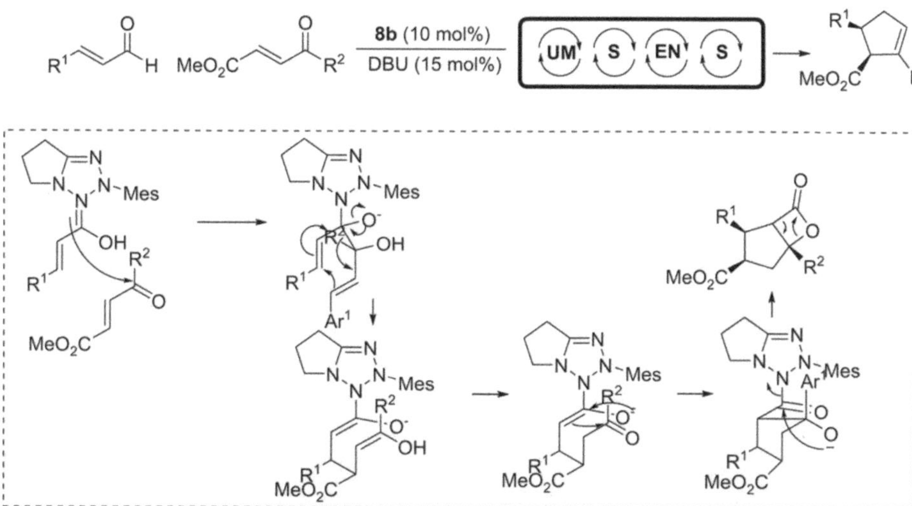

Within his studies on NHC-promoted reactions, Bode et al. also developed a four step cascade started by a cross-benzoin/oxy-Cope reaction between enals and chalcone-derived imines. A substoichiometric achiral base turned the triazolium precatalyst into the active carbene able to condense with the aldehyde and generate the Breslow intermediate. Concertedly, 1,2-addition to the imine and oxy-Cope rearrangement led to an enolate species, which was the nucleophile for the subsequent Mannich reaction; nitrogen attack onto the carbonyl allowed the released of the catalyst and formation of ring fused β-lactams in 60–80% yield and 88–99% ee. These outstanding results were obtained thanks to a careful tuning of reaction conditions, especially in the choice of the precatalyst/base combination, as the desired transformation has to outcompete the highly probable enal dimerization and aza-Diels-Alder reaction [73].

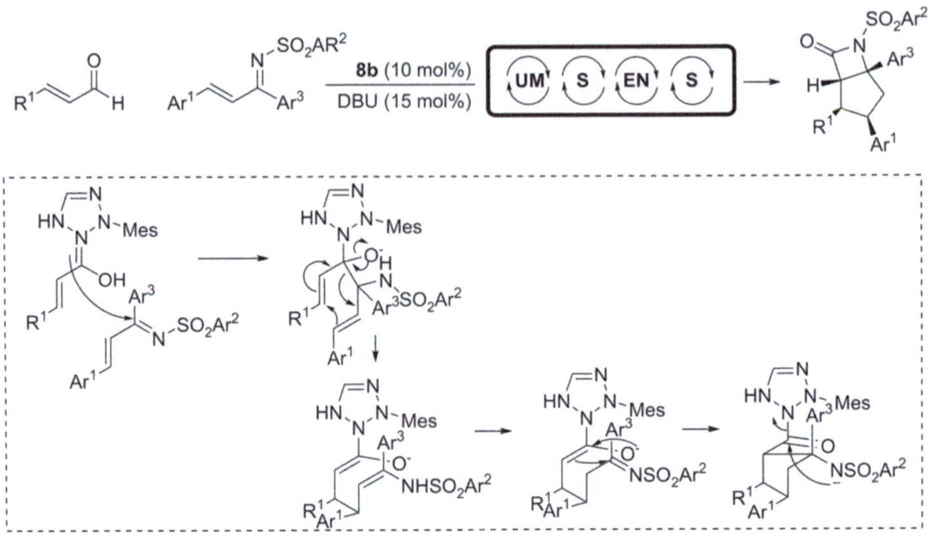

6.4.3 Miscellaneous

Reduction by Hantzsch ester was also exploited as the initial step of the hydrogenation/Michael/Michael/Aldol condensation sequence reported by Rueping et al. In this work, selective transfer hydrogenation of enals over nitroalkenes was achieved by the substrate activation as iminium ion. Hydride donation thus led to an enamine intermediate undergoing addition to the nitrostyrene derivative. Using an excess of the unsaturated aldehyde allowed a further 1,4-addition and a ring-closing crotonic step. The full four-step sequence occurred in about 50% yield and ee typically higher than 99% [74].

Hydrogenation/Michael/Michael/Aldol reaction sequence

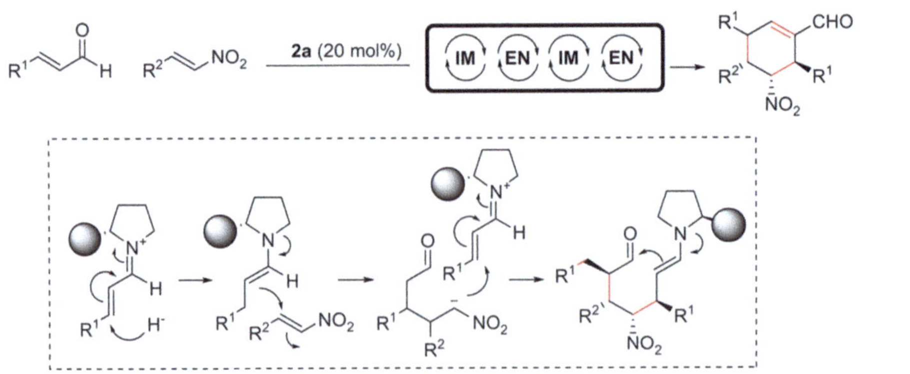

Oxidation was instead a key step in the synthesis of δ-lactone reported by the Bijou group. In this work, the Breslow intermediated formed by condensation between the NHC catalyst and an enal is turned into the corresponding chiral α,β-unsaturated acyl azolium by the bisquinone oxidant. The so formed intermediate undergoes an aza-Michael attack by a tailor made indole, featuring an enone group at the 7-position; this pendant becomes the protagonist of the subsequent intramolecular conjugate addition and lactonization. The tetracyclic products were obtained in yield typically higher than 80%, complete diastereoselectivity (dr >20:1) and excellent enantioselection (er >90:10) [75].

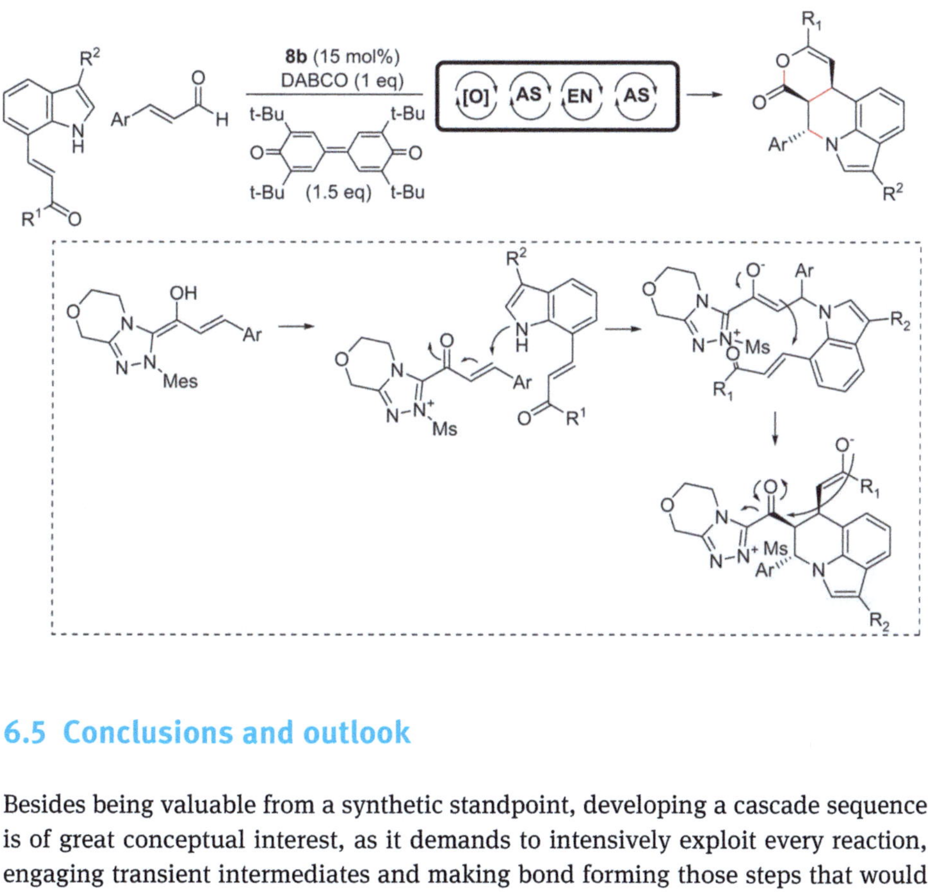

6.5 Conclusions and outlook

Besides being valuable from a synthetic standpoint, developing a cascade sequence is of great conceptual interest, as it demands to intensively exploit every reaction, engaging transient intermediates and making bond forming those steps that would otherwise have represented the quenching event.

As we have just shown, organocascades are meant to enable effective synthesis of structurally and stereochemically complex compounds. Even if tailor made substrates are often required, especially for long sequences and/or to obtain densely decorated products, their preparation is made worthy by the specificity of the connectivity achieved; (poli)cyclic compounds featuring several stereogenic centers are often prepared. For their efficacy organocascades find application as key steps in the synthesis natural products [76], including bioactive ones [77].

Abbreviations

PMB	-*ortho*-methoxybenzene
Boc	tert-butyloxycarbonyl
Ts	tosyl
Ms	mesyl
TFA	trifluoroacetic acid
TBA	tribromoacetic acid
ee	enantiomeric excess
er	enantiomeric ratio
dr	diastereoisomeric ratio
TMS	tetramethylsilyl
TPAP	tetrapropylammonium perruthenate
NMO	*N*-methylmorpholine N-oxide
TS	transition state

References

[1] Gaich T, Baran PS. Aiming for the ideal synthesis. J Org Chem. 2010;75:4657–73.
[2] (a) Newhouse T, Baran PS, Hoffmann RW. The economies of synthesis. Chem Soc Rev. 2009;38:3010–3021. (b) Wender PA, Verma VA, Paxton TJ, Pillow TH. Function-oriented synthesis, step economy and drug design. Acc Chem Res. 2008;41:40. (c) Wender PA, Miller BL Synthesis at the molecular frontier. Nature. 2009;460:197.
[3] Hayashi Y. Pot economy and one-pot synthesis. Chem Sci. 2016;7:866–80.
[4] (a) Nicolaou KC, Edmonds DJ, Bulger PG. Domino reactions: concepts for efficient organic synthesis. Angew Chem Int Ed. 2006;45:7134–86.
[5] (a) Tietze LF. Domino reactions in organic synthesis. Chem Rev. 1996;96:115. (b) Tietze LF, Beifuss U. Sequential transformations in organic synthesis: a synthetic strategy with a future. Angew Chem Int Ed. 1993;32:131.
[6] Mayer SF, Kroutil W, Faber K. Enzyme-initiated domino (cascade) reactions. Chem Soc Rev. 2001;30:332–9.
[7] Fogg DE, Dos Santos EN. Tandem catalysis: a taxonomy and illustrative review. Coord Chem Rev. 2004;248:2365.
[8] Chapman CJ, Frost CG. Tandem and domino catalytic strategies for enantioselective synthesis. Synthesis. 2007;1:1–21.
[9] Denmark SE, Thorarensen A. Tandem [4+2]/[3+2] cycloadditions of nitroalkenes. Chem Rev. 1996;96:137–65.
[10] Walji AM, MacMillan DW. Strategies to bypass the taxol problem. Enantioselective cascade catalysis, a new approach for the efficient construction of molecular complexity. Synlett. 2007;10:1477–89.

[11] (a) Wasilke J-C, Obrey SJ, Baker RT, Bazan GC. Concurrent tandem catalysis. Chem Rev. 2005;105:1001–1020. (b) Guo H-C, Ma JA. Catalytic asymmetric tandem transformations triggered by conjugate additions. Angew Chem Int Ed. 2006;45:354–66.
[12] Volla CM, Atodiresei I, Rueping M. Catalytic C–C bond-forming multi-component cascade or domino reactions: pushing the boundaries of complexity in asymmetric organocatalysis. Chem Rev. 2014;114:2390–431.
[13] Wende RC, Schreiner PR. Evolution of asymmetric organocatalysis: multi- and retrocatalysis. Green Chem. 2012;14:1821–49.
[14] Allen AE, MacMillan DW. Synergistic catalysis: a powerful synthetic strategy for new reaction development. Chem Sci. 2012;3:633–58.
[15] Ramon DJ, Yus M. Asymmetric multicomponent reactions (AMCRs): the new frontier. Angew Chem Int Ed. 2005;44:1602.
[16] Indu S, Kaliappan KP. A new and informative [a,b,c,d] nomenclature for one-pot multistep transformations: a simple tool to measure synthetic efficiency. RSC Adv. 2018;8:21292.
[17] Albrecht L, Jiang H, Jørgensen KA. A simple recipe for sophisticated cocktails: organocatalytic one-pot reactions—concept, nomenclature, and future perspectives. Angew Chem Int Ed. 2011;50:8492–509.
[18] For the pioneering work in the field see. Hajos ZG, Parrish DR. Asymmetric synthesis of bicyclic intermediates of natural product chemistry. J Org Chem. 1974;39:1615–21.
[19] Zhong G, Hoffmann T, Lerner RA, Danishefsky S, Barbas CF, III. Antibody-catalyzed enantioselective Robinson annulation. J Am Chem Soc. 1997;119:8131.
[20] Bui T, Barbas CF, III. A proline-catalyzed asymmetric Robinson annulation reaction. Tetrahedron Lett. 2000;41:6951.
[21] (a) Bahmanyar S, Houk KN. Transition states of amine-catalyzed aldol reactions involving enamine intermediates: theoretical studies of mechanism, reactivity, and stereoselectivity. J Am Chem Soc. 2001;123:11273. (b) Bahmanyar S, Houk KN, Martin HJ, List B. Quantum Mechanical Predictions of the Stereoselectivities of Proline-Catalyzed Asymmetric Intermolecular Aldol Reactions. J Am Chem Soc. 2003;125:2475. (c) Clemente FR, Houk KN. Theoretical studies of stereoselectivities of intramolecular aldol cyclizations catalyzed by amino acids. J Am Chem Soc. 2005;127:11294.
[22] Hepburn HB, Dell'Amico L, Melchiorre P. Enantioselective vinylogous organocascade reactions. Chem Rec. 2016;16:1787–806.
[23] Tian X, Melchiorre P. Control of remote stereochemistry in the synthesis of spirocyclic oxindoles: vinylogous organocascade catalysis. Angew Chem Int Ed. 2013;52:5360–3.
[24] Stiller J, Poulsen PH, Cruz D, Dourado J, Davis RL, Jørgensen KA. Organocatalytic [4+2] addition reactions *via* tetraenamine intermediate. Chem Sci. 2014;5:2052–6.
[25] Kaneko S, Yoshino T, Katoh T, Terashima S. Synthetic studies of huperzine A and its fluorinated analogues. 1. Novel asymmetric syntheses of an enantiomeric pair of huperzine A. Tetrahedron. 1998;54:5471.
[26] Cabrera S, Alemán J, Bolze P, Bertelsen S, Jørgensen KA. An unexpected organocatalytic asymmetric tandem Michael/Morita–Baylis–Hillman reaction. Angew Chem Int Ed. 2008;47:121.
[27] Lathrop SP, Rovis T. Asymmetric synthesis of functionalized cyclopentanones via a multicatalytic secondary Amine/*N*-Heterocyclic carbene catalyzed cascade sequence. J Am Chem Soc. 2009;131:13628–30.
[28] Hayashi Y, Okano T, Aratake S, Hazelard D. Diphenylprolinol silyl ether as a catalyst in an enantioselective, catalytic, tandem Michael/Henry reaction for the control of four stereocenters. Angew Chem Int Ed. 2007;46:4922.

[29] Reyes E, Jiang H, Milelli A, Elsner P, Hazell RG, Jørgensen KA. How to make five contiguous stereocenters in one reaction: asymmetric organocatalytic synthesis of pentasubstituted cyclohexanes Angew. Chem Int Ed. 2007;6:9202.

[30] Tan B, Chua PJ, Li Y, Zhong G. Organocatalytic asymmetric tandem Michael–Henry Reactions: A highly stereoselective synthesis of multifunctionalized cyclohexanes with two quaternary stereocenters. Org Lett. 2008;10:2437.

[31] Tan B, Lu Y, Zeng X, Chua PJ, Zhong G. Facile domino access to chiral bicyclo[3.2.1]octanes and discovery of a new catalytic activation mode. Org Lett. 2010;12:2682.

[32] Albertshofer K, Tan B, Barbas CF, III. Assembly of spirooxindole derivatives containing four consecutive stereocenters via organocatalytic Michael–Henry cascade reactions. Org Lett. 2012;14:1834.

[33] Dickmeiss AG, Acosta FC, Rodriguez-Escrich C, Davis RL, Jørgensen KA. Asymmetric organocatalytic formal [2 + 2]-Cycloadditions via Bifunctional H-bond directing Dienamine catalysis. J Am Chem Soc. 2012;134:2543–6.

[34] Talavera G, Reyes E, Vicario JL, Carrillo L. Cooperative Dienamine/ Hydrogen-Bonding catalysis: Enantioselective Formal [2+2] Cycloaddition of Enals with Nitroalkenes. Angew Chem Int Ed. 2012;51:4104–7.

[35] Kang G, Yamagami M, Vellalath S, Romo D. Enantioselective synthesis of medium-sized Lactams via Chiral α,β-Unsaturated Acylammonium salts. Angew Chem Int Ed. 2018;57:6527–31.

[36] Jin J-H, Li X-Y, Luo X, Deng W-P. Enantioselective synthesis of indolo[2,3-*b*]-dihydrothiopyranones via [3+3] cycloaddition of chiral *α,β*-unsaturated acylammonium salts. Tetrahedron. 2018;74:6804–8.

[37] Rueping M, Antonchick AP, Theissmann, TA. Highly Enantioselective Brønsted acid catalyzed cascade reaction: organocatalytic transfer hydrogenation of quinolines and their application in the synthesis of Alkaloids. Angew Chem Int Ed. 2006;45:3683–6.

[38] Rueping M, Antonchick AP. Organocatalytic enantioselective reductions of pyridines. Angew Chem Int Ed. 2007;46:4562–5.

[39] Nayak S, Panda P, Bhakta S, Mishra SK, Mohapatra S. Current advances of organocatalytic Michael-Michael cascade reaction in the synthesis of highly functionalized cyclic molecules. RSC Adv. 2016;6:96154–75.

[40] Wu LY, Bencivenni G, Mancinelli M, Mazzanti AB, Melchiorre P. Organocascade reactions of enones catalyzed by a chiral primary amine. Angew Chem Int Ed. 2009;48:7196–9.

[41] Massolo E, Benaglia M, Annunziata R, Palmieri A, Celentano G, Forni A. Stereoselective synthesis of functionalized Chiral 2-Nitrocyclohexanecarboxylic Esters *via* Catalytic Dienamine Addition to β-Substituted β-Nitroacrylates. Adv Synth Catal. 2014;356:493.

[41] Li JH, Du DM. Enantioselective cascade double Michael addition of 3-nitro-2*H*-chromenes and acyclic enones: efficient synthesis of functionalized tricyclic chroman derivatives. Org Biomol Chem. 2015;13:9600–9.

[43] Liang J, Chen Q, Liu L, Jiang X, Wang R. An organocatalytic asymmetric double Michael cascade reaction of unsaturated ketones and unsaturated pyrazolones: highly efficient synthesis of spiropyrazolone derivatives. Org Biomol Chem. 2013;11:1441–5.

[44] An Q, Shen J, Butt N, Liu D, Liu Y, Zhang W. Asymmetric domino double Michael addition of nitroolefins and Aldehyde esters with *trans*-Perhydroindolic acid as an organocatalyst. Synthesis. 2013;45:1612–23.

[45] Zu L, Li H, Xie H, Wang J, Jiang W, Tang Y, et al. Synthesis of highly functionalized chiral cyclopentanes by catalytic enantio- and diastereoselective double michael addition reactions angew. Chem Int Ed. 2007;46:3732–4.

[46] Miyabe H, Takemoto Y. Discovery and application of asymmetric reaction by multi-functional thioureas. Bull Chem Soc Jpn. 2008;81:785–95.

[47] Wang XF, An J, Zhang XX, Tan F, Chen JR, Xiao WJ. Catalytic asymmetric aza-michael addition cascade: enantioselective synthesis of polysubstituted 4-aminobenzopyrans. Org Lett. 2011;13:808–11.

[48] Ramachary DB, Chowdari NS, Barbas CF, III. Organocatalytic asymmetric domino knoevenagel/diels-alder reactions: a bioorganic approach to the diastereospecific and enantioselective construction of highly substituted spiro[5,5]undecane-1,5,9-triones. Angew Chem Int Ed. 2003;42:4233.

[49] Monaco MR, Prévost S, List B. Catalytic asymmetric synthesis of thiols. J Am Chem Soc. 2014;136:16982–5.

[50] Jones SB, Simmons B, MacMillan DW. Nine-step enantioselective total synthesis of (+)-Minfiensine. J Am Chem Soc. 2009;131:13606.

[51] Jones SB, Simmons B, Mastracchio A, MacMillan DW. Collective synthesis of natural products by means of organocascade catalysis. Nature. 2011;475:183.

[52] Abbasov ME, Hudson BM, Tantillo DJ, Romo D. Acylammonium salts as dienophiles in Diels-Alder/lactonization organocascades. J Am Chem Soc. 2014;136:4492–5.

[53] Enders D, Hüttl MR, Grondal C, Raabe G. Control of four stereocenters in a triple cascade organoctalytic reaction. Nature. 2006;441:861–3.

[54] Penon O, Carlone A, Mazzanti A, Locatelli M, Sambri L, Bartoli G, et al. Quaternary stereogenic carbons in complex molecules by an asymmetric, organocatalytic, triple cascade reaction. Chem Eur J. 2008;14:4788.

[55] (a) Carlone A, Cabrera S, Marigo M, Jørgensen KA. A new approach for an organocatalytic multicomponent domino asymmetric reaction. Angew Chem Int Ed. 2007;46:1101. (b) Enders D, Jeanty M, Bats JW. Organocatalytic asymmetric triple domino reactions of Nitromethane with α,β-unsaturated Aldehydes. Synlett. 2009;3175. (c) Companyó X, Zea A, Alba A-N, Mazzanti A, Moyano A, Rios R. Organocatalytic synthesis of spiro compounds via a cascade Michael-Michael-aldol reaction. Chem Commun. 2010; 46:6953. (d) Alba A-N, Zea A, Valero G, Calbet T, Font-Bardia M, Mazzanti A, Moyano A, Rios R. Highly stereoselective synthesis of spiropyrazolones. Eur J Org Chem. 2011;1318.

[56] Wang Y, Han R-G, Zhao Y-L, Yang S, Xu P-F, Dixon DJ. Asymmetric organocatalytic relay cascades_ catalyst-controlled stereoisomer selection in the synthesis of functionalized cyclohexanes. Angew Chem Int Ed. 2009;48:9834–9838.

[57] Mao Z, Jia Y, Xu Z, Wang R. Asymmetric organocatalytic cascade Michael/Michael/Henry reaction sequence: control of all stereocenters in one Cyclohexane skeleton. Adv Synth Catal. 2012;354:1401.

[58] Córdova A, Notz W, Barbas CF, III. Proline-catalyzed one-step asymmetric synthesis of 5-hydroxy-(2E)-hexenal from acetaldehyde. J Org Chem. 2002;67:301.

[59] Chowdari NS, Ramachary DB, Córdova A, Barbas CF, III. Proline-catalyzed asymmetric assembly reactions: enzyme-like assembly of carbohydrates and polyketides from three aldehyde substrates. Tetrahedron Lett. 2002;43:9591.

[60] Liu K, Chougnet A, Woggon W-D. A short route to a-Tocopherol. Angew Chem Int Ed. 2008;47:5827–9.

[61] Chen W-B, Wu Z-J, Pei Q-L, Cun L-F, Zhang X-M, Yuan W-C. Highly enantioselective construction of Spiro[4H-pyran-3,3′-oxindoles] through a domino Knoevenagel/Michael/Cyclization sequence catalyzed by cupreine. Org Lett. 2010;12:3132–5.

[62] Chen W-B, Wu Z-J, Pei Q-L, Cun L-F, Zhang X-M, Yuan W-C. Highly Enantioselective construction of Spiro[*4H*-pyran-3,3′-oxindoles] through a domino Knoevenagel/Michael/Cyclization sequence catalyzed by cupreine. Org Lett. 2010;12:3132.

[63] For a Knoevenagel/Michael/Michael reaction sequence see. Ramachary DB, Venkaiah C, Krishna PM. Discovery of 2-aminobuta-1,3-enynes in asymmetric organocascade catalysis:

construction of drug-like spirocyclic cyclohexanes having five to six contiguous stereocenters. Chem Commun. 2012; 48: 2252.
[64] For a L-proline-mediated Knoevenagel/Michael/acetalization reaction see: Rueping M, Merino E, Bolte M. Efficient proline and prolinol ether mediated 3-component synthesis of 3- and 3,4-substituted chromenone derivatives. Org Biomol Chem. 2012; 10: 6201.
[65] Rueping M, Dufour J, Bui L. Convergent catalysis: asymmetric synthesis of dihydroquinolines using a combined metal catalysis and organocatalysis approach. ACS Catal. 2014;4:1021–5.
[66] Kotame P, Hong B-C, Liao J-H. Synthesis of 2-chromanones. Tetrahedron Lett. 2009;50:4555.
[67] Zhang F-L, Xu A-W, Gong Y-F, Wei M-H, Yang X-L. Asymmetric organocatalytic four-component quadruple domino reaction initiated by Oxa-Michael addition of alcohols to acrolein. Chem Eur J. 2009;15:6815.
[68] Enders D, Krüll R, Bettray W. Microwave-assisted organocatalytic quadruple domino reaction of acetaldehyde and nitroalkenes. Synthesis. 2010;567.
[69] Enders D, Wang C, Mukanova M, Greb A. Organocatalytic asymmetric synthesis of polyfunctionalized 3-(cyclohexenylmethyl)-indoles *via* a quadruple domino friedel–crafts-type/Michael/Michael/aldol condensation reaction. Chem Commun. 2010;46:2447–2449.
[70] For an example of a Brønsted-acid-catalyzed quadruple cascade (two new C–C and two new C–N bonds), see. Rueping M, Volla CM. Brønsted-acid catalyzed condensation-michael reaction-pictet–spengler cyclization—highly stereoselective synthesis of indoloquinolizidines. RSC Adv. 2011; 1:79.
[71] Van KN, Romo D. Multicomponent, enantioselective Michael-Michael-aldol-β-lactonizations delivering complex β-lactones. J Org Chem. 2018;83:632–43.
[72] Chang YP, Gurubrahamam R, Chen K. Enantioselecetive synthesis of functionalized polycarbocycles via a three-component organocascade quadruple reaction. Org Lett. 2015;17:2908–11.
[73] He M, Bode JW. Enantioselective, NHC-catalyzed bicyclo-β-lactam formation via direct annulations of enals and unsaturated *N*-Sulfonyl ketimines. J Am Chem Soc. 2008;130:418.
[74] Rueping M, Haack KL, Ieawsuwan W, Sunden H, Blanco M, Schoepke FR. Nature inspired cascade catalysis: reaction control through substrate concentration-double vs quadruple domino reactions. Chem Commun. 2011;47:3828.
[75] Mukherjee S, Shee S, Poisson T, Besset T. Biju. Enantioselective N-heterocyclic carbene-catalyzed cascade reaction for the synthesis of Pyrroloquinolines via N–H functionalization of Indoles. Org Lett. 2018;20:6998–7002.
[76] Gronda C, Jeanty M, Enders D. Organocatalytic cascade reactions as a new tool in total synthesis. Nature Chem. 2010;167.
[77] Ishikawa H, Suzuki T, Hayashi Y. High-yielding synthesis of the anti-influenza neuramidase inhibitor (-)-oseltamivir by three "one-pot" operations. Angew Chem Int Ed. 2009;48:1304–1307.

Manuel Orlandi

7 Basic principles of substrate activation through non-covalent bond interactions

Abstract: In the last twenty years, chiral Brønsted acid and chiral counteranion catalysis have emerged as a fundamental area of organocatalysis. The development of chiral acidic catalysts has allowed extending many known Brønsted catalyzed reactions to the stereoselective domain. Moreover, the controlled conditions under which these catalysts can be used, allowed accessing reactivity of increasing complexity with extraordinary selectivity levels. However, compared to the explosion of this branch of organocatalysis in an applicative direction, only little has been done to understand and rationalize the observed reaction outcomes. This is due, in part, to the complex nature of the weak interactions (H-bonds, electrostatic, and dispersion interactions) governing this class of reactions. Here we review relevant mechanistic analyses from both chiral Brønsted acid and chiral counteranion directed catalysis. Both experimental and computational work is included that aimed at unveiling the nature of the interactions governing the a number of reactions. These include the: enantioselective reduction of ketoimines with Hantzsch esters; ring opening reactions of epoxides, oxetanes, aziridinium, and sulfonium ions; stereoselective fluorination of allylic alcohols; oxidative aminations of benzylic thioethers (enantioselective Pummerer reaction). These case studies are analyzed and discussed in order to highlight key features and similarities across the different catalytic systems.

Keywords: chiral Brønsted acid catalysis, chiral counteranion catalysis, mechanistic studies, H-bonding, non-covalent interactions

7.1 Introduction

Brønsted acid catalysis has been known from a long time as a powerful tool for the promotion of chemical transformations. In particular, acidic compounds have been employed primarily as catalysts for the formation and cleavage of C-O and C-N bonds, as in hydrolysis and formation of esters, acetals, imines, and other simple functional groups. However, during the first years 2000, Brønsted acids, and especially Chiral Phosphoric Acids (CPAs), emerged as efficient catalysts for a range of more valuable transformations involving the formation of C-C bonds. Indeed, today

This article has previously been published in the journal Physical Sciences Reviews. Please cite as: Orlandi, M. Basic principles of substrate activation through non-covalent bond interactions *Physical Sciences Reviews* [Online] 2020, 5. DOI: 10.1515/psr-2018-0090

https://doi.org/10.1515/9783110590050-007

Chiral Brønsted Acids (CBAs) are known to activate carbonyls, imines, alkenes, alkynes, and hydroxyl groups towards the attack of nucleophilic species in stereoselective fashion. Due to these important developments, an increasing attention has been devoted to the development of new CBAs, resulting in dozens of new catalysts able to promote hundreds of stereoselective transformations. However, compared to the explosion of this branch of organocatalysis in an applicative direction, only little has been done to understand and rationalize the observed reaction outcomes. This is due, in part, to the complex nature of the weak interactions (H-bonds, electrostatic and dispersion interactions) governing this class of reactions.

This chapter is a collection of the most relevant studies performed in the areas of CBA and chiral counter-anion catalysis. All of the experimental work aimed at determining the catalysts acidity and at understanding the chemical activation of electrophilic substrates is included in Section 7.2.1. Section 7.2.2 collects NMR and electrochemical studies that revealed the nature of the H-bond coordination between CPAs and imines. It also contains a summary of the theoretical investigations of the stereoselective reduction of imines with Hantzsch esters from several authors. From Sections 7.2.3 through Sections 7.2.5 summarize the computational studies performed to rationalize the stereochemical outcome of other reactions including: ring opening reactions, pericyclic reactions and oxidations. Sections 7.3.1 through Sections 7.3.4 collect mechanistic studies of counter-anion directed catalysis that is reactions which do not involve Brønsted acidic activation of the substrate, but that proceed in stereoselective fashion due to tight pairing of the chiral anionic catalyst and a cationic reaction intermediate. The case studies included are (i) ring opening of *meso*-aziridinium and sulfonium ions, (ii) stereoselective fluorination of allylic alcohols, (iii) oxidative aminations of benzylic thioethers (enantioselective Pummerer reaction).

7.2 Chiral Brønsted Acid (CBA) catalysis

7.2.1 H-bond activation and pK_a

The reaction rate of Brønsted acid catalyzed reactions have been shown long ago to correlate with the pK_a value of the reaction catalyst, in agreement with the Brønsted catalysis law (Figure 7.1) [1]. Importantly, the proportionality constant $\boldsymbol{\alpha}$ provides useful information as its magnitude suggests the degree of proton transfer at the transition state (TS) level thus distinguishing between reactions proceeding via H-bond catalysis (H^+ still located on the catalyst anion A^- in the TS, Figure 7.1) or ion pairing (H^+ almost completely transferred to the substrate S in the TS, Figure 7.1). In both cases, activation of the substrate occurs via lowering of the substrate's Lowest Unoccupied Molecular Orbital (LUMO) energy, even though the effect is more pronounced in the case of complete proton transfer (ion pairing).

Brønsted equation

$$\log(k) = \alpha \log(K_a) + \text{const.}$$
$$= -\alpha\, pk_a + \text{const.}$$

k, reaction rate
α, proportionality constant
$K_a = [H^+][A^-]/[HA]$, dissociation constant of an acid HA

H-bonding

$pK_a(HA) > pK_a(HS^+)$

$A\text{-}H + S \rightleftharpoons A\text{-}H\text{---}S$

Ion pairing

$pK_a(HA) < pK_a(HS^+)$

$A\text{-}H + S \rightleftharpoons A^{\ominus}\text{---}H\text{-}S^{\oplus}$

Figure 7.1: Brønsted equation, H-bonding and Ion pairing.

The degree of proton transfer does not only depend on the catalyst acidity, but also on the substrate basicity. As a rule of thumb, the pK_a of the catalyst has to be lower or at least of similar order of magnitude than the substrate's pK_a (acidity of the substrate's conjugated acid) in order to access a protonated (or activated) intermediate for catalysis (Figure 7.1). Knowing the pK_a values of Brønsted acid catalysts is therefore of high importance in order to establish *a priori* whether a certain acid will be suitable for substrate activation in a certain reaction.

The pK_a values of a wide range of acidic compounds in water are known. However, since most organic reactions are typically run in non-aqueous solvents, several groups established pK_a scales in media such as DMSO, MeCN or DCE. The most recent and comprehensive pK_a tables of strong Brønsted acids were collected by Leito and coworkers [2, 3]. The acidities of dozens of compounds within a pK_a range of 33 units (from 28 to −5) in MeCN were determined via spectroscopic titration methods [2]. Later, a similar *relative* scale of superacidic compounds was established in DCE [3].

In the context of stereoselective Brønsted acid catalysis, the pK_a values of a set of BINOL-based CBAs were first determined by O'Donoghue and Berkessel in 2011 by using UV-Vis methods and several indicators such as 4-nitrophenol, 2,4-dinitrophenol, 4-chloro-2,6-dinitrophenol and 2,4-dinitronaphthol in anhydrous DMSO [4]. The acidities of a similar set of compounds was determined in MeCN two years later by Leito and Rueping [5]. Some selected data from these two contributions are listed in Figure 7.2(a). These values suggest a clear acidity trend that mostly depends on the acidic functional group in compounds **1-3** and only marginally on the 3,3′-aryl substituents or on the BINOL scaffold (BINOL vs. $[H_8]$-BINOL). In particular, bis-sulfurylimides **3** developed by Berkessel were found to be the most acidic compounds, followed by N-triflylphosphoramides **2** (NTPs) and by phosphoric acids **1**. This trend was also confirmed by the measurement of the reaction rates for the Nazarov cyclization catalyzed by several compounds chosen from the set, which were found to correlate with the catalysts' pK_a values in accordance with the Brønsted equation (Figure 7.2(b)) [5].

Because of the rapid development of computational chemistry, the pK_a values of organic acids in solution can currently be calculated with similar accuracy than those determined experimentally [6, 7]. In 2013, a theoretical work appeared where Cheng

A

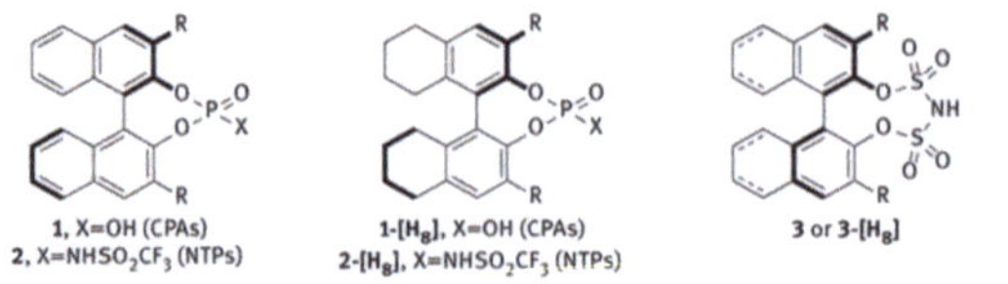

Acid	R	pK_a^{DMSO}	pK_a^{MeCN}	Acid	R	pK_a^{DMSO}	pK_a^{MeCN}
1a	H	3.37		**2a**	Ph		6.4
1a-[H$_8$]	H	4.12		**2a-[H$_8$]**	Ph		6.7
1b	Br	2.90		**2b**	4-OMe-Ph		6.4
1c	Ph	3.86	12.7	**2b-[H$_8$]**	4-OMe-Ph		6.9
1d	4-OMe-Ph	3.49		**2c-[H$_8$]**	4-F-Ph		6.3
1e	4-F-Ph		12.5	**2d**	2,4,6-*i*Pr$_3$-Ph	3.34	
1f	4-Ph-Ph	3.50		**3a**	H	1.74	
1g	3,5-(CF$_3$)$_2$-Ph	2.63		**3a-[H$_8$]**	H	1.85	
1h	2,4,6-*i*Pr$_3$-Ph	4.22	13.6	**3b**	3,5-(CF$_3$)$_2$-Ph		5.2
1i	2-Naphthyl	3.77					
1j	9-Phenanthryl		13.3				
1j-[H$_8$]	9-Phenanthryl		14.0				

B

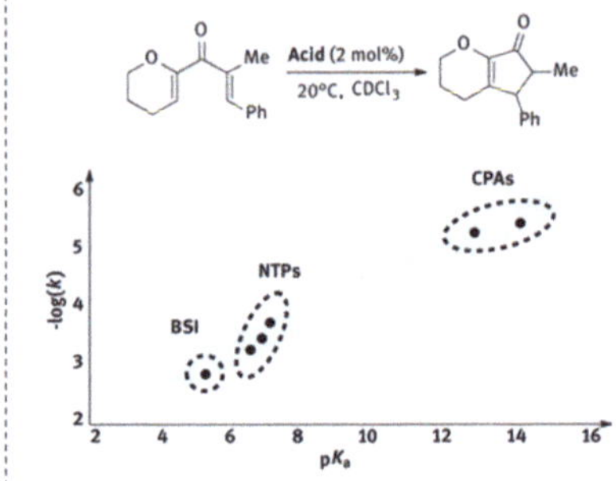

Figure 7.2: Common CBAs pK_a values, and correlation between catalyst's pK_a and reaction rate in a CBA-catalyzed Nazarov reaction.

and Li calculated the pK_a of a huge number of BINOL-derived phosphoric acids in DMSO [6]. The data reported by Berkessel were chosen as reference values, obtaining calculated pK_a values for additional 36 compounds. In particular, the pK_a values were found to range between 1.5 and 5.1 for CPAs, −3.1 and 1.9 for thiophosphoric acids (P(S)OH), and between −3.0 and −4.2 for dithiophosphoric acids (PS$_2$H). Despite in this first publication by Cheng and Li no information about NTPs or bis-sulfurylimides was provided, in a second paper published in 2014 by the same authors, new data completed the computational study [7]. Here the pK_a values of many additional acidic compounds were provided; in particular NTPs and bis-sulfurylimides were calculated to have pK_a of −3.9 to −2.2 and −3.7 to −2.3, respectively. Hence, the calculations present a trend that is in agreement with the experimental data provided by Rueping and Leito in acetonitrile (i. e. pK_a bis-sulfurylimides ≤ pK_a NTPs ≪ pK_a CPAs) [5].

7.2.2 Stereoselective reduction of imines

Since the pioneer reports by Akiyama and Terada in 2004 [8, 9], which introduced the use of CPAs as organocatalysts, imines proved the most widely used substrates in CBA catalysis. This is likely due to the relatively high pK_a of this class of compounds, which allows for the addition of a huge range of nucleophiles. It is therefore not surprising that the very few experimental mechanistic studies performed in order to understand the chemical–physical basis of Brønsted acid catalysis involved phosphoric acids in combination with imines.

The very first work in this context is due to Gschwind and Rueping [10], who in 2011 reported the NMR characterization of the acid-base complex between phosphoric acid **4** and imines **5**. Elucidating the exact mechanisms involved in chiral phosphoric acid catalyzed systems is not an easy task due to the large number of possible interactions that could occur between the catalyst and the substrate. However, by performing an NMR characterization of the ^{15}N-labeled adduct **6** (in toluene-d_8 as the solvent), the authors found three different species in solution: the H-bonded complex **6a**, the ion-pair complex **6b**, and the dissociated ion pair **6c** (Figure 7.3). Specifically, while at 300 K only one broad ^{1}H-NMR signal was detectable at ca. 16.2 ppm, by lowering the temperature this signal split in three broad signals. These signals became sharper as the temperature was further decreased to 240 K, at which temperature a singlet at $\delta = 18.16$ ppm and two doublets at $\delta = 15.50$ and 11.87 ppm with $^{1}J_{H,N}$ coupling constants of 86.01 and 69.5 Hz were observed. On the basis of further characterization (1D ^{31}P, ^{1}H INEPT, 2D ^{1}H, ^{15}N HMQC and ^{1}H DOSY), these signals were assigned respectively to **6a**, **6b** and **6c** (Figure 7.3).

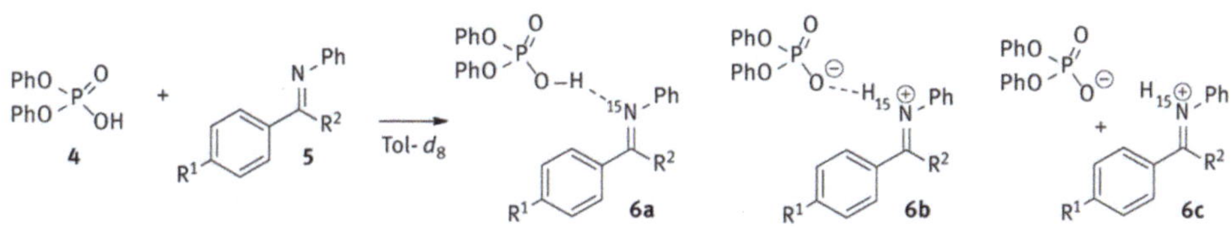

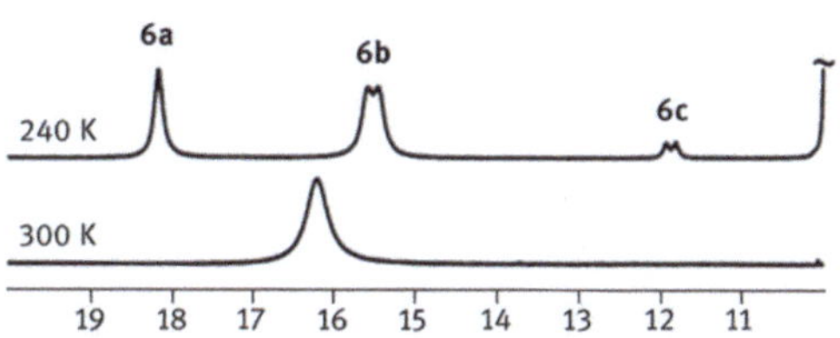

Relative integrals of species 6a, 6b and 6c for three different imines

R^1	R^2	6a	6b	6c
Me	H	0.38	0.62	0.08
Me	Me	0.42	0.58	0.18
CF_3	Me	0.51	0.49	0.11

Figure 7.3: Speciation of CBA/imine complexes observed by ^{1}H NMR. Adapted with permission from *Angew. Chem. Int. Ed.* **2011**, *50*, 6364.

As expected, the authors found that reducing the nitrogen basicity of **5** by changing the imine's substituent R^1 and R^2, the **6a:6b** ratio slightly decreases. The tested substrates are reported in Figure 7.3 together with the **6a:6b:6c** relative amount of each species for each salt. From the table it can be observed how the ratio between the H-bond complex **6a** and the ion pair complex **6b** changes toward **6b** quite accordingly with the imines' basicity. On the other hand, the relative amount of **6c** with respect to **6a** and **6b** can be hardly rationalized.

Being able to evaluate the ability of a catalyst to protonate, coordinate or activate a substrate is of crucial importance. Since an iminium ion is more reactive than the relative imine, it can be pointed out that the greater is the **6a:6b** ratio, the more efficient is the activation of the substrate by the catalyst. At the same time, a tight ion pair is necessary to achieve higher stereoselectivities. Hence, the optimum conditions for the development of an efficient catalysis (both in terms of chemical and stereochemical activity) are related to the relative amount of the tight ion pair

6b with respect to the H-bond complex **6a** and to the dissociated ion pair **6c**. In this context, Hunger and coworkers characterized and quantified the complex formation of **4** with 2-methylquinoline (of similar basicity than imines) by combining dielectric spectroscopy, quantum chemistry, and ^{1}H NMR techniques [11]. For a series of different solvents, the interaction between substrate and catalyst was found to be dominated by ion-pairing, rather than hydrogen bonding.

Despite the useful insights gained by Gschwind and coworkers, the studies were complicated by peak broadening of the H-bond signals, which indicated broad conformational space and multiple aggregation states for the achiral catalyst **4**, which is poorly representative of the much bulkier structures of the most widely used catalysts (**1-2**). Thus, for a more detailed analysis, the more sterically hindered CPA **1h** (TRIP) was used [12]. Characterization of complexes **7** (Figure 7.4), revealed a slightly different scenario than previously observed with **6**. Specifically, the neutral H-bond and the dissociated ion pair complexes were not observed (with the only exception of one strongly electron-deficient imine). Additionally, two peaks for the hydrogen bonded ion pair **7** appeared upon temperature lowering in the ^{1}H, ^{15}N and ^{31}P spectra. These were assigned to the salts of the *Z*- and *E*- isomers of the imine (***E*-7** and ***Z*-7**, Figure 7.4) based on NOESY experiments. This speciation was consistent through the set of imines used (10 examples) even though the *E*:*Z* ratio was found to be dependent on the nature of the imine aryl substituents.

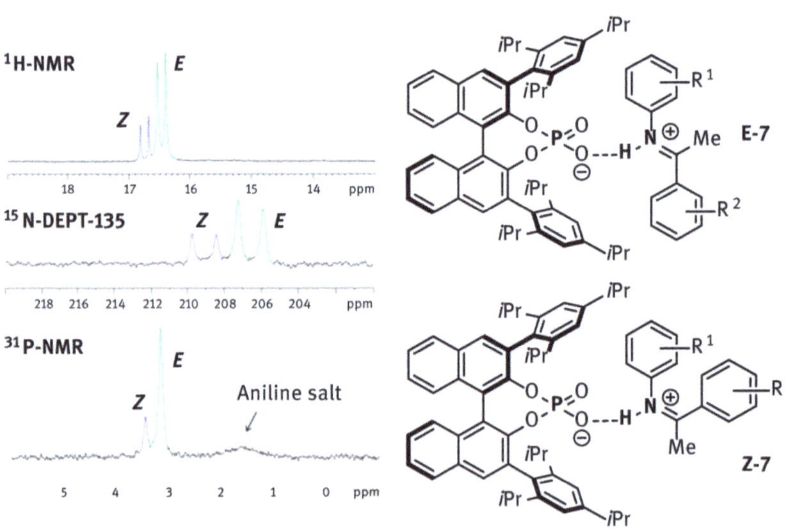

Figure 7.4: Speciation of TRIP/imine complexes observed by 1H NMR. Adapted with permission from *J. Am. Chem. Soc.* **2016**, *138*, 16345.

Having established a methodology for accessing reliable NMR measurements for these complexes, the authors turned their attention to a more detailed study of the properties of the H-bond structure in solution. Based on the determination of ^{1}H and ^{15}N chemical shifts obtained from several combinations of 10 imines and 8 acids (phenols, carboxylic acids, **1h** and HBF_4•OEt_2), the authors were able to determine the geometry and the strength of the complexes' H-bonds using a methodology previously developed by Limbach and Denisov [13–16].

The formation of strong H-bonds is associated with a proton-transfer reaction between H-bond donor and acceptor. Hence, the faster the proton exchange between the two basic sites (O and N), the stronger the H-bond. In other words, the H-bond results to be weaker when the proton steadily resides on one of the two partners resulting into a more likely dissociation of the two basic partners (one of the two taking the proton). In ^{1}H NMR spectra, protons in *strong* or *low-barrier* H-bonds are usually observed for chemical shifts above 16 ppm. The plot in Figure 7.5 shows the Steiner-Limbach correlation of the ^{1}H chemical shift δ(O$\underline{H}$N) against the ^{15}N chemical shift δ(OH$\underline{N}$) of several complexes reported by Gschwind and coworkers. The parabolic dependence of the proton chemical shifts δ(OHN) in combination with the linear dependence of the nitrogen chemical shifts δ(OHN) allows the determination of the valence bond orders (p_{NH} and p_{OH}) by employing empirical correlations. From these, important geometric features can be derived, like the NH and OH bond length (r_{NH} and r_{OH}). The experimentally derived r_{NH} and r_{OH} were then compared with the computed NH and OH distances in the condensed phase (DCM), which were found to be in excellent agreement.

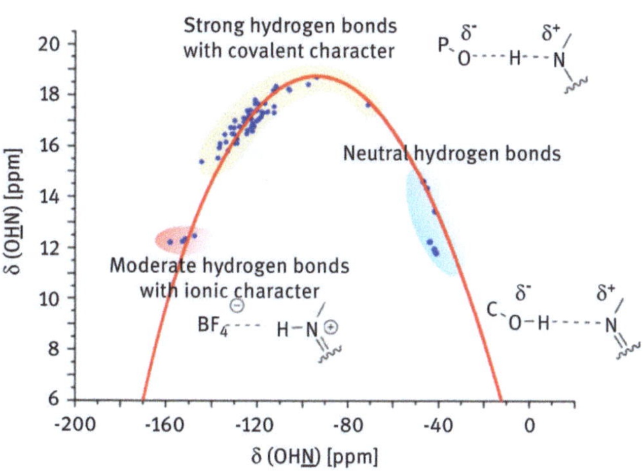

Figure 7.5: Steiner-Limbach correlation of the ^{1}H chemical shift δ(O$\underline{H}$N) against the ^{15}N chemical shift δ (OH$\underline{N}$) of several H-bond complexes. Adapted with permission from *Acc. Chem. Res.* **2017**, *50*, 2936.

Altogether, the collected data suggested the formation of a strong ionic H-bond between the catalyst **1h** and imine substrates **5**. This is explanatory of the good performance shown in terms of stereoselectivity when these catalysts and substrates are combined.

It is reasonable that the tight contact between **1h** and **5**, due to the presence of a strong H-bond, allows for other intermolecular interactions to take place. Gschwind and coworkers performed a conformational study of complexes ***E*-7** and ***Z*-7** separately using 1D $^{1}H,^{1}H$ selective NOESY and $^{1}H,^{19}F$ HOESY experiments and theoretical calculations [17]. Both complexes ***E*-7** and ***Z*-7** gave NOE patterns connecting both imine aryl groups and the catalyst binaphthyl backbone, while only for ***Z*-7** there were detectable contacts between the catalyst aryl substituents and the imine α-methyl group. The observed NOE patterns are highlighted with green ($^{19}F,^{1}H$) or purple ($^{1}H,^{1}H$) arrows in Figure 7.6, which shows the four conformations computed for the binary complexes ***E*-7** (two conformations, *Type I E* and *Type II E*) and **Z-7** (two conformations, *Type I Z* and *Type II Z*). Based on experiments and computations, the authors conclude that the energy barrier for the *Type I-Type II* transition is very low and allows free interconversion even at 180 K.

Interestingly this study showed that molecular recognition in these binary complexes is due not only to a strong ionic H-bond between the acid and the imine, but also to non-covalent interactions (NCIs) such as CH-π and T-shaped π contacts between the imine aryl groups and the catalyst binaphthyl backbone. Sigman and Toste found a similar catalyst-substrate interaction occurring in the asymmetric fluorination of allylic alcohols as will be discussed in Section 7.3.4 of this chapter [18].

Information about the structure of pre-reacting binary complexes is extremely important for mechanistic insights. Nevertheless, the ultimate understanding of a stereo- or regioselective reaction outcome is given by the knowledge of the TSs, especially in Curtin–Hammett scenarios and kinetically controlled transformations. TSs can hardly be studied experimentally due to their evanescent nature. However, their structure can be computed and experimental insights about their properties can be obtained via kinetic probing of the reaction under exam. Gschwind et al. used their acquired knowledge in the field of CBA catalysis to study the reduction of imines catalyzed by CPAs with Hantzsch ester **8** previously reported by Rueping and List [19, 20].

This reaction was proposed to proceed through a ternary complex of the catalyst with the two reagents according with the cycle depicted in Figure 7.7(a). As previously described, the coordination of the imine to **1h** leads to complexes ***E*-7** and ***Z*-7**, both of which are eligible to react with the nucleophile **8** to give either the *R* (the observed major product) or the *S* product **9** according with one of the four possible pathways depicted in Figure 7.7(b) (**TS-*E/R***, **TS-*E/S***, **TS-*Z/R***, **TS-*Z/S***). To experimentally determine which one of these TSs is actually operating in catalysis, the authors developed a method that exploits light as an external stimulus (orthogonal to the reaction mechanism) to modulate the equilibrium of slowly interchanging complexes ***E*-7** and ***Z*-7** (Figure 7.7(c)). In other words, irradiation of the reaction mixture at a proper wavelength would induce isomerization of the imine from the *E* to the *Z* configuration thus altering the thermodynamic equilibrium of the reactants

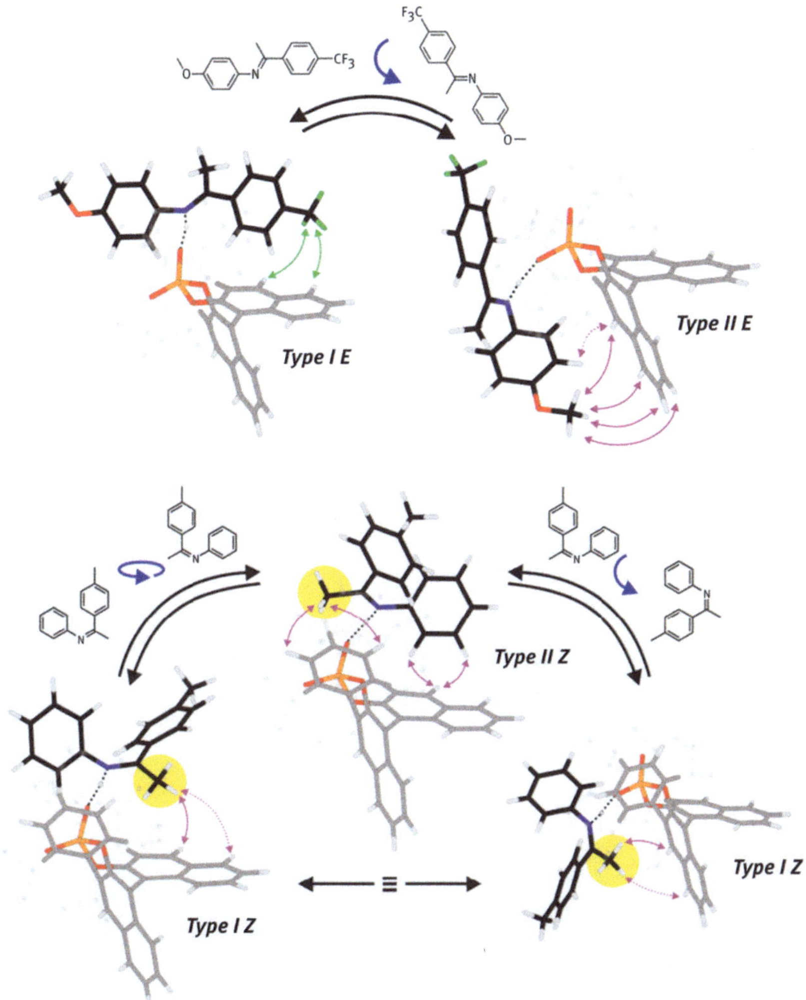

Figure 7.6: Computational/Spectroscopyc (NMR) conformational analysis of two TRIP/imine complexes. Reproduced with permission from *J. Am. Chem. Soc.* **2016**, *138*, 15965.

in favor of intermediate ***Z*-7**. If isomerization occurred at comparable or lower rate with respect to the reaction itself, the Curtin–Hammett boundaries would not be valid anymore. As a result, both the reaction outcomes (*ee* and reaction rate) could be affected by the change in the pre-reaction equilibrium in different ways depending on which reaction pathways are operating in catalysis. Thus, comparison of the reaction outcomes of the reaction performed in the dark and under light irradiation would reveal the operating TSs leading to each enantiomers of **9** according to the scheme in Figure 7.7(d). Since the reaction under light irradiation provided *identical* *ee's but higher reaction rates* with respect to the control experiments in the dark, the authors concluded that the reaction proceeds via **TS-*Z/R*** and **TS-*Z/S***, the former being the favorite one due to the observed absolute configuration of product **9**.

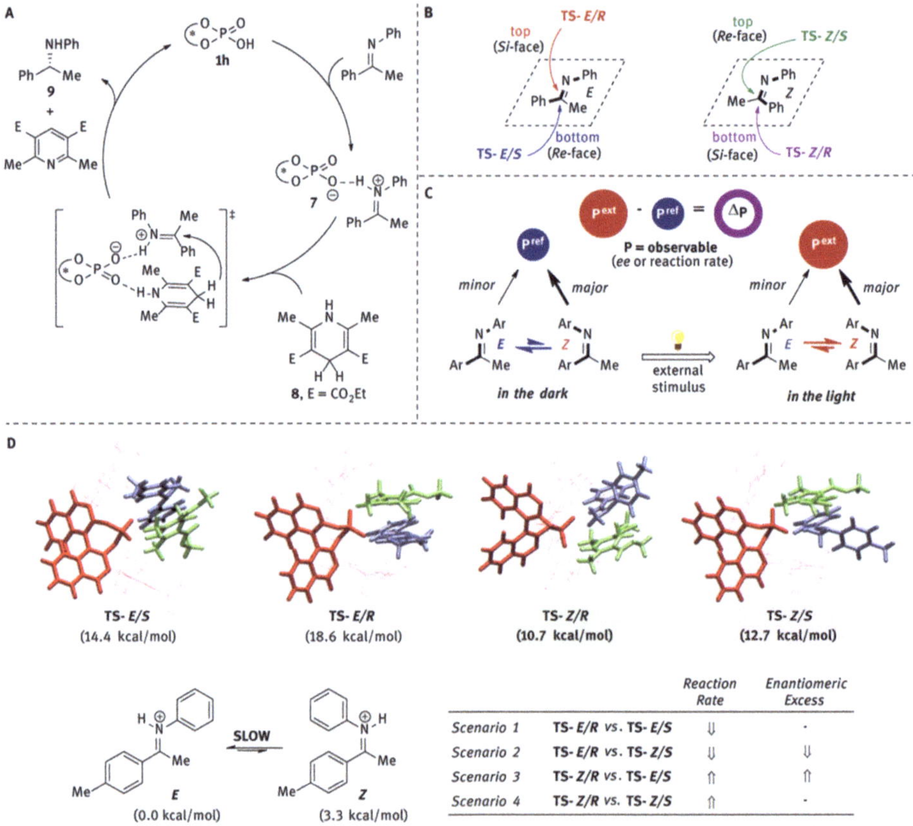

		Reaction Rate	Enantiomeric Excess
Scenario 1	TS- *E/R* vs. TS- *E/S*	⇓	-
Scenario 2	TS- *E/R* vs. TS- *Z/S*	⇓	⇓
Scenario 3	TS- *Z/R* vs. TS- *E/S*	⇑	⇑
Scenario 4	TS- *Z/R* vs. TS- *Z/S*	⇑	-

Figure 7.7: Experimental/computational study the reduction of imines catalyzed by CPAs with Hantzsch ester. Adapted with permission from *J. Am. Chem. Soc.* **2017**, *139*, 6752.

In order to provide theoretical support to the observed experimental outcomes, **TS-*E/R***, **TS-*E/S***, **TS-*Z/R*** and **TS-*Z/S*** were computed. As the Curtin–Hammett principle was shown not to apply by the experiments (*vide infra*), the energy of the TSs should not be directly compared, but the activation barrier from the corresponding *E*- or *Z*-imine should be considered instead. The computed structures are reported in Figure 7.7(d) together with their corresponding Gibbs free energy barriers calculated from the corresponding imine energy level (0.0 vs. 3.3 kcal/mol for the *E* and *Z* imine respectively, Figure 7.7(d)). **TS-*Z/R*** and **TS-*Z/S*** were found to be the TSs with the lowest energy barrier leading to (*R*)-**9** and (*S*)-**9** (10.7 vs. 12.7 kcal/mol). The energy difference between the two TSs accounts for the experimental *ee* and, since both of them derive from the imine in the *Z* configuration, the computations also account for the increase in reaction rate observed under light irradiation.

These calculations by Gschwind and coworkers were found to be in line with previous contributions by Himo and Goodman [21–23]. However, the experimental work here reviewed allowed accessing experimental information about the actual operating TSs thus providing support to the computed reaction mechanism.

This mechanistic study from the Gschwind group was mainly focused on the use of catalyst **1h** (TRIP). Recent theoretical investigations of the role of the BINOL 3,3′ aryl substituents in the reduction of imines was recently provided by Goodman [24], whose team showed that the hydrogenation mechanism with Hantzsch esters is dependent on the proximal and distal steric hindrance of the catalyst substituents. In a correlated study of the CPA catalyzed Strecker reaction [25], Goodman showed that the configuration of the reacting imine is also dependent on the sterics of the substrate's α-alkyl group. It was shown that while for ketimines the reaction typically proceeds with the substrate in the *Z* configuration, aldimines prefer to react in configuration *E*.

Altogether, the studies hereby reviewed show that the catalytic activity of CPAs is affected by a plethora of factors. These include:

i. the catalyst's acidity (low pK_a values ensure substrate activation);
ii. the ability of the PO_2H group to effectively coordinate the imine N atom to form a strong H-bond;
iii. the ability of the catalyst's scaffold to engage the substrate in NCIs;
iv. the size of the catalyst's pocket (which can affect the substrate's conformation in the binary complex).

As we will see through this chapter, the ability to control these factors proved to be crucial for developing efficient stereoselective transformations. Unfortunately, the coordination mode and the structure of the competing TSs is not easily accessible *a priori* for any reaction. Thus, establishing precedents and developing new physical-organic methodologies for gaining such information remains a task of great importance.

7.2.3 Ring opening reactions

7.2.3.1 Epoxides

A general rule of thumb states that Brønsted acid catalysis proceeds via coordination/activation of the reaction electrophile. The CPA-catalyzed epoxide and aziridine ring opening reactions (Figure 7.8(a)) developed by List and coworkers [26–30] are an exception to this rule. In these transformations, the catalyst coordinates the carboxylic acid **10**, the nucleophile, rather than the electrophilic epoxide **11** [26]. Specifically, the ^{1}H and ^{31}P NMR spectra of **1h** showed considerably shifted signals upon addition of a stoichiometric amount of benzoic or acetic acid. The translational diffusion coefficients of such mixtures were also found to be substantially different from those of the single components, thus suggesting complexation between the two partners. Additionally, the ^{31}P NMR titration experiments of **1h** with these acids resulted in the determination of the binding isotherm of association, which in turned allowed calculation of the binding constants ($K_a^{BzOH} = 3981 \pm 98\,M^{-1}$, $K_a^{AcOH} = 1948 \pm 26\,M^{-1}$, Figure 7.8(b)). Method of continuous variations and co-crystallization of AcOH and **1h** (X-ray analysis), revealed the formation of an equimolar complex (1:1 ratio) with structure **12**.

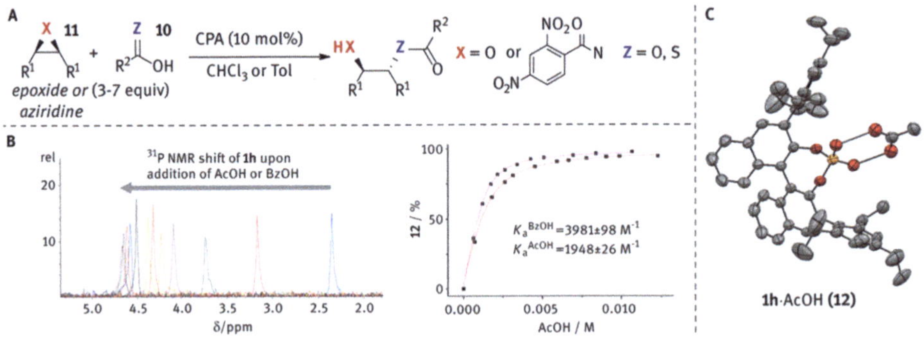

Figure 7.8: CPA-catalyzed epoxide and aziridine ring opening reaction. Adapted with permission from *Angew. Chem. Int. Ed.* **2014**, *53*, 7063.

Reactivity in CBA catalysis based on the electrophile activation is easily rationalized by invoking the lowering of the LUMO energy upon protonation by the catalyst. How do these reactions, proceeding via coordination of the nucleophile, work then? List and coworkers noticed that the ^{31}P chemical shift of **12** was shifted downfield with respect to **1h** alone, consistent with partial positive charge building up on the phosphate group. This is counterintuitive, as one would expect the more acidic compound (**1h**) to protonate the carboxylic acid and not *vice versa*. In other words, these NMR experiments seemed to suggest that the CPA behaves as a base within the complex, thus making the phosphate group more acidic and the carboxylic acid more nucleophilic in a synergistic activation. This was confirmed by computational evaluation of the HOMO and LUMO energies of free AcOH and of AcOH within complex **12**. While the LUMO energy change was calculated to be small ($\Delta E = +\,0.08$ eV), the HOMO energy was significantly increased upon coordination ($\Delta E = +\,0.25$ eV) [29].

The kinetic relevance of dimer **12** in catalysis was also established by determination of the epoxide opening reaction's rate law. The reaction rate was found to be first order in **13** and **1h**, but zeroth order in the carboxylic acid. This is consistent with the catalytic cycle depicted in Figure 7.9, where complex **12** is the catalyst resting state and reacts with epoxide **13** in the reaction rate- and stereo-determining step [29].

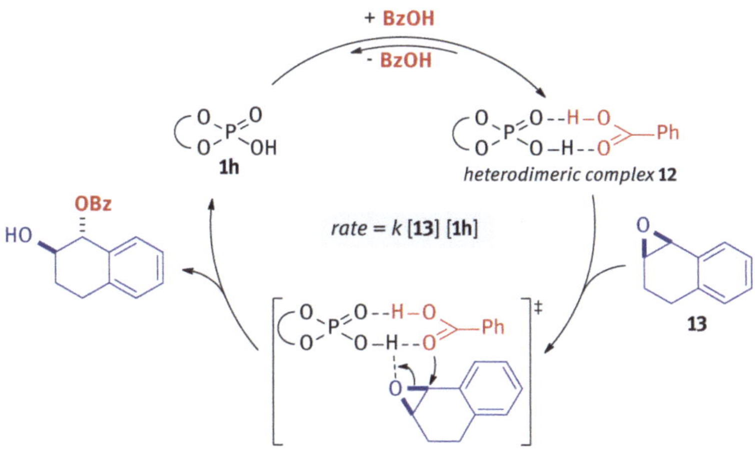

Figure 7.9: Mechanism of the CPA-catalyzed epoxide ring opening reaction.

The influence of the catalyst's structure on the enantioselectivity was investigated by DFT calculation of the reaction TSs for the opening reaction of the model system BzOH + **14** [29]. Catalysts **1h** and **1k** were considered for the calculations as they provided different catalytic activities in the opening reaction of cyclohexene oxide, (*S*)-**1k** being superior to (*S*)-**1h** (93 *vs.* 57% *ee*). The computed TSs showed the same general arrangement of the reagents within the catalyst chiral pocket. Specifically, the reagents complex are coordinated by the catalyst via two H-bonds and a strong electrostatic interaction between the electron-rich phosphoryl oxygen and the partially positively charged C-H moiety of the epoxide as previously found by Wheeler for a similar system [31]. Confined in the pocket by virtue of these attractive interactions, the electrophile also experiences steric repulsion from the isopropyl groups of (*S*)-**1h** and (*S*)-**1k**. The difference in selectivity between the two catalysts is due to a balance of these positive and negative contacts with the substrate. **(*S,S*)-TS1k-14** and **(*S,S*)-TS1h-14** in Figure 7.10 are the low lying energy TSs leading to both the (*R,R*)- and (*S,S*)-product for each catalyst. The computed $\Delta\Delta G^{\ddagger}$ was higher for catalyst **1k** (1.9 vs. 0.74 kcal/mol in the case of **1h**), thus reproducing the experimental relative performance of the two CPAs. A rationalization to the different $\Delta\Delta G^{\ddagger}$ was given based on the distance between the epoxide methyl substituent and the catalyst 3,3′ groups, which was shorter in **(*R,R*)-TS1k-14** (4.56 Å) than in **(*R,R*)-TS1h-14** (4.78 Å). Specifically, the shortest distance (i. e. the strongest repulsive interaction) was observed in **(*R,R*)-TS1k-14** between the epoxide and the catalyst's polycyclic substituent in the *meta* position of the 3,3′ aryl groups, which distinguish the structure of **1k** from **1h.**

This study by List and coworkers showed that CPAs strongly coordinate carboxylic acids via H-bonding. This coordination leads to: (i) rising of the carboxylic HOMO energy, thus making it more nucleophilic and, consequently, to (ii) building

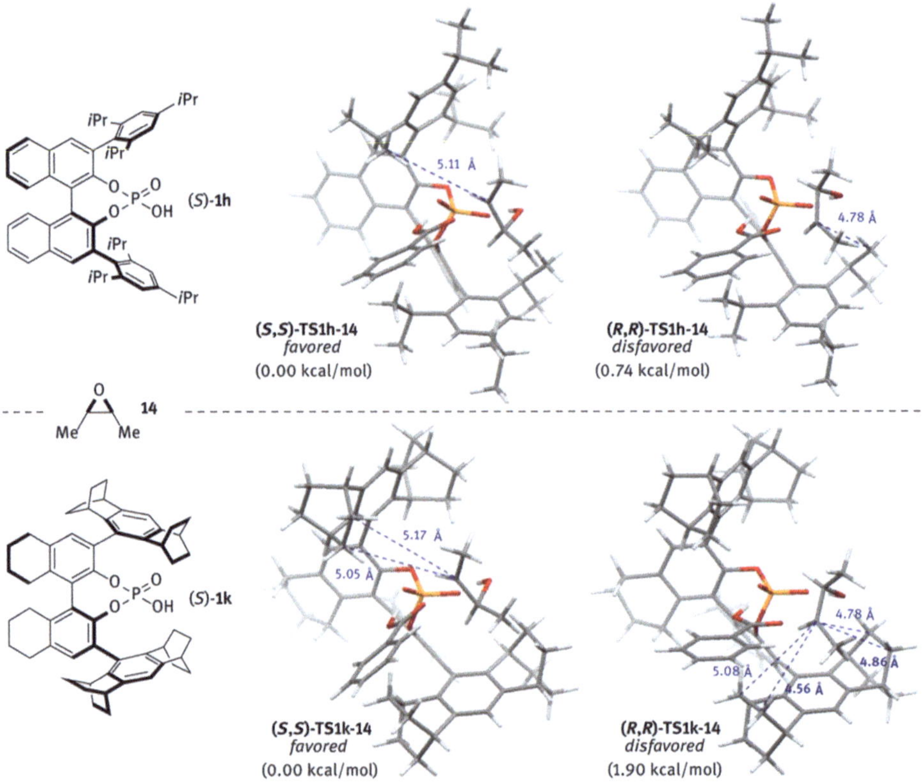

Figure 7.10: Computed TSs of the ring opening reaction of **14** catalyzed by **1h** and **1k**. Adapted with permission from *J. Am. Chem. Soc.* **2016**, *138*, 14740.

up of positive charge on the phosphate moiety, thus making the catalyst more acidic. The binary complex thus formed is able to react with electrophiles like epoxides and aziridines to give diols and aminoalcohols, respectively. The reaction enantioselectivity was shown to be highly dependent on the steric environment provided by the catalyst's 3,3′ groups. This was rationalized based on the balance of a *plethora* of attractive and repulsive NCIs that interplay at the TS level, sharing many general similarities with the case studies presented in the previous section.

7.2.3.2 Oxetanes

Oxetanes are strained four-membered rings that readily open in the presence of suited nucleophiles under CBA catalysis conditions as shown by Sun and coworkers (Figure 7.11) [32–34] The intramolecular version of this transformation was developed first, which employs the SPINOL based CBA **15a** as the optimal catalyst (Figure 7.11(a)).

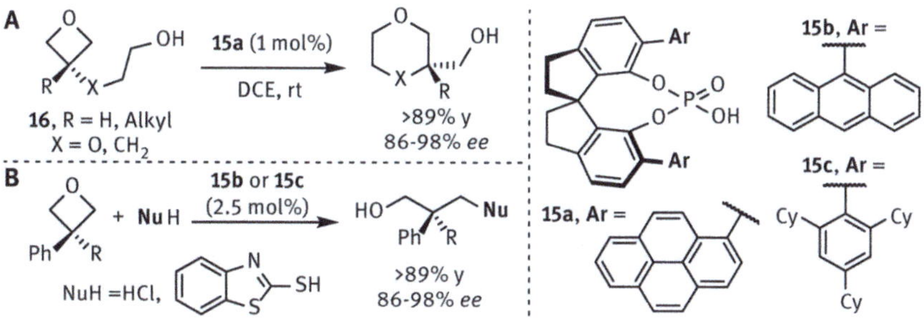

Figure 7.11: CPA-catalyzed oxetanes ring opening reaction.

The stereochemical outcome of this reaction was rationalized via DFT TS analysis by the groups of Wheeler and Houk [35], who showed that oxetanes **16** display a different mode of activation with respect to other common electrophiles in the presence of CPA **15a**. Specifically, it was found that in the favored TS leading to the major enantiomer **(*S*)-TS-16** the catalyst's acidic OH group coordinates/activates the oxetane O atom, the catalyst's basic site P=O is involved in an interaction with the four-membered ring H atoms. Thus, the catalyst is involved in the electrophilic activation of the substrate, while the nucleophilic OH group is only involved with an OH•••π interaction with the catalyst's pyrenyl groups (Figure 7.12(a)). On the other hand, in the TS leading to the disfavored product **(*R*)-TS-16**, such a stabilizing interaction was weaker (2.58 vs 2.48 Å), and showed steric clashes between the ring substituents and the catalyst's 3,3′ -aryl groups (Figure 7.12(b)). These structural differences between the diastereomeric TSs account for the computed $\Delta\Delta G^{\ddagger} \approx 2$ kcal/mol, in good agreement with the experimental selectivity.

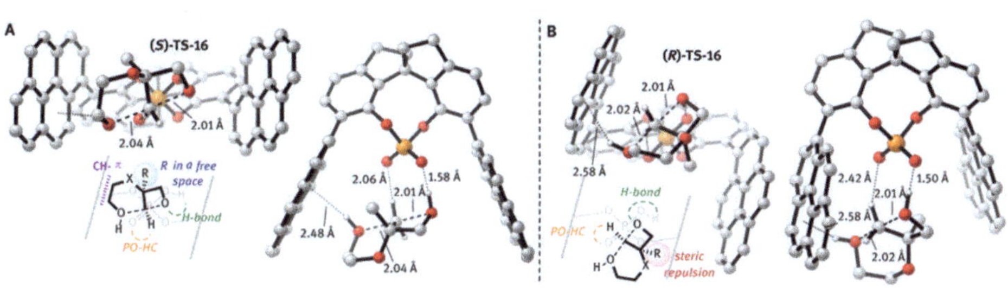

Figure 7.12: Computed TSs of the ring opening reaction of oxetane **16** catalyzed by CPA **15a**. Adapted with permission from *ACS Catal.* **2017**, *7*, 7332.

In 2016, Wheeler and Houk separately performed the computational study of the two intermolecular versions of the CPA catalyzed oxetane ring opening reaction in Figure 7.11(b) [31, 35]. The two studies show that, independently from the nucleophile, the oxetane ring coordination mode is preserved also in the case of intermolecular reactivity. However, differently from the intramolecular reaction, the attack of the external nucleophile is assisted by coordination of the P=O basic site (Figure 7.13(a)). Computation of the TSs leading to both the enantiomeric products for several substrates showed that a general stereochemical model is in place. Specifically, the catalyst's P=O group binds via PO-HC interaction to the oxetane methylene group that undergoes the addition. However, in the chiral environment provided by the catalyst, one of the two enantiotopic groups is coordinated preferentially due to the steric clash between the oxetane ring substituent R and the catalyst aryl group leading to the observed enantioselectivity (Figure 7.13(b)).

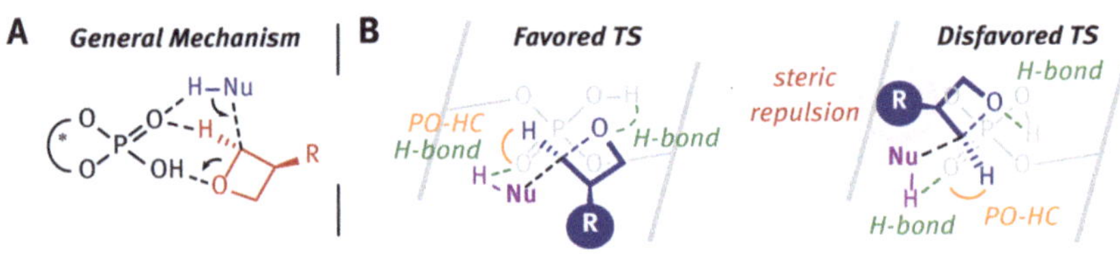

Figure 7.13: General stereochemical model for the CPA-catalyzed oxetanes ring opening reaction.

7.2.4 Asymmetric pericyclic reactions

7.2.4.1 [3±2]-cycloaddition of hydrazones

The [3+2]-cycloaddition between hydrazones and alkenes reported by Rueping and Houk is an example of N-triflylphosphoramide (NTP) catalyzed reaction (Figure 7.14(a)) [37]. NTPs are more acidic (see Section 7.2.1) and able to activate a wider range of substrates than CPAs. The triflyl group (CF_3SO_2) offers a greater number of possible coordination sites and orientations than the simple OH group. Additionally, the presence of two different substituents at the P atom reduces the symmetry of the catalyst from C2 (for CPAs) to C1. Consequently, the number of possible conformers and coordination modes increases considerably, making the computational study of reactions catalyzed by NTP more difficult. Therefore, Houk and coworkers began their study with a conformational analysis of the complex **19** formed between the model NTP **17** and hydrazone **18** (Figure 7.14(b)). Among all of the possible conformers, **19a–19d** were found to have the lowest energies. Therefore, these coordination modes were investigated as possible reaction partners of ethyl vinyl sulfide. **19c** was calculated to be the most reactive conformer toward ethylene, with a $\Delta G^{\ddagger} = 28.6$ kcal/mol.

Substitution of **17** with **2d-[H$_8$]** allowed computing the $\Delta\Delta G^{\ddagger}$ associated to the reaction enantioselectivity for the attack of ethyl vinylthioether to **18**. The coordination mode **19c** was conserved and a $\Delta\Delta G^{\ddagger} = 4.4$ kcal/mol in favor of product (3*R*,5*S*)-**20** was computed in agreement with the generally high selectivity observed in this

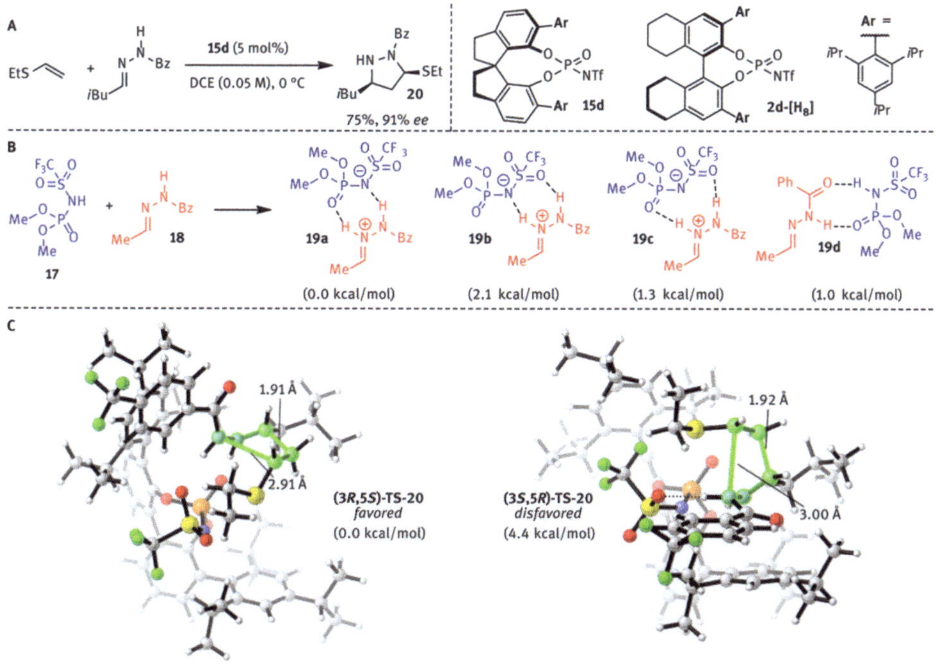

Figure 7.14: [3+2]-cycloaddition between hydrazones and alkenes catalyzed by NTPs. Adapted with permission from *J. Am. Chem. Soc.* **2014**, *136*, 13769.

reaction. The structure of the two diastereomeric TSs **(3*R*,5*S*)-TS-20** and **(3*S*,5*R*)-TS-20** are reported in Figure 7.14(c). Even though the origin of the computed $\Delta\Delta G^{\ddagger}$ is not clear at a first glance, removing the nucleophile from the structures and comparing the energy of the two remaining hydrazone/**2d-[H$_8$]** complexes results in an energy difference of 5.5 kcal/mol. This value is comparable to the computed $\Delta\Delta G^{\ddagger}$ and therefore the authors concluded that the selectivity in this reaction mostly originates from the interactions occurring between the catalyst's substituents and the hydrazone.

7.2.4.2 Nazarov reaction

The enantioselective Nazarov cyclization is another example of reaction catalyzed by NTPs. This transformation was also used by Rueping and coworkers as a benchmark reaction to gain insights into the Brønsted law for CBAs-catalyzed transformations (see Section 7.2.1, Figure 7.2). The stereochemical model for this catalytic system was first rationalized computationally by Krenske and coworkers [38]. As discussed in the previous paragraph, the computational treatment of the conformational flexibility of NTPs is a key point. Divinylketone and NTP **21** were used as model systems for a conformational analysis of a general catalyst/substrate complex **22** (Figure 7.15(a)). Interestingly, it was found that, differently from the [3+2] hydrazone cycloaddition reaction, the

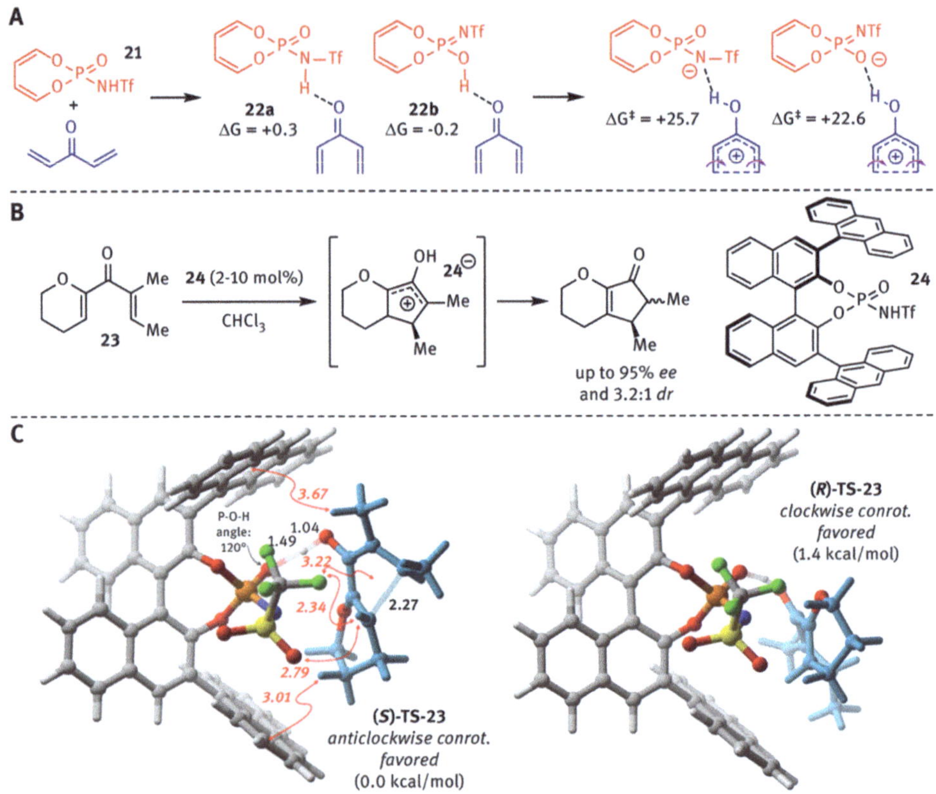

Figure 7.15: NTPs-catalyzed enantioselective Nazarov reaction.
Adapted with permission from *ACS Catal.* **2017**, *7*, 3466.

coordination of divinylketones occurs through either one of the tautomeric forms O=P-NH **22a** or TfN=P-OH **22b** of the NTP. However, while the Gibbs free energy difference between the two conformers at the ground state was negligible (0.5 kcal/mol in favor of **22b**), this energy difference was found to increase at the TS level (3.1 kcal/mol in favor of the TfN=P-OH form, Figure 7.15(a)). With evidence for the involvement of a tautomeric species like **22b** at the TS level, Kriske and coworkers studied the reaction of substrate **23** in the presence of NTP **24** (Figure 7.15(b)). The two diasteromeric TSs **(*S*)-TS-23** and **(*R*)-TS-23** for the conrotatory cyclization step, are reported in Figure 7.15(c). A $\Delta\Delta G^{\ddagger} = 1.4$ kcal/mol in favor of **(*S*)-TS-23** was calculated, which is in agreement with the experimental selectivity. Further interaction/distortion analysis of the two TSs allowed accessing more detailed information about the origin of the computed $\Delta\Delta G^{\ddagger}$. Distortion of the catalyst was found to be pivotal, being much stronger in the disfavored structure **(*R*)-TS-23**. Specifically, distortion of the N-triflyl phosphoramide group alone was computed to account for a $\Delta\Delta E_{distTf}^{\ddagger} = 4.8$ kcal/mol. Interestingly, the discrepancy between the computed $\Delta\Delta E^{\ddagger}$ and $\Delta\Delta E_{distTf}^{\ddagger}$ suggests that interactions between the catalyst and the substrate are in favor of **(*R*)-TS-23**. This is in contrast with the case studies examined so far where the interaction energy is responsible for the enantiodiscrimination towards the favored product. Thus, this example shows how in CBA catalysis the selectivity is often dependent on many factors which go beyond the ability of these catalysts to engage the substrate in non-covalent interaction networks.

7.2.4.3 Benzidine rearrangement

The enantioselective benzidine rearrangement of dinaphthylhydrazines catalyzed by CPAs was reported simultaneously by the groups of List and Kürti [39, 40]. They addressed the synthetic utility of this reaction that allows accessing enantioenriched BINAM derivatives from substrates **25** (Figure 7.16(a)). The two groups also undertook studies that provided interesting insights into the reaction mechanism from different perspectives. List and coworkers reasoned about the stoichiometry of the ionic pairing generated by the substrate and catalyst **1g**. Hydrazines **25** have two basic N atoms which could be protonated by the CPA. Thus, even though the second protonation would be less favorable, two possible mechanistic pathways can be depicted in which only one or two phosphate molecules are involved in the sigmatropic rearrangement (Figure 7.16(b)). In order to gain insights about the hypothesized mechanisms the authors undertook a non-linear effect experiment, in which the product *ee* was measured with **1g** at different optical purities. The plot of the product *ee* against the catalyst *ee* resulted in a non-linear correlation, implying the actual presence of a reaction pathway involving more than one phosphate molecules at the stereodetermining step (Figure 7.16(b)). It is worth noticing, that the observed non-linear effect does not exclude the presence of a one-phosphate mechanism. In fact, the interplay of multiple mechanisms or the presence of dimeric pathways where the rearrangement occurs within H-bonded aggregates of more catalyst/substrates molecules is consistent with the experiment outcome.

Kürti and coworkers performed a computational analysis of the one-phosphate mechanism. The computed TSs that involves the rearrangement of **25** in the presence of **[H$_8$]-1g** (**(*S*)-TS-25** and **(*R*)-TS-25**) are depicted in Figure 7.16(c). The $\Delta\Delta G^{\ddagger}$ associated to the reaction enantioselectivity is 2.2 kcal/mol. The energy difference was ascribed to the presence of CH-π interactions between the hydrazine aryl CH and the catalyst's electron-poor aryl substituents.

7.2.5 Stereoselective sulfoxidations

The enantioselective sulfoxidation reaction catalyzed by CBAs was first reported by List and coworkers in 2012 [41]. A wide set of CPAs was found to activate H_2O_2 toward thioanisoles to access sulfoxides albeit in low *ee* values (e. g. only 16% *ee* was obtained

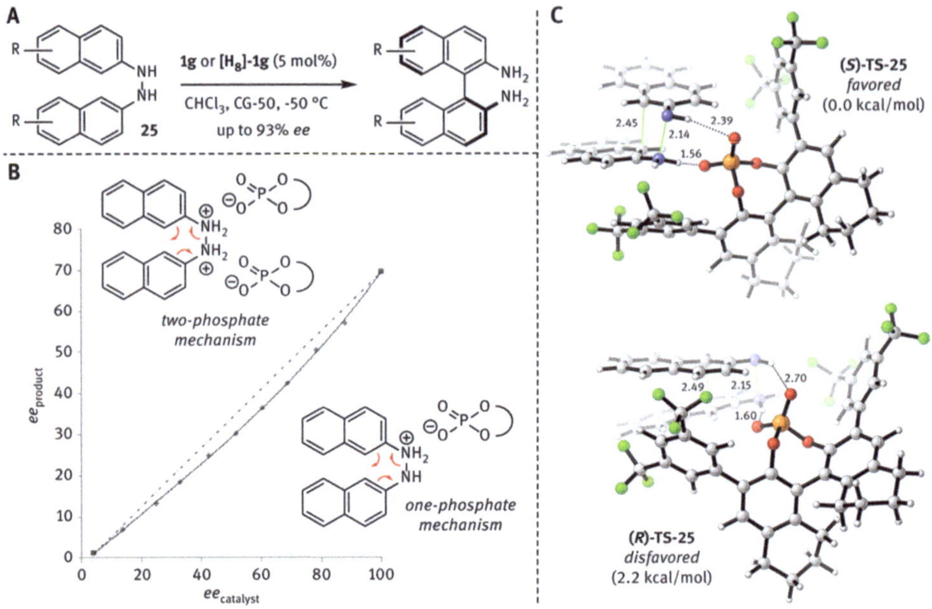

Figure 7.16: Enantioselective benzidine rearrangement of dinaphthylhydrazines catalyzed by CPAs. Adapted with permission from *Angew. Chem. Int. Ed.* **2013**, *52*, 9293 and *J. Am. Chem. Soc.* **2013**, *135*, 7414.

with catalyst **1H**). Higher selectivity was obtained by means of imidodiphosphoric acid **26** (Figure 7.17(a)), which provided oxidation of thioanisole **27** in 98% *ee*. In order to understand the mode of action of this new class of CBAs, computational studies were undertaken by the group of Sunoj [42]. Calculations showed that the reaction proceeds by coordination of H_2O_2 by the catalyst and consequent reaction of this complex with the substrate according with the mechanism in Figure 7.17(b). Computational prediction of the reaction enantioselectivity with (*S*)-**1H** resulted in structures **(*S*)-TS1h-27** and **(*R*)-TS1h-27**, with an energy difference $\Delta\Delta G^{\ddagger} = 0.5$ kcal/mol in favor of **(*R*)-TS1h-27**, in agreement with the low *ee* obtained with this catalyst. On the other hand, the computed $\Delta\Delta G^{\ddagger}$ for the reaction catalyzed by **26** was as high as 3.9 kcal/mol, and consistent with the ideal selectivity obtained experimentally. Analysis of structures **(*S*)-TS25-27** and **(*R*)-TS25-27** compared to **(*S*)-TS1h-27** and **(*R*)-TS1h-27** highlighted the reasons for the higher efficiency of **26** with respect to **1H** (see Figure 7.17(c), grey = catalyst, red = O atoms, blue = substrate). While **1H** has an open pocket that results in conformational freedom at the TS level with consequent low selectivity, the chiral pocket of catalyst **26** is closed and resembles the active site of an enzyme.

Further comparative analysis of **(*S*)-TS25-27** and **(*R*)-TS25-27** using the Van der Waals radii visualization mode, shows a better fitting of the substrate molecule into the catalyst's pocket in **(*R*)-TS25-27** (Figure 7.17(c), poor fitting for **(*S*)-TS25-27**

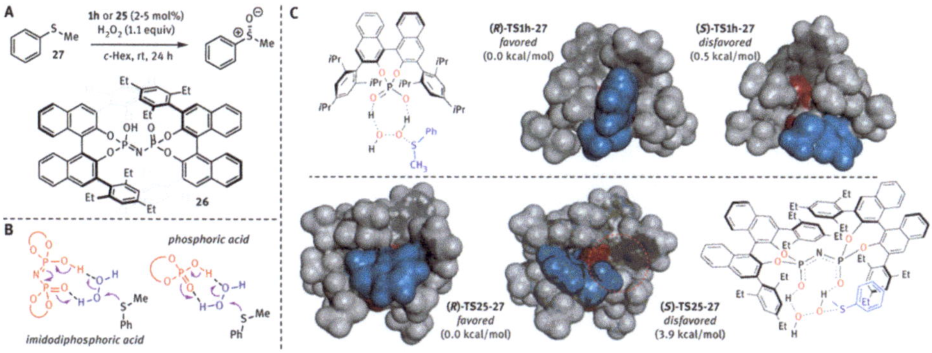

Figure 7.17: Enantioselective sulfoxidation reaction catalyzed by CBAs. Computational comparison between CPAs and imidophosphoric acids. Adapted with permission from *Angew. Chem. Int. Ed.* **2014**, *53*, 4432.

highlighted by a red circle). This better fitting allows maximization of non-covalent interactions at the TS level, which results in the observed preference for the (*R*)-product.

7.3 Chiral counter-anion catalysis

7.3.1 Meso-Aziridinium and meso-episulfonium ring opening reactions

Counter-anion directed catalysis is based on the formation of a tight ion pair between a positively charged reaction intermediate and a chiral counter-anion from the catalyst [43]. The formation of such ion pair is dependent on several factors, reaction temperature and solvent properties being the most important. Indeed, for the ion pair to be formed, the energy associated with the electrostatic attraction between the two partners has to be larger than the thermal energy available to separate them. For this reason, chiral counter-anion directed reactions are performed in non-polar solvents (toluene, DCM, $CHCl_3$, alkanes, etc.) and at low/moderate temperatures (typically not higher than r.t.). The first computational report assessing these concepts in the field of asymmetric catalysis is due to Duarte and Paton [44]. These authors studied the aziridinium and episulfonium ring opening reactions catalyzed by the phosphate anion of **1H** (Figure 7.18(a)). In these reactions the transient cationic reaction intermediates **28a-28b** are formed by abstraction of a chloride anion by Ag_2CO_3 from the starting material. Subsequent nucleophilic attack by an alcohol nucleophile leads to the formation of the desired aminoether or thioether in high enantioselectivity (91–94% *ee*). The study began with the evaluation by means of molecular dynamics (ΔG_{MD}) and DFT (ΔG_{DFT}) of the ion pair Gibbs free energy of association in various solvents (toluene, DCM, MeCN and H_2O). Not surprisingly, the calculated values were found to correlate with the corresponding dielectric constants and are listed in Figure 7.18(b). The coordination mode of **28a** and **28b** by **1H** is reminiscent of the complexation of the epoxides in **(*S*,*S*)-TS1h-14** and **(*R*,*R*)-TS1h-14** (Figure 7.10). Also in this case, the three membered ring H atoms were found to be involved in a direct electrostatic interaction with the phosphate anion, leading to a rigid and tight ionic complex (Figure 7.18(b)).

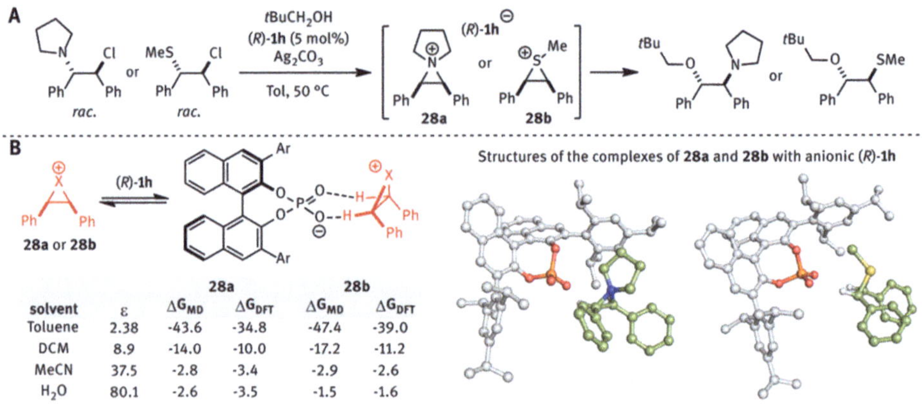

solvent	ε	28a ΔG_{MD}	28a ΔG_{DFT}	28b ΔG_{MD}	28b ΔG_{DFT}
Toluene	2.38	-43.6	-34.8	-47.4	-39.0
DCM	8.9	-14.0	-10.0	-17.2	-11.2
MeCN	37.5	-2.8	-3.4	-2.9	-2.6
H_2O	80.1	-2.6	-3.5	-1.5	-1.6

Figure 7.18: *meso*-aziridinium and *meso*-episulfonium ring opening reactions by counter-anion directed catalysis. Adapted with permission from *J. Am. Chem. Soc.* **2017**, *139*, 8886.

Due to the *meso* relative configuration of intermediates **28**, attack of the nucleophile to one of the two C atoms leads to opposite enantiomers of the desired product. Due to the chiral environment provided by catalyst (*R*)-**1H**, the attack occurs at the (*R*)-carbon leading to the preferential formation of the (*S*,*S*)-product. The structure of the diastereomertic TSs for the attack to **28a** (**(*S*,*S*)-TS-28a** and **(*R*,*R*)-TS-28a**) and **28b** (**(*S*,*S*)-TS-28b** and **(*R*,*R*)-TS-28b**) are reported in Figure 7.19. The $\Delta\Delta G^{\ddagger}$ associated to the reaction enantioselectivity is 2.1 kcal/mol for **28a** and 5.4 kcal/mol for **28b**, in agreement with the high *ee* obtained experimentally. Interaction/distortion analysis revealed that the selectivity is almost exclusively due to a stronger substrate distortion in **(*R*,*R*)-TS-28b**.

7.3.2 Stereoselective fluorination of allylic alcohols

In 2014, Toste and coworkers developed an in situ directing group strategy to access the enantioselective fluorination of allylic alcohols in high enantioselectivity [45]. In this reaction, **Selectfluor** was used as the fluorinating agent under chiral anion phase transfer regime with a chiral phosphate as catalyst. An arylboronic acid **BA** was added as additive to the reaction, which in the presence of the substrate **29** was shown to form a mixed boronic ester **30**. Such intermediate is the actual substrate that

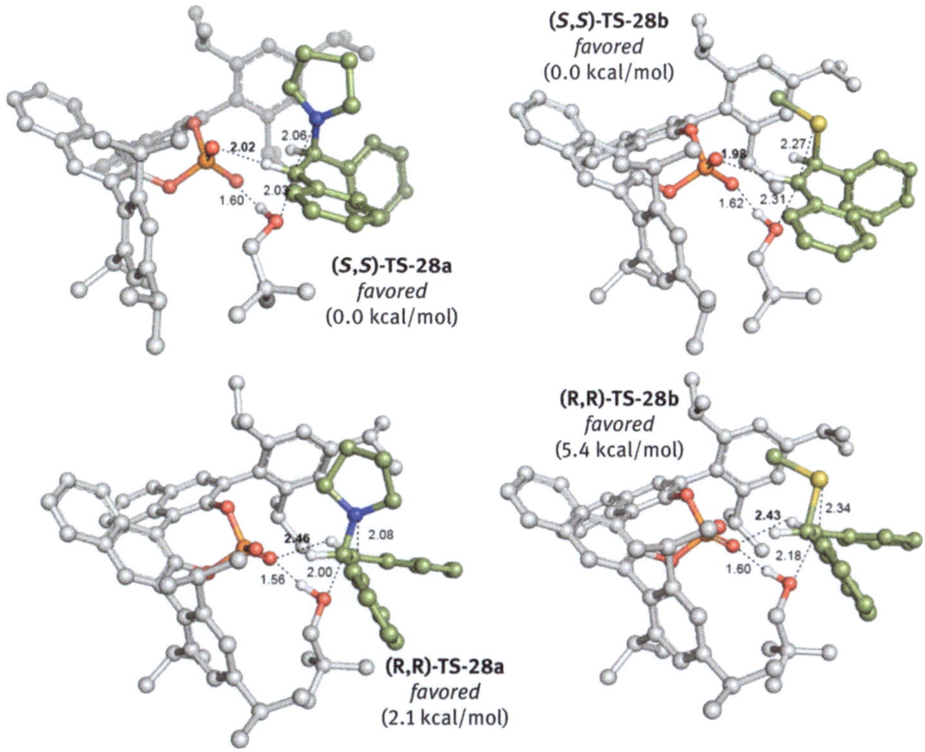

Figure 7.19: Computed TSs for the TRIP-counterion catalyzed ring opening reaction of meso-aziridinium and meso-episulfonium. Reproduced with permission from *J. Am. Chem. Soc.* **2017**, *139*, 8886.

undergoes fluorination to give product **31** after a hydrolytic workup. The pattern substitution of **BA** strongly affected the reaction selectivity, which was shown to reverse upon substitution of the *meta*-positions of the aryl group (Figure 7.20(a)). In order to gain mechanistic understanding of the manifold at the basis of this reaction, the groups of Toste and Sigman undertook a mechanistic study [46]. Non-linear effect experiments showed that chiral phosphates with *ortho*-substituents at the 3,3′ -aryl substituents (**1H, 1l, 1m** and **1H,** Figure 7.20(a)) behaved differently form unsubstituted ones. Specifically, the former ones did not give non-linear effects, differently from other catalysts. Thus, only catalysts **1H, 1l, 1m** and **1H** were selected for further studies, as these showed to work with the same mechanism. A stereochemical model for this transformation was developed [18] by means of DFT calculations and Multidimensional Correlation Analysis (MCA) [47, 48]. In this approach, experimental *ee* values are correlated through multidimensional regression analysis to a set of computationally acquired parameters. The resulting polynomial equation contains those parameters that describe the features of the catalyst/substrate that affect the reaction stereochemical outcome. Therefore, mechanistic information can be obtained by the analysis of such descriptors.

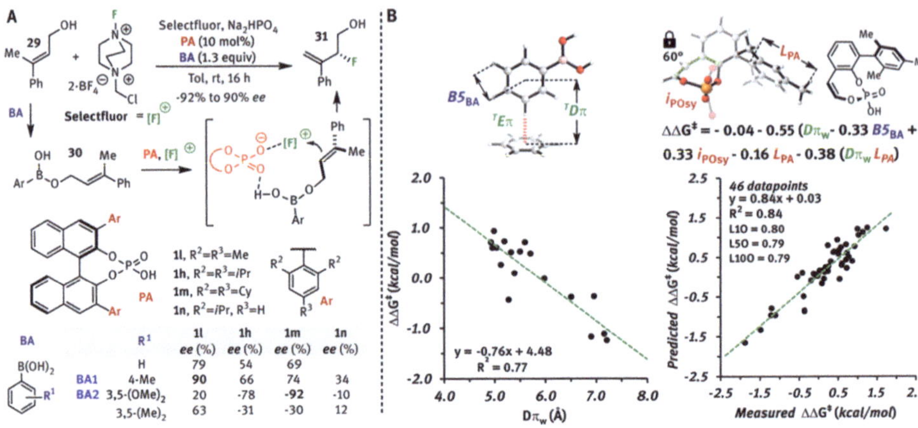

Figure 7.20: Computational analysis of the enantioselective fluorination of allylic alcohols catalyzed by TRIP. Reproduced with permission from *J. Am. Chem. Soc.* **2017**, *139*, 6803.

Application of MCA to a dataset of 46 catalyst/**BA** combinations resulted in the model depicted in Figure 7.20(c). The parameters include $^{\mathbf{T}}\boldsymbol{D\pi}_{\mathbf{w}}$, accounting for geometric requirements derived from the apparent presence of a π interaction (Figure 7.20(b)); $\mathbf{B5_{BA}}$ and $\mathbf{L_{PA}}$, Sterimol values describing steric effects from **BA** and the catalyst, respectively; and, $\boldsymbol{i}_{\mathbf{POsy}}$, the phosphate symmetric stretching intensity, which represents the ability of each phosphate to engage in H-bonding/electrostatic interactions. The coefficients of the model equation suggest the T-shaped π-stacking interaction engaged by the *meta*-position of BA to be important. Furthermore, simple correlation of $^{\mathbf{T}}\boldsymbol{D\pi}_{\mathbf{w}}$ against the experimental enantioselectivity (Measured $\Delta\Delta G^{\ddagger}$ expressed in kcal/mol) showed a quantitative trend with $R^2 = 0.77$ (Figure 7.20(b)). Thus, based on this parameters analysis, a TS involving i) a π-stacking between the **BA** *meta*-susbtituent and the catalyst BINOL scaffold, and ii) H-bond/electrostatic interaction between the catalyst phosphate moiety, **Selectfluor** and **BA** was hypothesized.

In order to access visualization of the hypothesized TSs, DFT computations were undertaken combining catalyst **1H** and **BA1** or **BA2**. Remarkably, this DFT study revealed that TSs presenting a T-shaped interaction between the aryl ring of **BA1** and the BINOL scaffold of **1H** are the lowest in energy. The favored TSs leading to the *R* and *S* products are depicted in Figure 7.21(a), with **(*R*)-TSBA1-30** leading to the major observed product (computed $\Delta\Delta G^{\ddagger} = 1.0$ kcal/mol; experimental $\Delta\Delta G^{\ddagger} = 0.9$ kcal/mol). Location of low-lying TSs was also performed for the **1H**/**BA2** system. The analysis showed that **(*R*)-TSBA2-30** and **(*S*)-TSBA2-30** (Figure 7.4(b)) are the lowest in energy, and the formation of the *S* enantiomer as the major product is consistent with the observed results (computed $\Delta\Delta G^{\ddagger} = 2.6$ kcal/mol; experimental $\Delta\Delta G^{\ddagger} = 1.2$ kcal/mol). Supporting the hypothesis resulting from the parameters found in the MCA analysis, T-shaped π-interactions between the *meta* substituent of **BA2** and the BINOL moiety of **1H** are present in both systems (Figure 7.21(b)). Moreover, H-bonding (**a**), π stacking (**b**), and electrostatics (**c**) interactions were found in the computed TSs (Figure 7.21), which can be connected to the parameters $\boldsymbol{i_{POsy}}$ (**a** and **c**), and $\boldsymbol{^{T}D\pi_{w}}$ and $\mathbf{B5_{BA}}$ (**b**) of the MCA.

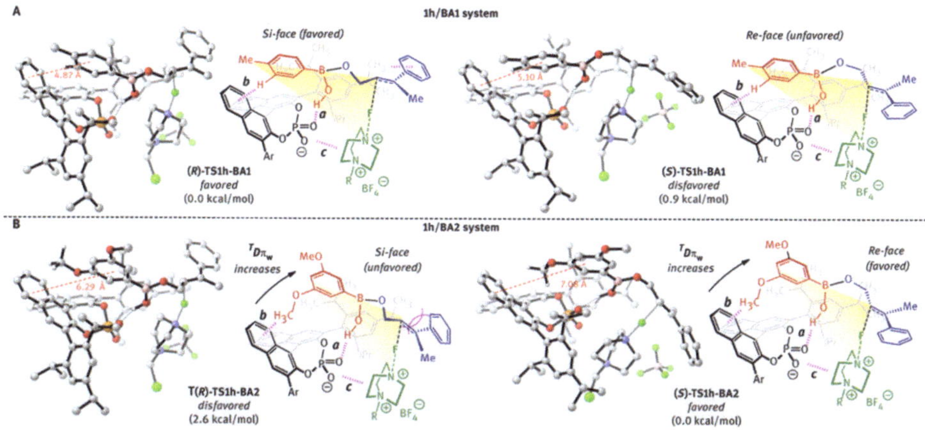

Figure 7.21: Computational analysis of the enantioselective fluorination of allylic alcohols catalyzed by TRIP. Reproduced with permission from *J. Am. Chem. Soc.* **2017**, *139*, 6803.

7.3.3 Enantioselective Pummerer reaction

In the 2017, Sigman and Toste developed and studied an enantioselective Pummerer reaction based on a chiral anion phase transfer process [49]. In this reaction, a benzylic thioether **32** is oxidized in *situ in* by an oxammonium salt **33** in the presence of chiral phosphate **1o**. The resulting thionium cation forms a tight ion pair **34** with the phosphate **1o**, and undergoes a stereoselective cyclization to give the desired product **35** (Figure 7.22(a)). The reaction proceeded with moderate to good enantioselectivity over a wide range of substrates and catalysts, which prompted the authors to study the reaction via a combined MCA and DFT approach. MCA was performed separately on a *ee* values dataset obtained with substrate **32** and 12 different catalysts (Figure 7.22(b)). Three Sterimol parameters appeared in the equation: **B5**, maximum width of the entire catalysts' aryl substituent; $\mathbf{B1_2}$ and $\mathbf{B1_4}$, minimum width of the substituents in the *ortho*-positions and of the substituent in the *para*-position of the aryl group, respectively. These parameters describe the requirement for a proper steric environment into the catalyst's chiral pocket. Additional insights into the stereochemical recognition were obtained by MCA of the ion pair cationic counterpart. Descriptors for a set of nine substrates with different aryl groups at the sulfonamide moiety were calculated from the TS of the corresponding thionium cation in the uncatalyzed cyclization. The analysis resulted in the statistical model in

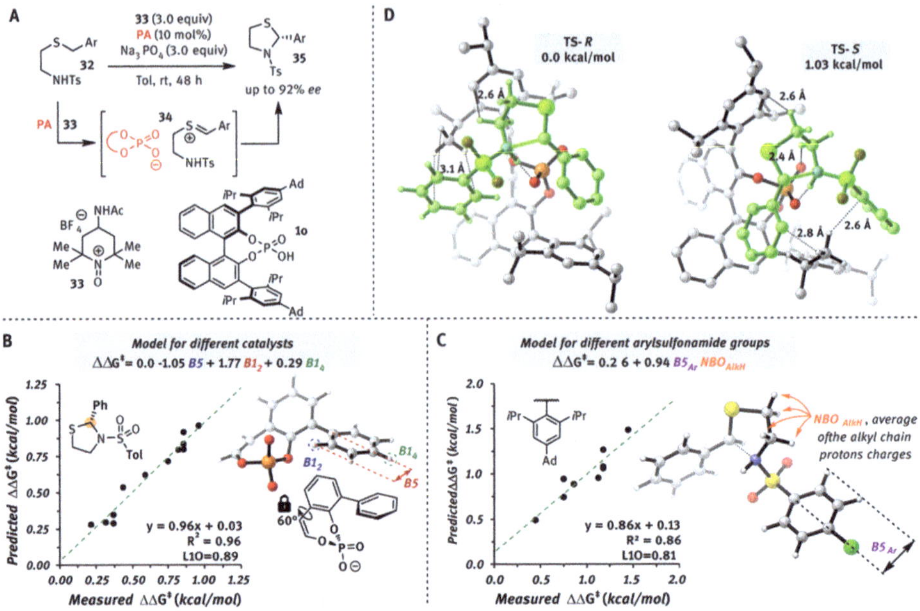

Figure 7.22: Enantioselective Pummerer reaction via Chiral Anion Phase Transfer Catalysis, and its computational/MCA analysis. Adapted with permission from *Angew. Chem. Int. Ed.* **2018**, *57*, 589.

Figure 7.22(c), which contains two parameters condensed into one term. These parameters accounts for either the steric hindrance or the shape of the aryl substituent (**B5$_{Ar}$**), and the average charge of the thionium alkyl chain (**NBO$_{AlkH}$**). The presence of this latter descriptor in the model suggested the presence of non-covalent interactions at the TS level involving the alkyl chain next to the thionium moiety.

Further information about the interaction mode between the catalyst and the substrate was gained by DFT computation of the reaction TSs involving **32** and **1H.** The low-lying TSs leading to the *R*- and *S*-configured products (**(*R*)-TS-32** and **(*S*)-TS-32**, respectively) are depicted in Figure 7.22(d) (substrate highlighted in green). **(*R*)-TS-32** is favored and consistent with the experimentally observed product configuration, and the relative energy of **(*S*)-TS-32** (1.03 kcal/mol) matched with the selectivity measured experimentally (56% *ee*, 0.75 kcal/mol). The two TSs showed several non-covalent interactions between the phosphate and the substrate. However, **(*R*)-TS-32** displayed a better accommodation of the benzenesulfonamide group through CH–π interactions, which involved the iPr substituents and the binaphthyl backbone of the catalyst. This representation provides an interpretation for the statistical model in Figure 7.22(b), where strict requirements about the catalyst's substituents were highlighted. Additionally, a stronger CH–π interaction between the aryl substituent or **1H** (iPr$_3$Ph) and the electron-deficient alkyl chain of the cationic intermediate was present in **(*R*)-TS-32** (three CH–π contacts, while

(S)-TS-32 presented only one CH–π contact). The presence of these interactions in the computed TSs is fully consistent with the model equation obtained by MCA in Figure 7.22(c).

7.4 General considerations

In the last twenty years, CBA and chiral counteranion catalysis have emerged as a fundamental area of organocatalysis. The development of chiral acidic catalysts has allowed extending many known Brønsted catalyzed reactions to the stereoselective domain. Moreover, the controlled conditions under which these catalysts can be used, allowed accessing reactivity of increasing complexity with extraordinary selectivity levels. For instance, the development of bisimidophosphoric acids by List allowed accessing reactivity of new substrates within the tight catalyst pocket, which is reminiscent of enzymatic catalysis [50]. Exploitation of the solubility of bulky CPAs allowed Toste and coworkers to develop the chiral counteranion phase transfer catalysis [51], where the low solubility of the starting material allows high control of background reactions of highly reactive reagents. However, compared to the fast growth of these fields in an applicative fashion, only little has been done in order to understand the mode of action of these new catalysts. Part of the work reported in this direction was reviewed in this chapter, with examples from both CBA and counteranion directed catalysis. The contributions summarized here allow outlining several key points about the use of chiral phosphates in catalysis.

Studies by Leito and Rueping highlighted the importance of the catalyst's acidity for accessing desirable levels of reactivity. Indeed, for the Nazarov cyclization, it has been shown that the reaction rate follows the Brønsted law. However, acidity alone is not sufficient for achieving a good control of the reaction selectivity. In fact, Gschwind and coworkers showed that the success of CPAs with respect of other possible acidic functional groups is due to the properties of the H-bond that phosphates can form. In particular, it was shown that phosphates can engage H-bonds with iminium ions that are stronger with respect to those formed by other acidic compounds. The same team showed that BINOL-based phosphates are able to coordinate iminium ions by non-covalent interactions thanks to their extended π-systems. Computational analysis of a variety of CBA catalyzed reactions by Goodman, Houk, Wheeler, Sunoj, and others allowed assessing the importance of these features for optimal molecular recognition at the TS level, rationalizing the high levels of selectivity experimentally accessed by CPAs. On the other hand, it was shown that the activation mode and the coordination of the substrate by these catalysts is often difficult to predict. Indeed, differently from covalent catalysis, the interactions interplaying at the TS level are strongly dependent on the features of the electrophile, including steric, electronic and substitution patterns. For this reason, the continuous development of new methodologies and physical organic techniques (for instance as pioneered by Gschwind and Sigman) aiming the study and understanding of these catalytic systems is a topic of major importance in the field of organocatalysis.

References

[1] Brønsted JN, Pedersen KJ. Die katalytische Zersetzung des Nitramids und ihre physikalisch-chemische Bedeutung. Zeitschrift für Phys Chemie Stöchiometrie und Verwandtschaftslehre. 1924;108:185–235.

[2] Kütt A, Leito I, Kaljurand I, Sooväli L, Vlasov VM, Yagupolskii LM, et al. A comprehensive self-consistent spectrophotometric acidity scale of neutral Brønsted acids in acetonitrile. J Org Chem. 2006;71:2829–38.

[3] Kütt A, Rodima T, Saame J, Raamat E, Mäemets V, Kaljurand I, et al. Equilibrium acidities of superacids. J Org Chem. 2011;76:391–5.

[4] Christ P, Lindsay AG, Vormittag SS, Neudörfl JM, Berkessel A, O'Donoghue AC. K_a values of chiral brønsted acid catalysts: phosphoric acids/ amides,sulfonyl/sulfuryl imides, and perfluorinated TADDOLs (TEFDDOLs). Chem Eur J. 2011;17:8524–8.

[5] Kaupmees K, Tolstoluzhsky N, Raja S, Rueping M, Leito I. On the acidity and reactivity of highly effective chiral Brønsted acid catalysts: establishment of an acidity scale. Angew Chem Int Ed. 2013;52:11569–72.

[6] Yang C, Xue XS, Jin JL, Li X, Cheng JP. Theoretical study on the acidities of chiral phosphoric acids in dimethyl sulfoxide: hints for organocatalysis. J Org Chem. 2013;78:7076–85.

[7] Yang C, Xue XS, Li X, Cheng JP. Computational study on the acidic constants of chiral brønsted acids in dimethyl sulfoxide. J Org Chem. 2014;79:4340–51.

[8] Uraguchi D, Terada M. Chiral Brønsted acid-catalyzed direct mannich reactions via electrophilic activation. J Am Chem Soc. 2004;126:5356–7.

[9] Akiyama T, Itoh J, Yokota K, Fuchibe K. Enantioselective Mannich-type reaction catalyzed by a chiral Brønsted acid. Angew Chem Int Ed. 2004;43:1566–8.

[10] Fleischmann M, Drettwan D, Sugiono E, Rueping M, Gschwind RM. Brønsted acid catalysis: hydrogen bonding versus ion pairing in imine activation. Angew Chem Int Ed. 2011;50:6364–9.

[11] Kim H, Sugiono E, Nagata Y, Wagner M, Bonn M, Rueping M, et al. Role of ion-pairs in Brønsted acid catalysis. ACS Catal. 2015;5:6630–3.

[12] Sorgenfrei N, Hioe J, Greindl J, Rothermel K, Morana F, Lokesh N, et al. NMR spectroscopic characterization of charge assisted strong hydrogen bonds in Brønsted acid catalysis. J Am Chem Soc. 2016;138:16345–54.

[13] Sharif S, Denisov GS, Toney MD, Limbach H. Coupling of functional hydrogen bonds in pyridoxal-5'-phosphate–enzyme model systems observed by solid-state NMR spectroscopy. J Am Chem Soc. 2007;129:6313.

[14] Limbach HH, Pietrzak M, Sharif S, Tolstoy PM, Shenderovich IG, Smirnov SN, et al. NMR parameters and geometries of OHN and ODN hydrogen bonds of pyridine–acid complexes. Chem Eur J. 2004;10:5195.

[15] Benedict H, Shenderovich IG, Malkina OL, Malkin VG, Denisov GS, Golubev NS, et al. Nuclear scalar spin–spin couplings and geometries of hydrogen bonds. J Am Chem Soc. 2000;122:1979.

[16] Limbach HH, Pietrzak M, Benedict H, Tolstoy PM, Golubev NS, Denisov GS. Empirical corrections for quantum effects in geometric hydrogen bond correlations. J Mol Struct. 2004;706:115.

[17] Greindl J, Hioe J, Sorgenfrei N, Morana F, Gschwind RM. Brønsted acid catalysis—structural preferences and mobility in imine/phosphoric acid complexes. J Am Chem Soc. 2016;138:15965–71.
[18] Orlandi M, Coelho JA, Hilton MJ, Toste FD, Sigman MS. Parametrization of non-covalent interactions for transition state interrogation applied to asymmetric catalysis. J Am Chem Soc. 2017;139:6803–6.
[19] Rueping M, Sugiono E, Azap C, Theissmann T, Bolte M. Enantioselective Brønsted acid catalyzed transfer hydrogenation: organocatalytic reduction of imines. Org Lett. 2005;7:3781–3.
[20] Hoffmann S, Seayad AM, List B. A powerful Brønsted acid catalyst for the organocatalytic asymmetric transfer hydrogenation of imines. Angew Chem Int Ed. 2005;44:7424–7.
[21] Simón L, Goodman JM. Theoretical study of the mechanism of Hantzsch ester hydrogenation of imines catalyzed by chiral BINOL-phosphoric acids. J Am Chem Soc. 2008;130:8741–7.
[22] Marcelli T, Hammar P, Himo F. Hantzsch esters employed for asymmetric imine reduction. Adv Synth Catal. 2009;351:525–9.
[23] Marcelli T, Hammar P, Himo F. Phosphoric acid catalyzed enantioselective transfer hydrogenation of imines: a density functional theory study of reaction mechanism and the origins of enantioselectivity. Chem Eur J. 2008;14:8562–71.
[24] Reid JP, Goodman JM. Goldilocks catalysts: computational insights into the role of the 3,3′ substituents on the selectivity of BINOL-derived phosphoric acid catalysts. J Am Chem Soc. 2016;138:7910–17.
[25] Simón L, Goodman JM. Mechanism of BINOL–phosphoric acid-catalyzed strecker reaction of benzyl imines. J Am Chem Soc. 2009;131:4070–7.
[26] Monaco MR, Poladura B, Diaz de Los Bernardos M, Leutzsch M, Goddard R, List B. Activation of carboxylic acids in asymmetric organocatalysis. Angew Chem Int Ed. 2014;53:7063–7.
[27] Monaco MR, Prévost S, List B. Organocatalytic asymmetric hydrolysis of epoxides. Angew Chem Int Ed. 2014;53:8142–5.
[28] Monaco MR, Prévost S, List B. Catalytic asymmetric synthesis of thiols. J Am Chem Soc. 2014;136:16982–5.
[29] Monaco MR, Fazzi D, Tsuji N, Leutzsch M, Liao S, Thiel W, et al. The activation of carboxylic acids via self-assembly asymmetric organocatalysis: a combined experimental and computational investigation. J Am Chem Soc. 2016;138:14740–9.
[30] Liao S, Leutzsch M, Monaco MR, List B. Catalytic enantioselective conversion of epoxides to thiiranes. J Am Chem Soc. 2016;138:5230–3.
[31] Seguin TJ, Wheeler SE. Competing noncovalent interactions control the stereoselectivity of chiral phosphoric acid catalyzed ring openings of 3-substituted oxetanes. ACS Catal. 2016;6:7222–8.
[32] Yang W, Sun J. Organocatalytic enantioselective synthesis of 1,4-dioxanes and other oxa-heterocycles by oxetane desymmetrization. Angew Chem Int Ed. 2016;55:1868–1871.
[33] Wang Z, Chen Z, Sun J. Catalytic asymmetric nucleophilic openings of 3-substituted oxetanes. Org Biomol Chem. 2014;12:6028–6032.
[34] Yang W, Wang Z, Sun J. Enantioselective oxetane ring opening with chloride: unusual use of wet molecular sieves for the controlled release of HCl. Angew Chem Int Ed. 2016;55:6954–6958.
[35] Maji R, Champagne PA, Houk KN, Wheeler SE. Activation mode and origin of selectivity in chiral phosphoric acid-catalyzed oxacycle formation by intramolecular oxetane desymmetrizations. ACS Catal. 2017;7:7332–9.
[36] Champagne PA, Houk KN. Origins of selectivity and general model for chiral phosphoric acid-catalyzed oxetane desymmetrizations. J Am Chem Soc. 2016;138:12356–9.

[37] Hong X, Küçük HB, Maji MS, Yang YF, Rueping M, Houk KN. Mechanism and selectivity of N-triflylphosphoramide catalyzed (3+2) cycloaddition between hydrazones and alkenes. J Am Chem Soc. 2014;136:13769–80.
[38] Lovie-Toon JP, Tram CM, Flynn BL, Krenske EH. Mechanisms of carbonyl activation by BINOL *N*-triflylphosphoramides: enantioselective Nazarov cyclizations. ACS Catal. 2017;7:3466–76.
[39] Li GQ, Gao H, Keene C, Devonas M, Ess DH, Kürti L. Organocatalytic Aryl–Aryl bond formation: an atroposelective [3,3]-rearrangement approach to BINAM derivatives. J Am Chem Soc. 2013;135:7414–17.
[40] De CK, Pesciaioli F, List B. Catalytic asymmetric benzidine rearrangement. Angew Chem Int Ed. 2013;52:9293–5.
[41] Liao S, Čorić I, Wang Q, List B. Activation of H2O2 by chiral confined Brønsted acids: a highly enantioselective catalytic sulfoxidation. J Am Chem Soc. 2012;134:10765–8.
[42] Jindal G, Sunoj RB. Axially chiral imidodiphosphoric acid catalyst for asymmetric sulfoxidation reaction: insights on asymmetric induction. Angew Chem Int Ed. 2014;53:4432–6.
[43] Mahlau M, List B. Asymmetric counteranion-directed catalysis: concept, definition, and applications. Angew Chem Int Ed. 2013;52:518–33.
[44] Duarte F, Paton RS. Molecular recognition in asymmetric counteranion catalysis: understanding chiral phosphate-mediated desymmetrization. J Am Chem Soc. 2017;139:8886–96.
[45] Zi W, Wang YM, Toste FD. An *in Situ* directing group strategy for chiral anion phase-transfer fluorination of allylic alcohols. J Am Chem Soc. 2014;136:12864–7.
[46] Neel AJ, Milo A, Sigman MS, Toste FD. Enantiodivergent fluorination of allylic alcohols: data set design reveals structural interplay between achiral directing group and chiral anion. J Am Chem Soc. 2016;138:3863–75.
[47] Sigman MS, Harper KC, Bess EN, Milo A. The development of multidimensional analysis tools for asymmetric catalysis and beyond. Acc Chem Res. 2016;49:1292–301.
[48] Santiago CB, Guo J-Y, Sigman MS. Predictive and mechanistic multivariate linear regression models for reaction development. Chem Sci. 2018;9:2398–412.
[49] Biswas S, Kubota K, Orlandi M, Turberg M, Miles DH, Sigman MS, et al. Enantioselective synthesis of N,S-acetals by an oxidative pummerer-type transformation using phase-transfer catalysis. Angew Chem Int Ed. 2018;57:589–93.
[50] Čorić I, List B. Asymmetric spiroacetalization catalysed by confined Brønsted acids. Nature. 2012;483:315.
[51] Rauniyar V, Lackner AD, Hamilton GL, Toste FD. Asymmetric electrophilic fluorination using an anionic chiral phase-transfer catalyst. Science. 2011;334:1681–4.

Sara Meninno, Rosaria Villano and Alessandra Lattanzi

8 Recent developments in stereoselective organocatalytic oxyfunctionalizations

Abstract: In this chapter, asymmetric at carbon oxidations using organocatalytic systems reported from 2012 up to 2018 have been illustrated. Asymmetric epoxidations and oxidation of heteroatom-containing molecules were not included. The processes selected encopass alpha-hydroxylation of carbonyl compounds, dihydroxylation and dioxygenation of alkenes, Baeyer-Villiger and oxidative desymmetrization reactions.

Keywords: asymmetric oxidation, organocatalysis, green chemistry

8.1 Introduction

The area of asymmetric oxidations is one of the most actively investigated by organic chemists, being, the introduction of "oxygen-containing groups" and changing the oxidation state of functional groups and atoms onto a molecule, fundamental and recurrent transformations [1]. The development of these processes in a stereoselective fashion is often prompted by the need to produce chiral molecules, which are ubiquitous in the industry, e. g. pharmaceuticals, agrochemicals, food products, materials science to cite the most relevant. Over the years, the attention of the synthetic community steadily increased toward the development of environmentally-sustainable and convenient to apply processes of asymmetric oxidation [1]. In this endeavor, an important place is occupied by the organocatalytic oxidative methods, which have two major advantages: (1) use of readily available catalysts, often derived from natural sources, (2) mild and manageable reaction conditions, which overcome product contamination problems, associated to residual metal traces, typical of metal-catalysis [2]. Several methodologies are available to stereoselectively oxyfunctionalize organic molecules by catalytic metal-based systems, whereas the organocatalytic tool can be considered still underdeveloped. The aim of the chapter is to highlight the progress achieved from 2012 to 2018 in fundamental stereoselective at carbon oxidation reactions, where the introduction of oxygen atom(s) onto the molecule has been achieved by using organocatalytic systems. Recent reviews and book chapters reported on organocatalytic asymmetric epoxidation and sulfoxidation reactions [1c, 3], which are amongst the most developed. Accordingly, they will not be included, as well as oxidative coupling and dearomatization reactions. The processes selected

This article has previously been published in the journal Physical Sciences Reviews. Please cite as: Meninno, S., Villano, R., Lattanzi, A. Recent developments in stereoselective organocatalytic oxyfunctionalizations *Physical Sciences Reviews* [Online] 2021, 2. DOI: 10.1515/psr-2018-0095

https://doi.org/10.1515/9783110590050-008

encopass α-hydroxylation of carbonyl compounds, dihydroxylation and dioxygenation of alkenes, Baeyer–Villiger (BV) and oxidative desymmetrization reactions.

8.2 α-Oxygenation of carbonyl compounds

α-Oxygenated carbonyl compounds are a common motif in biologically active natural products or pharmaceuticals. The direct stereoselective α-oxygenation of carbonyl compounds promoted by organocatalysts represents a powerful strategy for the construction of optically active α-oxygenated targets and, in the field of asymmetric organocatalysis, chiral H-bonding catalysts and phase-transfer ion-pairing catalysts are extremely useful.

8.2.1 α-Hydroxylation

In a pioneering work on the enantioselective hydroxylation of 3-substituted-2-oxindoles with molecular oxygen, the Itoh's group [3] demonstrated that the cinchonidine-derived phase-transfer catalyst **1** accelerated the reaction considerably (Figure 8.1). High chemical yields and moderate to good enantioselectivities were obtained, above all by using 4 equiv of triethyl phosphite as additive to reduce the peroxide intermediate, even though the examples reported were restricted to oxindole.

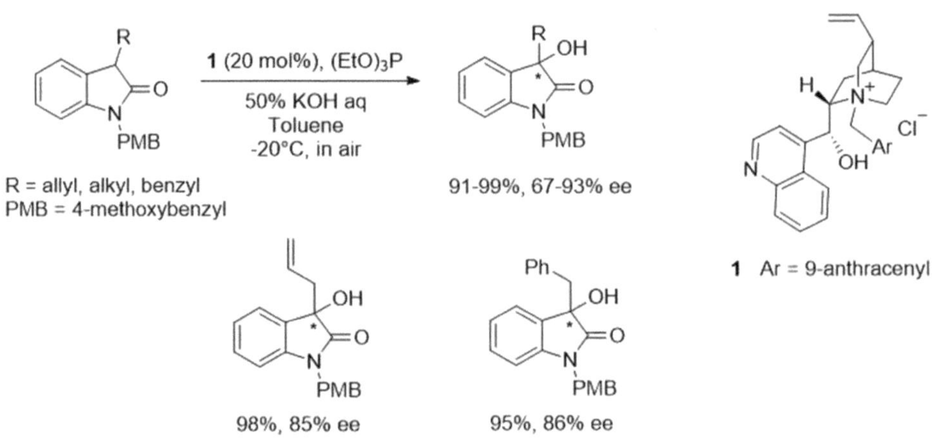

Figure 8.1: Catalytic asymmetric hydroxylation of oxindoles by using a phase-transfer catalyst.

More recently, the use of the cinchona alkaloid-derived dimeric phase-transfer catalyst **2** has been reported by Zhao and co-workers in the oxidation of acyclic as well as cyclic ketones [4], affording α-hydroxylated products in high yields and good stereoselectivities using molecular oxygen (Figure 8.2). Different solvents and P-based additives were screened and the best results were obtained by using 1,2-bis(diphenylphosphino)ethane (DPPE) as reductant.

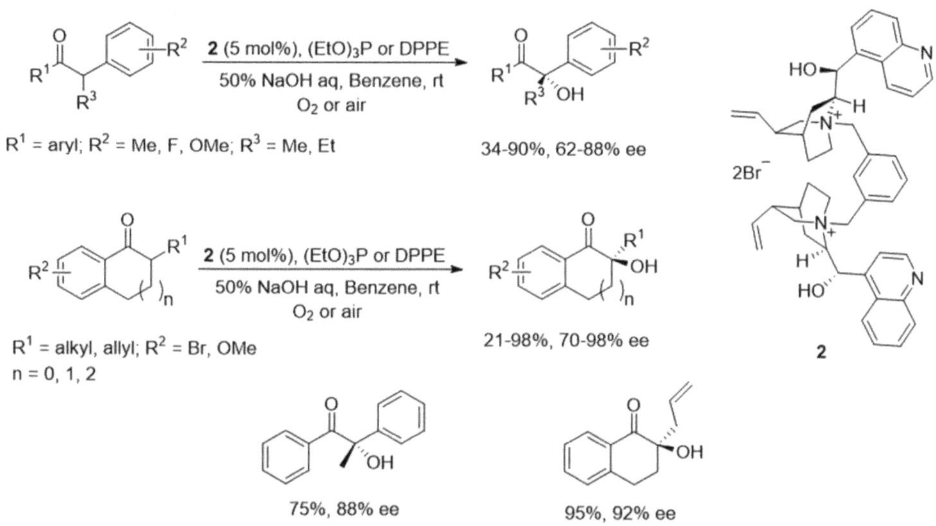

Figure 8.2: Asymmetric α-hydroxylation of acyclic and cyclic ketones by using a phase-transfer catalyst.

Molecular oxygen has also been used as an efficient oxidant in the enantioselective α-hydroxylation of 2-oxindoles catalyzed by pentanidiums as phase transfer catalysts. Contrary to all previous reports, in this methodology no reductant (triethyl phosphite or DPPE) was required (Figure 8.3) [5]. Several 3-hydroxyl-2-oxindoles were obtained with excellent enantioselectivities (86–98 % ee) and the authors proposed a two-step mechanism for the reaction: the enantioselective formation of hydroperoxide oxindole controlled by pentanidium, followed by its kinetic resolution, *via* reduction by the enolates of 3-substituted-2-oxindoles *in situ* generated under the basic reaction conditions. Under the same reaction conditions, the amount of hydroperoxide oxindoles obtained was related to the amount of molecular oxygen used in the reaction as well as to the addition of a strong inorganic base such as NaOH or KOH (Figure 8.3) [6].

As a further development of the strategy promoted by chiral phase-transfer catalysts using molecular oxygen as oxidant, Meng and co-workers explored the photo-oxygenation of β-keto esters under mild conditions [7]. The effect of different light sources was evaluated, as well as the influence of different sensitizers on the reaction process (Rose Bengal, methylene blue, tetraphenylporphyrin TPP). The optimized conditions depicted in Figure 8.4 were applied to a range of substrates and the products were isolated in high yields and moderate ees. Moreover, the free OH group at the C9 position of the organocatalyst proved to be fundamental for the enantiocontrol. The authors provided evidence of the involvement of $^1O_2$ in the hydroxylation process and proposed a plausible mechanism for the reaction (Figure 8.4). They also reported a

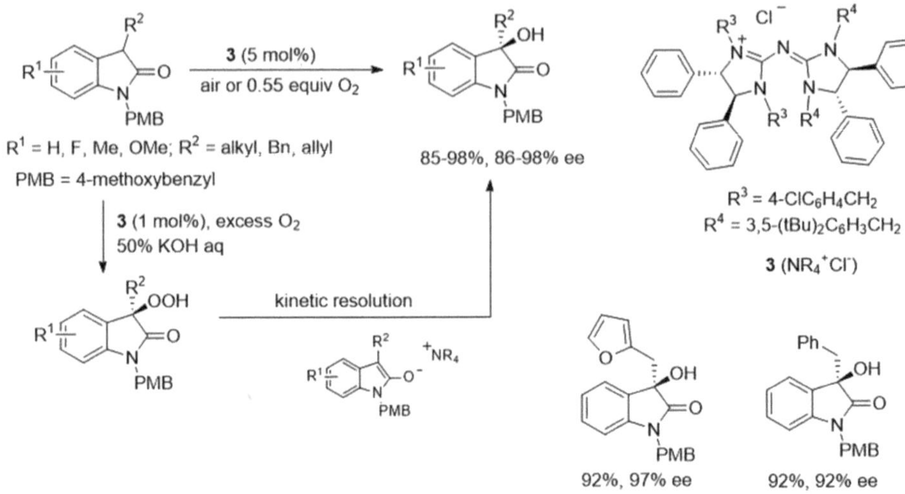

Figure 8.3: Pentanidium-catalyzed enantioselective α-hydroxylation of oxindoles with O_2.

different methodology for the α-hydroxylation of β-keto esters through catalysis by methylhydrazine and cinchona alkaloid, using O_2 as the oxygen source.

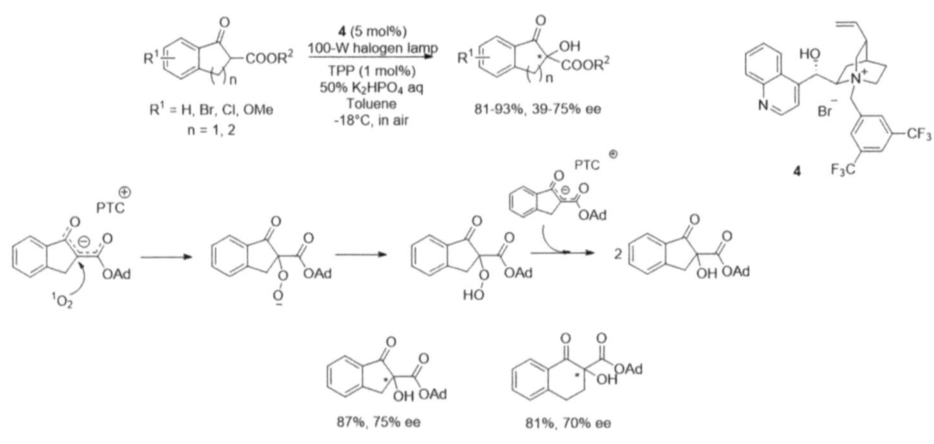

Figure 8.4: Asymmetric photo-oxygenation of β-keto esters and proposed mechanism for the reaction.

Enantioenriched α-hydroxy β-dicarbonyl compounds were obtained with excellent yields (up to 95 %) and interesting level of enantioselectivity (ee up to 85 %) [8].

Recently, a modification of the chiral phase-transfer catalyst to include a N-oxide moiety [9] (**5**, Figure 8.5) has been used in another photo-oxidative methodology with improved results compared to the modest enantioselectivity reported in Figure 8.4 for adamantyl α-hydroxy-β-keto esters

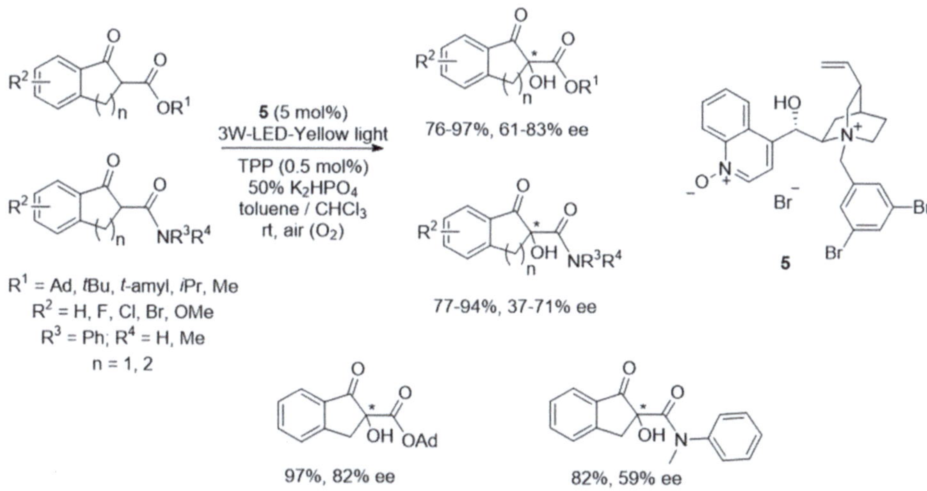

Figure 8.5: Enantioselective photo-oxygenation of β-dicarbonyl compounds catalyzed by **5**.

Furthermore, thanks to this cinchona-derived N-oxide asymmetric phase-transfer organocatalyst, the elaborated methodology could also be extended to other β-keto esters and β-keto amides (Figure 8.5), different from the adamantyl derivatives, with excellent yields and good enantioselectivities. Moreover, these catalysts could be easily recovered at the end of the reaction and reused without significant loss of reactivity or enantioselectivity.

The choice of the oxidant is fundamental for the sustainability of an oxidation reaction and undoubtedly molecular oxygen is the ideal choice but recently, hydrogen peroxide and organic hydroperoxides have emerged as useful alternatives for oxidations under metal-free conditions.

In 2017, Ooi and co-workers elaborated an efficient methodology for the asymmetric α-hydroxylation of oxindoles catalyzed by chiral 1,2,3-triazolium salts as phase-transfer catalysts (**6**), using hydrogen peroxide and trichloroacetonitrile under basic conditions (Figure 8.6) [10]. The authors suggested the *in situ* formation of peroxy trichloroacetimidic acid as the active oxygenating agent and in fact no oxidation products were detected in the absence of trichloroacetonitrile. The elaborated protocol tolerated the presence of both electron-donating and electron-withdrawing substituents on the oxindoles and the products were generally isolated in good yields and excellent enantioselectivities.

Cumene hydroperoxide (CHP) has been used as alternative oxygen-transfer reagent in stereoselective α-hydroxylations of carbonyl compounds. In 2012, Meng and co-workers reported the enantioselective phase-transfer catalyzed direct oxidation of 1-adamantyl β-oxo esters with CHP (Figure 8.7). The corresponding products were isolated with good yields and moderate to good ees [11]. Interestingly, the authors found that when modifying the hydroxyl and quaternary nitrogen groups of alkaloid catalysts

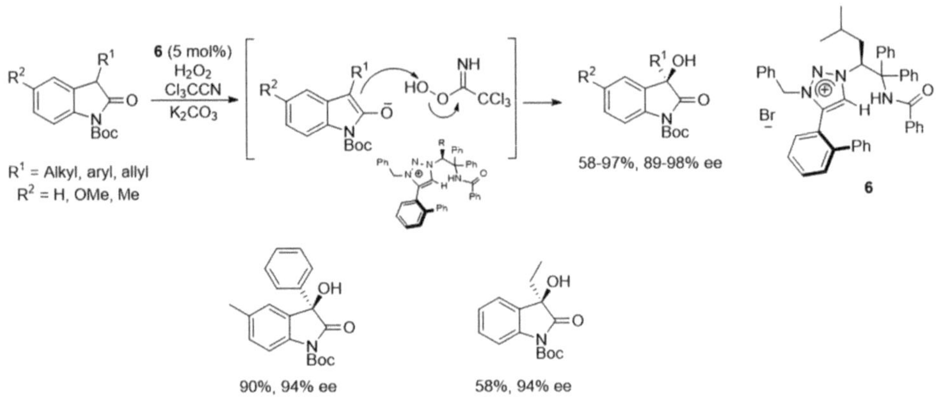

Figure 8.6: Asymmetric α-hydroxylation of 3-substituted oxindoles.

7 using bulky groups, the enantioselectivity reached 90 % but only for 1-indanone-derived β-keto esters (1-tetralone-derived β-oxo esters were also examined but they gave <35 % yields and moderate ees).

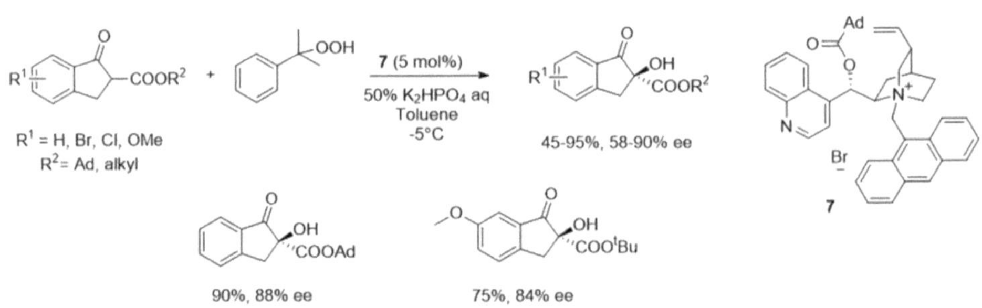

Figure 8.7: Asymmetric α-hydroxylation of β-oxo esters catalyzed by **7**.

Only one year later, Odagi and co-workers [12] achieved a highly enantioselective catalytic oxidation of 1-tetralone-derived β-keto-esters by using the guanidine-bisurea bifunctional organocatalyst **8** in the presence of CHP (Figure 8.8). The corresponding α-hydroxy-β-keto-esters were obtained in 82–99 % yields with high enantioselectivities (85–95 % ee). It was suggested that interactions between guanidine and β-keto ester and between the urea group and oxidant were important to achieve high enantioselectivity (Figure 8.8). Oxidants such as tBuOOH and oxaziridine were also tested, but the selectivity drastically decreased. The usefulness of this reaction was confirmed by applying it to the synthesis of a key-intermediate of *ent*-daunorubicin.

The origin of stereocontrol in guanidine-bisurea bifunctional organocatalyzed α-hydroxylation of tetralone-derived β-ketoesters in the presence of CHP was examined by DFT calculations [13]. The identified TS structure indicated that the β-keto

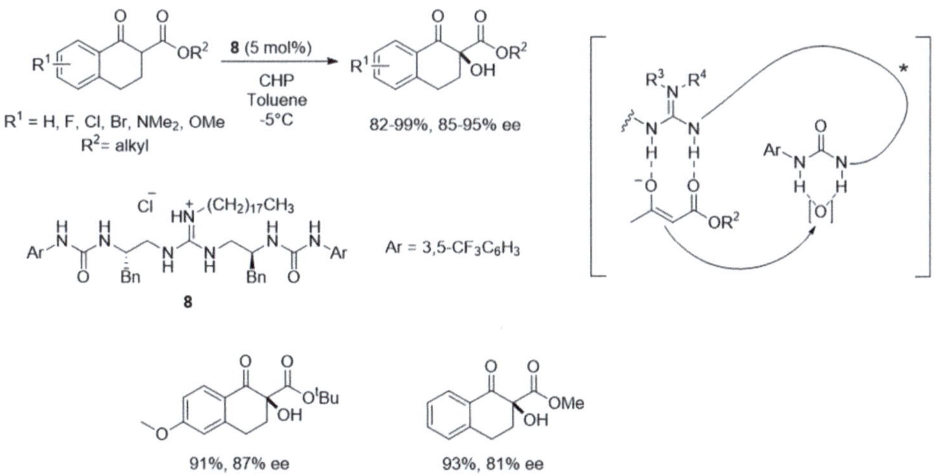

Figure 8.8: Catalytic asymmetric α-hydroxylation of β-keto esters by guanidine-bisurea **8** and proposed transition state model.

ester group of tetralone interacts with guanidine and urea groups in the catalyst, and oxidant coordinates with the remainder of the urea group. Based on the insights provided by this TS model, the authors designed a new methodology for asymmetric synthesis of β- and γ-substituted tetralones by oxidative kinetic resolution reaction of tetralone-derived β-ketoesters, using guanidine-urea bifunctional organocatalyst and CHP. The kinetic resolution reaction of various tetralone derivatives proceeded with high selectivity. This methodology was successfully applied to the synthesis of (+)-linoxepin and rishirilide B [14].

In 2013, Lattanzi and co-workers [15] described the first enantioselective α-hydroxylation reaction of α-substituted β-keto amides by using the commercially available hydroquinine **9**/TBHP system (Figure 8.9). The products were obtained in good to high yields and up to 83 % ee, which could be improved by a single crystallization (up to 99 %).

CHP and H_2O_2, tested as alternative oxygen sources in the presence of hydroquinine **9** gave inferior results. Halogen atoms and electron-donating groups on the aromatic ring of the indane scaffold were well tolerated, whereas substrates with an aliphatic amido group reacted sluggishly at room temperature, affording the products in modest yields and ees. Interestingly, the authors investigated the nature of interactions between hydroquinine and β-ketoamide *via*[1] H-NMR, recording a number of spectra of **9** in $CDCl_3$ at room temperature, adding different amounts of substrate. The findings suggested that the catalyst OH group might be involved in hydrogen bonding interaction with the enolate of β-ketoamide.

Cinchona alkaloid derivatives have also been reported as efficient catalysts for the asymmetric α-hydroxylation of β-keto esters in the presence of CHP [16]. The

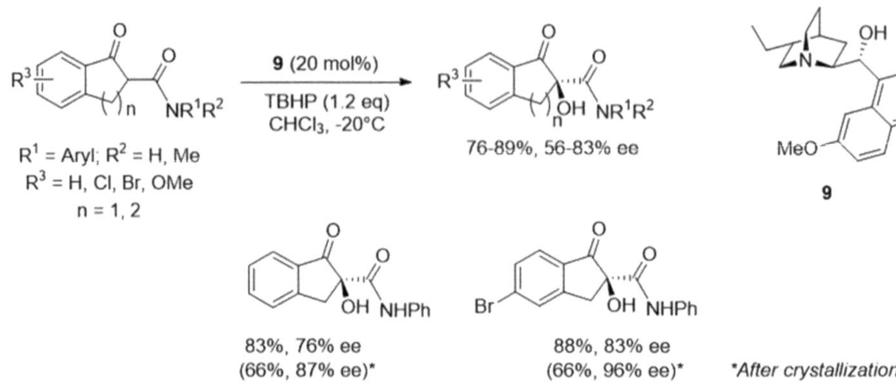

Figure 8.9: Asymmetric α-hydroxylation of β-keto amides.

reaction was successfully scaled up to a gram scale, without reduction in yield and ee. Then, this oxidative methodology was extended by Meng and co-workers [17] to different β-keto esters and β-keto amides using safe, commercially available peroxides (CHP, TBHP and H_2O_2) as the oxidant (Figure 8.10). Even in this case, compounds with aliphatic amido groups provided modest yields and ees. Moreover, under optimized conditions, non-cyclic β-keto esters did not react. Three possible transition states were proposed by the authors to elucidate the origin of the stereoselectivity of this reaction.

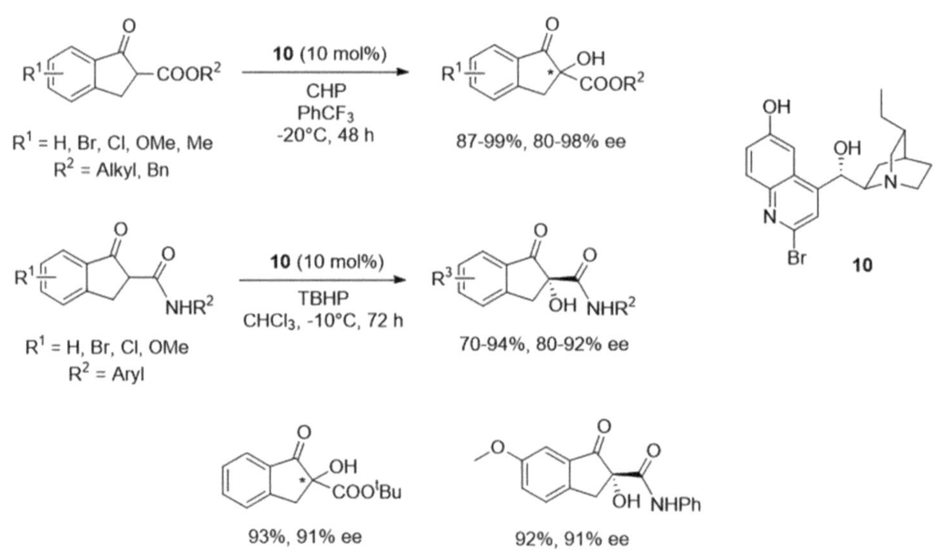

Figure 8.10: Asymmetric α-hydroxylation of β-indanone esters and amides.

Chiral bifunctional urea-containing ammonium salts were found to be very efficient catalysts for asymmetric α-hydroxylation reactions of β-keto esters, using racemic oxaziridines rac-**11** as oxygen-transfer reagents under base-free conditions, with enantiomeric ratios up to 99:1 [18]. This transformation was accompanied by a simultaneous kinetic resolution of the oxaziridine (Figure 8.11).

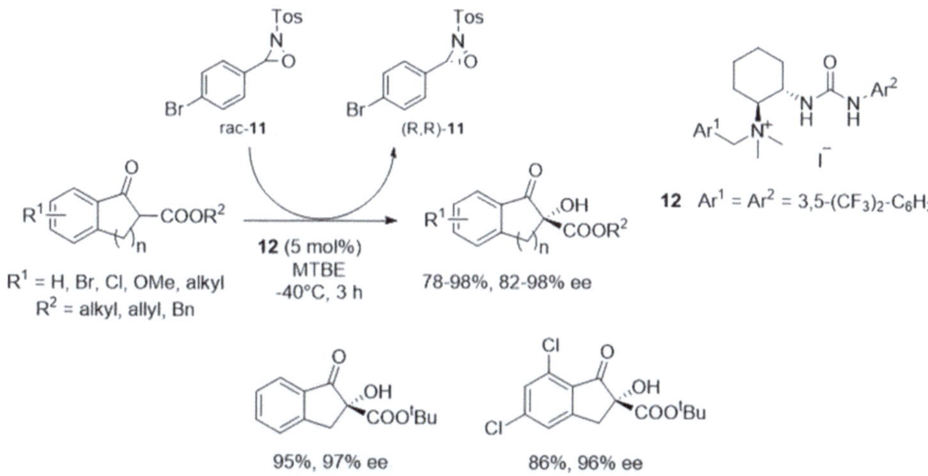

Figure 8.11: Asymmetric α-hydroxylation of β-keto esters by simultaneous kinetic resolution of rac-**11**.

Concurrently, Liu and co-workers developed a chiral bifunctional guanidine-catalyzed synchronous α-hydroxylation of β-keto esters and kinetic resolution of oxaziridines. The resolution could easily be performed on a gram-scale, without reduction in selectivity or yield [19]. In the presence of catalyst **14**, a variety of optically active oxaziridines and chiral α-hydroxy-β-keto esters were satisfactorily generated (Figure 8.12).

8.2.2 α-Benzoyloxylation

In 2012, Bencivenni and co-workers [20] explored the reactivity of cyclic ketones for the enantioselective synthesis of α-benzoyloxy derivatives using 9-amino(9-deoxy) epi-hydroquinidine **15** in combination with salicylic acid as cocatalyst, using dibenzoyl peroxide as the oxidant (Figure 8.13). Interestingly, salicylic acid was the best choice thanks to a fruitful hydrogen bond interaction between the hydroxyl group of the acid and the carbonyl groups of the benzoyl peroxide. This methodology was also effectively applied to the direct enantioselective synthesis of α-benzoyloxylated 1-indanone derivatives.

Recently, List and coauthors reported [21] the direct asymmetric α-benzoyloxylation of α-branched aldehydes and α-branched enals *via* enamine and dienamine

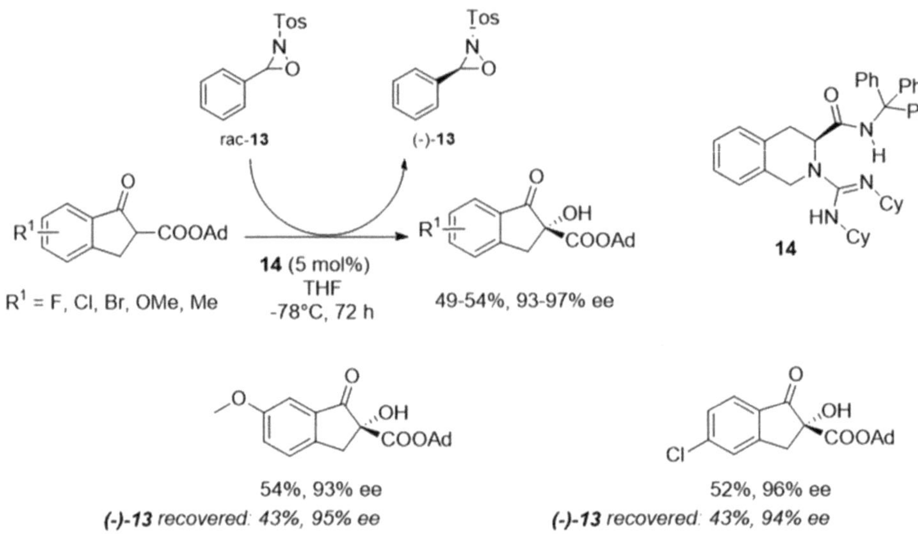

Figure 8.12: Asymmetric α-hydroxylation of β-keto esters by simultaneous kinetic resolution of **13**.

catalysis, employing cinchona alkaloid-derived primary amine as the catalyst, benzoyl peroxide (BPO) as a readily available and inexpensive hydroxylation reagent and 10 mol% of the radical inhibitor 2,6-di-*tert*-butyl-4-methylphenol BHT (Figure 8.14). This methodology gave access to valuable aldehydes containing a quaternary stereocenter. The authors suggested that an external acid additive was unnecessary and that the Brønsted acid involved in the catalytic cycle could be the benzoic acid generated *in situ* after the first catalytic turnover, even though chiral acids **TRIP** (3,3'-bis(2,4,6-triisopropylphenyl)-1,1'-bi-2-naphthol cyclic monophosphate) affected the enantioselectivity.

Their hypothesis on the role of benzoic acid was supported by the autocatalytic behavior of the reaction as confirmed by the sigmoidal curve observed plotting the results of the conversion of the aldehyde over time.

The modest enantioselectivity observed in some examples was attributed to the formation of equilibrium mixtures of E- and Z-enamine intermediates, which are similar in energy but which furnish opposite enantiomers of the product.

In 2017, Maruoka and coauthors [22] reported an efficient synthesis of chiral tritylpyrrolidine derivatives as single stereoisomers (their precursors were mixtures of diastereoisomers which could be separated into single stereoisomers by simple silica gel column chromatography). The potential of these molecules as novel chiral catalysts was evaluated in the asymmetric benzoyloxylation of 3-phenylpropanal with BPO, with results comparable in term of reactivity and selectivity to those reported by using other catalysts.

Luo and co-workers [23] developed an elegant direct asymmetric α-benzoyloxylation of β-ketocarbonyls catalyzed by primary amines. The reaction worked more

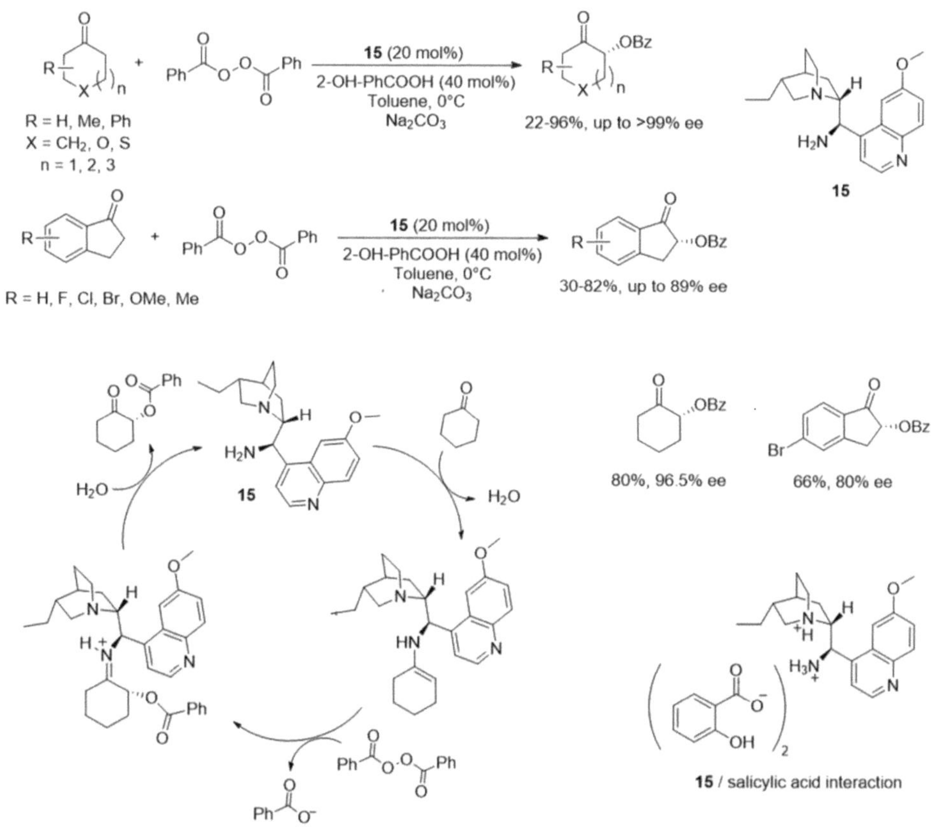

Figure 8.13: Asymmetric α-benzoyloxylation of ketones and proposed catalytic cycle for the reaction.

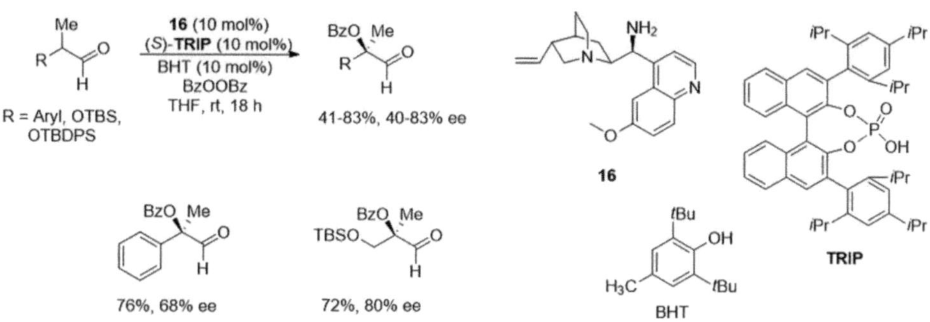

Figure 8.14: Enantioselective α-benzoyloxylation of α-branched aldehydes.

cleanly under N_2, thus all the reactions were conducted under anaerobic atmosphere, in the presence of butylated hydroxytoluene (BHT) as a radical inhibitor and benzoyl peroxide as the hydroxylation reagent (Figure 8.15). Acyclic and cyclic ketoesters in the presence of benzoyl peroxide afforded the desired α-benzoyloxylation products in high yields and excellent enantioselectivities. A number of α-substituents on acetoacetates were well tolerated as well as various ester moieties, including those with sterically bulky groups. To further explore the versatility of the reaction, β-ketoamides were also examined, achieving good results. From a mechanistic point of view, the authors proposed three transition states: an H-bonding TS **I** to account for the R-selectivity, a N-oxide addition pathway **II** to give R-selectivity and an S-selective enamine N-oxide pathway **III**.

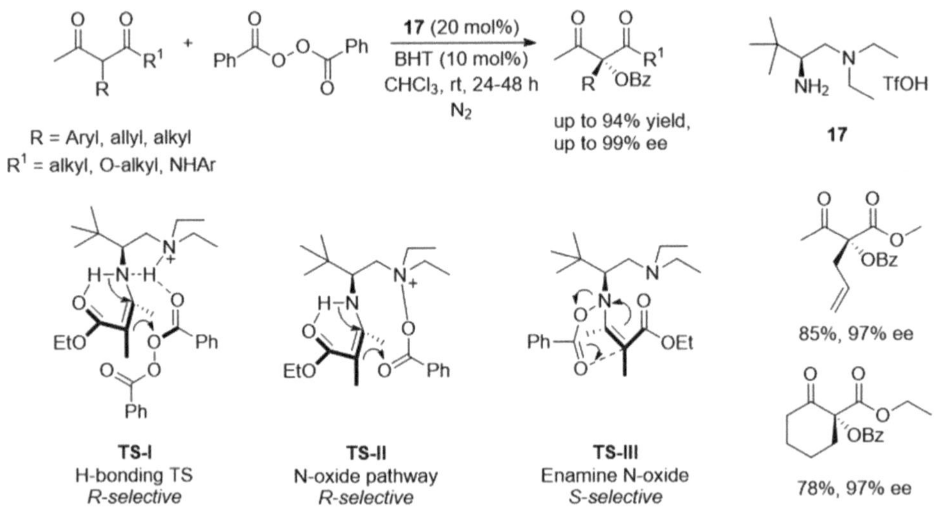

Figure 8.15: Asymmetric α-benzoyloxylation of β-ketocarbonyls and proposed transition states.

They suggested that the N-oxide pathways should be minor and disfavored and that the H-bonding directed addition would be the major productive pathway under the optimized experimental conditions (Figure 8.15).

One year later, Kanemitsu and Itoh published [24] the first enantioselective α-benzoyloxylation of α-monoalkylated malonic diesters promoted by a phase-transfer catalyst (Figure 8.16). Under the optimized conditions, the reaction was carried out in the presence of aqueous KOH and N-(9-anthracenylmethyl)cinchoninium chloride **18** by using BPO as the oxidant and afforded the corresponding products in excellent yields (up to 99 %) and high ees (up to 96 %). The utility of this methodology was demonstrated by its application in the synthesis of a mineralocorticoid receptor antagonist.

Very recently, Meng and co-workers [25] have successfully applied a cinchona-derived N-oxide phase-transfer catalyst also in the α-benzoyloxylation of β-keto esters (Figure 8.17). In this way a range of substituted 1-indanone and 1-tetralone derivatives were synthesized with excellent yields and enantioselectivities under mild conditions. The methodology was also used at gram-scale, without reduction in efficiency

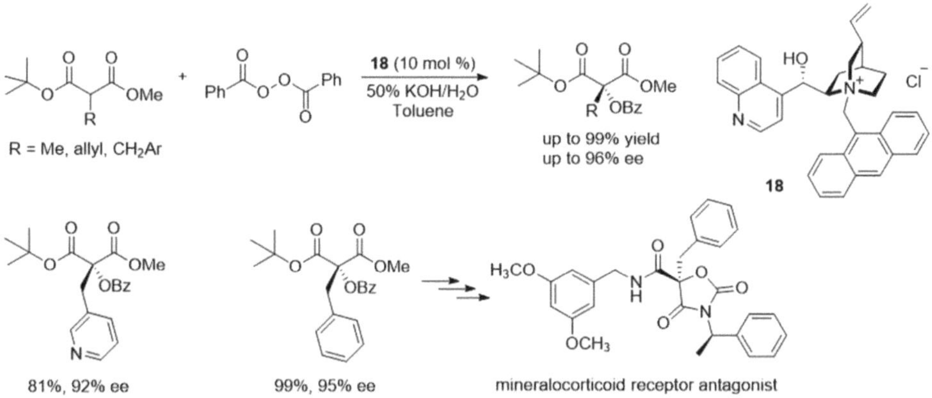

Figure 8.16: Enantioselective α-benzoyloxylation of malonic diesters and its application in the synthesis of a mineralocorticoid receptor.

and enantioselectivity and the catalyst can be easily separated at the end of the reaction and reused without loss of activity.

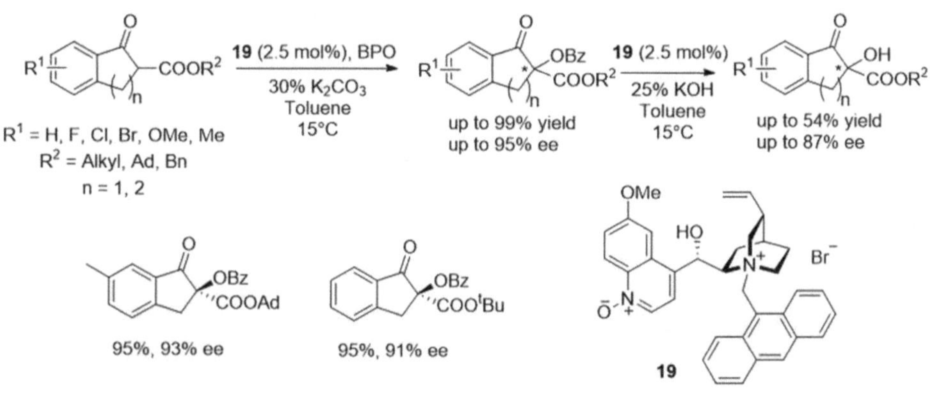

Figure 8.17: Enantioselective α-benzoyloxylation of β-keto esters.

From a mechanistic point of view, the authors proposed the deprotonation of the β-keto ester in the presence of a base followed by its reaction with the active BPO with formation of the α-benzoyloxylation product. In a possible transition state proposed by the authors, only the *Si*-face of the enolate was available for the reaction. Under the phase-transfer conditions, the α-benzoyloxylation product could be converted to the α-hydroxy derivative by kinetic resolution of the benzoates, for which only the *R* or *S* product was preferentially hydrolyzed.

8.2.3 α-Tosyloxylation

In 2012, Legault and co-workers [26] developed chiral iodoaryl oxazolines **20** as new catalysts for the α-tosyloxylation of propiophenone. In this case, *meta*-chloroperoxybenzoic acid (*m*-CPBA, slow addition) was used as the stoichiometric oxidant and *p*-toluenesulfonic acid (*p*-TsOH) as a source of tosylate nucleophile (Figure 8.18). The same authors reported [27] that the stereoinduction is exclusively controlled by the stereogenic center alpha to the oxazoline oxygen, as the epimer **20b** of catalyst **20a** led to an inversion of the sense of stereoinduction. Computational chemistry was used to rationalize the stereoinduction of the process, in particular, due to the excess of *p*-TsOH with respect to the catalyst, the oxazoline was driven to its protonated state. The protonation furnished the necessary activation to favor the formation of the reactive iodane intermediate (Figure 8.18), a much more electrophilic bis-cationic intermediate and in this structure the stereogenic center, near to the oxygen atom influenced the approach of the substrate. The introduction of an alkyl group *ortho* to the iodine center in the catalysts led to a drastic improvement in activity. Moreover, the best, although modest, selectivity was obtained after the introduction of a chlorine atom *para* to iodine (**20c**).

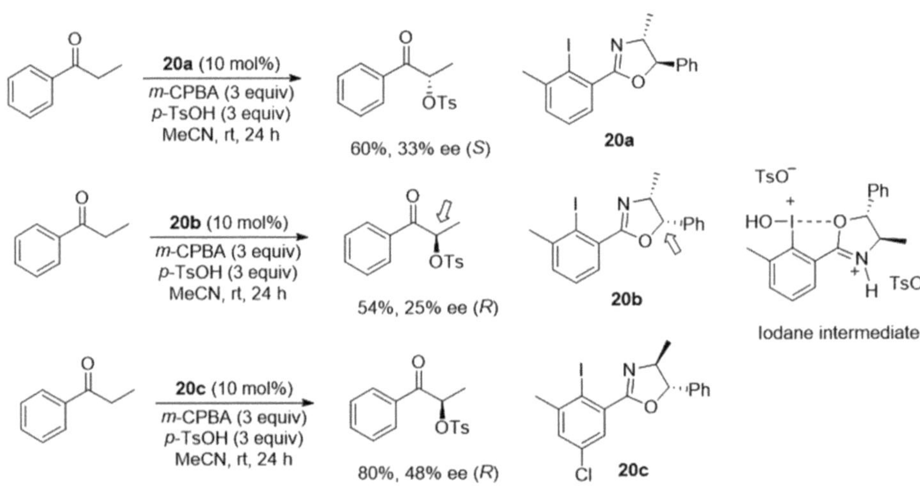

Figure 8.18: Enantioselective α-tosyloxylation of propiophenone.

Later, Berthiol et al. [28, 29] synthesized 3,3′-diiodo-BINOL-fused maleimides **21** as hypervalent iodine (III) organocatalysts, which were used in the α-tosyloxylation of ketones. With these catalysts, yields and ees were found to be comparable to the best results previously achieved using other organocatalysts (Figure 8.19).

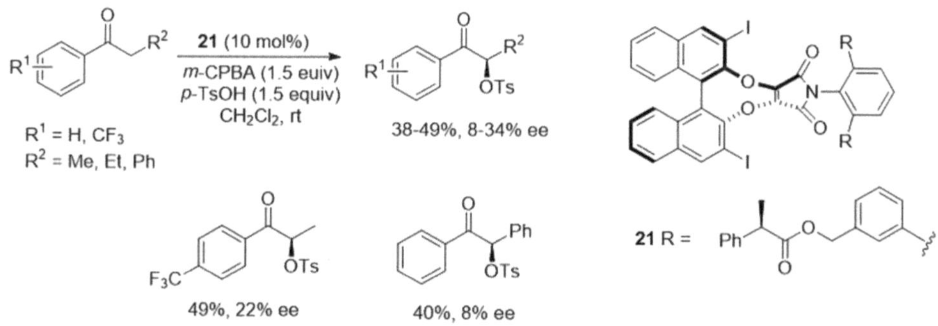

Figure 8.19: Asymmetric α-tosyloxylation of ketones.

The authors highlighted that the maleimide moiety was idoneous for obtaining moderate enantioselectivities, and the investigation of the reaction mechanism revealed that the steric crowding around the iodine center improved the enantioselectivity.

Recently, new chiral non-C_2-symmetric iodoarenes **22** have been found to be effective catalysts for the asymmetric α-tosyloxylation of ketones (Figure 8.20) [30]. In a plausible mechanism, proposed by Masson et al., the enol form of propiophenone reacted with the aryl-λ^3-iodane generated *in situ* from reaction of the chiral organocatalyst, *m*-CPBA and sulfonic acid (path B) to provide a C-bonded intermediate which was able to react with the oxygenated nucleophile in order to give the final product. However, an alternative mechanism involving an O-bonded iodane intermediate could be plausible [31] (path A) and the low stereocontrol in the reaction was attributed to the competition between these two different paths (Figure 8.20). AcOEt was found to be the best reaction medium and *m*-CPBA was the most efficient oxidant for the reaction. Furthermore, the same authors described an interesting strategy [32] in order to access highly atropoenriched 2-iodobiphenyls by means of Andersen's reagent under kinetic resolution conditions.

To avoid the low selective path A (O-bonded intermediate) and favor path B (C-bonded intermediate), Legault and Basdevant used enol esters and iodoarenes **23** as chiral iodine(III) reagents for the synthesis of α-tosyloxy ketones under stoichiometric and catalytic conditions [33] (Figure 8.21). The authors pointed out that while the reaction with ketones proceeded preferentially through a S_N2'-type mechanism (O-bonded intermediate), this pathway was inaccessible for enol derivatives (in this case the reaction proceeded through a C-bonded intermediate).

Iodoarenes **23** were also used by Coeffard and co-workers [34] in the asymmetric α-oxidation of α-substituted β-keto esters in the presence of *m*-CPBA as the oxidant (Figure 8.22). After a careful screening of experimental conditions, modest yields and satisfactory enantiomeric excesses were obtained for the synthesis of products with a quaternary stereogenic center.

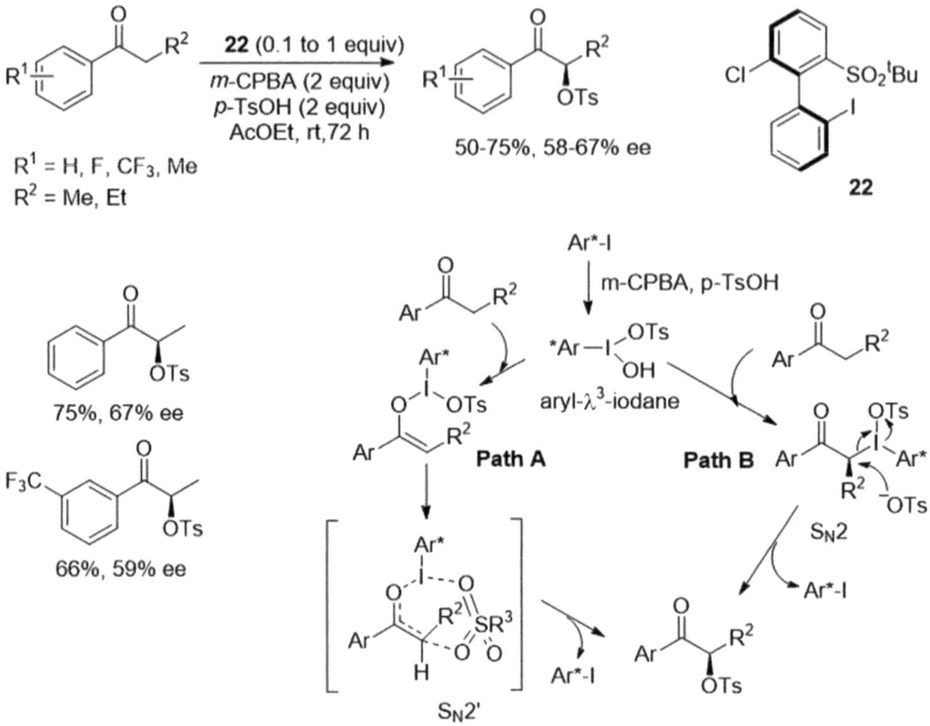

Figure 8.20: Enantioselective α-tosyloxylation of ketones and proposed mechanism for the reaction.

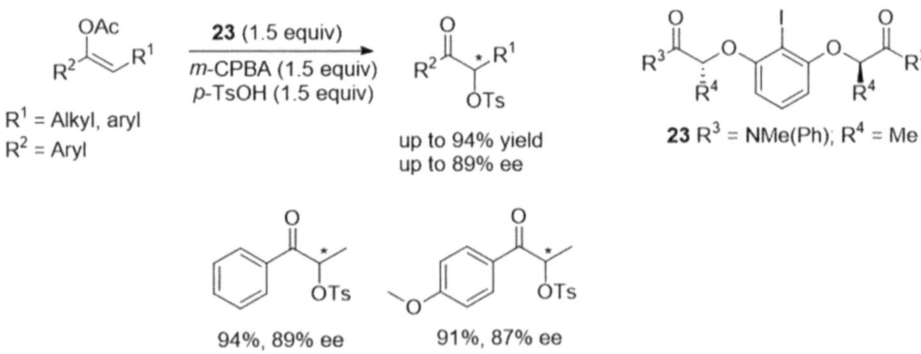

Figure 8.21: Enantioselective α-tosyloxylation of enol esters.

8.3 Dihydroxylation and dioxygenation of alkenes

One of the most useful and exploited oxyfuntionalization of the carbon-carbon double bond is the oxidation to 1,2-diols or their protected derivatives. This transformation is often reported as key-step in total synthesis of bioactive and natural compounds [35].

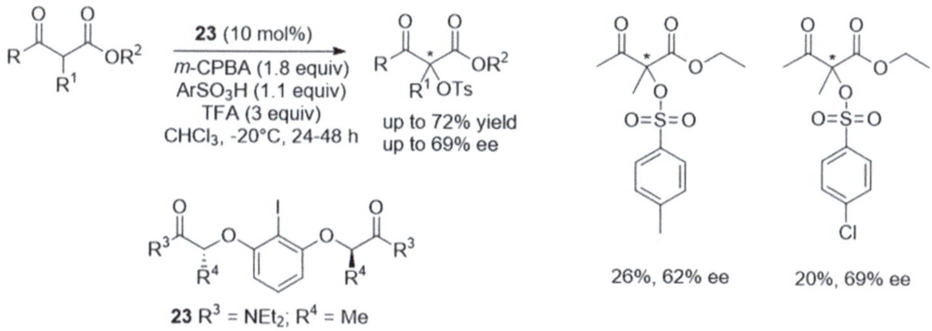

Figure 8.22: Enantioselective α-oxysulfonylation of α-substituted β-keto esters.

8.3.1 Dihydroxylation of alkenes

The stereoselective oxidation of alkenes enables the preparation of 1,2-diols, which are important derivatives for the synthesis of every-day life compounds such as sugars and a fundamental class of chiral ligands for metal-based catalysis. Among the metal-based systems, the Sharpless asymmetric dihydroxylation (SAD) provides a simple and readily available catalytic protocol for the preparation of *cis*-1,2-diols with high stereocontrol [36].

Chiral anion-controlled ion-pairing catalysis is an area of research showing brilliant examples of highly stereocontrolled transformations, exploiting noncovalent interactions provided prevalently by binaphthyl-derived chiral phosphate anion [37]. A complementary and still greatly underdeveloped approach foresees the use of chiral cations for the control of metal anions. In this context, a pioneering example of dihydroxylation of alkenes has been reported by Brown in 2002, using stoichiometric amount of a cinchonidinium salt in the presence of MnO_4^- anions [38]. Although the enantioselectivity was low and the catalyst proved to be unstable under the oxidative conditions, this system attested the feasibility of ion-pairing to induce the stereoselective processes with metal containing anions.

In 2015, Wang, Tan and coauthors, disclosed stable dicationic bisguanidinium salts **24**, as catalysts for the enantioselective dihydroxylation and oxohydroxylation of α-aryl acrylates and α-aryl-β-alkyl acrylates, using $KMnO_4$ as the terminal oxidant (Figure 8.23) [39].

When using catalysts **24**, readily accessed from a commercially available optically pure 1,2-diamine, the reactions proceeded at −60 °C in ethereal/water mixtures with only 2 mol % of the chiral salt to give diols, bearing electron-donating groups in the aromatic moiety, in moderate to satisfactory yield and higher ees. The presence of the α-ketoester as by-product was observed. α-Alkyl substituted acryates were transformed to diols with lower level of enantioselectivity. Interestingly, when employing mixtures of *Z*, *E*-trisubstituted enoates, the initially formed mixture of

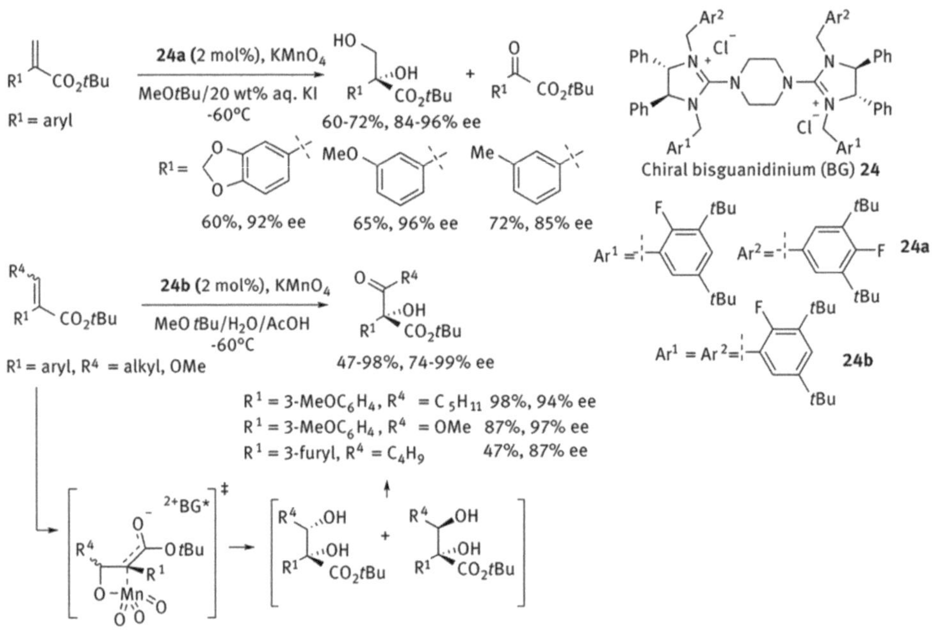

Figure 8.23: Enantioselective dihydroxylation and oxohydroxylation of acrylates.

diastereoisomeric diols could be transformed under acidic conditions, into the oxohydroxy products with good to excellent ees. This outcome demonstrated that *Z*- and *E*-trisubstituted enoates afforded the same enantiomer of the product. The chiral catalyst **24**, being dicationic in character, should form stronger electrostatic interaction with the enolate anion derived from the attack of MnO_4^- toward the C-C double bond. At this stage, the chiral environment of the dication would dictate the stereoselectivity in the formation of the quaternary stereocenter of the cyclic mangate ester. The protocol can be considered a useful new tool which somewhat complements the SAD.

An interesting enantioselective diboration of alkenes has been reported by Morken and coauthors, using readily available carbohydrate-derived glycols [40]. The process is based on the knowledge that bis(neopentylglycolato)diboron reagent $B_2(neo)_2$ in the presence of Lewis basic alkoxides (YO^-) forms an ate-complex that would promote the reaction with unactivated alkenes. The relevant process for the stereocontrol relies on the reversible replacement of neopentyl glycol ligands with appropriate chiral alcohols to give optically pure ate complex. Specifically, the combination of catalytic loading of diol **25** or its pseudoenantiomer **26** and DBU as base in THF, after the one-pot oxidative treatment afforded the 1,2-diols in good yields and good to high enantioselectivity (Figure 8.24(a)).

The transformation is of note as the chiral diols **25** and **26** derives from low cost D-glucal and L-rhamnal and both enantioenriched products are available by choosing the proper catalyst. Terminal alkenes, except styrenes, generally led to the final

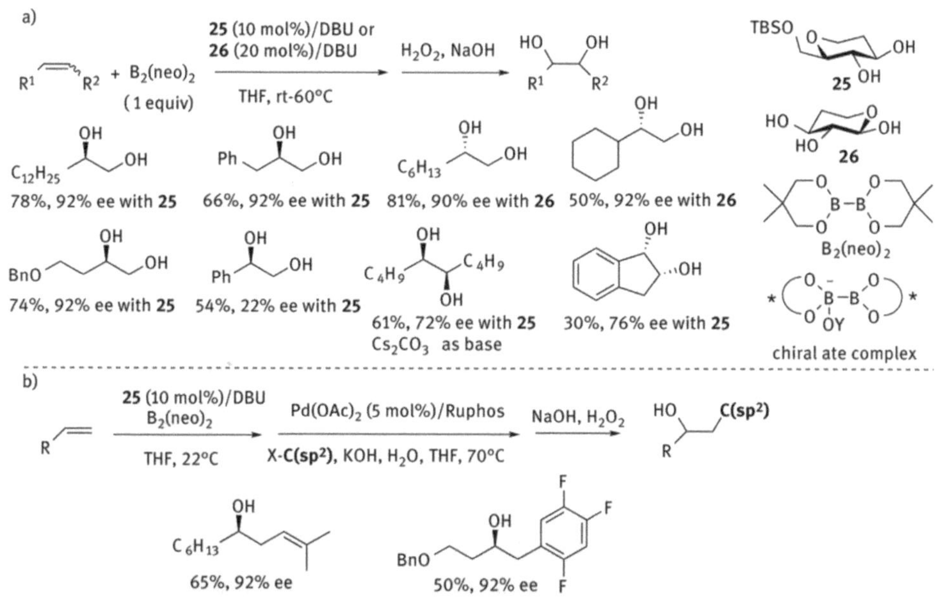

Figure 8.24: (a) Enantioselective diboration/oxidation of alkenes to 1,2-diols and (b) Cascade diboration/cross-coupling/oxidation sequence.

diols with high ees and the process could be scaled up to 10 g of starting alkene. More challenging internal alkenes, furnished the 1,2-diols in excellent diastereoselectivity and good enantiocontrol. The procedure could be coupled with Pd-catalyzed cross-coupling reactions in a cascade fashion, performing the subsequent step on the crude mixture of the previous step, useful to prepare secondary non racemic alcohols (Figure 8.24 (b)).

8.3.2 Dioxygenation of alkenes

An appealing approach for the asymmetric oxidative difunctionalization of carbon-carbon double bond relies on the employment of optically pure hypervalent iodine compounds [41]. These molecules have been increasingly used over the years as an environment-friendly alternative to toxic heavy metal oxidants. The dioxygenation of alkenes by this class of reagents was applied in intramolecular manner to construct heterocycles containing one or two contiguous stereocenters and more recently in the intermolecular version.

In 2012, Fujita illustrated a first catalytic dioxygenation of alkenes mediated by chiral hypervalent iodine precatalyst **27**, trifluoroacetic acid as an activator and *m*-CPBA as the terminal oxidant (Figure 8.25) [42]. Precatalyst **27** is oxidized by the terminal oxidant to give a chiral organoiodane(III) species able to attack the carbon–carbon double bond to form a iodonium intermediate. Intramolecular nucleophilic attacks of the OH and the carbomethoxy groups occur with inversion of configuration at the carbon atoms to give the dihydrofuran-fused isochromanones with complete diastereoselectivity in moderate yields and moderate to high enantioselectivity.

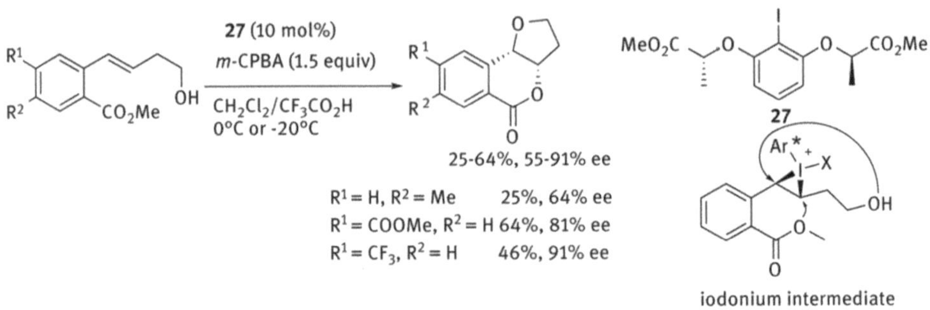

Figure 8.25: Asymmetric oxylactonization of alkenes with C_2-symmetric iodoarene.

The selectivity of the reaction is not complete due to the competitive epoxidation of the double bond by the terminal oxidant. The same group reported a trifluoroacetoxy-lactonization of *ortho*-alk-1-enylbenzoates to 3-alkyl-4-hydroxyisochroman-1-ones catalyzed by a set of chiral iodo-based precatalysts (Figure 8.26) [43]. This transformation is related to that one reported in Figure 8.25. In this case, trifluoroacetic acid represents the external nucleophile attacking the chiral iodonium intermediate to give the trifluoroacetoxylactone, easily hydrolyzed to products **29** and **30**. The process catalyzed by one of the most effective compounds **28**, provided four lactones in a variable ratio, with the enantioenriched *cis*-isochromanone **29** as the most abundant with up to 98 % ee, regioselectively obtained via a 6-*endo* ring closure. A parallel detrimental racemic epoxidation, followed by ring-opening to *anti*-derivatives **31** and **32** occurred, thus suppressing the final yield of lactone **29**.

In 2017, Masson and coauthors illustrated a tosyloxy- and phosphoryloxy-lactonization of 4-pentenoic acid derivatives mediated by precatalyst **33** under stoichiometric amount of *m*-CPBA and sulfonic or phosphoric acids at 0 °C (Figure 8.27) [44]. The γ-(hydroxymethyl)-γ-butyrolactones were isolated in acceptable yields and good ee values. Only one example of reactivity, leading toward the corresponding δ-butyrolactone, proceeded with lower conversion and enantiocontrol.

Optimized reaction conditions prevented the epoxidation of the terminal carbon–carbon double bond, leading to a more selective process. Mechanistically, the oxidation of precatalyst **33** by *m*-CPBA would form a chiral λ^3-aryliodane species, which would be attacked by the alkene moiety to give a chiral iodonium intermediate.

This in turn would undergo 5-*exo*-closure to a iodonium species, followed by sulfonyl or phosphoryl nucleophiles substitution. This approach represents a straightforward

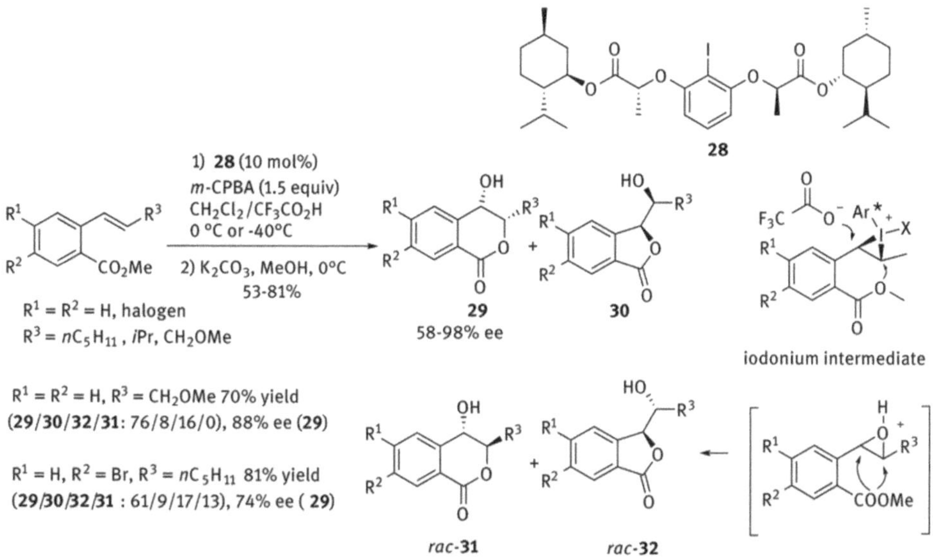

Figure 8.26: Stereoselective oxylactonization of *ortho*-alk-1-enylbenzoates.

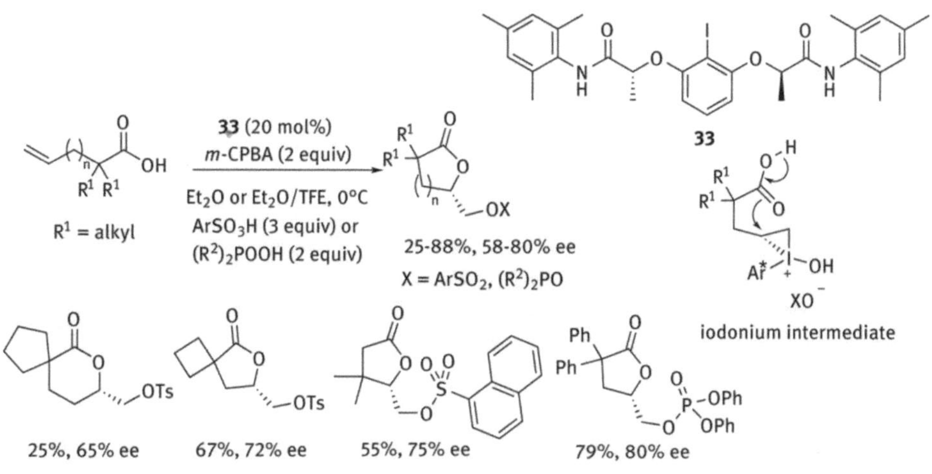

Figure 8.27: Enantioselective tosyloxy- and phosphoryloxy-lactonization of 4-pentenoic acid.

route to γ-(hydroxymethyl)-γ-butyrolactones, useful intermediates for the synthesis of natural products [45].

The chiral lactones were produced with low to moderate optical purities by alternative organocatalyzed BV reaction (30–42 % ee) [46]. The first intermolecular example of asymmetric dioxygenation of alkenes mediated by hypervalent iodine organocatalysts has been illustrated by Muñiz and coauthors in 2016 (Figure 8.28) [47].

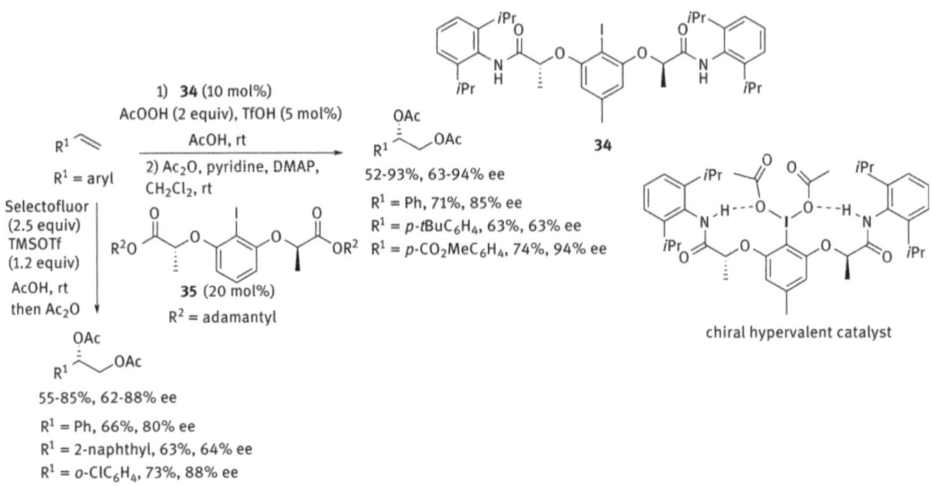

Figure 8.28: Enantioselective diacetoxylation of styrenes.

A variety of stryrenes was reacted using precatalyst **34** at 10 mol % loading, peracetic acid and catalytic loading of triflic acid in acetic acid as solvent, leading to the enantioselective formation at room temperature of two regioisomeric acetoxy alcohols.

The mixture was then treated with acetic anhydride to obtain the diacetoxylated products in good yields and up to 94 % ee. The protocol has the advantage to use a more convenient oxidant source, whose by-product is the reaction solvent. The X-ray structure of a few chiral active hypervalent catalysts was determined for the first time, showing a supramolecular chiral helical ensemble around the iodine atom. This feature has been ascribed to H-bonding interactions of the amide moieties with the oxygen atoms of the acetoxy ligands, which persisted also in solution. Hence, the helical chirality was suggested to play a role in the control of the asymmetric induction in carbon-carbon double bonds oxidation generally mediated by hypervalent catalysts having this structure. The same group shortly after showed that bislactate precatalyst **35** promoted the same process using Selectfluor as an alternative terminal oxidant, and trimethylsilyl triflate (TMSOTf) as the activator (Figure 8.28) [48]. Moreover, *trans*-cinnamyl alcohols were reported as suitable internal alkenes for the diacetoxylation to give, under the same conditions, the triacetoxylated products in good to high diastereo- and enantioselectivity (dr > 80/20 and ee from 78 % to 96 %).

8.4 BV oxidation and oxidative desymmetrization

The BV oxidation is one of the fundamental transformations in organic synthesis useful to directly transform ketones to the corresponding esters or lactones. The most traditional approach to carry out this transformation uses organic peracids as oxygen source [49]. However, over the last years, more attention has been paid to the development of new green chemistry methods that exploit atom-efficient and mild oxidants such as hydrogen peroxide or molecular oxygen [1b-c, 51]. Particular efforts have been directed toward the development of enantioselective processes and, among metal-free protocols, excellent enantioselectivity has been achieved by enzymatic systems, such as mono-oxygenases [50]. Despite the considerable efforts, the development of organocatalytic BV oxidation protocols using H_2O_2 is considered to be one of the most challenging issues in synthetic chemistry [1b-c, 53]. The main reason could be ascribed to its low activity and strong acids or metal promoters generally required [50e, 54].

Before 2012, few examples of organocatalytic asymmetric BV reactions have been reported essentially focused on prochiral cyclobutanones. The first organocatalyzed BV oxidation, reported by Murahashi [51], showed the use of planar-chiral C_2-symmetric bisflavinium perclorate catalyst **36** in the oxidation of 3-aryl-substituted cyclobutanones with hydrogen peroxide. The resulting chiral lactones were isolated in moderate yield and up to 74 % ee. (Figure 8.29). In 2008, Miller reported two preliminary examples of oxidative desymmetrization of cyclic ketones promoted by the oligopeptide catalyst **37**, using the combined action of carbodiimide and hydrogen peroxide to generate transient peracids upon exposure to carboxylic acid catalyst (Figure 8.29) [47].

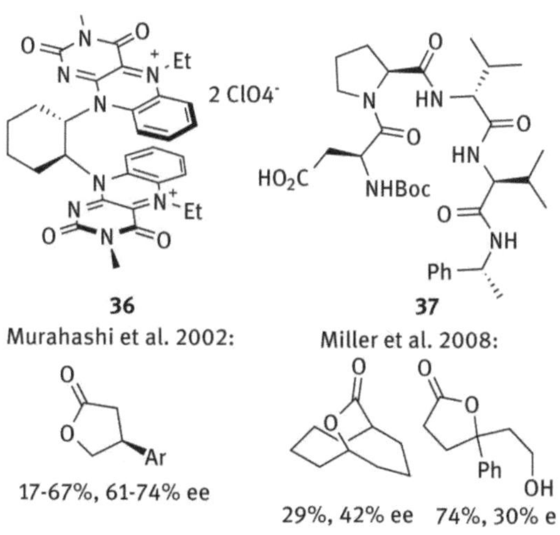

Figure 8.29: Organocatalysts used for enantioselective Baeyer–Villiger oxidations with H_2O_2 and relative chiral products reported before 2012.

The most remarkable results in the field of asymmetric organocatalyzed BV reactions were achieved by Ding and coauthors who, from 2008 to 2011, published their studies on enantioselective BV oxidations promoted by chiral BINOL- and H_8-BINOL-derived phosphoric acids of the type **38** of 3-substituted and 2,3,4-trisubstituted tricyclic cyclobutanones to γ-lactones using aqueous hydrogen peroxide as oxidant [52]. Exploiting chiral phosphoric acid/H_2O_2 systems, the γ-lactones were isolated in quantitative yield and good to high enantioselectivity (55–95 %) (Figure 8.30(a)). The reaction mechanism was studied based on experimental and theoretical findings and the authors proposed that the bifunctional activation of the reagents through hydrogen-bonding interactions is central to the catalysis. The catalyst simultaneously increases the electrophilicity of the carbonyl carbon, acting as a general acid, and the nucleophilicity of hydrogen peroxide in the addition step **39.** It promotes the dissociation of the hydroxyl group from the Criegee intermediate in the rearrangement step **40** (Figure 8.30(c)). The authors extended the use of H_8-BINOL-derived phosphoric acid **38**/H_2O_2 system also to a divergent kinetic resolution of racemic 2,3-disubstituted bicyclic cyclobutanones via asymmetric BV oxidation. Two chiral lactones were obtained in moderate regioisomeric ratio and generally low ees for the major regioisomer and much higher enantioselectivity for the minor one (74– > 99 % ee) (Figure 8.30(b)).

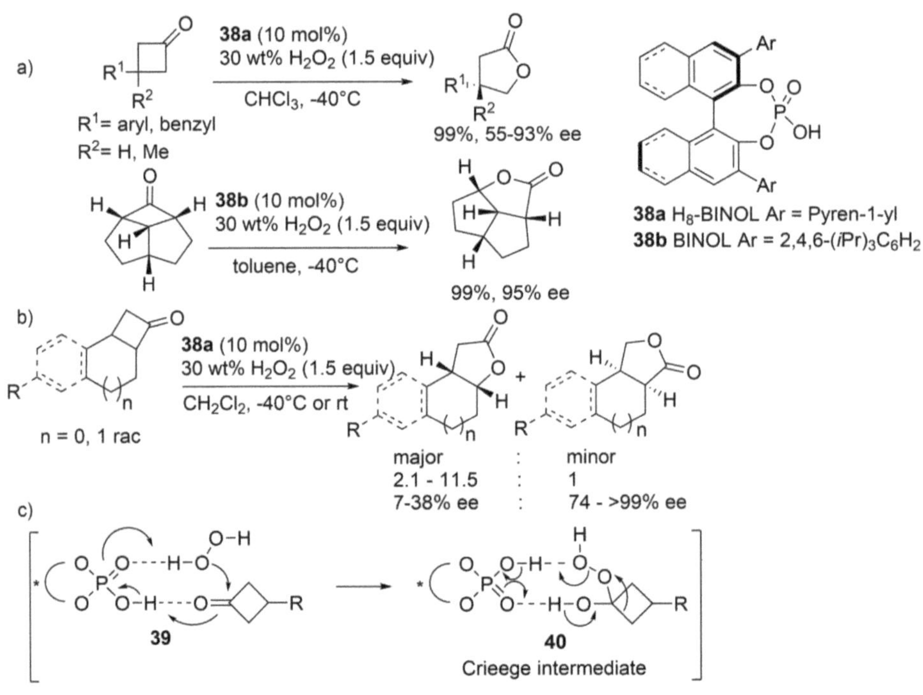

Figure 8.30: Chiral phosphoric acid catalyzed: (A) enantioselective BV reaction and (B) divergent kinetic resolution via BV oxidation. (C) Mechanism of the chiral phosphoric acid catalyzed asymmetric BV reaction of 3-substituted cyclobutanones.

In 2012, Kotsuki and coauthors reported a new environmentally friendly and mild method, organocatalyzed by thioureas, for the BV oxidation of cyclobutanones using H_2O_2 in Et_2O as the oxidant, working in toluene at room temperature (Figure 8.31) [53]. A variety of aryl and alkyl substituted cyclobutanones were oxidized to γ-lactones in high yields, observing the normal mode of migratory aptitude of the ketonic alkyl moiety. Unfortunately, other cyclic ketones such as cyclopentanones and cyclohexanones gave only the corresponding peroxide Criegee intermediates. By ^{13}C NMR experiments it was demonstrated that catalyst coordinates with cyclobutanone via hydrogen-bond interactions. The authors reported preliminary attempts to develop an asymmetric version of the methodology using chiral thioureas, without observing any asymmetric induction.

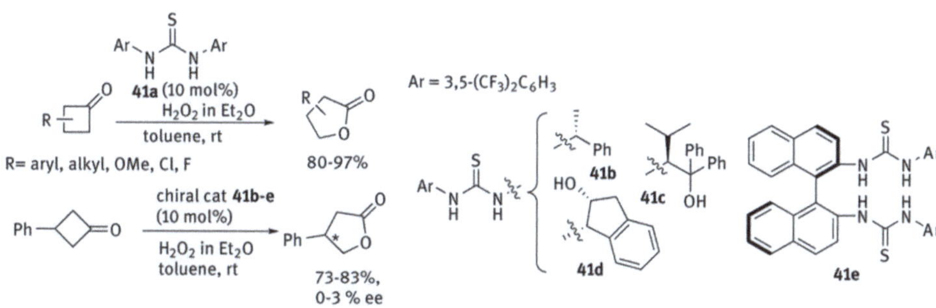

Figure 8.31: BV oxidation of cyclobutanones catalyzed by thioureas.

As cited before, in 2008 Miller and coauthors developed a peptide-catalyzed BV oxidation exploiting an acid/peracid catalytic cycle which had already proven particularly effective in the electrophilic epoxidation [52]. Even if in the BV oxidation the peracid has opposite functionality, acting as a nucleophile, this catalysis concept proved to be applicable in the oxidation of ketones. The catalytic acid/peracid cycle for the BV oxidation is depicted in Figure 8.32. The side-chain of a peptide-embedded aspartic acid **42** shuttles between the peracid oxidation state **44** and a carboxylic acid resting state **42**. In more detail, the carboxyl moiety of the catalyst is activated by diisopropyl carbodiimide (DIC) generating the intermediate **43**. Capture of intermediate **43** by H_2O_2 leads to the formation of peracid **44** which, via the classical BV reaction, leads to product esters, regenerating **42** for a new catalytic cycle. The nucleophilic cocatalyst 4-dimethylamino pyridine (DMAP) is necessary to guarantee catalyst turnover, since peracid **44** may react with excess DIC or O-acyl urea **43** to give inactive diacyl peroxide **45** [54]. DMAP enabled perhydrolysis of the diacyl peroxide **45** to regenerate the active peracid **44**.

In 2014, the authors exploited a combinatorial approach in order to find the optimal peptide sequence able to catalyse the oxidization of relatively unstrained ketones, such as cyclohexanones and cyclopentanones, to afford products with both

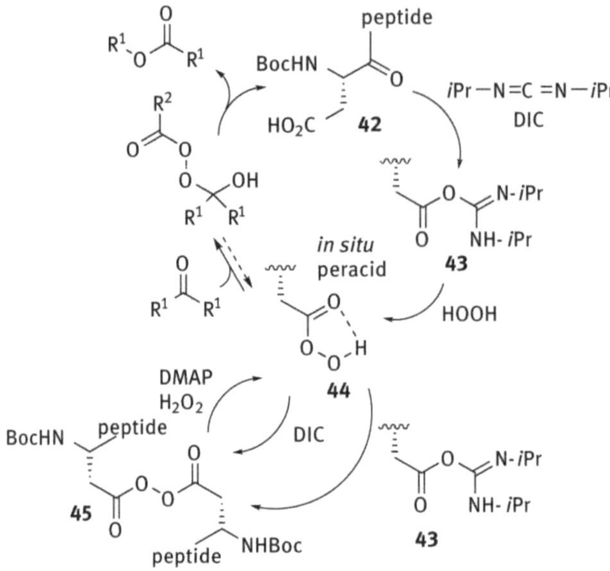

Figure 8.32: Acid/peracid catalytic cycle for the BV oxidation promoted by aspartate-containing peptide.

regio- and enantioselectivity [55]. An on-bead library of catalysts was synthesized and tested in a model kinetic resolution via BV reaction, identifying the most suited peptide motif sequence. Finally, the optimization and identification of mechanistic roles of the different regions of the catalyst led to the choice of the best catalyst **46** to promote the regio- and enantioselective BV oxidation of cyclic ketones bearing amide, urea or sulfonamide functional groups (Figure 8.33 and Figure 8.34). These catalysts, by means of directing group interactions, showed the ability to overturn the regioselectivity exhibited by *m*-CPBA, that is, the intrinsic migratory aptitude of substrates, giving access to generally disfavored BV products (Figure 8.33).

They suggested that both types of selectivity arise from the interactions between the peptide and H-bonding functionality of the substrate. Moreover, depending on the substrate, the catalyst is able to promote different reaction mechanisms, from classical kinetic resolution (KR) to parallel kinetic resolution (PKR) [56] (Figure 8.34).

The first example in Figure 8.34 (equation a) represents a classical kinetic resolution in which only the "*m*-CPBA-disfavored" lactone was isolated in 38 % yield and 86 % ee and the starting ketone was recovered in 50 % yield and 72 % ee (k_{rel} = 27). The second oxidation (Figure 8.34(b)) is a case in the middle between KR/PKR, since each substrate enantiomer affords a different regioisomer but at different rates, allowing the isolation of the "*m*-CPBA-disfavored" lactone in 24 % yield (regioisomeric ratio > 58:1) and 94 % ee. In the third example (Figure 8.34(c)), the cyclohexanone undergoes an oxidation with PKR behavior, in which the product "*m*-

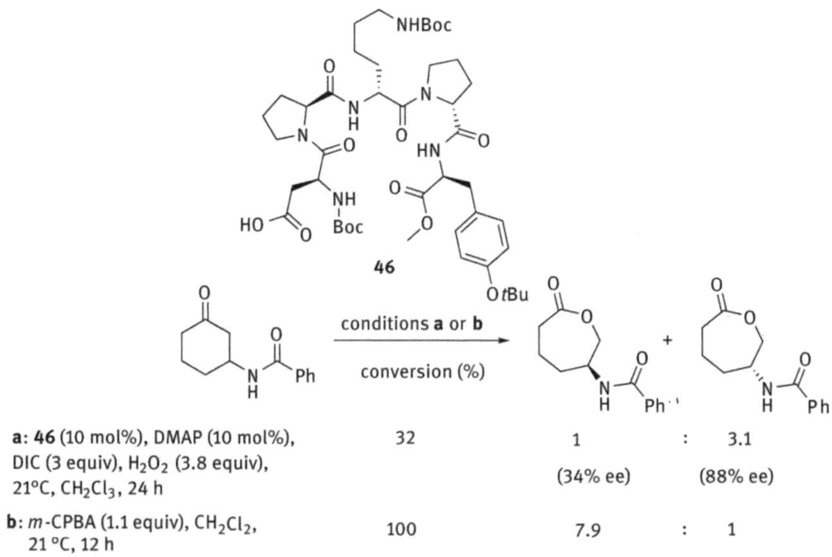

Figure 8.33: Effect of peptide catalyst **7a** on regioselectivity of BV oxidation.

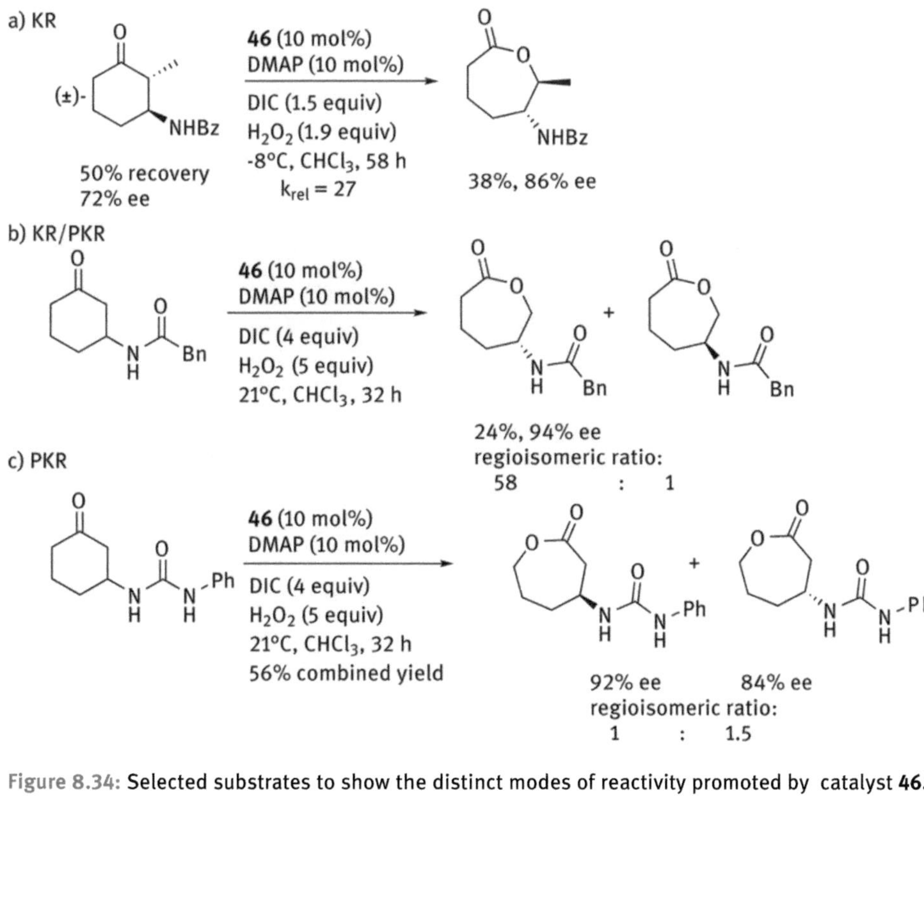

Figure 8.34: Selected substrates to show the distinct modes of reactivity promoted by catalyst **46.**

CPBA-disfavored" and the "normal" lactone, result of the intrinsic migratory aptitude, are formed in near equal quantities with similar ees.

Later, Miller and Anslyn reported a synergistic bioinformatics-inspired combinatorial and chiroptical study of peptide catalysts to set up the enantioselective BV oxidation of substrates lacking directing groups [57].

Investigation on protein loop sequences which have been observed to interact with tetrahedral anions, structurally similar to the Criegee intermediate, led to the identification of two families of peptide catalysts able to promote the enantioselective BV oxidation of a sterically challenging *cis*-2,6-diphenylcyclohexanone. This substrate is a minimum model of highly encumbered ketones found in terpene and polyketide natural products (Figure 8.35(a)).

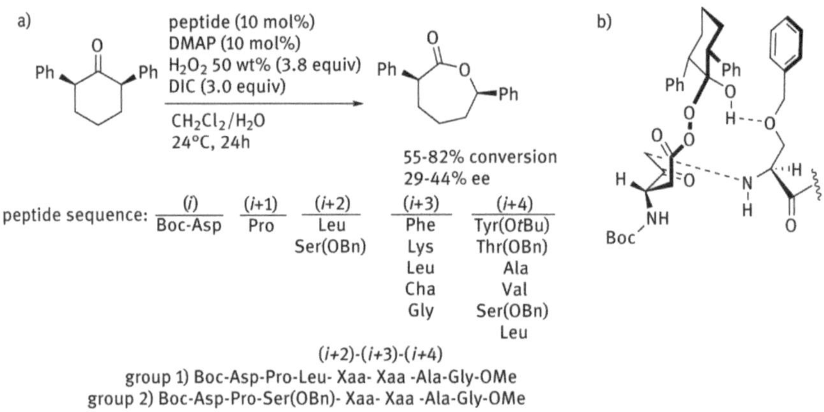

Figure 8.35: (A) Peptide catalysts for the BV oxidation of encumbered directing group-free ketone. (B) Hypotesized H-bonding interactions between Criegee intermediate and Ser(OBn) side-chain.

One family universally has an O-benzyl serine residue in the $(i+2)$ position, while the second group has an $(i+2)$ leucine residue, where (i) is the position of N-terminal residue N-(*t*-butoxycarbonyl)-L-aspartic acid followed by an L-proline residue at $(i+1)$ position. The authors suggested a mode of asymmetric induction where the H-bond interactions between the $(i+2)$ residue and the catalyst-bound Criegee intermediate are crucial for the asymmetric induction (Figure 8.35(b)).

In 2016, Yamamoto and coauthors, reported the first application in asymmetric BV reactions of a supramolecular assembly, an ion-pair catalyst derived by *in situ* assembly of a chiral flavinium and a cinchona alkaloid dimer [58]. The idea was the generation of a supramolecular catalyst by self-assembly via non-covalent interactions, of a chiral base, that induces the stereoselectivity, and a catalytically active molecule, an active flavinium specie. Flavin derivatives are known promoters of BV oxidation, using molecular oxygen or hydrogen peroxide. Nevertheless, they provided limited stereoselectivities, due to their conformational changes during the redox cycle. A variety of 3-substituted cyclobutanones were oxidized with good to excellent enantioselectivities, up to 96 % ee (Figure 8.36). The authors proposed a model of the transition state depicted in Figure 8.37. The chiral flavinium is linked with the matched chiral cinchona dimer by π–π and ionic interactions, setting a selective orientation in which there is no steric interference between the ethyl group of the DHQ and the phenyl group next to N^1. The hydrogen peroxide, deprotonated by the other basic site, attacks the iminium of the flavinium generating a hydroperoxy specie which reacts with the less-hindered face of the cyclobutanone. The latter likely being activated by H-bonding with ammonium of cinchona alkaloid, producing the Criegee intermediate that rearranges to produce the enantioenriched γ-lactone.

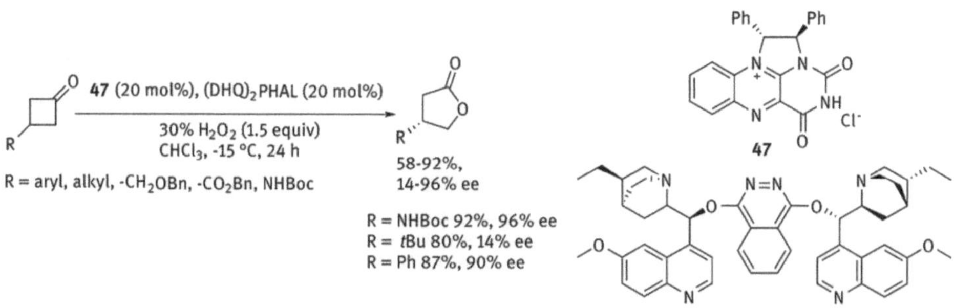

Figure 8.36: Self-assembled flavinium-cinchona alkaloid dimer organocatalyst for enantioselective BV oxidation.

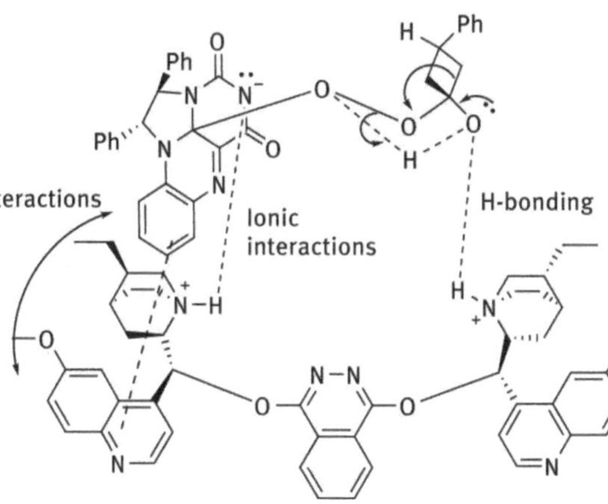

Figure 8.37: Hypothesized transition state of BV oxidation of 3-phenylcyclobutanone catalyzed by flavinium-cinchona alkaloid dimer.

Oxidative enantioselective desymmetrization of prochiral and *meso*-compounds has been also reported. Zhao and coauthors described the enantioselective oxidation of 1,2-diols using N-bromophthalimide (NBP), organocatalyzed by quinine derived urea **48** at room temperature, in a mixed solvent system ($CHCl_3$/PhCl 10:1) (Figure 8.38(a)) [59]. The method gave access to the corresponding α-hydroxy ketones in good yields (65–94 %) and good to excellent enantioselectivity (66–95 % ee) starting from diaryl-substituted *meso*-1,2-diols. The catalytic process was selective toward diol mono-oxidation and the overoxidized dione was produced in traces. When starting from dialkyl-substitued diols, the resulting α-hydroxy ketone was isolated in lower yield and enantioselectivity (Figure 8.38(a)).

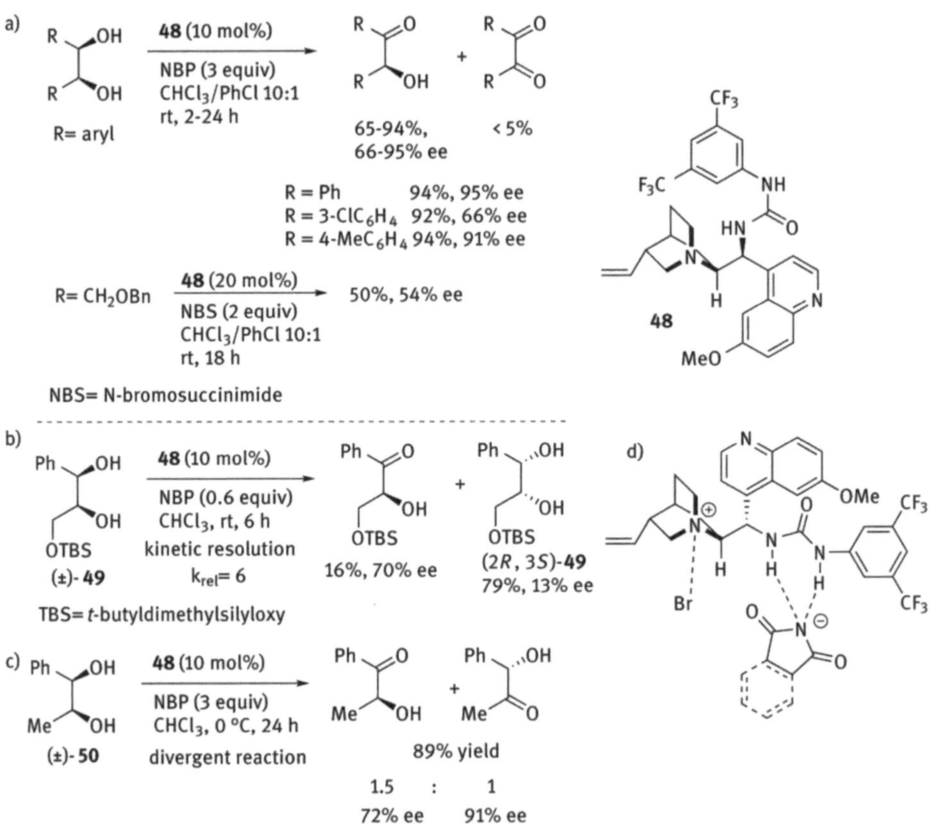

Figure 8.38: (A) Enantioselective oxidative desymmetrization of meso-diols organocatalyzed by quinine-derived thiourea. (B) Kinetic resolution and (C) divergent reaction on racemic diols. (D) Model of Lewis base and H-bond donor catalytic abilities of quinine-thiourea.

The less efficient oxidation of dialkyl-substituted diols was exploited in the resolution of racemic *syn*-diols via classical kinetic resolution or divergent reaction, depending on the substrate (Figure 8.38(b, c)).

The authors speculated that H-bonding interactions among the catalyst urea moiety, the phthalimide and diol are pivotal in the organization of a rigid transition state accounting for the high enantioselectivity. Moreover, the quinuclidine nitrogen of the catalyst is involved in complexation with NBP (Figure 8.38(d)).

An interesting strategy for the enantioselective organocatalyzed desymmetrization of 2-substituted and 2,2-disubstituted 1,3-diols was reported in 2014 by Houk and Zheng [60]. The desymmetrization of 2-substituted 1,3-diols represents a more challenging task given the pro-stereogenic center distance from hydroxyl group and primary alcohols are stronger nucleophiles. The 1,3-diols were protected with benzylidene acetals, able to undergo a selective oxidation by dimethyldioxirane (DMDO), forming an "ortho ester" intermediate **51** (Figure 8.39). Since a proton transfer process is involved to transform intermediate **51** into the final product, the authors envisioned that chiral phosphoric acids, could promote this step inducing enantioselectivity. A wide variety of 2-substituted and 2,2-disubstituted 1,3-diols were transformed into the corresponding products in high yields and enantioselectivity, using the chiral phosphoric acid **TRIP** (Figure 8.39).

The DFT study of the mechanism showed that the oxidation of acetal by DMDO is the rate-determining step and the proton transfer is significantly accelerated by the phosphoric acid. Attractive aryl–aryl interactions between benzylidene moiety of the substrate and 2,4,6-triisopropylphenyl group of the catalyst were identified as crucial for the enantiocontrol.

In 2018, Zheng applied this approach to the enantioselective synthesis of acyclic α-tertiary amines via desymmetrization of 2-substituted 2-nitro-1,3-diols (Figure 8.40) [61].

The protocol allowed to isolate a wide variety of chiral 2-nitro-1,3-diols in high yields and enantioselectivity, giving access to optically active valuable building blocks such as unnatural α-alkyl serines, precursors of hydroxymethyl glutamic acid (HMG) and natural product salinosporamide. Furthermore, the formal synthesis of chiral manzacidines, a family of bromopyrrole alkaloids isolated from sponges, was accomplished.

8.5 Miscellaneous

The development of sustainable methodologies which employ readily available molecular oxygen and visible light source received a lot of attention in the last years, becoming an important area of research in catalytic oxidation reactions [62]. The 2,2-disubstituted indolin-3-ones are molecular units found in pseudoindoxyl alkaloids and natural products. Therefore, simple and sustainable methods focused on their stereoselective approach are highly desirable. In this context, a first example on an enantioselective cascade aerobic oxidation and semipinacol rearrangement

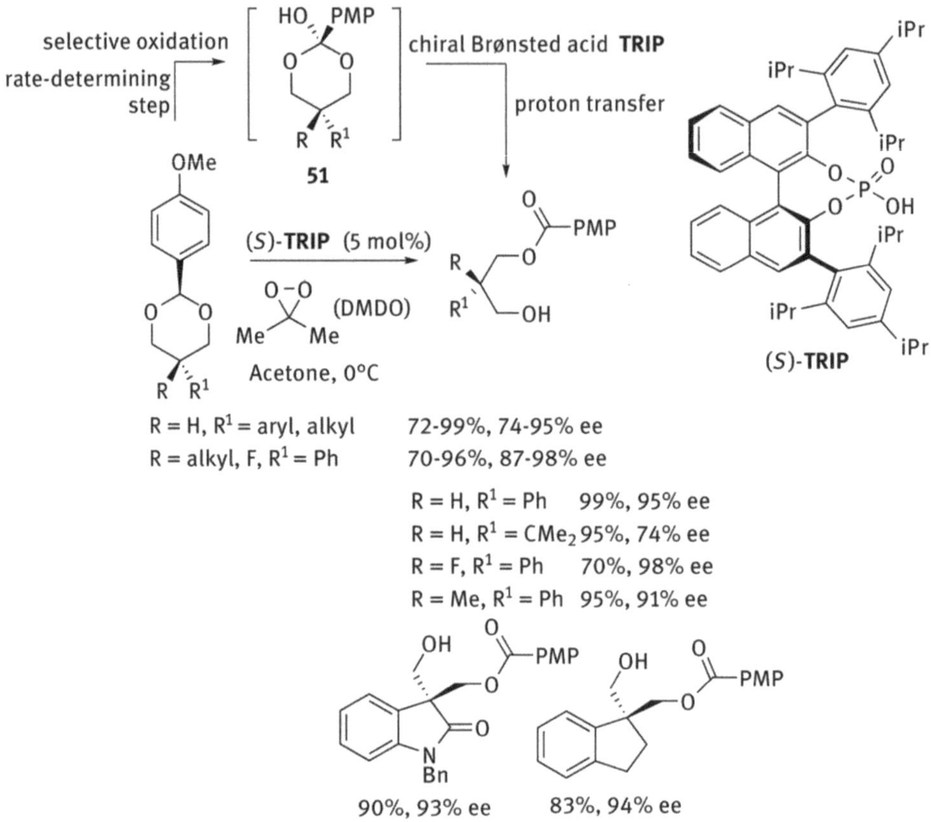

Figure 8.39: Chiral phosphoric acid catalyzed enantioselective desymmetrization of 2-substitued and 2,2-disubstituted 1,3-diols.

reaction of 2-aryl-3-alkylsubstituted indoles via photoredox and hydrogen-bonding catalysis was illustrated by Lu and coauthors in 2014 (Figure 8.41) [63].

Irradiation of the reaction mixture containing substituted indole, catalytic loadings of $Ru(byp)_3Cl_2$ as the photoredox catalyst and a BINOL-based phosphoric acid, afforded the 2,2-disubstituted indolin-3-one in 83 % yield and 60 % ee. This enantioselective cascade reaction was recently in-depth investigated by Jiang, Zhao et al. combining the dicyanopyrazine-derived chromophore (DPZ) as an organic photoredox catalyst and a SPINOL-based spirocyclic phosphoric acid **52** [64].

Under aerobic and slightly different reaction conditions, the 2-aryl-3-alkylsubstituted indoles, were converted into the 2,2-disubstituted indolin-3-ones, via the formation of intermediate 3-benzyl-2-phenyl-3H-indolin-3-ol, in good yields and good to high ee values (Figure 8.41). Single-electron transfer of DPZ as the photoredox catalyst to oxidize indole to a radical cation species, then captured by molecular oxygen, was demonstrated. Preliminary investigations on the stereoselective outcome of the

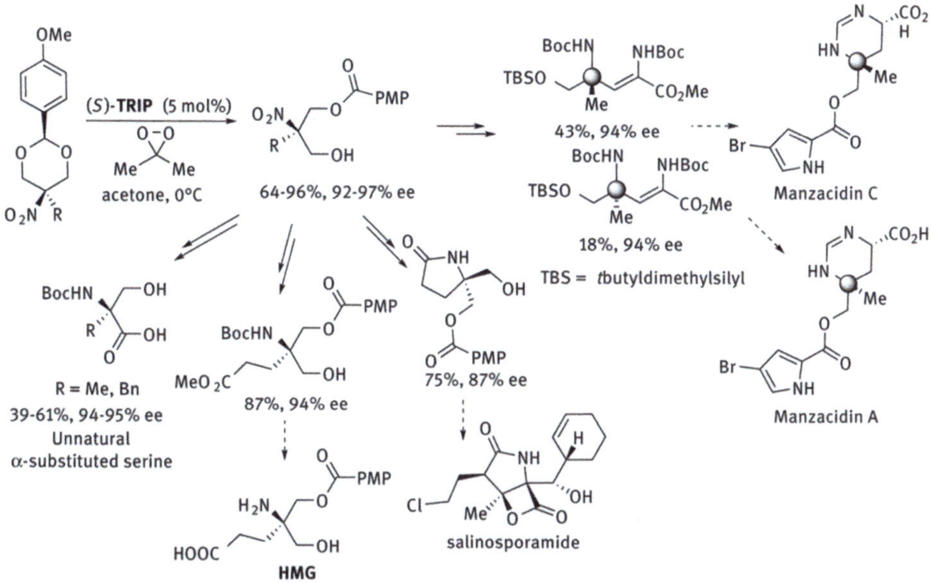

Figure 8.40: Chiral phosphoric acid catalyzed enantioselective desymmetrization of 2-substituted 2-nitro-1,3-diols for the synthesis of acyclic α-tertiary amines.

cascade process showed the primary role of catalyst **52** to foster the conversion to the product. Moreover, catalyst **52** would be involved in the enantiocontrol of the first oxidation, whereas the intermediate 3-benzyl-2-phenyl-3H-indolin-3-ol would act as a chiral promoter for the semipinacol rearrangement.

8.6 Outlook and perspectives

In the last 6 years, new methodologies to stereoselectively oxidize readily available carbonyl compounds, alkenes and heterocycles have been developed by using a variety of organocatalysts such as bifunctional or phase transfer catalysts derived from cinchona alkaloids, BINOL or SPINOL-based phosphoric acids, bisguanidinium salts, diols, peptides, hypervalent iodine-based compounds. These organocatalysts were employed in fundamental oxyfunctionalization reactions, thus becoming either alternative or complements to well-known metal-based catalyzed processes. Given the continuous request of a less-polluting and energetically convenient chemistry for the production of fine chemicals, it is likely to expect that future developments will focus on the synthesis and application of new organocatalysts suitable to work under greener oxidative and visible light conditions. Indeed, in this context, competitive stereoselective methodologies are truly awaited.

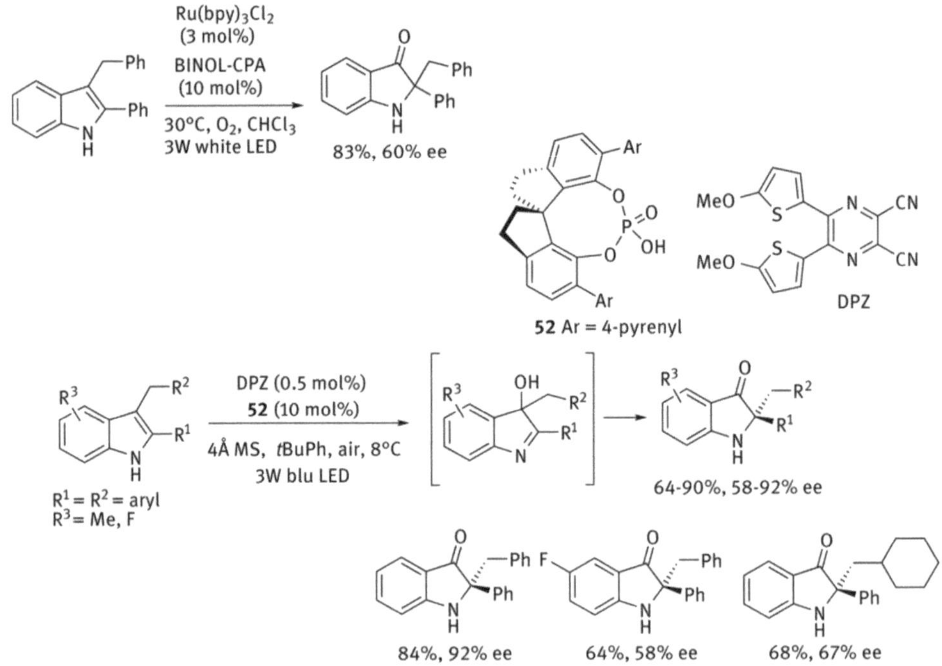

Figure 8.41: Enantioselective visible light-driven photo- and organocatalyzed access to 2,2-disubstituted indolin-3-ones under aerobic conditions.

References

[1] For reviews: (a) Wang C, Yamamoto H. Organocatalytic asymmetric oxidations with hydrogen peroxide and molecular oxygen. Chem Asian J 2015;10:2056–68. (b) Russo A, De Fusco C, Lattanzi A. Organocatalytic asymmetric oxidations with hydrogen peroxide and molecular oxygen. ChemCatChem. 2012;4:901–16. (c) Bryliakov PK. Catalytic asymmetric oxygenations with the environmentally benign oxidants H_2O_2 and O_2. Chem Rev. 2017;117:11406–59. (d) Mandal S, Bera T, Dubey G, Saha J, Laha JK. Uses of $K_2S_2O_8$ in metal-catalyzed and metal-free oxidative transformations. ACS Catal. 2018;8:5085–144. (e) Zhang X, Rakesh KP, Ravindar L, Qin H-L. Visible-light initiated aerobic oxidations: a critical review. Green Chem. 2018;20:4790–833.

[2] Albanese DC, Foschi F, Penso M. Sustainable oxidations under phase-transfer catalysis conditions. Org Process Res Dev. 2016;20:129–39.

[3] Sano D, Nagata K, Itoh T. Catalytic asymmetric hydroxylation of oxindoles by molecular oxygen using a phase-transfer catalyst. Org Lett. 2008;10:1593–5.

[4] Sim SB, Wang M, Zhao Y. Phase-transfer-catalyzed enantioselective α-hydroxylation of acyclic and cyclic ketones with oxygen. ACS Catal. 2015;5:3609–12.

[5] Yang Y, Moinodeen F, Chin W, Ma T, Jiang Z, Tan CH. Pentanidium-catalyzed enantioselective α-hydroxylation of oxindoles using molecular oxygen. Org Lett. 2012;14:4762–5.

[6] Zhou S, Zhang L, Li C, Mao Y, Wang J, Zhao P, et al. Pentanidium catalyzed enantioselective hydroperoxidation of 2-oxindole using molecular oxygen. Catalysis Commun. 2016;82:29–31.

[7] Lian M, Li Z, Gai Y, Meng Q, Gao Z. Enantioselective photooxygenation of β-keto esters by chiral phase-transfer catalysis using molecular oxygen. Chem Asian J. 2012;7:2019–23.
[8] Wang Y, Xiong T, Meng Q. Methylhydrazine-induced enantioselective α-hydroxylation of β-keto esters with molecular oxygen catalyzed by hydroquinine. Tetrahedron. 2015;71:85–90.
[9] Wang Y, Yin H, Tang X, Wu Y, Meng Q, Gao Z. A series of cinchona-derived N-oxide phase-transfer catalysts: application to the photo-organocatalytic enantioselective α-hydroxylation of β-dicarbonyl compounds. J Org Chem. 2016;81:7042–50.
[10] Ooi T, Ohmatsu K, Ando Y. In situ electrophilic activation of hydrogen peroxide for catalytic asymmetric α-hydroxylation of 3-substituted oxindoles. Synlett. 2017;28:1291–4.
[11] Yao H, Lian M, Li Z, Wang Y, Meng Q. Asymmetric direct α-hydroxylation of β-oxo esters catalyzed by chiral quaternary ammonium salts derived from cinchona alkaloids. J Org Chem. 2012;77:9601–8.
[12] Odagi M, Furukori K, Watanabe T, Nagasawa K. Asymmetric α-hydroxylation of tetralone-derived β-ketoesters by using a guanidine–urea bifunctional organocatalyst in the presence of cumene hydroperoxide. Chem Eur J. 2013;19:16740–5.
[13] Odagi M, Furukori K, Yamamoto Y, Sato M, Iida K, Yamanaka M, et al. Origin of stereocontrol in guanidine-bisurea bifunctional organocatalyst that promotes α-hydroxylation of tetralone-derived β-ketoesters: asymmetric synthesis of β- and γ-substituted tetralone derivatives via organocatalytic oxidative kinetic resolution. J Am Chem Soc. 2015;137:1909–15.
[14] Odagi M, Furukori K, Takayama K, Noguchi K, Nagasawa K. Total synthesis of rishirilide B by organocatalytic oxidative kinetic resolution: revision of absolute configuration of (+)-rishirilide B. Angew Chem Int Ed. 2017;56:6609–12.
[15] De Fusco C, Meninno S, Tedesco C, Lattanzi A. Enantioselective α-hydroxylation of β-ketoamides. Org Biomol Chem. 2013;11:896–9.
[16] Wang Y, Li Z, Xiong T, Zhao J, Meng Q. Enantioselective α-hydroxylation of β-keto esters catalyzed by cinchona alkaloid derivatives. Synlett. 2014;25:2155–60.
[17] Wang Y, Yin H, Qing H, Zhao J, Wu Y, Meng Q. Asymmetric α-hydroxylation of β-indanone esters and β-indanone amides catalyzed by C-2' substituted cinchona alkaloid derivatives. Adv Synth Catal. 2016;358:737–45.
[18] Novacek J, Izzo JA, Vetticatt MJ, Waser M. Bifunctional ammonium salt catalyzed asymmetric α-hydroxylation of β-ketoesters by simultaneous resolution of oxaziridines. Chem Eur J. 2016;22:17339–44.
[19] Lin X, Ruan S, Yao Q, Yin C, Lin L, Feng X, et al. Kinetic resolution of oxaziridines via chiral bifunctional guanidine-catalyzed enantioselective α-hydroxylation of β-keto esters. Org Lett. 2016;18:3602–5.
[20] Jadhav MS, Righi P, Marcantoni E, Bencivenni G. Enantioselective α-benzoyloxylation of ketones promoted by primary amine catalyst. J Org Chem. 2012;77:2667–74.
[21] Demoulin N, Lifchits O, List B. Organocatalytic asymmetric α-benzoyloxylation of α-branched aldehydes and enals: a useful approach to oxygenated quaternary stereocenters. Tetrahedron. 2012;68:7568–74.
[22] Shimogaki M, Maruyama H, Tsuji S, Homma C, Kano T, Maruoka K. Synthesis of chiral tritylpyrrolidine derivatives and their application to asymmetric benzoyloxylation. J Org Chem. 2017;82:12928–32.
[23] Wang D, Xu C, Zhang L, Luo S. Asymmetric α-benzoyloxylation of β-ketocarbonyls by a chiral primary amine catalyst. Org Lett. 2015;17:576–9.
[24] Kanemitsu T, Sato M, Yoshida M, Ozasa E, Miyazaki M, Odanaka Y, et al. Enantioselective α-benzoyloxylation of malonic diesters by phase-transfer catalysis. Org Lett. 2016;18:5484–7.
[25] Wang Y, Gao Q, Liu Z, Bai S, Tang X, Yin H, et al. Enantioselective α-benzoyloxylation of β-keto esters by N-oxide phase-transfer catalysts. J Org Chem. 2018;83:2263–73.

[26] Guilbault AA, Legault CY. Drastic enhancement of activity in iodane-based α-tosyloxylation of ketones: iodine(III) does the hypervalent twist. ACS Catal. 2012;2:219–22.
[27] Guilbault AA, Basdevant B, Wanie V, Legault CY. Catalytic enantioselective α-tosyloxylation of ketones using iodoaryloxazoline catalysts: insights on the stereoinduction process. J Org Chem. 2012;77:11283–95.
[28] Brenet S, Berthiol F, Einhorn J. 3,3'-diiodo-BINOL-fused maleimides as chiral hypervalent iodine(III) organocatalysts. Eur J Org Chem. 2013;8094–6.
[29] Brenet S, Minozzi C, Clarens B, Amiri L, Berthiol F. 3,3'-Diiodobinaphthol and 3,3'-diiodobiphenol derivatives as hypervalent iodine organocatalysts for the α-oxytosylation of ketones. Synthesis. 2015;47:3859–73.
[30] Levitre G, Dumoulin A, Retailleau P, Panossian A, Leroux FR, Masson G. Asymmetric α-sulfonyl- and α-phosphoryl-oxylation of ketones by a chiral hypervalent iodine(III). J Org Chem. 2017;82:11877–83.
[31] Beaulieu S, Legault C. Mechanistic insights on the iodine(III)-mediated α-oxidation of ketones. Chem Eur J. 2015;21:11206–11.
[32] Bortoluzzi J, Jha V, Levitre G, Fer MJ, Berreur J, Masson G, et al. Stereoselectivity switch in the trapping of polar organometallics with Andersen's reagent-access to highly stereoenriched transformable biphenyls. J Org Chem. 2018;83:7751–61.
[33] Basdevant B, Legault CY. Enantioselective iodine(III)-mediated synthesis of α-tosyloxy ketones: breaking the selectivity barrier. Org Lett. 2015;17:4918–21.
[34] Feng Y, Huang R, Hu L, Xiong Y, Coeffard V. Chiral C_2-symmetric iodoarene-catalyzed asymmetric α-oxidation of β-keto esters. Synthesis. 2016;48:2637–44.
[35] Heravi MM, Zadsirjan V, Esfandyari M, Lashaki TB. Applications of sharpless asymmetric dihydroxylation in the total synthesis of natural products. Tetrahedron Asymmetry. 2017;28:987–1043.
[36] Jacobsen EN, Marko I, Mungall WS, Schroeder G, Sharpless KB. Asymmetric dihydroxylation via ligand-accelerated catalysis. J Am Chem Soc. 1988;110:1968–70.
[37] (a) Mayer S, List B. Asymmetric counteranion-directed catalysis. Angew Chem Int Ed. 2006;45:4193–5. (b) Rauniyar V, Lackner AD, Hamilton GL, Toste FD. Asymmetric electrophilic fluorination using an anionic chiral phase-transfer catalyst. Science. 2011;334:1681–4.
[38] Bhunnoo RA, Hu Y, Lainé DI, Brown RC. An asymmetric phase transfer dihydroxylation reaction. Angew Chem Int Ed. 2002;41:3479–80.
[39] Wang C, Zong L, Tan CH. Enantioselective oxidation of alkenes with potassium permanganate catalyzed by chiral dicationic bisguanidinium. J Am Chem Soc. 2015;137:10677–82.
[40] Fang L, Yan L, Haeffner F, Morken JP. Carbohydrate-catalyzed enantioselective alkene diboration: enhanced reactivity of 1,2-bonded diboron complexes. J Am Chem Soc. 2016;138:2508–11.
[41] (a) Romero RM, Wöste TH, Muñiz K. Vicinal difunctionalization of alkenes with iodine (III) reagents and catalysts. Chem Asian J. 2014, 9, 972–83. (b) Claraz A, Masson G. Asymmetric Iodine catalysis-mediated enantioselective oxidative transformations. Org Biomol Chem. 2018;16:5386–402. (c) For a review on mechanistic aspects, see: Fujita M. Mechanistic aspects of alkene oxidation using chiral hypervalent iodine reagents. Tetrahedron Lett. 2017;58:4409–19.
[42] Fujita M, Mori K, Shimogaki M, Sugimura T. Asymmetric synthesis of 4,8-dihydroxyisochroman-1-one polyketide metabolites using chiral hypervalent iodine(III). Org Lett. 2012;14:1294–7.
[43] Shimogaki M, Fujita M, Sugimura T. Enantioselective oxidation of alkenylbenzoates catalyzed by chiral hypervalent iodine(III) to yield 4-hydroxyisochroman-1-ones. Eur J Org Chem. 2013;7128–38.

[44] Gelis C, Dumoulin A, Bekkaye M, Neuville L, Masson G. Chiral hypervalent iodine(III) catalyst promotes highly enantioselective sulfonyl- and phosphoryl-oxylactonizations. Org Lett. 2017;19:278–81.

[45] Seitz M, Reiser O. Synthetic approaches towards structurally diverse γ-butyrolactone natural-product-like compounds. Curr Opin Chem Biol. 2005;9:285–92.

[46] Peris G, Miller SJ. A nonenzymatic acid/peracid catalytic cycle for the Baeyer-Villiger oxidation. Org Lett. 2008;10:3049–52.

[47] Haubenreisser S, Wöste TH, Martínez C, Ishihara, Muñiz K. Structurally defined molecular hypervalent iodine catalysts for intermolecular enantioselective reactions. Angew Chem Int Ed. 2016;55:413–17.

[48] Wöste TH, Muñiz K. Enantioselective vicinal diacetoxylation of alkenes under chiral iodine(III) catalysis. Synthesis. 2016;48:816–27.

[49] (a) Krow GR. Baeyer-Villiger oxidation of ketones and aldehydes. In organic reactions, Vol. 43 (Ed.: L. A. Paquette). New York: Wiley, 1993:251–798. (b) Strukul G. Transition Metal catalysis in the Baeyer-Villiger oxidation of ketones. Angew Chem Int Ed. 1998;37:1198–209. (c) Renz M, Meunier B. 100 years of Baeyer-Villiger oxidations. Eur J Org Chem. 1999:737–50. (d) Jiménez-Sanchidrían C, Ruiz JR. The Baeyer-Villiger reaction on heterogeneous catalysts. Tetrahedron. 2008;64:2011–26. (e) Michelin RA, Sgarbossa P, Scarso A, Strukul G. The Baeyer-Villiger oxidation of ketones: a paradigm for the role of soft lewis acidity in homogeneous catalysis. Coord Chem Rev. 2010;254:646–60.

[50] Leisch H, Morley K, Lau PC. Baeyer-Villiger monooxygenases: more than just green chemistry. Chem Rev. 2011;111:4165–222.

[51] Murahashi SI, Ono S, Imada Y. Asymmetric Baeyer–Villiger reaction with hydrogen peroxide catalyzed by a novel planar-chiral bisflavin. Angew Chem Int Ed. 2002;41:2366–8.

[52] (a) Xu S, Wang Z, Zhang X, Zhang X, Ding K. Chiral brønsted acid catalyzed asymmetric Baeyer–Villiger reaction of 3-substituted cyclobutanones by using aqueous H_2O_2. Angew Chem Int Ed. 2008;47:2840–3. (b) Xu S, Wang Z, Li Y, Zhang X, Wang H, Ding K. Mechanistic investigation of chiral phosphoric acid catalyzed asymmetric Baeyer–Villiger reaction of 3-substituted cyclobutanones with H_2O_2 as the oxidant. Chem Eur J. 2010;16:3021–35. (c) Xu S, Wang Z, Zhang X, Ding K. Charge-transfer effect on chiral phosphoric acid catalyzed asymmetric Baeyer-Villiger oxidation of 3-substituted cyclobutanones using 30 % aqueous H_2O_2 as the oxidant. Chin J Chem. 2010;28:1731–5. (d) Xu S, Wang Z, Zhang X, Ding K. Asymmetric Baeyer–Villiger oxidation of 2,3- and 2,3,4-substituted cyclobutanones catalyzed by chiral phosphoric acids with aqueous H_2O_2 as the oxidant. Eur J Org Chem. 2011:110–16.

[53] Sasakura N, Nakano K, Ichikawa Y, Kotsuki H. A new environmentally friendly method for the Baeyer-Villiger oxidation of cyclobutanones catalyzed by thioureas using H_2O_2 as an oxidant. RSC Adv. 2012;2:6135–9.

[54] (a) Greene FD, Kazan J. Preparation of diacyl peroxides with N,N' dicyclohexylcarbodiimide. J Org Chem. 1963:28:2168–71. (b) Peris G, Jakobsche CE, Miller SJ. Aspartate-catalyzed asymmetric epoxidation reactions. J Am Chem Soc. 2007;129:8710–11.

[55] Romney DK, Colvin SM, Miller SJ. Catalyst control over regio- and enantioselectivity in Baeyer–Villiger oxidations of functionalized ketones. J Am Chem Soc. 2014;136:14019–22.

[56] (a) Vedejs E, Chen X. Parallel kinetic resolution. J Am Chem Soc. 1997;119:2584–5. (b) Dehli JR, Gotor V. Parallel kinetic resolution of racemic mixtures: a new strategy for the preparation of enantiopure compounds? Chem Soc Rev. 2002;31:365–70.

[57] Giuliano MW, Lin CY, Romney DK, Miller SJ, Anslyn EV. A synergistic combinatorial and chiroptical study of peptide catalysts for asymmetric Baeyer–Villiger oxidation. Adv Synth Catal. 2015;357:2301–9.

[58] Poudel PP, Arimitsu K, Yamamoto K. Self-assembled ion-pair oganocatalysis-asymmetric Baeyer-Villiger oxidation mediated by flavinium-cinchona alkaloid dimer. Chem Commun. 2016;52:4163–6.
[59] Rong ZQ, Pan HJ, Yan HL, Zhao Y. Enantioselective oxidation of 1,2-diols with quinine-derived urea organocatalyst. Org Lett. 2014;16:208–11.
[60] Meng SS, Liang Y, Cao KS, Zou L, Lin XB, Yang H, et al. Catalytically enantioselective synthesis of acyclic α-tertiary amines through desymmetrization of 2-substituted 2-nitro-1,3-diols. J Am Chem Soc. 2014;136:12249–52.
[61] Meng SS, Tang WB, Zheng WH. Catalytically enantioselective synthesis of acyclic α-tertiary amines through desymmetrization of 2-substituted 2-nitro-1,3-diols. Org Lett. 2018;20:518–21.
[62] Bian C, Singh AK, Niu L, Yi H, Lei A. Visible-light-mediated oxygenation reactions using molecular oxygen. Asian J Org Chem. 2017;6:386–96.
[63] Ding W, Zhou QQ, Xuan J, Li TR, Lu LQ, Xiao WJ. Photocatalytic aerobic oxidation/semipinacol rearrangement sequence: a concise route to the core of pseudoindoxyl alkaloids. Tetrahedron Lett. 2014;55:4648–52.
[64] Bu L, Li J, Yin Y, Qiao B, Chai G, Zhao X, et al. Organocatalytic asymmetric cascade aerobic oxidation and semipinacol rearrangement reaction: a visible light-induced approach to access chiral 2,2-disubstituted indolin-3-ones. Chem Asian J. 2018;13:2382–7.

Andrea Gualandi, Pier Giorgio Cozzi, Giacomo Rodeghiero, Thomas Paul Jansen and Rossana Perciaccante

9 Stereoselective synergystic organo photoredox catalysis with enamines and iminiums

Abstract: Application of small chiral organic molecules in catalysis has been dominated by formation of chiral enamines or iminium ions. Nucleophiles – electrophiles reactivity has been exploited in many papers. Now, the possibility to combine organocatalysis with photochemistry open new "exciting" possibilities and opportunities, in reactions that are mediated by radicals.

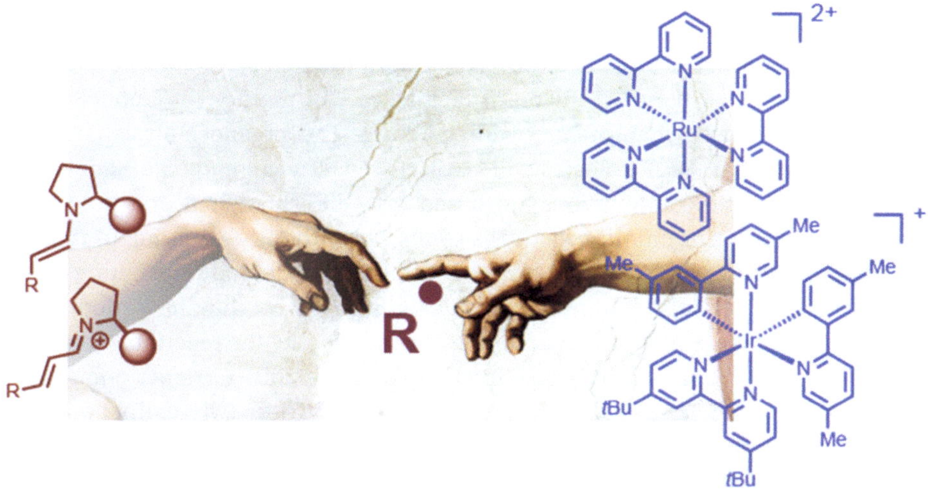

Keywords: photoredox catalysis, organocatalysis, stereoselective reactions, enamine catlysis, iminium catalysis

9.1 Introduction

Photoredox catalysis has grown in recent years as an important field in organic chemistry [1]. New photocatalysts (PCs) [2], reactions [3], devices [4] and technical solutions for large scale synthesis [5], have developed rapidly giving rise to new directions and improvements. However, one peculiar aspect of photoredox catalysis

This article has previously been published in the journal Physical Sciences Reviews. Please cite as: Gualandi, A., Cozzi, P., Rodeghiero, G., Jansen, T. P., Perciaccante, R. Stereoselective synergystic organo photoredox catalysis with enamines and iminiums *Physical Sciences Reviews* [Online] 2020, 11. DOI: 10.1515/psr-2018-0098

https://doi.org/10.1515/9783110590050-009

is the possibility to merge the photoredox process with other activation mode [6]. The application of synergistic photoredox catalysis, where one or more catalytic cycles work cooperatively with the photoredox cycle enables new interesting transformations. These multiple catalytic systems provide effective control of selectivity, and more importantly, of stereoselectivity. In fact, though enantioselectivity photoredox catalysis can also be controlled by chiral photoredox metal complexes [7], the employment of a synergistic activation mode can give excellent solutions to this challenge. The possibility to merge different activation modes with photoredox catalysis relies on the quite peculiar way in which the PCs are activated, using visible light irradiation, normally functional groups present in organic molecules do not absorb photons in this wavelength region (380–700 nm). In addition, Lewis acids, Lewis bases and many solvents do not dramatically influence the absorption of light by the PCs. In many cases, when organometallic photoredox catalysts are employed, they are coordinatively saturated and stable avoiding any ligand exchange with reaction partners. Finally, organic molecules used as PCs are quite robust to many reaction conditions [8]. In this chapter, we will discuss the application of photoredox catalysis when merged with stereoselective organocatalysis. Two distinct and particular methodologies will be discussed. The common methodologies, using a synergistic combination, are based on well-established and previously discussed enamine or iminium catalysis combined with the generation of radicals through photocatalytic cycles [9]. Another quite interesting and intriguing possibility of organocatalytic photoredox catalysis is based on the absorption of the intermediates (again iminium or enamines), generated by the reaction of an organocatalyst with the substrate. These intermediates are able to trigger photoredox events that takes place in a stereocontrolled manner. These two distinct methodologies will be discussed, with the prevalent catalysts used and the chemistry developed in recent years. A short introduction regarding properties and behavior of excited states precedes the discussion, for readers unfamiliar with these concepts. It is important to stress that an interdisciplinary combination of expertise is required to provide a clear view of this field of chemistry.

9.2 Photophysical and photochemical properties of organic photoredox catalysis: a guide for dummies

Photochemistry and photophysics of organic molecules are well-established fields, with dedicated journals, books and more than 100 years of investigations. Many applications of photoredox catalysis in organic chemistry have been reported, also in total synthesis. However, the field was addressed mainly to specialists, and common organic laboratories seldomly employed these interesting methods, hampered by technical problems. The renaissance of these studies in synthetic organic chemistry started in 2008, when MacMillan [10] and Yoon [11] reported the possibility to use a well-known coordination compound, $[Ru(bpy)_3]^{2+}$ (bpy = bipyridine) in promoting interesting transformations. In the same period, Stephenson [12] was also reporting the use of $[Ru(bpy)_3]^{2+}$ in another kind of reaction. Researchers soon realized the key role played by $[Ru(bpy)_3]^{2+}$ as a PC. Since then many other metal complexes, dyes, quantum dots, semiconductors were exploited taking advantage of their properties [13]. The ability of a PC to absorb light and reach a relatively stable excited state (*PC) is required to engage in different processes with present substrates: (a) by single electron transfer (SET), in which radicals are formed, or (b) energy transfer (ET), in which other excited states are generated by direct energy transfer from the PC to other molecules. Both these two modes create a starting event that requires a precise design in order to observe a successful reaction. Another important aspect is that the PC needs to absorb light, preferable visible light, at wavelengths where other species present in solution do not absorb light. Furthermore, the quantum yield of the PC should be as close to one as possible, where the quantum yield of luminescence is defined as the ratio of the number of photons emitted to the number of photons absorbed. In addition, the excited state, once efficiently reached, needs to exist long enough time to permit the exchange of electrons or energy with other species present. It is important to realize immediately, that the PC in excited state can donate (behaving as a reductant) or receive (behaving as an oxidant) one electron easily compared to the PC in the ground state (Figure 9.1).

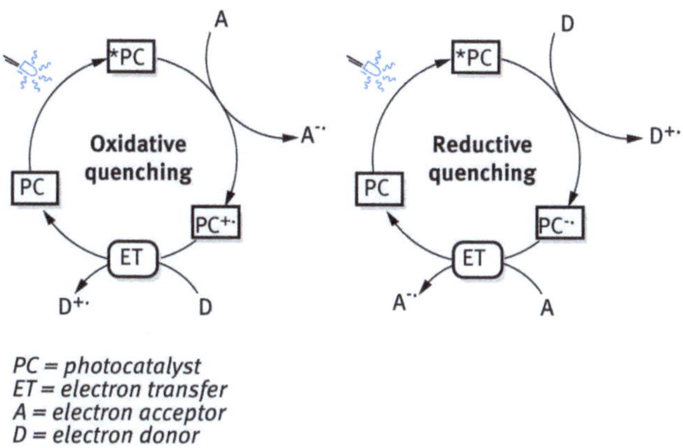

Figure 9.1: Quenching of excited states of a photocatalyst.

Therefore, the excited state of the PC is a stronger oxidant or reductant compared to the ground state of the PCs. This realization has led to a breakthrough in synergistic photoredox catalysis. The obtainment of reactive species is possible by the active role played by PC and by its extra energy gained after the absorption of light. After the electron transfer process, the PC is reduced or oxidized, and the ability to undergo successive electron transfer, can be evaluated by the ground state potential of the new specie produced. Oxidation and reduction potentials of many PC are well known [14] or can be determined by cyclic voltammetry. In organometallic complexes, such as $[Ru(bpy)_3]^{2+}$, oxidations and reductions are producing stable species, while other PC might produce quite unstable species. In other words, oxidation or reduction can be irreversible. Nevertheless, also molecules, for which the ground state potential is irreversible, can successfully engage in photoredox reactions. As the redox potentials of the excited state and ground state are known, it would be possible to establish if the reaction with the species present in solution is exothermic or weakly endothermic. Since redox potentials of organic species are known and were recently re-evaluated [15], it is also possible, using well-known concepts of electrochemistry, to quite firmly establish if the reaction is favored or not (Figure 9.2). It is also important to mention that redox potentials are always referred to a reference electrode, or redox couple (i. e. saturated calomel electrode [SCE] and ferrocenium/ferrocene). It is crucial to compare redox potentials using the same reference electrode, as it may sometimes happen that the potentials are compared using different references leading to wrong conclusions.

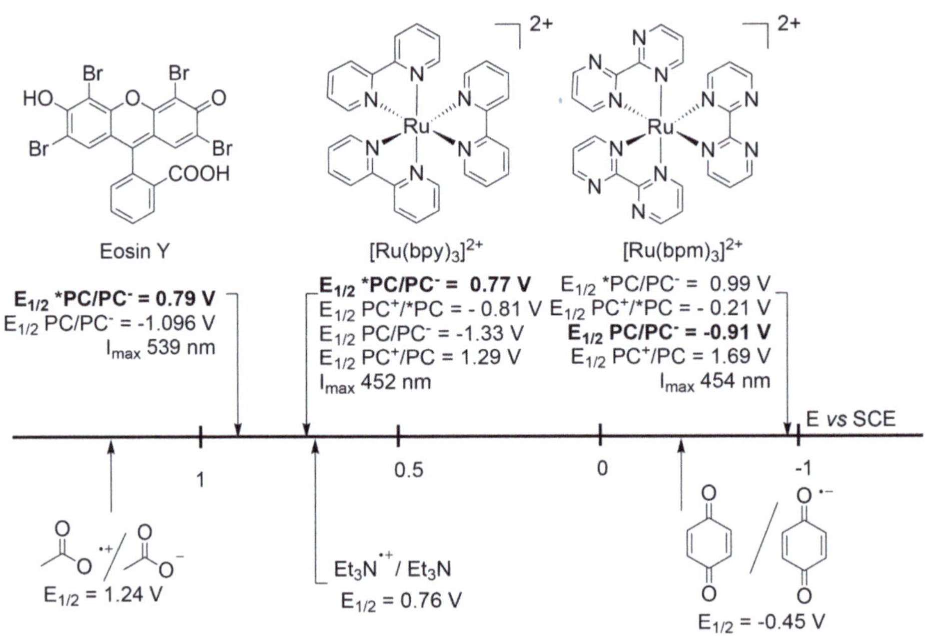

Figure 9.2: Electrochemical series of photocatalyst and organic compounds.

However, as many potentials used in these calculations are irreversible, care needs to be taken, and only experimental confirmation can ultimately give a direct assessment on the feasibility of a new reaction. In metal complexes a strong absorption maximum is observed near the 425 nm region. This region is known as the MLCT (metal to ligand charge transfer) region, in which electron transfer from a *d* orbital of the metal to a $*\pi$ of the ligand (bpy in the case of $[Ru(bpy)_3]^{2+}$) occurs. Excited states have their own terminology and energy, and we suggest reading more comprehensive and exhaustive texts for more detailed explanations [16]. In case of $[Ru(bpy)_3]^{2+}$, following excitation, a new state called ^{3}[MLCT] is quite rapidly reached. The symbol 3 indicates the coupling between the single electrons in the two semi-occupied states that formed after absorption of light. This state can relax back, through different pathways. Photophysicists describe these phenomena with rate constants k. The ^{3}MLCT can return to the ground state by emitting light (radiative process) with rate constant indicated as k_r or by emitting heat (not radiative process) with a rate constant indicated as k_{nr}. By definition, k observed (k_o) is the sum of $k_r + k_{nr}$. The observed quantum yield (ϕ_O) of the excited state can also be expressed by a ratio between rate constants of the two emission processes and, in this case, $\phi_O = k_r/k_r + k_{nr}$. It is important to realize that the light emitted by the $[Ru(bpy)_3]^{2+}$ is observed at a different wavelength with respect to the MLCT state, due to dissipation of energy. The emission maximum of the ^{3}MLCT is used to evaluate an important value, the E_{OO} energy, which is the energy difference between the ^{3}MLCT state and the ground state. This is the gained energy that the PC can exchange in electron transfer. As shown in in Figure 9.3, this energy value is measured in eV, and can be directly used for having, as first approximation, the values of the excited states potential [16].

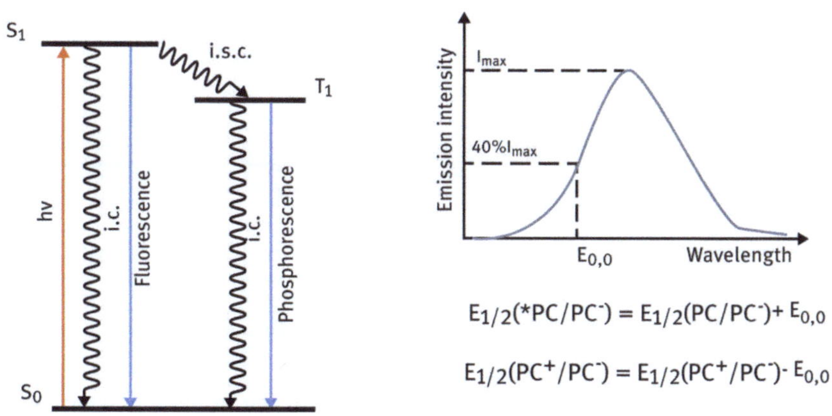

Figure 9.3: Jablonski diagram for a generic photocatalyst and evaluation of electrochemical potentials of its excited states.

This formalism is known as the Rehm–Weller equation and it needs to be stressed that other energy terms, important in some cases, are neglected. Anyway, by direct experiments, or by tabulated values, ground state potential and E_{OO} can be used to establish the series of the four redox potentials available for a PC in ground or excited states (Figure 9.2). At this point, another piece of information is missing. Which is the right potential that needs to be considered? Of course, many aspects as evaluation of organic species present in solutions with their redox potential and the excited or ground state potentials of the PC, need to be evaluated for understanding the kind of electron transfer process (reduction or oxidation) and the involved species. However, experimental data can give an indication if the excited states of the PC are involved. The simplest experiment that can be performed, called the Stern–Volmer quenching study [17], determines if a bimolecular reaction (electron or energy transfer) is occurring between the PC and an organic molecule present in the reaction mixture. The PC is excited with a laser light in the presence of one of the reaction partners in a measured concentration, and the emission of the PC is monitored at increased concentration of the reaction partners. If the emission of the PC is altered by the presence of the reaction partner, this partner is defined as quencher of the excited state. In other words, as electrons or energy are transferred between the quenchers and the PC in the excited state, the concentration of the excited state will be reduced, and, consequently, the intensity of the emitted light. Therefore, it is possible to monitor the phenomenon by plotting the measured intensity of the emitting light [18] in function of the quencher concentrations (Figure 9.4).

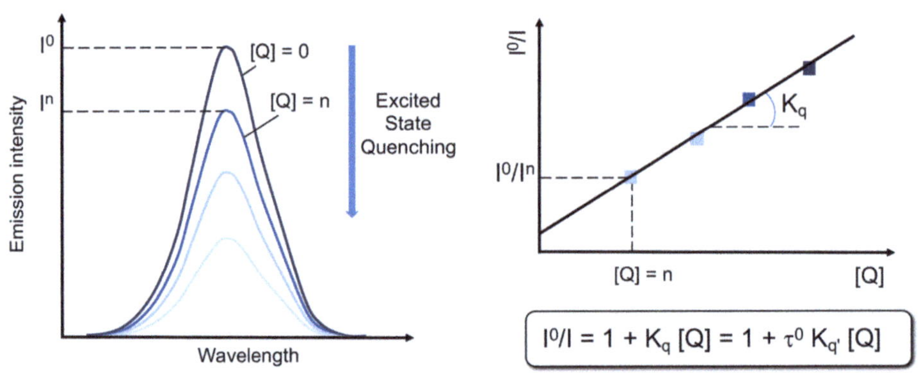

Figure 9.4: The Stern–Volmer experiment: determination of the effective quencher.

The Stern–Volmer experiment is also able to show which is the organic species in solution that is interfering with the excited state, providing the PC in its ground state by SET or energy transfer. At this point we have all the information necessary in order to understand the use of PC in stereoselective organic catalysis. It is important to mention that the PC needs to return in its original state, and therefore, if an electron transfer event (oxidation or reduction) is occurring, another complementary electron transfer through the species in the solution needs to restore the catalysts. Therefore, also this step needs to be exoergonic, or thermodynamically favored. This needs to be coupled with another possible event in which redox potentials are matched in the way that the thermodynamic of the process is favored. The Stern–Volmer experiment is also useful in evaluating energy transfer processes, which are only now starting to be considered in photocatalysis [18]. In the examples discussed in this chapter, only electron transfer events seem to be operative, and a further description of energy transfer will not be presented here.

9.3 Stereoselective alkylation of enamines through dual catalysis

As previously mentioned, the pioneering work published by MacMillan in 2008 reopened the field of photoredox catalysis. This work is based on merging enamine catalysis with photoredox catalysis. Enamine catalysis is recognized as a powerful activation mode [19] in organocatalysis [20]. Enamines of aldehydes and ketones are strong nucleophiles [21] activating the α-position of the aldehydes and ketones towards reaction with compatible electrophiles. As nucleophilicity of enamines is quantified by the Mayr scale [22], it is possible to predict if the reaction with electrophiles is successful if the Mayr relation ($\log k = S_N[N + E]$, where k is the rate constant of the reaction, S_N is the nucleophile-specific sensitivity parameter, E and N are the electrophilicity and nucleophilicity parameters) is satisfied [23]. In particular, if $E + N > -5$ it will be possible to observe the reaction to proceed in reasonable time (5–24 h). As enamine catalysis can be conducted in the presence of water, and water is produced by the formation of enamines, the presence of water needs to consider when designing a catalytic cycle. The ability of enamines to react through radicals, generated by oxidation, in so called SOMO (Single Occupied Molecular Orbitals) catalysis, proved a crucial observation to establish the photoredox cycle [24].

9.3.1 Ruthenium stereoselective alkylation of enamines

Radicals were found compatible with enamine catalysis, and in 2008 MacMillan reported a way to generate radicals and solve one decisive problem in enamine catalysis, the α-alkylation of aldehydes **1** (Figure 9.5) [10].

Enamines are electron rich compounds and can be reacted with electron poor radicals. In Figure 9.5 the proposed catalytic cycle is reported, requiring the presence and action of two distinct cycles, one controlled by the PC and one by the enamine. The PC used in this reaction was the $[Ru(bpy)_3]^{2+}$ discussed above. In order to elucidate the reaction mechanism, Stern–Volmer experiments were carried out, indicating that the enamine was able to quench the excited state of ruthenium and not the aldehydes, or, more importantly, the bromo derivative. As enamines get oxidized, from the reported redox potential it is possible to propose that enamines are able to reduce

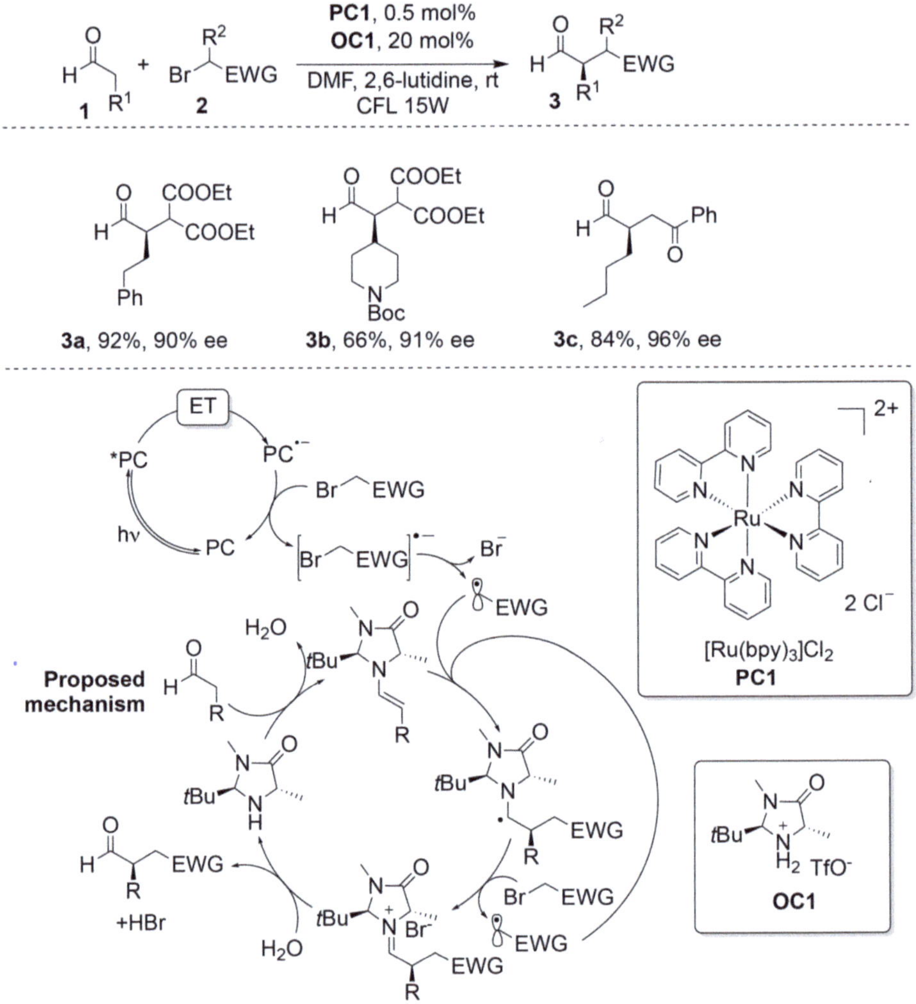

Figure 9.5: Photocatalytic stereoselective alkylation of aldehydes promoted by ruthenium(II) catalyst.

the ruthenium complex to form a ruthenium(I) species. This part of the catalytic cycle is not shown in Figure 9.5 and is the starting event that allows the production of active $[Ru(bpy)^{\bullet-}(bpy)_2]^+$. This intermediate is a strong reductant ($E_{1/2}[Ru^{2+}/Ru^+] = -1.33$ V vs. SCE), and the irreversible reduction of bromo derivatives substituted with an electron-withdrawing group is favored. Bromo derivatives **2**, described in the article, have reduction potentials between −0.49 and −0.9 V, and as mentioned, the formation of stabilized alkyl radicals is irreversible process and proceeds with elimination of bromide. The radicals formed react with the enamine giving rise to the formation of α-amine radicals. These are quite unstable and reactive with low reduction potential (−1.12 V vs. SCE) and are easily oxidized by an oxidant. Which species behaves as the oxidant? The excited state of $[Ru(bpy)_3]^{2+}$, as reported in Figure 9.2, has the calculated potential of $E_{1/2}[{}^*Ru^{2+}/Ru^+] = 0.77\,V$ and is able to oxidize this species. A sacrificial enamine starts the event producing sufficient quantities of Ru^+ that is then able to reduce the bromo derivatives present in solution forming Ru^{2+}. As soon as radicals are formed, they react with enamines producing the α-amine radical, the species that is able to restart the catalytic cycle by reducing the $^*Ru^{2+}$. The stereoselective process is controlled by the attack of the generated radical to the accessible face of the chiral enamine that depends from the chiral secondary amine used. It is quite important to mention that, among all the enamine catalysts described by the MacMillan group, the *trans* imidazolidinone **OC1** was the most effective catalyst. This catalyst is reasonable instable in solution and tends to equilibrate to the less effective *cis* form. Probably, other imidazolidinones, bearing benzylic groups in α–positions, can suffer from the generation of radicals, or formation of benzylic stabilized radicals. As the key event in this cycle is the formation of the stabilized radicals from bromo species, another PC able to form a strong reducing intermediate can be used. This strong reductant can be either formed by a PC in its excited state or can be obtained as an intermediate after the reductive quenching of the excited state.

9.3.2 Organic dyes in stereoselective alkylation of enamines

König and Zeitler reported the use of eosin Y (Figure 9.2) as PC for promoting the same reaction described by MacMillan in 2008, with similar substrate scope [25]. In the proposed mechanistic cycle of this reaction, the authors assumed that eosin Y acts as a photoredox catalyst after its excitation with visible light. Eosin Y acts as a reductant, and similarly to the ruthenium case, the sacrificial oxidation of a catalytic amount of the enamine as the initial event is assumed. The use of organic dyes in a MacMillan's imidazolidinones mediated photocatalytic alkylation was recently reported by Aleman (Figure 9.6) [26], employing the bifunctional photoaminocatalyst based on an imidazolidinone bearing a thioxanthone group **PC2**.

The preparation of these catalysts is quite straightforward. Good enantiomeric excesses and yields were obtained for the examined transformation. Contrary to the previous two examples, careful mechanistic analysis was carried out, indicating that the catalyst can work under visible light conditions. Laser flash photolysis experiments showed that the intramolecular quenching of the excited state (singlet or triplet) of the thioxanthone moiety, by the amino group in the bifunctional catalyst, is reduced after enamine formation. The thioxantone is then able to directly reduce the bromo derivatives by ET, forming a cationic species. The radical cation is then reduced by the radical intermediates formed by addition of the radical to the enamines, resulting in formation of the iminium intermediate. The calculated potential

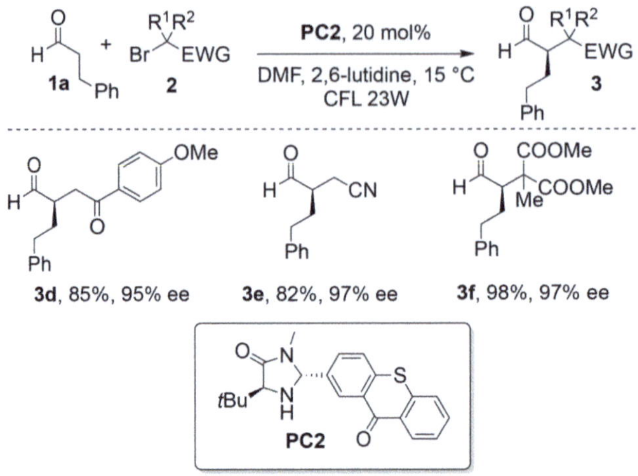

Figure 9.6: Photocatalytic stereoselective alkylation of aldehydes promoted by a imidazolidinone conjugated with an organo-photocatalyst.

$E_{1/2}{}^{ox}$ and $E_{1/2}{}^{red}$ for the designed catalysts are very interesting. The high reduction potentials allowed the use of bromonitrile species (bromoacetonitrile, $E^{red} = -1.23$ V vs. SCE in CH_3CN). A simple calculation of the energy ($\Delta G_{PET} = -2.3$ kcal/mol) guarantees the thermodynamic driving force of the process, and in fact, the reaction with nitriles was observed in high yields and excellent enantioselectivities. However, this reaction is also possible with the combination of organocatalysis and $[Ru(bpy)_3]^{2+}$ [27] as the Ru(I) species, obtained by sacrificial oxidation of enamine, has a reduction potential sufficient to induce the generation of radicals from bromonitriles. Again, the available values of potentials can give a rough idea on the thermodynamic driving force. Another strong organic reductant, a coumarin derivative that can reach $E^{red} = -1.9$ V (vs. SCE in CH_3CN) was recently reported as a PC for the alkylation of aldehydes [28]. Also in this case, the coumarin PC directly acts as the reductant of the bromo derivative and no sacrificial enamine starts the catalytic cycle. Yoon, measuring the quantum yield (molecules of product formed/number of photons absorbed) of ruthenium catalyzed process [29] obtained a value of 18. Although the photocatalytic cycle was proposed as a key feature of the reaction, the value obtained by Yoon is not compatible with cycles controlled by the PCs, indeed 18 molecules of products were formed per photon absorbed. Radical chain reactions seem to be a relevant issue in many photocatalytic processes. Careful and precise measurement of the quantum yield can give some indications to whether the process is effectively promoted by a PC. For an interesting discussion on radical chains and radicals produced under photoredox conditions, a font of inspiration can be found in a review published in Angewandte Chemie by Curran and Studer [30]. Furthermore, a recent theoretical investigation into the origin of enantioselectivity in asymmetric radical additions to the MacMillan imidazolidinone enamines was reported [31]. Interestingly, this calculation can shed some light on the role played by the E-*cis* enamine in enantiocontrol of the reaction.

9.3.3 Use of semiconductors and Earth abundant metals

The direct stereoselective α-alkylation of aldehydes with α-bromocarbonyl compounds can be obtained by using semiconductors, such as bismuth-based materials (Bi_2O_3) as low-band-gap PCs [32], in the presence of imidazolidinone **OC1** used for the above described processes. The authors did not conduct a deep photophysical analysis of the results and the suggested catalytic cycle is believed to occur via direct oxidation (oxidative quenching, Figure 9.1) of the semiconductor, that is able to directly transfer electrons from the conduction band to the bromo derivative. The low abundance and high cost of iridium and ruthenium as well as their use in many electronic devices, demands for new and effective PCs. Well-structured and characterized complexes of abundant earth metals can behave effectively as PC, as many of these complexes have excited-states reachable by visible light irradiation and with lifetimes between 10 ps and 1 ms in solution. New PCs based on Cr, Mn, Fe, Co, Ni, Cu, Zn, Zr, Mo, W and Ce in various oxidation states surrounded with diverse ligands, have been recently studied and applied to established photoredox processes [33]. As iron is an abundant metal, cheap and that presents low toxicity, the search for PCs based on Fe(II) complexes as $3d^6$ analogues of the emissive $(Ru[\alpha\text{-diimine}]_3)^{2+}$ compounds is a hot topic. The principal problems related to iron complexes are due to the non-emissive character and low lifetime of its excited state. For example, the lifetime of excited $[Fe(bpy)_3]^{2+}$ is about hundreds of picoseconds [34]. Surprisingly, $[Fe(bpy)_3]^{2+}$ is an effective and compelling PC for α-alkylation of aldehydes, in the presence of the imidazolidinone **OC1** (Figure 9.7) [35].

Visible light and $[Fe(bpy)_3]Br_2$ were both necessary to drive the reaction to completion. EPR studies with a radical trap evidenced the formation of alkyl radicals from bromo derivatives used in the reaction. A series of compounds was obtained with yields and enantiomeric excesses comparable to those obtained with the before mentioned methodologies. Furthermore, the method was applied to the synthesis of (–)-isodehydroxypodophyllotoxin. A radical chain mechanism was proposed for the reaction, with the iron photosensitizer capable of promoting, upon excitation, a chain radical reaction in which the photochemical event is only the starting step. The photophysical process was studied carefully by femtosecond laser absorption spectroscopy, showing the lifetime of the excited states to be around 600 ps. By evidence obtained with Stern–Volmer experiments, the $[Fe(bpy)_3]Br_2$ photosensitizer acts as a reductant for initiating the chain mechanism. Remarkably, although similar, the iron based PC is substantially different from the ruthenium cases reported, as the PC in the excited state is able to directly generate the radical species from bromo derivative, without the assistance of the enamine to generate the catalytic active

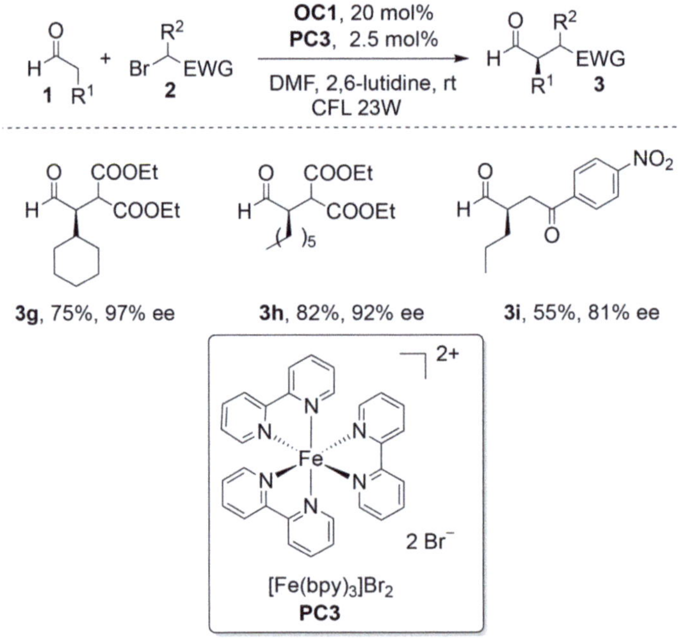

Figure 9.7: Photocatalytic stereoselective alkylation of aldehydes promoted by $[Fe(bpy)_3Br_2]$ catalyst.

species. However, the quite short life of the PC-excited state poses an intriguing problem, as it is significantly shorter than the diffusion rate, which is the limit in encountering processes in solution. Probably, further work is necessary in order to better understand the potential of these PCs, investigating if other processes induced by light (spin transitions) are responsible for the observed reactivity. In addition to the use of abundant iron complexes, commercially available aluminum based luminescent complexes are also quite effective PC for the catalytic reactions [36].

9.3.4 Iridium complexes in alkylation of enamines

One of the problems associated with this example of synergistic catalysis is the limited scope of the reaction. In fact, just to mention, benzyl bromides were found completely unreactive. This is not surprising, as the limitation is determined by the redox potential of the PC in its excited or reduced states. In the case of $[Ru(bpy)_3]^{2+}$, the produced Ru(I) is able to reduce organic molecules that have a potential greater or equal to ca. −1.3 V. In order to favor the reaction a stronger photoreductant is necessary. MacMillan was able to show that the alkylation of aldehydes with benzylic bromides is possible using iridium complexes. These complexes are cyclometallated iridium(ppy) (III), (ppy = 2-phenylpyridine) based and are commonly used as green triplet emitters in OLEDs. Their excited states are readily reached by visible light to generate the strong reductant specie *fac*-*Ir(ppy)$_3$ ($E_{1/2}$ = −1.73 V vs SCE in CH_3CN). Unfortunately, the reaction was ineffective with benzyl bromide and, only the introduction of EWG to lower the reduction potential, allowed the alkylation to take place. Anyway, interesting compounds **5** could be obtained using this methodology (Figure 9.8) [37].

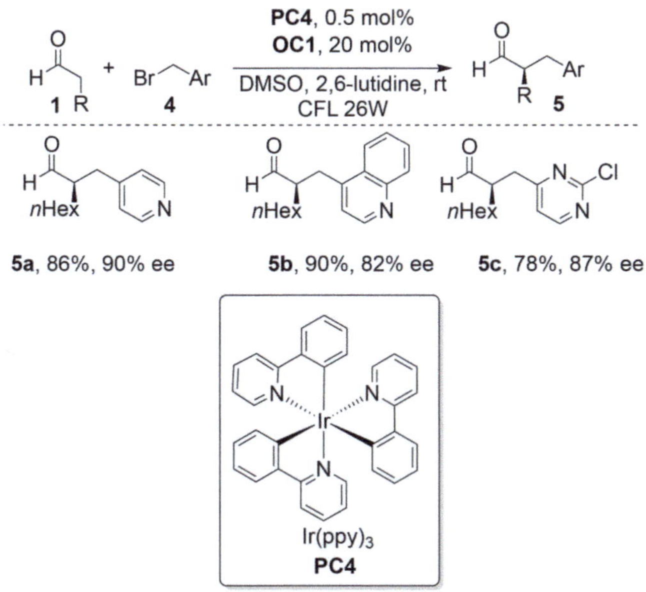

Figure 9.8: Photocatalytic stereoselective alkylation of aldehydes with benzylic bromides promoted by an iridium (III) catalyst.

Stern–Volmer experiments were performed to provide insight into the mechanistic details of the reaction. While *fac*-*Ir(ppy)$_3$ oxidation of the enamine (Figure 9.8) is not observed, aryl and heteroaryl methylene bromides **4** were able to display efficient quenching of the PC, suggesting that the electron transfer event occurs between the PC and organic halides, as shown in the illustrated cycle. The organocatalytic α-trifluoromethylation and α-perfluoroalkylation of aldehydes were reported by MacMillan, by using the imidazolidinone catalyst and an iridium (III) complex [38].

Although Ir(ppy)$_3$ was able to catalyze the reaction, the authors found that another Ir(III) complex, Ir(ppy)$_2$(dtbbpy)$^+$ (dtbbpy = 4,4′-di-*tert*-butyl-2,2′-bipyridine), that has a reduction potential in its reduced Ir(II) state of −1.51 V (vs. SCE in CH_3CN), was more effective. In fact, a SET with trifluoromethyl iodide (−1.22 V vs. SCE in DMF) is favored if the reduction potential of the species is more negative. The asymmetric trifluoromethylation was compatible with a wide range of functional groups including ethers, esters, amines, carbamates, and aromatic rings (61–86 % yield, 93–98 % ee), and the use of hindered aldehydes was also possible. The catalytic cycle started with reduction of the *Ir(III) complex from the enamine generating the strong reductant Ir(II) species able to produce the CF_3 radical from CF_3I. After the addition of the radical to the enamine, the α-aminyl radical obtained is the reductant for Ir(III) in its excited state, restoring and maintaining the catalytic cycle. As pointed out, the catalyst behaves like $[Ru(bpy)_3]^{2+}$ in the photocatalytic alkylation of aldehydes (Figure 9.5). The ruthenium PC could not be employed in this photoredox reaction due to an inferior (–1.33 vs. –1.55, vs SCE) reduction potential.

9.3.5 Stereoselective photoredox catalysis and HAT

The synergistic merger of three different catalytic processes (photoredox, enamine and hydrogen-atom transfer) was shown by MacMillan being quite effective and allows the direct alkylation of aldehydes with aryl alkenes **6** (Figure 9.9) [39].

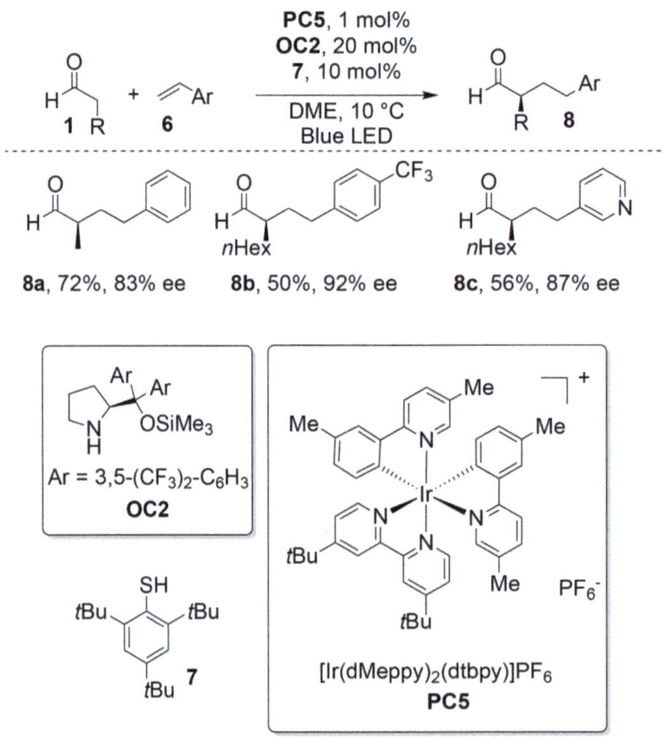

Figure 9.9: Photocatalytic stereoselective alkylation of aldehydes with alkene by hydrogen-atom transfer (HAT) catalysis.

The reaction uses a chiral imidazolidinones or prolinols as the organocatalysts, in combination with a hindered thiophenol **7** and the iridium photoredox catalyst **PC5**. The process needs visible light, through a complex series of independent events connected by three catalytic cycles. The iridium PC used in the process is Ir(III)(Fmppy)$_2$(dtbbpy)PF$_6$ [Fmppy = 2-(4-fluorophenyl)-4-(methylpyridine), dtbbpy = 4,4′-di-*tert*-butyl-2,2′-bipyridine] (**PC5**), able to absorb visible light, with a long-lived triplet (τ = 1.2 µs) excited-state. The PC is assumed to behave as an oxidant, ($E_{1/2}$ [*Ir(III)/Ir(II)] = + 0.77 V vs. SCE), capable of oxidizing the enamine. This electron transfer reaction produces the Ir(II) complex and the enaminyl radical species. The enamine radical species is an electrophilic radical, capable of intercepting in a stereoselective manner the alkene present in solution. This C–C bond forming reaction generates another nucleophilic radical species. This species is therefore able to react in a radical manner with the electrophilic species. The hindered thiol added to the reaction mixture is this specie, and the S-H bond now reacts by a HAT (hydrogen atom transfer) mechanism forming a thiol radical. The sterical hindrance of the thiol is crucial to stabilize the thiyl radical and allows an electron transfer from Ir(II). This electron transfer is highly favorable ($E_{1/2}$[PhS•/PhS$^-$] = + 0.02 V vs. SCE, DMSO) and regenerates both the ground state Ir(III) **PC5** as well as the HAT thiol catalyst, after thiolate protonation. Another remarkable process in which a proton and a spin transfer are operative is the alkylation of aldehydes, in the presence of organocatalyst with benzylic alcohols **9** derived from pyridine and quinolines (Figure 9.10) [40].

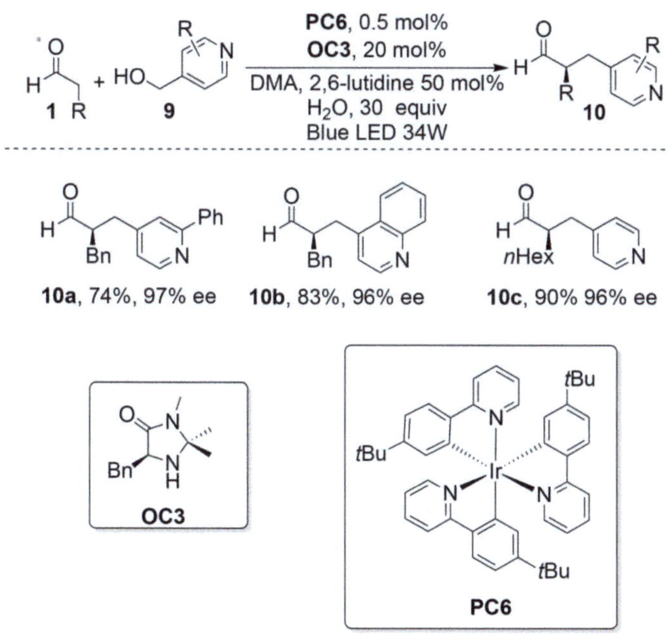

Figure 9.10: Photocatalytic stereoselective alkylation of aldehydes with methanol azines.

The single-electron reduction of a pyridinemethanol can be obtained by a photoredox catalyst, as pyridine is an electron withdrawing group. The HAT process and the spin rearrangement of the obtained species give a benzylic C–O bond cleavage, facilitated by the protonation of the alcohol intermediate. The electron poor radical, produced by the rearrangement, is able to react with electron rich enamines in a stereoselective manner. The strong reducing PC in its excited state Ir(III)(ppy)$_3$ was employed in the reaction. The formed Ir(IV) intermediate **PC6** ($E_{1/2}$[Ir(IV)/(III)] = + 0.77 V vs. SCE in CH_3CN) is capable of oxidizing the resulting α-amino radical (Figure 9.5) $E_{1/2}$= −1.12 to −0.92 V vs. SCE in CH_3CN), restoring the PC to the ground state.

9.3.6 Stereoselective formation of C–N bonds with photoredox enamine catalysis

The enantioselective construction of C–N bonds for the synthesis of key intermediates or natural products is an important goal in modern stereoselective catalysis. Photoredox stereoselective catalysis can also be applied to form C–N bonds. In particular, as α-amino aldehydes are quite useful starting materials, the approach to asymmetric α-amination of aldehydes by photoredox enamine catalysis, can solve some problematic issues related to the introduction of nitrogen on acidic substrates. In this perspective, MacMillan has reported the formation of N-centered radicals that undergo enantioselective α-addition to chiral enamines (Figure 9.11) [41].

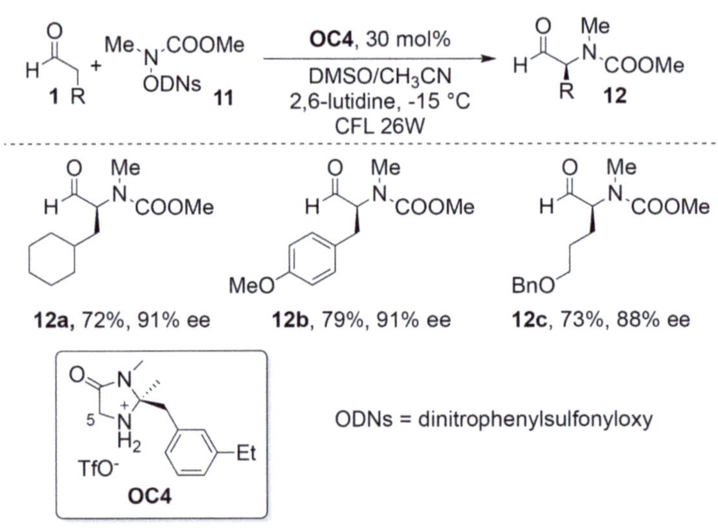

Figure 9.11: Photocatalytic stereoselective amination of aldehydes with a tailored organocatalyst.

By absorption of light from amine **11**, it is possible to generate a N-based radical under mild conditions. The cycle starts with the generation of an amynyl radical, induced by

UV light. The light source for the reaction is a CFL (Compact Fluorescent Lamp), and although the lamp has normally a broad emission (white light) some UV components are also emitted. The aminyl radical formed reacts with the enamine giving rise to the formation of a α-amino radical species. This species is oxidized by a SET process that involves the amino reagent **11** in its excited state. The scope of the reaction is broad, and many functional groups present on the aldehydes (ethers, amines, alkenes, and aromatic rings) are tolerated, with products isolated in 71–79 % yields, and 88–91 % ees. It is also quite important to mention that a new type of imidazolidinone catalyst (**OC4**) was introduced for the transformation. A chiral imidazolidinone catalyst less hindered in position 5 was considered important. Careful analysis of the remaining catalyst clarified that the imidazolidinone was not very stable under the reaction conditions as it underwent H radical abstraction at the aminal C2 position. Therefore, the need to reduce sterical hindrance and eliminate the undesired decomposition pathway led to the design of a novel organocatalyst with a fully substituted carbon stereocenter at the aminal positions The trial and error design was justified by density functional theory (DFT) of the corresponding enamine intermediate, with the position of the electron-rich enamine away from the fully substituted carbon center on the imidazolidinone framework. The presence of a *meta* substituent and the orientation of the arene group are important to effectively cover the enamine stereoface.

9.3.7 Other stereoselective processes with enamine catalysis

The photocatalytic stereoselective α-alkylation of β-ketocarbonyls **13** by synergistic photoredox catalysis with enamine catalysis was described (Figure 9.12) [42].

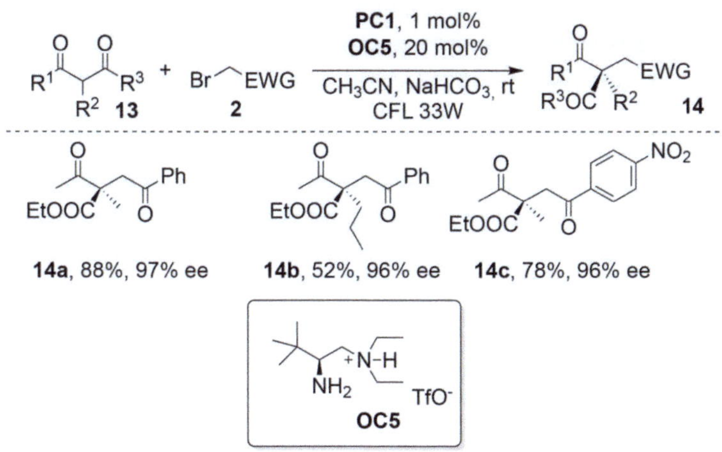

Figure 9.12: Photocatalytic stereoselective alkylation of ketoesters with a chiral primary amine organocatalyst.

In this example all-carbon quaternary stereocenter with high enantioselectivity was prepared from acyclic and cyclic ketoesters. Acyclic acetoacetates bearing either steric bulky *tert*-butyl or benzyl ester groups are useful substrates giving products in high yields and excellent enantioselectivities. In addition, the described catalysis is quite successful with cyclic β-ketoesters. Photophysical studies were not reported, as the reaction employed $[Ru(bpy)_3]^{2+}$ as the PC and the reaction mechanism is quite similar to the previously discussed proposal by MacMillan. The organocatalyst **OC5** needs a brief comment. As the formation of ketone enamine is involved, the catalysis by primary amine is quite crucial [43], in order to reach a sufficient concentration of enamine able to quench the excited state and interact with the formed radical.

9.4 Stereoselective alkylation of enamines through EDA

One of the main issues related to stereoselective photoredox processes, discussed in the previous sections, was the demand of a PC (either a metal complex, semiconductor, or organic molecule) able to intercept light, reaching the excited state suitable for the electron transfer event. As chiral enamines are strong donors and enamines are the key catalyst for the stereoselective formation of C–C bonds, it would be possible to induce photon-absorption by chiral electron donor – acceptor (EDA) complexes. The energy necessary to favor the electron transfer between the two partners is then furnished by the absorbed light. The strong radical couple formed could then collapse trough radical-radical coupling, forming a new bond in a stereoselective manner. EDA complexes are well known and reported in literature [43]. Also the ability of tertiary amines to form EDA complexes with electron-accepting molecules of high electron affinity is well established [44]. In enamines, the nitrogen lone pair is suitable to form EDA complexes with an acceptor compound or an electron poor compound. This behavior can easily be confirmed, as EDA complexes are characterized by the appearance of a weak absorption band, called the charge-transfer band, associated with an ET transition from donor to acceptor [45]. In many cases, the absorption is visible to the naked eye, with the appearance of a colored solution. This means that the energy associated with this absorption is within the visible wavelength range. If it would be possible to produce an EDA complex using a chiral enamine, the ET event can drive the stereoselective α-alkylation of aldehydes. This possibility and the processes associated with a rational design of EDA complexes in enamine catalysis were introduced in literature by Melchiorre (Figure 9.13(a)) [46].

The alkylation was realized using suitable alkyl halides **4**, such as benzyl bromides carrying a nitro group on the aromatic ring. They have – a low reduction potential and they are capable of forming EDA complexes with enamines, by a n-π* type interaction [47]. Interestingly, the commercially available diarylprolinol silylether catalyst **OC6** was selected. This selection agreed with the better nucleophilic properties of

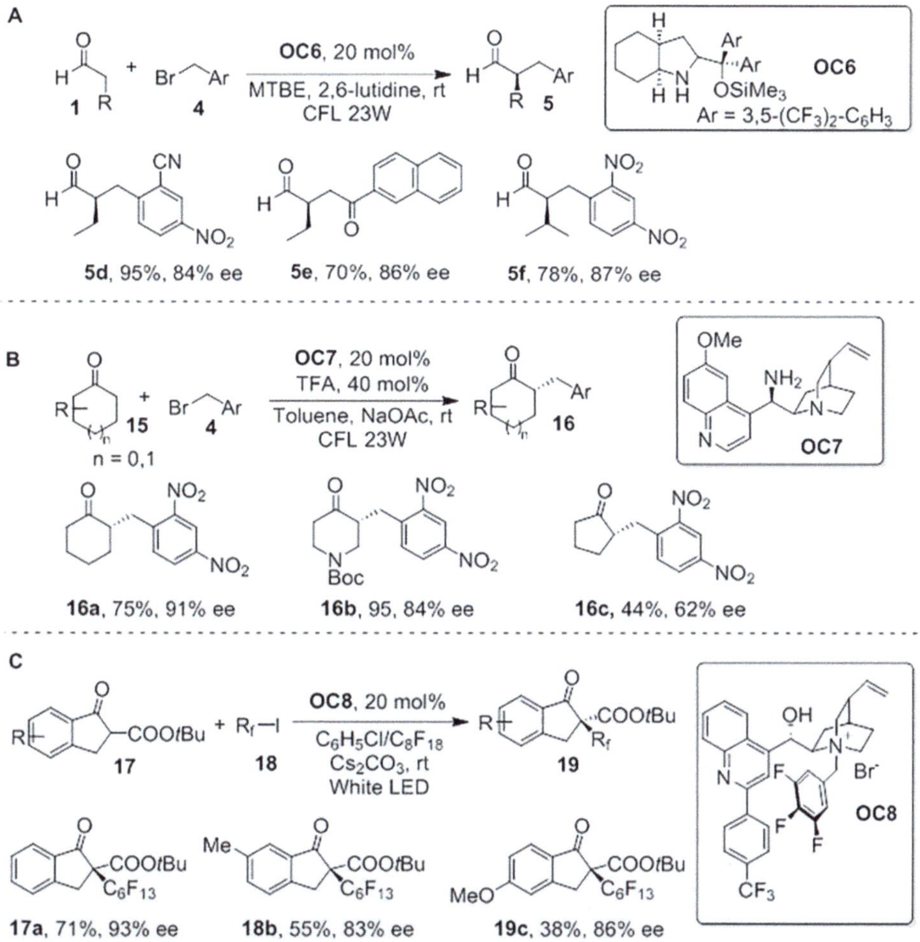

Figure 9.13: Examples of EDA mediated organocatalytic stereoselective alkylation.

the enamines derived from Hayashi–Jørgensen catalyst **OC2**. In addition, the pyramidalization of the nitrogen atom in the MacMillan catalysts [21] can reduce the availability of the nitrogen lone pair.

For the stereoselective reaction, a more hindered Hayashi–Jørgensen type of catalyst was prepared, and with this catalyst, good yields, broad scope and high stereoselectivities were observed. Not only benzyl bromides bearing electron-withdrawing groups were useful substrates, but also aromatic and heteroaromatic α-bromo ketones. The reaction also proceeded using branched aldehydes and ketones [48] in case primary amine obtained from the cinchona alkaloid **OC7** was used as organocatalyst (Figure 9.13(b)). The photochemical enantioselective perfluoroalkylation of β-ketoesters was developed using a similar concept (Figure 9.13(c)) [49]. An EDA complex arises from the association of a perfluoroalkyl iodide **18** and the chiral enolate of the ketoester. The latter was generated by phase transfer catalysis (PTC) [50], where the phase transfer catalyst is a chiral quaternary ammonium salt **OC8**, commercially available or obtained in a in straigthforward manner from a cinchona alkaloid. The reaction proceeded in the presence of Cs_2CO_3 (2 equiv) and under irradiation with white LEDs at room temperature for 16 h in good yields and high enantiomeric excesses. Due to the scarce solubility of perfluorinated compound in organic solvents, a mixture chlorobenzene/perfluorooctane (2:1 ratio) was used as the reaction solvent to allow the reaction to proceed. Investigation on a model capable of explaining the stereochemical outcome of the reaction, by DFT calculations, was reported by Li [51]. The calculations showed that stereoinduction was not determined by π–π stacking interaction between the cinchona catalyst and the ketoester, but rather multiple hydrogen-bonding interactions play a decisive role.

An interesting formation of quaternary stereogenic centers by EDA process, in a photoinduced Michael type reaction, was reported quite recently by Melchiorre (Figure 9.14) [52] illustrating the concepts of careful design of a photoactive organocatalyst [53].

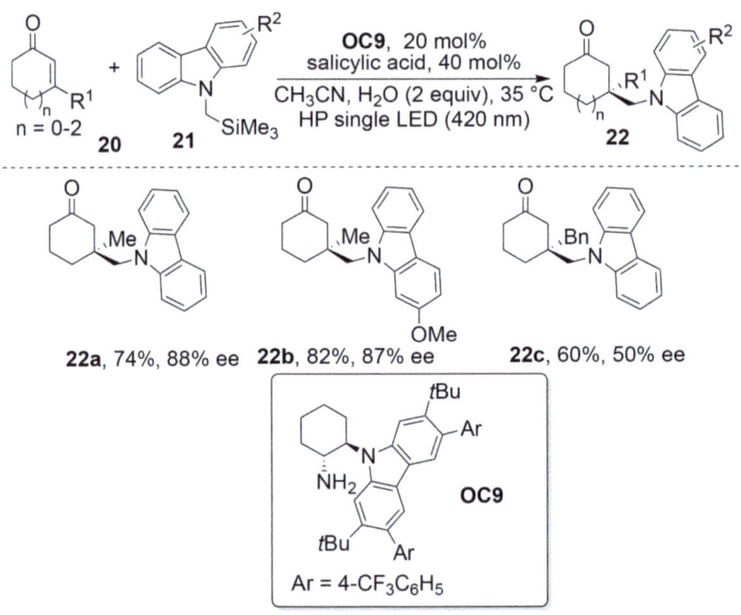

Figure 9.14: Formation of quaternary stereogenic centers by photoinduced type Michael EDA.

The strategy is based on the formation of an intramolecular EDA complex, capable of triggering a photochemical catalytic enantioselective radical process. A radical conjugate addition to β-substituted cyclic enones, challenging substrates for Michael reactions, was described. Although the scope is limited to certain substrates, the chiral iminium ion formed in these reactions testify to the possibility to expand the repertoire of photocatalytic processes using both iminium and enamine catalysis. As a result of conjugation and the presence of a carbazole moiety in the design of the organocatalyst **OC9**, the chiral iminium ions formed show a broad absorption band in the visible region. This absorption band originates from an intramolecular charge transfer π–π interaction between the electron-rich carbazole and the electron-deficient iminium double bond. The catalyst is able to transfer an electron to the double bond, by irradiation, forming a radical couple. Crucial to the design of the catalyst is the carbazole moiety, as the long-lived carbazole radical cation formed by electron transfer is a persistent species and can behave as an oxidant. After formation of the radical couple, the carbazole radical cation can oxidize suitable species forming the corresponding radical that undergoes the coupling with the enaminyl radical intermediate.

9.5 Stereoselective alkylation through iminium intermediate and a PC

The stereoselective Michael-type conjugate addition to unsaturated carbonyl substrates is a powerful methodology to access interesting molecules and introduce a variety of moietes. Remarkably, stereoselective radical additions were also investigated using chiral auxiliaries or ligands, to provide a good level of stereoselection [54]. As photoredox catalysis can be used to access radicals and iminium catalysis is a well-established activation mode in organocatalytic methodologies [55], it could beg the question if the effective combination of photocatalytic production of radicals with the effective iminium activation mode, can give powerful and selective methodologies. MacMillan in 2013 reported an elegant strategy for the β-arylation of carbonyl compounds by combining photoredox and organocatalysis using a chiral amine derived from cinchona alkaloid **OC10** and cyclohexanone (Figure 9.15(a)) [56].

However, the enantioselectivity was only moderate (50% ee), and no substrate scope was reported. This limitation is due to the fast ET that is occurs by the generated short-lived, highly reactive α-iminyl radical cation, which has a high tendency to undergo radical elimination (β-scission) to re-form the more stable iminium ion. Probably, for this reason, the stereoselective photoredox iminium-type reactions were underdeveloped. However, moving towards the formation of quaternary centers and providing a better stabilization/protection of the radical, Melchiorre [57] (Figure 9.15(b)) reported a remarkable Michael type addition of benzodioxole **25** to β-alkylated cyclohexenones **20**. In order to allow the reaction to proceed , different design elements were considered. Due to the high reactivity of the α-iminyl radical cation, a very rapid SET reduction is necessary to stabilize the adduct as an enamine adduct before β-scission occurs to re-form the more stable iminium ion. An electron rich carbazole moiety, capable of ET, was inserted in the chiral primary amine used as the

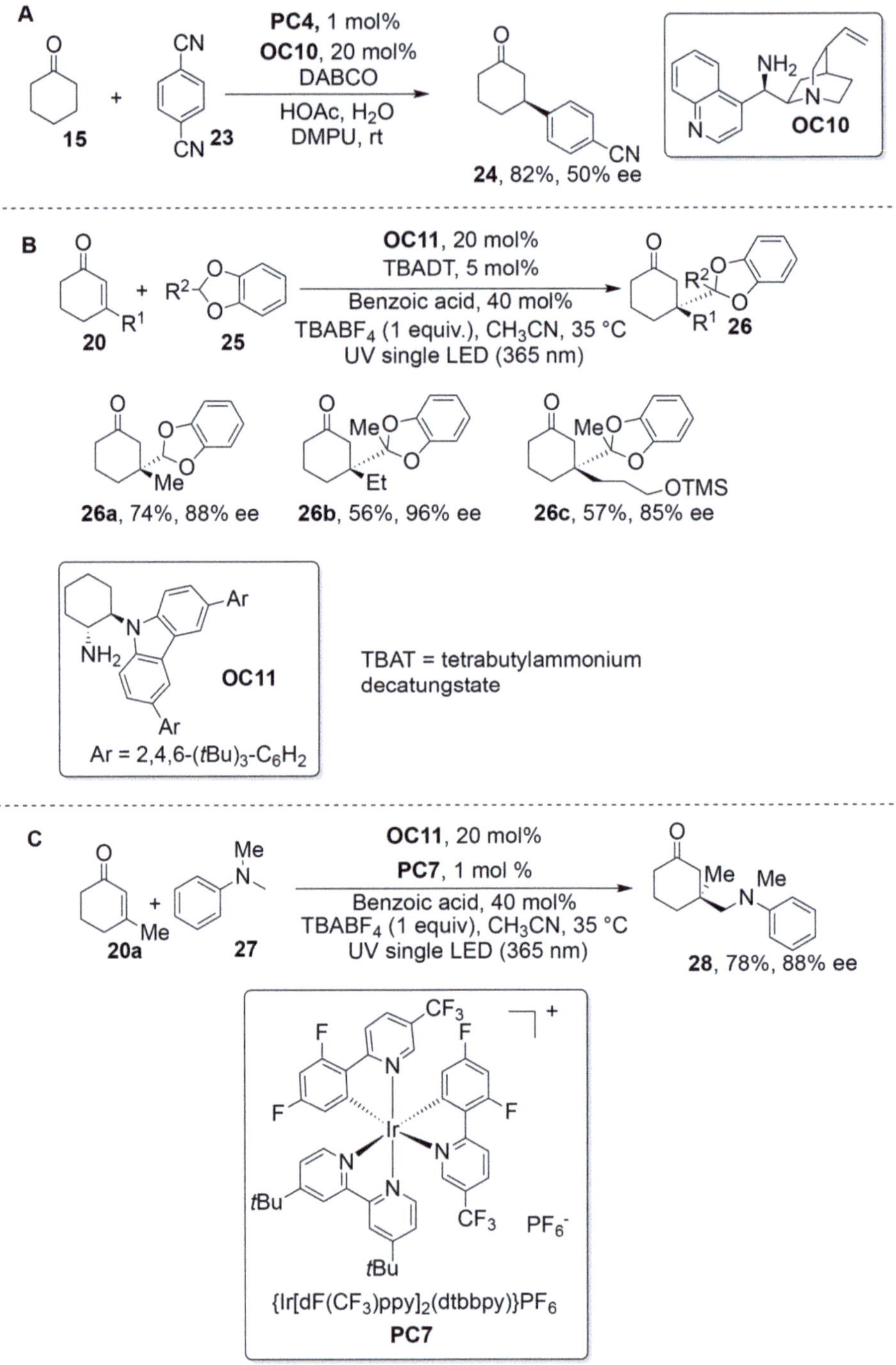

Figure 9.15: Michael and arylation in β-position of a carbonyls by organo-photoredox catalysis.

catalyst, to favor iminium formation. Finally, in order to generate the benzodioxole radical, a PC capable of abstracting a hydrogen atom, tetrabutylammonium decatungstate (TBADT, 5 mol%) was employed. However, this catalyst needs a (UV)-light emitting diode (UV LED, $\lambda_{max} = 365$ nm) to be excited. Five, six and seven membered ketones, with different alkyl substituent in the β positions gave the observed products, in good yields and moderate to good selectivity. The photoredox reactions can also be promoted by **PC7** Ir[dF(CF_3)ppy]$_2$(dtbbpy)PF_6, that was used (Figure 9.15(c)) to generate α-amino radicals directly from tertiary amines. For these reactions, due to the more favorable photophysical properties of Ir(III) catalyst, a blue LED was employed. The combination of electrochemical, spectroscopic, computational, and kinetic studies was also employed to investigate the reaction mechanism of the reaction [58]. Another interesting process that coupled stereoselective iminium catalysis with a photoredox event is the photoenolization/Diels – Alder sequence, to produce benzannulated carbocyclic molecule (Figure 9.16) [59].

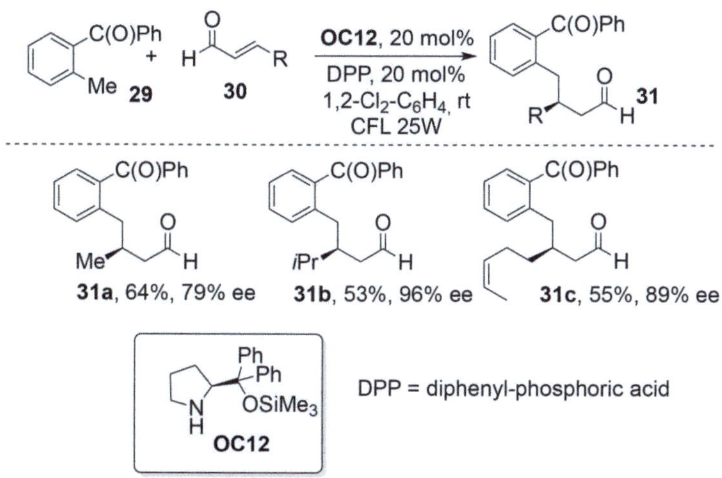

Figure 9.16: Photoenolization/Diels–Alder sequence for the synthesis of benzannulated carbocyclic molecules.

This approach exploits the light-triggered enolization of 2-alkyl-benzophenones **29** forming in situ the transient hydroxy-*o*-quinodinomethanes, highly reactive intermediates that can behave as a Michael donor towards electron-poor alkenes. If the electron poor alkene is a chiral iminium, a conjugate stereoselective addition is observed. This chemistry provides a straightforward method for the direct β-benzylation of enals. Using the optimized reaction conditions, a variety of diversely substituted enals can be used for the reaction and sterically hindered groups are also tolerated. In order to perform the reaction, and allow the formation of the key intermediate, a BLB (black light bulb $\lambda_{max} = 365$ nm) was employed. Beside enals, also 3-substituted-

2-cyclohexenones can be employed [60]. Since ketones were used in the reaction, chiral primary amines derived from amino acids were employed as organocatalysts, to promote the formation of the chiral iminium ion.

9.6 Stereoselective alkylation through enamine as the PC

An interesting feature of the enamines is their ability to reach an electronically excited state upon light absorption. As the other organic species present in the reaction mixture are not capable to absorb light at wavelength where enamines are absorbing, the formation of catalytic amount of enamine can possibility provide the in situ PC preparation, without the need of adding other organic molecules or metals that behave as PC. In addition, reactive radicals formed by the photocatalytic event could intercepted in a stereoselective fashion by the chiral enamine. This general activation mode, that can be added to the list of enamine catalysis, was first reported by Melchiorre (Figure 9.17(a)) [61].

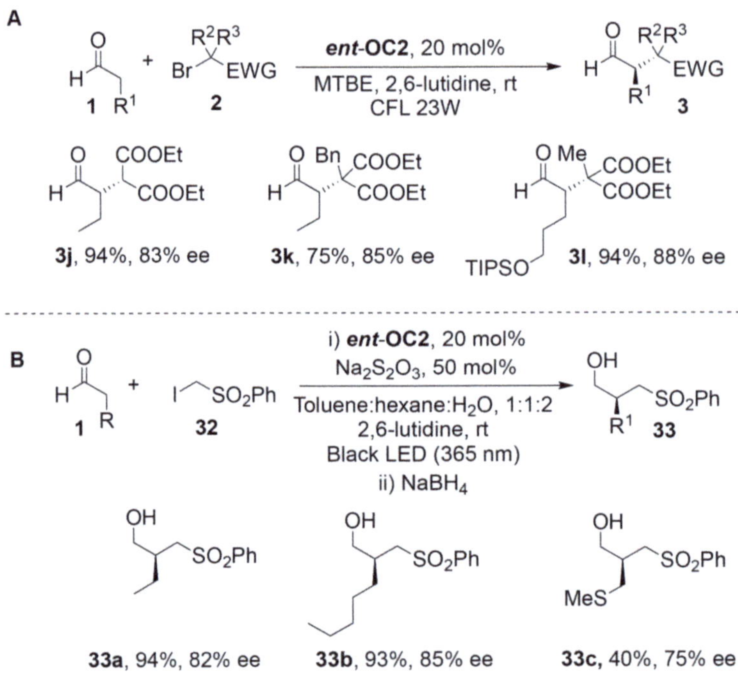

Figure 9.17: Stereoselective alkylation through enamine as the photocatalyst.

Quite remarkably, the enamine used for the direct photocatalytic alkylation is based on the Hayashi–Jørgensen catalysts ***ent*-OC2** probably because of their nucleophilicity.

The absorption and emission of light by the enamine prepared from condensation of the chiral Hayashi-Jørgensen catalyst with 2-phenylacetaldehyde was demonstrated [61]. Furthermore, the Stern–Volmer quenching experiment performed between chiral enamine and bromomalonate indicated the electron transfer between the two. Based on this finding, a radical chain mechanism was proposed, in which starting radicals are formed by the enamine PC. Remarkably, the calculated redox potential of the enamine in its excited state was estimated to be −2.50 V (vs Ag/AgCl, NaCl sat.) based on electrochemical and spectroscopic measurements. Differently substituted bromomalonates can be used for the enantioselective alkylation and a variety of aldehydes bearing diverse functional groups are compatible. A substantial difference is observed in the reaction conditions employed. Normally, photocatalytic alkylation of aldehydes using the MacMillan catalyst in the presence of ruthenium or iridium are conducted in DMF, while in the current case, *tert*-butyl methyl ether (MTBE) is employed. Contrary to the reactions catalyzed by ruthenium and iridium complexes, the use of a CFL is necessary due to the absorption of enamines in the near UV region. It is also possible to apply this chemistry to dienamine catalysis [62] by introducing stereocenters in remote γ-position. What is also important to mention, is that a comprehensive and authoritative mechanistic analysis of the process, including kinetic studies and full photochemical investigations, was published [63]. The high reduction potential of enamines in their excited states can be exploited to provide a general stereocontrolled α-alkylation of carbonyl compounds. Alkyl halides could not be employed as electrophiles in enamine catalysis, apart from rare examples of S_N2-type alkylation [64]. Only one effective procedure was published by List, limited to branched aldehydes with benzyl bromide derivatives [65]. More successful methodologies for the alkylation of aldehydes are based on S_N1-type reactions [66]. An organo-photocatalytic methodology to install either a methyl or a benzyl moiety onto simple aldehydes with high stereoselectivity was reported by Melchiorre (Figure 9.17(b)) [67]. Because of the ability of chiral enamines to be strong reductant in their excited states, SET reduction of the (phenylsulfonyl)alkyl iodides **32** gave (phenylsulfonyl)alkylated intermediates that can easily be desulfonylated. As the reduction potential of iodosulfones is −1.49 V ($E_{1/2}$ vs. Ag/Ag + in CH_3CN), the reaction is thermodynamically favored. The desulfonation is performed by activated magnesium in anhydrous MeOH.

9.7 Stereoselective reactions of iminium ions as the PC

As enamines can directly absorb light and become stronger reductants, conversely, iminium ions in their excited states can behave as strong oxidants. Starting from suitable precursors, iminium ions could be used in effective stereoselective Michael reactions. This process is also observed in nature where iminium ions are involved in visual processing invertebrates responsible for light excitation of the formed iminium ion from condensation of 11-*cis*-retinal with the ε-amino group of a lysine residue occurs within opsins [68]. This strategy was implemented by Melchiorre, (Figure 9.18) detailing the stereoselective addition of benzylic substrates **35** to cinnamyl aldehydes **34** [69].

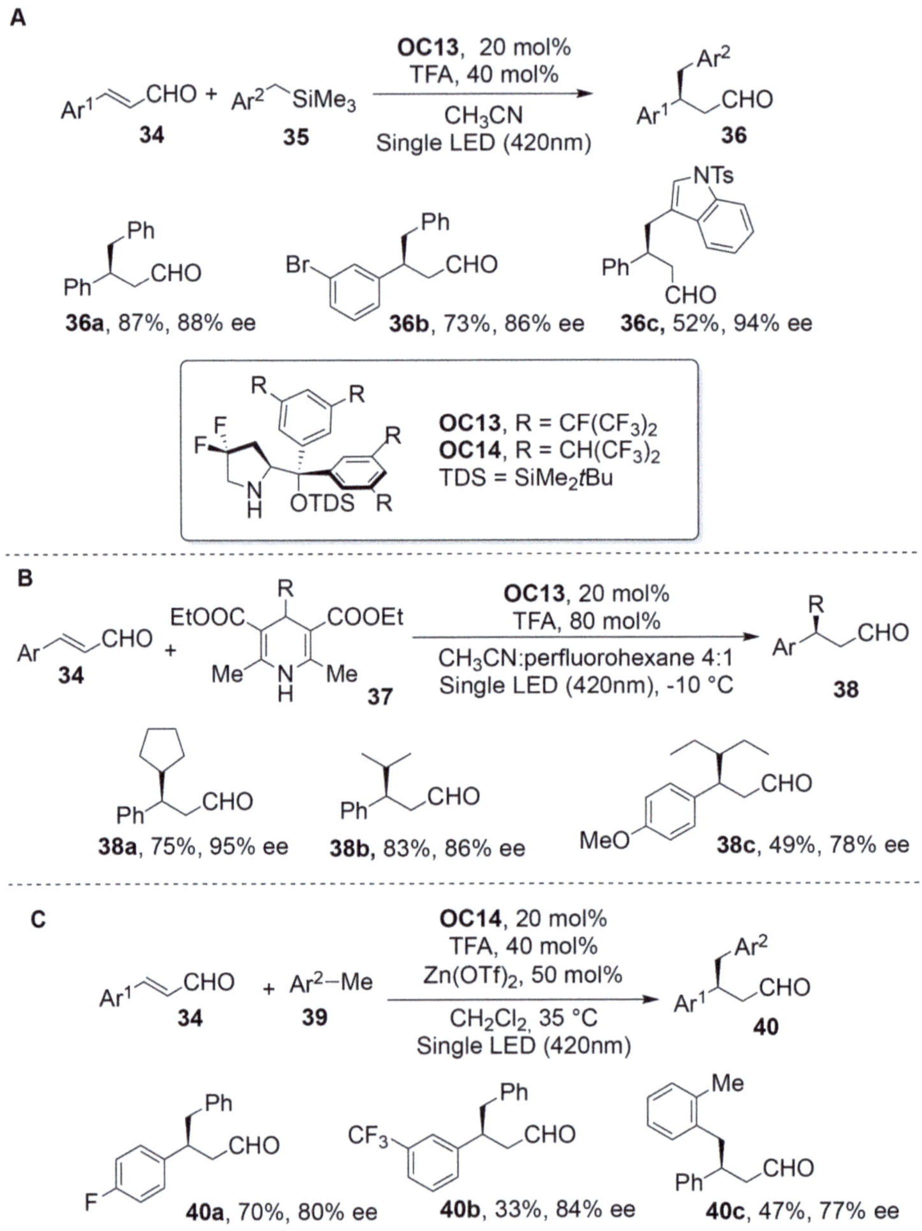

Figure 9.18: Stereoselective addition of benzylic substrates to cinnamyl aldehydes.

The potential of the excited iminium ion ($E_{1/2}$ [*Ia/Ia^-], was estimated to be + 2.3 V [vs. Ag/Ag^+]) and therefore the oxidation of benzylic trimethylsilane derivative **35** ($E_{1/2}$ = + 1.74 vs. Ag/Ag^+ for benzyl trimethylsilane) is a thermodynamically favored process. As iminium ions are capable of absorbing visible light, the excitation was performed using blue LEDs (λ_{max} = 420 nm). Stern–Volmer experiments were able to demonstrate that benzylsilane effectively quenched the excited state of the iminium ion. The reaction scope was investigated with different cinnamyl derivatives as well as benzylic silanes. Unsaturated aliphatic aldehydes did not undergo any reaction, probably due to the low molar extinction coefficient of the corresponding iminium ions. Importantly the iminium ion in its excited state reaches a very high oxidation potential. This strong oxidant is also capable of decomposing the catalyst in solution. A modified diarylprolinol catalyst in which fluorine atoms were introduced to guarantee stability towards oxidations was prepared and evaluated. The *gem*-difluorinated catalyst retains strong oxidation properties in its excited state, ($E_{1/2}$ [I^+·/ *I] = + 2.20 V vs. Ag/Ag^+ in CH_3CN) and proved to be a productive catalyst. Several β-benzylated aldehydes were obtained in high yields and stereoselectivities through this methodology. In order to introduce alkyl groups in cinnamyl aldehydes, other suitable radical precursors were investigated, identifying the alternative precursors **37**. By using the *gem*-difluorinated diarylprolinol silyl ether catalyst **OC13** in the formation of chiral iminium ions, the reaction was effective in the presence of potassium alkyl tetrafluoroborates and substituted dihydropyridine **37**. Due to the higher solubility and easy preparation of substituted dihydropyridine, the reaction was described with these substrates (Figure 9.18(b)) [70]. In order to improve the stereoselectivity of the reaction, a new chiral amine bearing bulkier perfluoro-isopropyl groups on the arene scaffold was prepared. The new alkylation methodology was general and was investigated with different 4-alkyl-1,4-dihydropyridines, both substituted with linear or cyclic fragments. The enantiomeric excesses obtained for the reaction were all under 90 % with yields from moderate to good. More interesting is the possibility to directly form radicals by oxidation of cheap commercially available substrates. Because of the strong oxidation power of the iminium ion in its excited state, an organocatalytic direct C–H functionalization of toluene with Michael acceptors was developed (Figure 9.18(c)) [71]. The iminium ion in its excited state can trigger the formation of radicals through SET oxidation of inactivated benzyl substrates **39**. In fact, as redox properties of toluene can be evaluated ($E_{1/2}$ = + 2.26 V vs. SCE), an oxidation of toluene mediated by the iminium acting as *PC seems feasible. The radical cation obtained by oxidations of benzylic C(sp^3)–H with an enhanced acidity (pK_a of the radical cation is estimated as–13 in CH_3CN) and the presence of a weak basic anion could generate the corresponding benzylic radical allowing the photochemical alkylation to take place. The authors, in order to facilitate the formation of iminium, and, at the same time, introduce a non-coordinating weak anion, used zinc triflates in sub-stochiometric amount. With the optimized conditions, the reaction was shown to be general and suitable to introduce several benzylic derivatives directly when 10 equivalents of the toluene derivative are employed. The reaction was again limited to the employment of cinnamyl aldehydes, and the enantiomeric excesses obtained were in all cases below 84%. However, the reaction tolerates the presence of functional groups in the benzylic position.

Iminium-enamine catalysis can be a suitably coupled for cascade reactions [72]. Radical cascade processes can easily be realized exploiting the capabilities of iminium ions in their excited states. Melchiorre described a two sequential radical-based bond-forming events, using inactivated olefins and α,β-unsaturated aldehydes to provide chiral products in a single step (Figure 9.19(a)) [73].

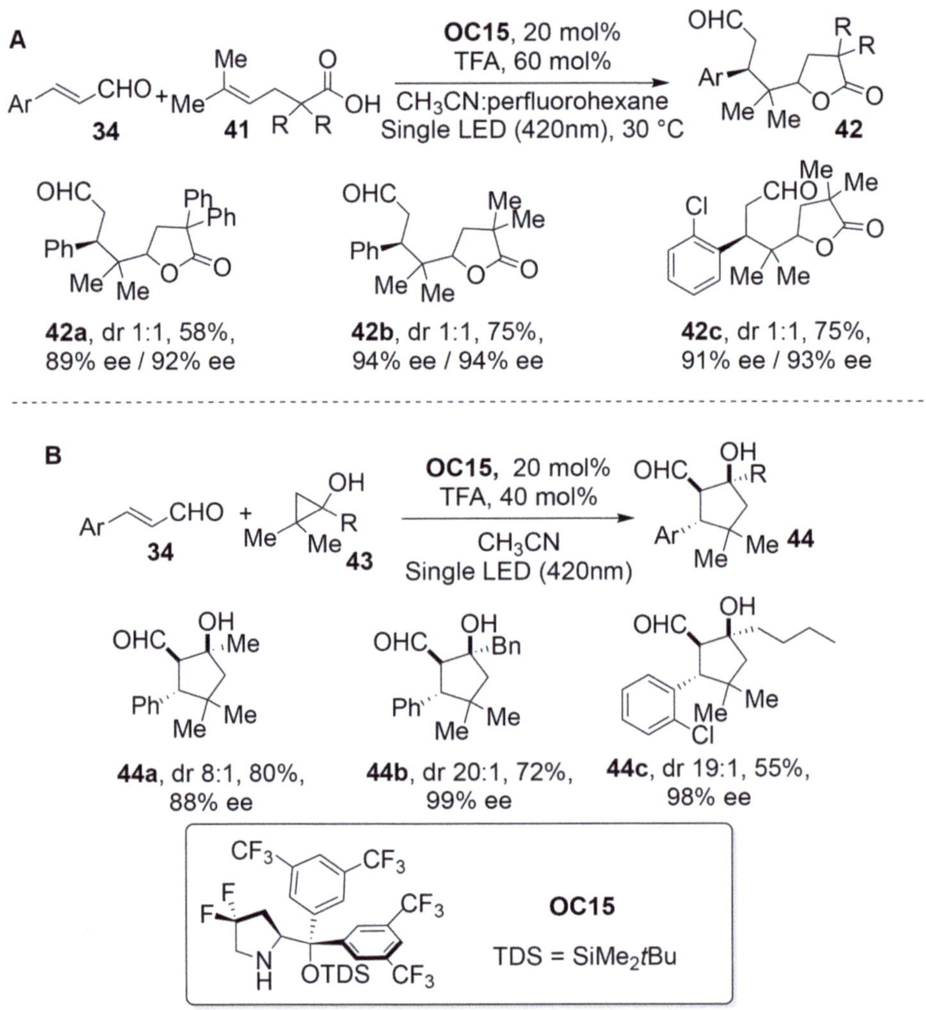

Figure 9.19: Photocatalytic annulations based on sequential radical bond-forming events.

Again, the chemistry works due to the powerful oxidizing properties of the iminium ion in its excited state and its ability to oxidize an alkene substrate bearing a carboxylate group in γ position. Upon oxidation driven by the iminium, the alkene was forming a radical cation that was intercepted by the carboxylate oxygen forming a five membered intermediate. This intermediate, by a quite rapid loss of a proton was forming an alkyl radical. The formed alkyl radical can then react with the iminium radical resulting from the electron transfer event. This radical recombination is followed by hydrolysis of the intermediate enamine. Differently substituted olefins were suitable substrates, affording the products in good yields and stereoselectivities. Not only carboxylic acids but also alcohols can be used to trigger the formation of the radical adduct. Again, only cinnamyl aldehydes can be used, but many substitutions are tolerated on the aromatic moiety. As a last example of this chemistry, another photochemical cascade process that combines the excited-state and ground-state reactivity of chiral organocatalytic intermediates was described by Melchiorre (Figure 9.19(b)) [74]. The methodology used racemic cyclopropanols as starting material that, by oxidation, can be converted to the β-alkyl radical ketones. As racemic cyclopropanols have oxidation potentials around 1.60 V versus Ag/Ag^+ in CH_3CN, oxidation of these substrates by the excited state of the enamine is thermodynamically possible. As the intermediate of the radical coupling between the ketone radical and the enamyl radical is an enamine, the carbonyl obtained by fragmentation engages in a stereoselective reaction with the enamine to give quite interesting cyclopentanols. The reaction is quite tolerant toward structural and electronic variations of the coupled substrates, and many complex cyclopentanols, in high yields and ees were obtained. Interestingly also spirocyclic compounds could be prepared.

9.8 Outlook and perspectives

Enamine and iminium activation modes can be effectively combined with PC, enhancing the possibility offered by organocatalysis to control the stereochemistry in product formation. As photoredox catalysis is a clean way to produce active radicals, many new radical reactions will be introduced in the future. In addition, cascade and domino reactions are available combining Michael radical reactions with enamines. As photoredox catalysis will be also possible in the presence of transition metals, the combination of enamine or iminium catalysis with transitions metal catalysis triggered by photoredox event, will be certainly investigated. Finally, all investigated processes, involving iminium ions or enamines (with or without PC), were conducted in non-aqueous conditions. Merging water compatible photocatalysis with the chemistry already developed for enamine or iminium catalysis in water, or in the presence of surfactants, will be surely investigated, providing novel clean and green stereoselective processes.

References

[1] Marzo L, Pagire SK, Reiser O, König B. Visible-light photocatalysis: does it make a difference in organic synthesis? Angew Chem Int Ed. 2018;57:10034–72.
[2] Wenger OS. Photoactive complexes with Earth-abundant metals. J Am Chem Soc. 2018;140:13522–33.
[3] Shaw MH, Twilton J, MacMillan DW. Photoredox catalysis in organic chemistry. J Org Chem. 2016;81:6898–926.
[4] http://chemlabs.princeton.edu/macmillan/photoredox/.
[5] Douglas JJ, Sevrin MJ, Stephenson CR. Visible light photocatalysis: applications and new disconnections in the synthesis of pharmaceutical agents. Org Process Res Dev. 2016;20:1134–47.
[6] Allen AE, MacMillan DW. Synergistic catalysis: A powerful synthetic strategy for new reaction development. Chem Sci. 2012;3:633–58.
[7] Prier CK, Rankic DA, MacMillan DW. Visible light photoredox catalysis with transition metal complexes: applications in organic synthesis. Chem Rev. 2013;113:5322–63.
[8] Romero NA, Nicevicz DA. Organic photoredox catalysis. Chem Rev. 2016;116:10075–166.
[9] Wallentin CJ, Nguyen JD, Stephenson CR. Radical carbon-carbon bond formations enabled by visible light active photocatalysts. Chimia. 2012;66:394–8.
[10] Nicevicz, MacMillan DW. Merging photoredox catalysis with organocatalysis: the direct asymmetric alkylation of aldehydes. Science. 2008;322:77–80.
[11] Ischay MA, Anzovino ME, Du J, Yoon TP. Efficient visible light photocatalysis of [2 + 2] enone cycloadditions. J Am Chem Soc. 2008;130:12886–7.
[12] Narayanam JM, Tucker JW. Stephenson CRJ. Electron-transfer photoredox catalysis: development of a tin-free reductive dehalogenation reaction. J Am Chem Soc. 2009;131:8756–7.
[13] Mills A, Le Unte S. An overview of semiconductor photocatalysis. J Photochem Photobiol A. 1997;108:1–35.
[14] http://vrg-uofsc.com/images/719_Merck-Photocatalysis-Chart.pdf.
[15] Roth HG, Romero NA, Nicewicz DA. Experimental and calculated electrochemical potentials of common organic molecules for applications to single-electron redox chemistry. Synlett. 2016;27:714–23
[16] Balzani V, Ceroni P, Juris A. Photochemistry and photophysics: concepts, research, applications. Weinheim, Germany: Wiley-VCH Verlag GmbH & Co. KGaA, 2014
[17] Castanho MA, Prieto MJ. Fluorescence quenching data interpretation in biological systems: the use of microscopic models for data analysis and interpretation of complex systems. Bba. 1998;1373:1–16.
[18] Boaz H, Rollefson GK. The quenching of fluorescence. Deviations from the Stern-Volmer law. J Am Chem Soc. 1950;72:3435–43.
[19] MacMillan DW. The advent and development of organocatalysis. Nature. 2008;455:304–8.
[20] Mukherjee S, Woon Yang S, Hoffmann S, List B. Asymmetric enamine catalysis. Chem Rev. 2007;107:5471–569.
[21] Mayr H, Lakhdar S, Maji B, Ofial AR. A quantitative approach to nucleophilic organocatalysis. Beilstein J Org Chem. 2012;8:1458–78.
[22] Kempf B, Hampfel N, Ofial AR, Mayr H. Structure-nucleophilicity relationships for enamines. Chem Eur J. 2003;9:2209–18.
[23] Mayr H. Reactivity scales for quantifying polar organic reactivity: the benzhydrylium methodology. Tetrahedron. 2015;71:5095–111.

[24] Beeson TD, Mastracchio A, Hong J-B, Ashton K, MacMillan DW. Enantioselective organocatalysis using SOMO activation. Science. 2007;316:582–5.

[25] Neumann M, Füldner S, König B, Zeitler K. Metal-free, cooperative asymmetric organophotoredox catalysis with visible light. Angew Chem Int Ed. 2011;50:951–4.

[26] Rigotti T, Casado-Sánchez A, Cabrera S, Pérez-Ruiz R, Liras M, de la Peña O'Shea VA, et al. A bifunctional photoaminocatalyst for the alkylation of aldehydes: design, analysis, and mechanistic studies. ACS Catal. 2018;8:5828–40.

[27] Juris A, Balzani V, Barigelletti F, Campagna S, Belser P, von Zelewsky A. "Ru(II) polypyridine complexes - photophysics, photochemistry, electrochemistry, and chemiluminescence. Coord Chem Rev. 1988;84:85–277.

[28] Gualandi A, Rodeghiero G, Della Rocca E, Bertoni F, MArchini M, Perciaccante R, et al. Application of coumarin dyes for organic photoredox catalysis. Chem Comm. 2018;54:10044–7.

[29] Cismesia MA, Yoon TP. Characterizing chain processes in visible light photoredox catalysis. Chem Sci. 2015;6:5426–34.

[30] Studer A, Curran DP. Catalysis of radical reactions: a radical chemistry perspective. Angew Chem Int Ed. 2016;55:58–102.

[31] Um JM, Gutierrez O, Schoenebeck F, Houk KN, MacMillan DW. Nature of intermediates in organo-SOMO catalysis of α-arylation of aldehydes. J Am Chem Soc. 2010;132:6001–5.

[32] Riente P, Matas Adams A, Palomares E, Pericàs MA. Light-driven organocatalysis using inexpensive, nontoxic Bi_2O_3 as the photocatalyst. Angew Chem Int Ed. 2014;53:9613–6.

[33] Larsen CB, Wenger OS. Photoredox catalysis with metal complexes made from earth-abundant elements. Chem Eur J. 2018;24:2039–58.

[34] McCusker JK, Walda KN, Dunn RC, Simon JD, Magde D, Hendrickson DN. Subpicosecond $^1MLCT\ ^5T_2$ intersystem crossing of low-spin polypyridyl ferrous complexes. J Am Chem Soc. 1993;115:298–307.

[35] Gualandi A, Marchini M, Mengozzi L, Natali M, Lucarini M, Ceroni P, et al. Organocatalytic enantioselective alkylation of aldehydes with $[Fe(bpy)_3]Br_2$ catalyst and visible light. ACS Catal. 2015;5:5927–31.

[36] Gualandi A, Marchini M, Ceroni P, Cozzi PG Manuscript in preparation.

[37] Shih H-W, Vander Val MN, Grange RL, MacMillan DW. Enantioselective α-benzylation of aldehydes via photoredox organocatalysis. J Am Chem Soc. 2010;132:13600–3.

[38] Nagib DA, Scott ME, MacMillan DW. Enantioselective α-trifluoromethylation of aldehydes via photoredox organocatalysis. J Am Chem Soc. 2009;131:10875–7.

[39] Carpacci AG, Malinowski JT, McAlpine N, Kuhne J, MacMillan DW. Direct, enantioselective α-alkylation of aldehydes using simple olefins. Nat Chem. 2017;9:1073–7.

[40] Nacsa ED, MacMillan DW. Spin-center shift-enabled direct enantioselective α-benzylation of aldehydes with alcohols. J Am Chem Soc. 2018;140:3322–30.

[41] Cecere G, König C, Alleva JL, MacMillan DW. Enantioselective direct α-amination of aldehydes via a photoredox mechanism: a strategy for asymmetric amine fragment coupling. J Am Chem Soc. 2013;135:11521–4.

[42] Zhu Y, Zhang L, Luo S. Asymmetric α-photoalkylation of β-ketocarbonyls by primary amine catalysis: facile access to acyclic all-carbon quaternary stereocenters. J Am Chem Soc. 2014;136:14642–5.

[43] Melchiorre P. Cinchona-based primary amine catalysis in the asymmetric functionalization of carbonyl compounds. Angew Chem Int Ed. 2012;51:9748–70.

[44] Foster R. Electron donor–acceptor complexes. J Phys Chem. 1980;84:2135–41.

[45] Rosokha SV, Kochi JK. Fresh look at electron-transfer mechanisms via the donor/acceptor bindings in the critical encounter complex. Acc Chem Res. 2008;41:641–53.

[46] Arceo E, Jurberg ID, Alvarez-Fernandez A, Melchiorre P. Photochemical activity of a key donor–acceptor complex can drive stereoselective catalytic α-alkylation of aldehydes. Nat Chem. 2013;5:750–6.
[47] Husch T, Seebach D, Beck AK, Rehiner M. Rigorous conformational analysis of pyrrolidine enamines with relevance to organocatalysis. Hel Chim Acta. 2017;100:e1700182
[48] Arceo E, Bahamonde A, Bergonzini G, Melchiorre P. Enantioselective direct a-alkylation of cyclic ketones by means of photo-organocatalysis. Chem Sci. 2014;5:2438–42.
[49] Woźniak L, Murphy JJ, Melchiorre P. Photo-organocatalytic enantioselective perfluoroalkylation of β-ketoesters. J Am Chem Soc. 2015;137:5678–81.
[50] Shirakawa S, Maruoka K. Recent developments in asymmetric phase-transfer reactions. Angew Chem Int Ed. 2013;52:4312–48.
[51] Yang C, Zhang W, Li YH, Song XS, Li X, Cheng J–P. Origin of stereoselectivity of the photoinduced asymmetric phase-transfer-catalyzed perfluoroalkylation of β-ketoesters. J Org Chem. 2017;82:9321–7.
[52] Silvi M, Melchiorre P. Enhancing the potential of enantioselective organocatalysis with light. Nature. 2018;554:41–9.
[53] Cao Z–Y, Ghosh T, Melchiorre P. Enantioselective radical conjugate additions driven by a photoactive intramolecular iminium-ion-based EDA complex. Nature Comm. 2018;9:3274.
[54] Srikanth GS. Castle S L Advances in radical conjugate additions. Tetrahedron. 2005;61:10377–441.
[55] Erkkilä A, Majander I, Pihko PM. Iminium catalysis. Chem Rev. 2007;107:5416–70.
[56] Pirnot MT, Rankic DA, Martin DB, MacMillan DW. Photoredox activation for the direct β-arylation of ketones. Science. 2013;339:1593–6.
[57] Murphy JJ, Bastida D, Paria S, Fagnoni M, Melchiorre P. Asymmetric catalytic formation of quaternary carbons by iminium ion trapping of radicals. Nature. 2016;532:218–22.
[58] Bahamonde A, Murphy JJ, Savarese M, Brémond E, Cavalli A, Melchiorre P. Studies on the enantioselective iminium ion trapping of radicals triggered by an electron-relay mechanism. J Am Chem Soc. 2017;139:4559–67.
[59] Dell'Amico L, Vega-Peñaloza, Caudrados S, Melchiorre P. Enantioselective organocatalytic diels–alder trapping of photochemically generated hydroxy-o-quinodimethanes. Angew Chem Int Ed. 2016;55:3313–7.
[60] Yuan X, Dong S, Liu Z, Wu G, Zou C, Ye J. Enantioselective Michael addition of photogeneratedo-quinodimethanes to enones catalyzed by chiral amino acid esters. Org Lett. 2017;19:2322–5.
[61] Silvi M, Arceo E, Juberg ID, Cassani C, Melchiorre P. Enantioselective organocatalytic alkylation of aldehydes and enals driven by the direct photoexcitation of enamines. J Am Chem Soc. 2015;137:6120–23.
[62] Ramachary DB, Reddy YV. Dienamine catalysis: an emerging technology in organic synthesis. Eur J Org Chem. 2012:865–87.
[63] Bahamonde A, Melchiorre P. Mechanism of the stereoselective α-alkylation of aldehydes driven by the photochemical activity of enamines. J Am Chem Soc. 2016;138:8019–30.
[64] Hodgson DA, Charlton A. Methods for direct alkylation of α-alkyl-sybstituted aldehydes. Tetrahedron. 2014;70:2207–36.
[65] List B, Čorić´ I, Grygorenko OO, Kaib PS, Komarov I, Lee A, et al. The catalytic asymmetric α-benzylation of aldehydes. Angew Chem Int Ed. 2014;53:282–5.
[66] a) Cozzi PG, Benfatti F, Zoli L. Organocatalytic asymmetric alkylation of aldehydes by S_N1-type reaction of alcohols. Angew. Chem Int Ed. 2009, 48, 1313–6. b) Gualandi A, Emer E, Guiteras Capdevila M, Cozzi PG. Highly enantioselective α-alkylation of aldehydes with 1,3-

benzodithiolylium tetrafluoroborate: a formal organocatalytic α-alkylation of aldehydes by the carbenium ion. Angew Chem. 2011;123:7988–92.
[67] Filippini G, Silvi M, Melchiorre P. Enantioselective formal α-methylation and α-benzylation of aldehydes by means of photo-organocatalysis. Angew Chem Int Ed. 2017;56:4447–51.
[68] Nathans J, Thomas D, Hogness DS. Molecular genetics of human color vision: the genes encoding blue, green, and red pigments. Science. 1986;232:193–202.
[69] Silvi M, Verrier C, Rey YP, Buzzetti L, Melchiorre P. Visible-light excitation of iminium ions enables the enantioselective catalytic β-alkylation of enals. Nat Chem. 2017;9:868–73.
[70] Verrier C, Alandini N, Pezzetta C, Moliterno M, Buzzetti L, Hepburn HB, et al. Direct stereoselective installation of alkyl fragments at the β-carbon of enals via excited iminium ion catalysis. ACS Catal. 2018;8:1062–6.
[71] Mazzarella D, Crisenza GE, Melchiorre P. Asymmetric photocatalytic C–H functionalization of toluene and derivatives. J Am Chem Soc. 2018;140:8439–43.
[72] Hepburn HB, Dell'Amico L, Melchiorre P. Enantioselective vinylogous organocascade reactions. Chem Rec. 2016;16:1787–806.
[73] Bonilla P, Rey YP, Holden CH, Melchiorre P. Photo-organocatalytic enantioselective radical cascade reactions of unactivated olefins. Angew Chem Int Ed. 2018;57:12819–23.
[74] Woźniak L, Magagnano G, Melchiorre P. Enantioselective photochemical organocascade catalysis. Angew Chem Int Ed. 2018;57:1068–72.

Alessandra Puglisi and Sergio Rossi

10 Stereoselective organocatalysis and flow chemistry

Abstract: Organic synthesis has traditionally been performed in batch. Continuous-flow chemistry was recently rediscovered as an enabling technology to be applied to the synthesis of organic molecules. Organocatalysis is a well-established methodology, especially for the preparation of enantioenriched compounds. In this chapter we discuss the use of chiral organocatalysts in continuous flow. After the classification of the different types of catalytic reactors, in Section 2, each class will be discussed with the most recent and significant examples reported in the literature. In Section 3 we discuss homogeneous stereoselective reactions in flow, with a look at the stereoselective organophotoredox transformations in flow. This research topic is emerging as one of the most powerful method to prepare enantioenriched products with structures that would otherwise be challenging to make. Section 4 describes the use of supported organocatalysts in flow chemistry. Part of the discussion will be devoted to the choice of the support. Examples of packed-bed, monolithic and inner-wall functionalized reactors will be introduced and discussed. We hope to give an overview of the potentialities of the combination of (supported) chiral organocatalysts and flow chemistry.

Keywords: stereoselective synthesis, stereoselective catalysis, organocatalysis, supported catalysts, flow chemistry, catalytic reactors, microreactors, mesoreactors

10.1 Introduction

Continuous-flow processes were recently rediscovered as an enabling technology [1] to be applied to the synthesis of organic molecules [2, 3]. Kirschining has defined Enabling technologies as “various traditional as well as new techniques which have been developed to speed up synthetic transformations and importantly ease workup as well as isolation of products” [4]. The merits of flow chemistry both on a micro- and on a meso-scale have been reviewed [5–10] a number of times, but can be shortly summarized here: (1) *efficient heat and mass transfer*: this is particularly emphasized in narrow-channel system, where heat and concentration profiles are homogeneous along the reactor. (2) An easy *combination with other enabling technologies* such as microwave irradiation, inductive heating,

This article has previously been published in the journal Physical Sciences Reviews. Please cite as: Puglisi, A., Rosi, S. Stereoselective organocatalysis and flow chemistry *Physical Sciences Reviews* [Online] 2021, 2. DOI: 10.1515/psr-2018-0099

https://doi.org/10.1515/9783110590050-010

photochemistry, electrochemistry, which opens the door to efficient and fast synthetic transformations. (3) An *improved safety* of the process: this is mainly due to the accurate control of the reaction parameters (temperature, pressure) in the (micro)reactor. In addition, the possibility of using and reacting small amounts of potentially toxic and/or dangerous reagents at a given time drastically expands the pool of reagents available to the synthetic organic chemist. Safety issues are particularly important in the case of massive production and are generally concerns of big industries. Pharmaceutical companies have recently started to adopt continuous technologies thanks to the great benefits associated: the possibility to develop a fully automatic process while analyzing the data *in continuo*, using low-cost equipment. Food and Drug Administration (FDA) and European Medicines Agency (EMA) on their side recommend the massive use of continuous-flow technologies to meet high-quality standards and green chemistry principles [11, 12]. The equipment necessary to run a reaction *in continuo* is by far more flexible and cheaper than the common equipment used to run the same process in batch. The immediate benefit is that the optimization campaign for the active pharmaceutical ingredient (API) synthesis is much shorter and the synthetic organic chemist can produce the desired amount of the API in a very short time. Consequently, the risk and the cost associated with the development and production of and API is greatly reduced, due to reduced investment. The specific topic of the application of continuous-flow processes to the synthesis of pharmaceuticals intermediates was broadly reviewed in the past few years and will not be specifically discussed in this chapter [13–21].

All these discussed features contribute to have faster reactions and lead to an implementation of the existing processes. Some common problems related to the reaction scale-up can be more easily tackled in flow rather than in batch: this is the case of runaway reactions, inefficient mixing, and byproduct formation that benefits from the use of microreactors. The small dimension of microreactors allows to better control the reaction parameters and, therefore, possible problems that may arise. Accordingly, the reaction scale-up in small reactors is easier than in batch, at least in principle. There are three reported methods to produce large amounts of a compound *in continuo*: (i) the scaling-out, which is the easiest method and consists on running the process for long time, (ii) the numbering up, which is used in many reactors in a parallel manner, and (iii) the scaling-up, where the process is performed on larger continuous reactors [22].

Finally, we can observe a considerable increase in the complexity of the continuous-flow processes published in the literature; this is particularly true for the applications toward APIs. We assisted to a rapid evolution of flow chemistry, from single-step reactions to multistep processes, often completely integrated, comprising advanced downstream purification and formulation [23].

10.2 Catalytic flow reactors: classification, tools and parameters

"Continuous-flow chemistry" is the common and wide term used to describe the performance of a reaction in a continuous manner, within the channels of a fluidic reactor. It collects a variety of process technologies, tools and strategies now routinely used in many research groups both for explorative and preparative purposes. Intuitively, we can describe a flow reactor as a vessel with an inlet and an outlet, in which the reagents are fed continuously and the products are continuously removed. Figure 10.1 depicts the different types of flow reactors that will be discussed in this chapter.

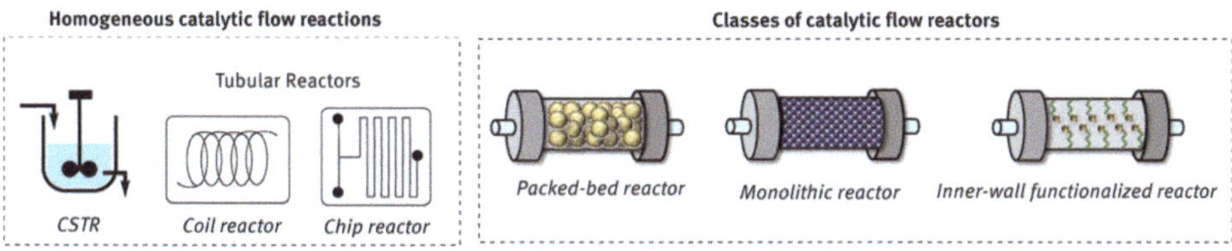

Figure 10.1: Types of flow reactors discussed in this chapter.

There are two main classes of flow reactors: the continuous stirred-tank reactor (CSTR) and the tubular reactor (TR). On a laboratory scale, TRs are commonly used.

Catalytic reactors can be defined as those in which the catalyst permanently resides inside. They can be divided into three main classes, according to the method used to incorporate the immobilized (organo)catalyst into the device: (i) packed-bed; (ii) monolithic; and (iii) inner wall-functionalized. These types of catalytic reactors will be discussed in Section 4.

Johnston et al. [24] have reported an example of a homogeneous Brønsted acid–base catalyzed aza-Henry reaction performed in a CSTR. This work was developed in collaboration with Eli Lily researchers to testify the interest of big pharmaceutical industries to the development of continuous-flow processes. Due to the absence of (micro)channels, CSTRs are more suitable for handling the solids in flow with respect to TRs; thus, heterogeneous reactions with solid precipitation can be run in flow using a CSTR. The initial goal of the project was to translate the enantioselective aza-Henry reaction from batch to continuous flow, to accelerate the large-scale production of diamine building blocks. During the reaction, the formation of a less-soluble product prevented the use of a typical tubular flow reactor for the initial investigations. The presence of solids usually leads to plugging and clogging of the (micro)reactors. The authors selected an automated intermittent flow-stirred tank reactor, an equivalent process to continuous flow. Under the optimized operating conditions, the aza-Henry product continuously crystallized from the reaction mixture; this solid was intermittently filtered, while the mother liquor, containing the soluble catalyst, returned to the reactor. The sequence separation-catalyst recycling allowed to achieve

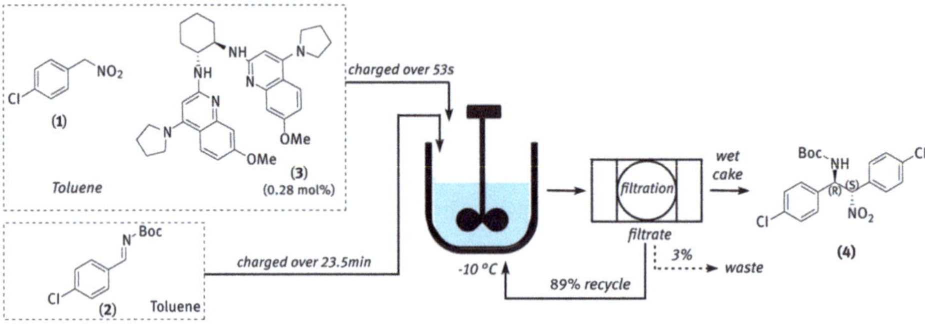

Figure 10.2: Homogeneous Brønsted acid–base catalyzed aza-Henry reaction performed in a CSTR.

a high-catalyst concentration in the reactor, thus leading to a faster organocatalytic reaction. The reaction setup is illustrated in Figure 10.2. The benchmark reaction was the aza-Henry reaction between 1-chloro-4-(nitromethyl)benzene **1** and *N*-Boc-4-chloro-benzaldimine **2** promoted by the diamine catalyst **3** to afford compound **4**. Overall, the development of an automated process allowed to improve the safety of the process, since small amounts of reactive nitroalkane were present at any given time. In comparison to the batch reaction, the reaction in flow was run with a higher output with a reduced time (40 min for a single cycle in flow *vs* 22 h for the batch process). Moreover, it turned out that the aza-Henry reaction was highly reproducible, thanks to the precise control over the parameters, and the scale-up required minimum optimization. The authors applied this process to the multigram synthesis of chiral nitro amines, direct precursors to differentially protected diamines.

Apart from few exceptions, in the course of this chapter, when we refer to "flow reactor," we usually consider a TR. TRs are typically characterized by narrow and well-defined channels with internal dimensions in the 10^2–10^3 μm range, and by internal volumes ranging from a few microliters to several milliliters. Based on internal diameters and volumes, they are referred to as micro- or mesofluidic reactors. They can be "chips" or "coils" according to the geometry (see Figure 10.1).

In a TR, no mixing is performed along its length, but usually the reagents are premixed before entering the reactor. There are many flow systems available on the market, with their costs variable from high to exorbitant, depending on the features and equipment. The high price of the commercially available systems has encouraged the researchers to engineer their own homemade flow setup, using some common laboratory tools. It is worth mentioning here that 3D-printing technology is becoming increasingly available and is significantly contributing to the field [25].

A list of tools for developing flow chemistry devices is reported here. (For a complete dissertation, see ref [10].)

PUMP: It ensures a constant flow rate. There are several possibilities, but it is commonly a syringe pump or an HPLC pump.

TUBING: It represents the connections and the real TR. There are several materials that can be used; the most common materials are PTFE, (stainless) steel, glass, and PEEK.

TEE: It is usually made by PEEK and it serves to connect different streams. It is used also for the reactants mixing at the inlet of the reactor.

The combination of these three pieces is sufficient to build a basic flow setup. Other devices that help the operator and complete the system are listed below.

CHECK VALVE: It is a device that ensures a unidirectional flow.

BACK-PRESSURE REGULATOR (BPR): It is a device that allows having and maintaining pressure in the reactor. It is used with gases and low-boiling solvents, to avoid the formation of bubbles.

LIQUID-LIQUID SEPARATOR: It is a device used for in-line reaction work-up that allows to separate two liquids based on their polarity.

IN-LINE MONITORING: It is an instrument that can be connected to the flow setup. It allows obtaining real-time information about the reaction progress by operating on the flow stream, without stopping the system. Commonly, it is a FLOW-IR, although other devices are becoming available (FLOW-NMR).

When dealing with continuous-flow systems, some important parameters need to be considered, besides the common parameters used to describe batch reactions such as temperature, pressure, concentration and molar ratios. Vr is the reactor volume; the flow rate ϕ is the rate at which the reagents are fed into the reactor. The flow rates are additive, so, if two reagents are pumped independently, the flow rate inside the reactor will be the sum of the two flow rates. The residence time τ is the time in which the reagents reside into the reactor. Intuitively, it corresponds to the reaction time for batch reactions. The relationship that connects the three parameters is:

$$\tau = \mathrm{Vr}/\phi \tag{10.1}$$

In the case of packed-bed reactor (see below) the volume to consider is the void volume V_0, determined by Vr minus the volume occupied by the catalyst:

$$V_0 = \mathrm{Vr} - \mathrm{Vcat} \tag{10.2}$$

In case of catalytic reactors, it is important to assess the efficiency of the catalyst to evaluate the overall convenience of the flow process. Different parameters have been proposed and are reported in the literature for a given transformation. However, it must be pointed out that it is not trivial to compare two different flow processes or a flow process to its corresponding batch version. The most common parameters found in the literature are listed below.

Mole hourly space velocity (MHSV): It represents the substrate feed to the reactor per unit mole of the catalyst and time unit.

Space–time yield (STY): It is defined as the amount of product obtained per unit time and reactor (or catalyst bed) unit volume.

Productivity: It is defined as the amount of product (obtained per amount of catalyst) in the unit time.

Time on stream (TOS): It is the overall duration of the process; it indicates the temporal stability of the catalyst inside the reactor.

Total turnover number (TTON): It is given by the molar ratio between the substrate converted into the expected product and the initial catalyst amount; it indicates the total amount of product that can be obtained by the process.

Accumulated TON: it is given by the product of catalyst TOF by operation time.

In a catalytic flow process, the catalyst can be homogeneous, so it flows through the reactor (chip or coil) together with the reactants; in this case, at the end of the reaction, a separation step of the product from the catalyst (and possible by-products) is required. On the contrary, the catalyst can be heterogeneous, that is, supported on a solid matrix. The solid catalyst resides into the flow reactor while the reactants pass by; at the end of the reaction there is no additional separation step required. The types of catalytic reactors are discussed in Section 4.

10.3 Homogenous organocatalysis in flow

Despite the great popularity of continuous-flow processes in the last decades, the use of homogeneous stereoselective organocatalysts in flow is still underdeveloped. This is mostly due to the fact that, in a homogenous catalytic continuous-flow process, the catalyst flows in the (micro)reactor together with the reactants, leading to a process in which, at the end, a separation step is required in order to purify the product and, possibly, recover and recycle the catalyst. This technique does not seem very appealing if compared to the use of supported catalysts that will be discussed in the next section. However, homogeneous catalysts are usually more active than their supported counterpart, so a continuous-flow process with an integrated catalyst separation and recovery would be amenable (see below).

Microfluidic technology allows for a fast screening of different reaction parameters, such as temperature, solvent, reactants concentration, reaction times, catalyst loading, and catalyst/reagents combinations just to name a few. The great benefit is the quick identification of the best reaction conditions that guarantee high chemical and stereochemical efficiency. The reactions developed under continuous-flow conditions favorably compare with those carried out in a flask only if reduced reaction times and improved productivity can be achieved.

Among the possible advantages of the use of continuous process, there is the possibility of heating the reaction, in a very controlled manner, in order to speed up the reaction; however, in the presence of a homogeneous chiral organocatalyst, heating while maintaining a high level of enantioselectivity can be challenging. Odedra and Seeberger [26] reported the 5-(pyrrolidin-2-yl)tetrazole-catalyzed aldol reaction between acetone and different aromatic aldehydes in a glass chip reactor. The reaction was run in a 1:1 acetone:DMSO mixture at 60 °C; with a residence time of 40 min, they obtained the desired product in shorter time than the corresponding reaction in batch, with similar yield and enantiomeric excess. With 4-nitrobenzaldehyde **6** yield and enantioselectivity were superior than those of the same reaction performed in batch in the same conditions (60 °C, 20 min, 79 % yield, 75 % ee, Figure 10.3(a)). They extended the scope to different aldehydes and cyclohexanone. They also applied the same catalyst **5** and the same flow set-up for studying the Mannich reaction between cyclohexanone and *N*-PMP protected α-imino ethylglyoxylate. Performing the reaction at 60 °C allowed to obtain the desired product in short reaction time maintaining high yield and high diastereo- and enantioselectivity. The authors showed that the continuous process allowed to lower the catalyst loading with respect to the batch (5–10 mol %).

Luisi et al. [27] reported the use of the same catalyst **5** to promote the Michael addition of cyclohexanone to differently substituted β-nitrostyrenes to yield the products with moderate to good levels of enantioselectivity. The system was then implemented by adding a second microreactor and the set-up was thus modified: the first chip was dedicated to perform a Michael addition, leading to an intermediate that react in a second chip with a different Michael acceptor (e.g. an α,β-unsaturated carbonyl compound). This second organocatalyzed process afforded the corresponding adduct with up to four consecutive stereogenic centres. As an example, cyclohexanone **8** and *trans*–β–nitrostyrene **9** reacted in the presence of catalyst **5**; the output of the first microreactor, containing **5** and adduct **10**, was fed to the second microreactor with a solution of cinnamaldehyde **11** and DIPEA. The multifunctionalized adduct **12**, with six stereogenic centers, was obtained in 73 % yield, 66:34 dr and 84 % ee for the major isomer, thus demonstrating the ease of realizing multi-step stereoselective organocatalytic continuous-flow processes (Figure 10.3(b)).

The same catalyst **5** was used by Nakashima and Yamamoto [28] in a completely different approach for the use of nonsupported organocatalyst in flow. They packed proline tetrazole catalyst **5** into an empty column, thus realizing a reactor to be used with low-polarity organic solvents. The resulting heterogeneous system was studied in continuo without consumption and loss of catalytic activity. The column was used in the aldol reaction between ketones and different aromatic aldehydes using ethyl acetate as nonpolar solvent; the addition of 3 % water was beneficial for improving the ee. Notably, the packed-bed column was repeatedly used for reactions on 10 mmol scale and for different reactions, such as Mannich and *ortho*-nitro aldol reactions. Products were obtained in high yield (generally more than 70 %) and ee. Although the nonsupported proline tetrazole showed lower efficiency with respect to other solid-supported organocatalysts, this approach proved to be simple and efficient in the aldol reactions (Figure 10.3(c)).

Benaglia et al. [29] reported the stereoselective synthetic strategy for the preparation of trifluoromethylamine mimics of retro thiorphan, an inhibitor of metalloproteinase NEP (neutral endopeptidase). The crucial step is the stereoselective, catalytic reduction of a fluorinated enamine with trichlorosilane as a reducing

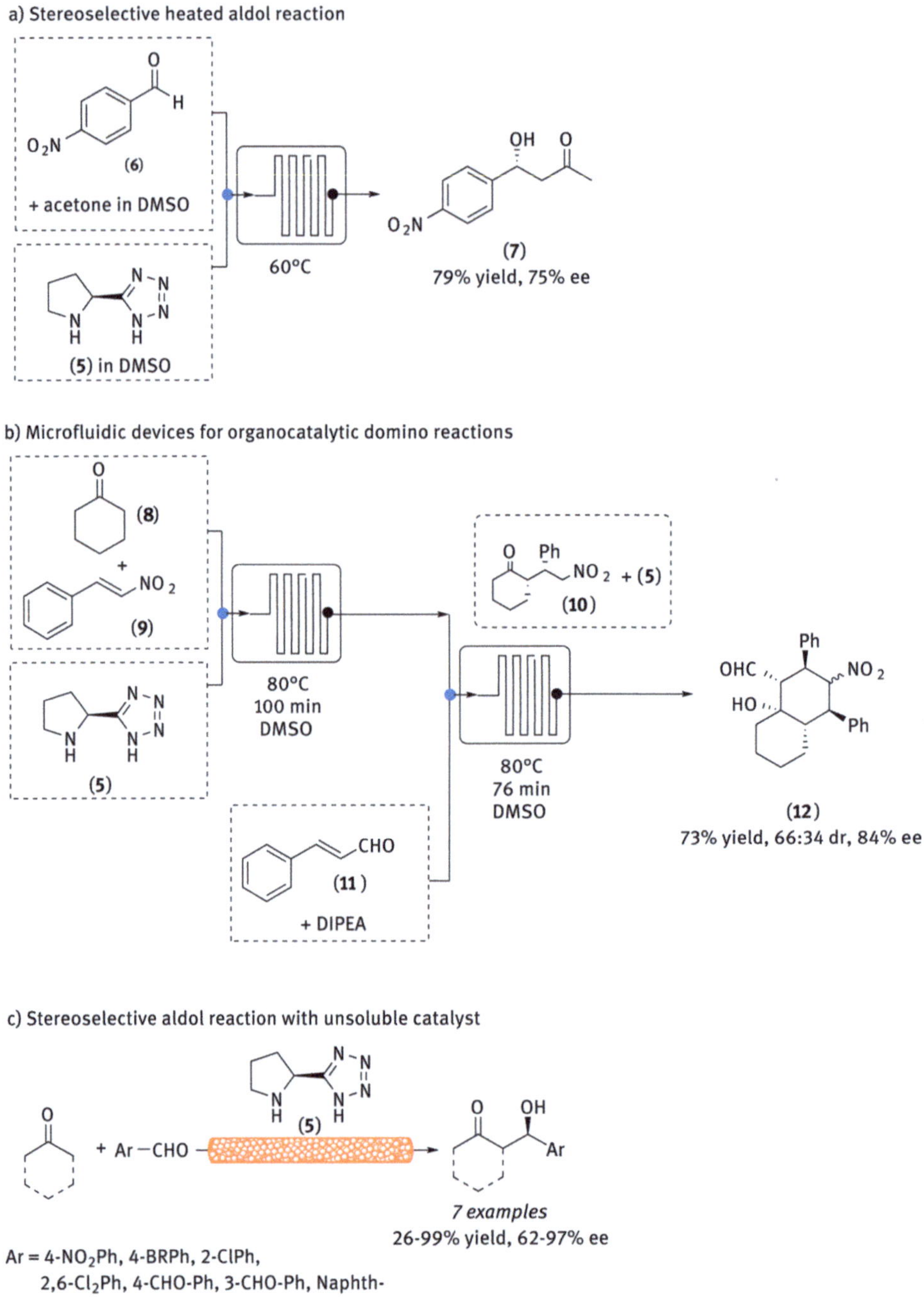

Figure 10.3: Examples of homogenous organocatalytic reaction under flow conditions using chip reactors (examples a and b) and packed bed reactor (example c).

agent in the presence of a chiral Lewis base. The whole synthesis was then investigated under continuous-flow conditions: an advanced intermediate of the target molecule was synthesized in only two in-flow synthetic modules, avoiding isolation and purifications of intermediates, leading to the isolation of the target chiral fluorinated amine in up to an 87:13 diastereoisomeric ratio. The entire flow sequence is illustrated in Figure 10.4. The commercially available fluorinated alkyne **13** was reacted with the O-allyl protected phenyl alaninol ether **14** to afford enamine **15** with complete conversion after 10 min residence time. **15** was consequently subjected to the $HSiCl_3$-mediated diastereoselective reduction in the presence of the chiral Lewis base **16.** The reduction step required long residence times in order to have an acceptable yield: heating up the reactor to 35 °C, allowed to obtain **17** in up to 37 % isolated yield, but with lower d.r., while cooling to room temperature, the yield was below 20 %, with 95:5 d.r. Although yield and d.r. were not totally satisfactory, this work represents a multistep continuous-flow process for the synthesis of enantiomerically pure, fluorinated, pharmaceutically relevant products.

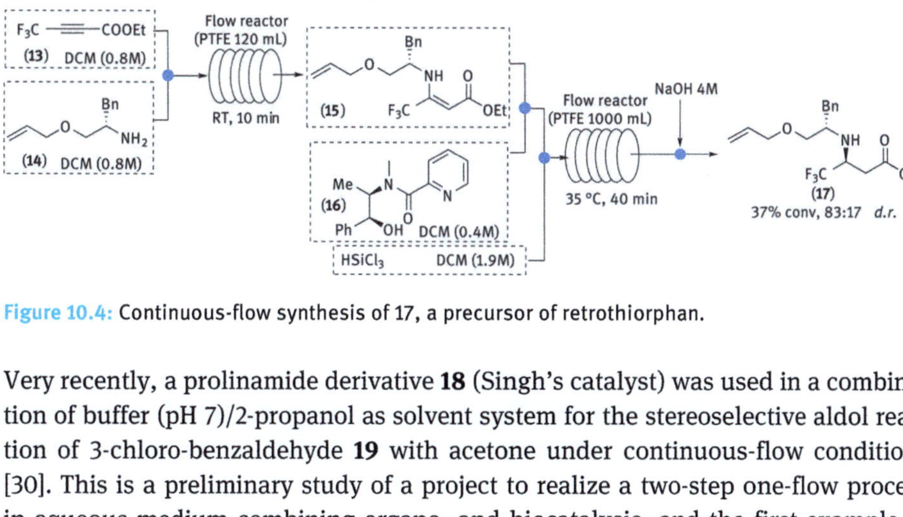

Figure 10.4: Continuous-flow synthesis of 17, a precursor of retrothiorphan.

Very recently, a prolinamide derivative **18** (Singh's catalyst) was used in a combination of buffer (pH 7)/2-propanol as solvent system for the stereoselective aldol reaction of 3-chloro-benzaldehyde **19** with acetone under continuous-flow conditions [30]. This is a preliminary study of a project to realize a two-step one-flow process in aqueous medium combining organo- and biocatalysis, and the first example of an organocatalytic reaction with a hydrophobic substrate in water running in flow mode. After extensive optimization, the authors were able to obtain yield and enantiomeric excess comparable to the batch reaction, opening the way to the second step. A solution of 3-chloro-benzaldehyde **19**, 2-propanol and phosphate buffer was mixed with a solution of Singh's catalyst **18** (3.6 mol %), 2-propanol, acetone and phosphate buffer and pumped into a coil Teflon tube reactor. After equilibration of the reaction, the eluted sample was collected and quenched by dropwise addition into a solution of CH_2Cl_2/2.0 M HCl; the organic materials were extracted with CH_2Cl_2, concentrated in vacuo and analysed, affording 74 % conversion and 89 % ee of product **20** (Figure 10.5).

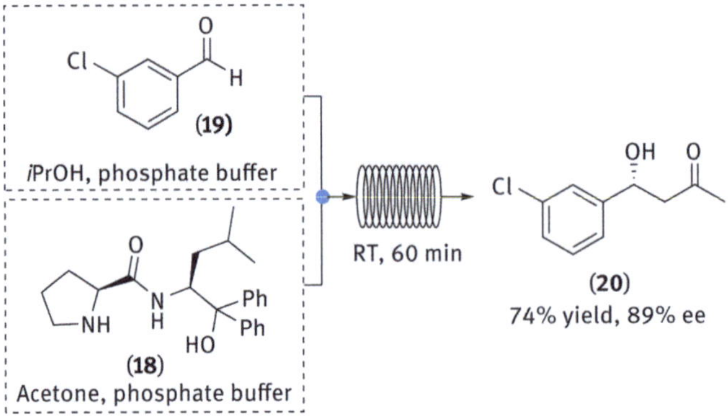

Figure 10.5: Stereoselective aldol reaction of 3-chloro-benzaldehyde 19 with acetone in aqueous media under continuous-flow conditions.

Another great advantage of microfluidic techniques is the possibility of integrating in-line analytical devices with the aim of real-time monitoring of the reaction without stopping the system. Flow IR is currently the most popular analytical device for in-line monitoring, thanks to its affordable price and ease of use. In parallel, due to the urgency of new systems for in-line reaction monitoring, also commercial Flow NMR was recently developed. In a flow NMR, the reaction mixture is continuously transferred from a vessel to the NMR magnet into the probe for the (fast) analysis and returned via an insulated line. Due to the popularity of NMR among organic chemists, it is not surprising that, over the past decades, the use of Flow-NMR has been increasingly reported [31]. In-line analysis by NMR spectroscopy is especially attractive for reaction monitoring and mechanistic investigation, since NMR provides sophisticated structural information that is usually distinctive and characteristic, for all species possessing the nuclide under observation and is inherently quantitative in nature. The availability of NMR cryoprobes has greatly enhanced the signal-to-noise ratio, substantially increasing the sensitivity; moreover, significant developments in solvent signal suppression methodology means that non-deuterated solvents can also be used, opening the way to use NMR for reaction monitoring in flow. We expect that Flow-NMR will become an in-line monitoring technique for catalyzed continuous processes in the near future.

In a study conducted by Rueping [32], a ReactIR flow cell was coupled with the microreactor and applied as an inline monitoring device to study the continuous-flow organocatalytic asymmetric transfer hydrogenation of a wide range of cyclic imines. The chiral phosphoric acid **21** was the organocatalyst and Hantzsch dihydropyridine **23** was the hydrogen source. The device allowed studying different reaction conditions and adjusting the reaction parameters until optimum conditions were

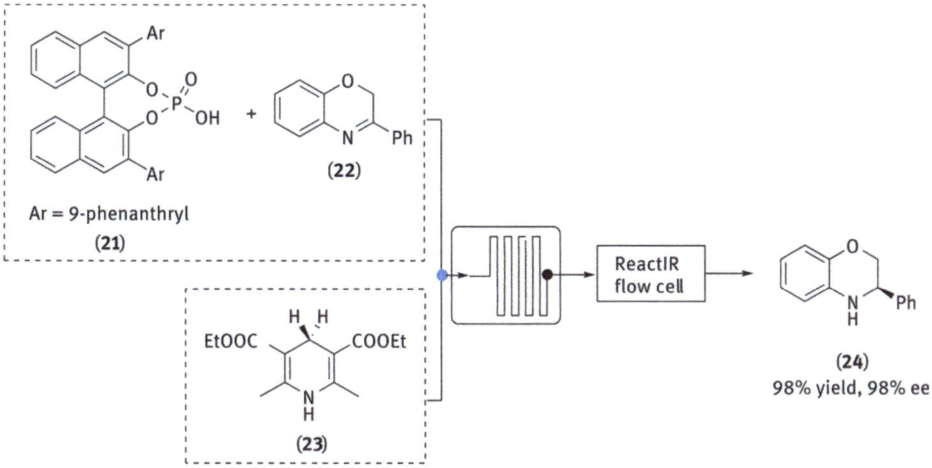

Figure 10.6: Continuous-flow stereoselective reduction of 3-Ph-benzoxazine promoted by Hantzsch ester in the presence of a chiral phosphoric acid.

found by recording a reaction temperature profile. As an example, hydrogenation of 3-Ph-benzoxazine **22** with 1 h residence time at 60 °C lead to the desired product **24** in 98 % isolated yield and 98 % ee (Figure 10.6). Noteworthy, the same reaction in batch lead to a noticeable drop in conversion (67 % yield). Better heat transfer in the microreactors compared to the glass flask of batch reactions may be responsible of the observed behavior. The work is a clear demonstration of how flow processes involving microdevices may be coupled with other "enabling technologies", techniques designed to speed up reactions, work up and isolation processes.

As mentioned at the beginning of this section, the use of homogenous organocatalyst in (micro)fluidic devices has the major issue of requiring a separation step of the precious chiral catalyst from the reaction mixture, in order to purify the product and, hopefully, to recycle the catalyst. This is the main reason why continuous-flow processes with an integrated catalyst separation and recovery started to appear in the literature. Integrated catalysis-separation system is a rapidly emerging field; the membrane-assisted recovery of homogeneous catalysts [33] is considered "sustainable" thanks to a low energy consumption. Two other features are the simple scale-up and the possibility to implement the process in continuous-flow quite easily [34]. The choice of the membrane is crucial to have an optimum retention of the catalyst. The other factor that influences the separation efficiency is the molecular weight gap between the catalyst and the other reaction components. In order to improve the separation efficiency, usually the catalyst undergoes a size-enlargement, and this can be done by embedding the catalyst in soluble polymers or conjugating it to dendrimers. A very recent example was reported by Kupai and Szekely: they prepared β-cyclodextrins decorated with cinchona-thiourea and squaramide catalysts as enlarged organocatalysts for the Michael addition of diketones to β-nitrostyrene [35]. The homogeneous catalysts were first tested in batch and the optimized conditions were used as the starting point for the evaluation of the model reaction in the continuous-flow reactor. The schematic process diagram for the continuous catalysis-separation platform is illustrated in Figure 10.7. Different membranes with different features were also tested, and the authors chose the most open membrane, that demonstrated 100 % rejection of catalyst **25** and less than 5 % rejection of the other species. This result was possible thanks to the large gap in the size of the catalyst and the reactants. During the integrated synthesis-separation process, the flow reactor outlet stream containing the crude reaction mixture was diverted to a cross-flow membrane cell. The permeate stream consisted of the highly concentrated product **27** (41 g L^{-1}) having a purity of 92 %. The retentate stream in situ recycled 100 % of catalyst **25** and 50 % of the 2-MeTHF solvent. A fresh solution of diketone **26** and *trans*–β–nitrostrene **29** was pumped into a dynamic mixing chamber where it met the recycled stream. The recirculation was defined as the ratio of the retentate flow rate and permeate flow rate. A sensitivity analysis was performed to reveal the effect of recirculation on the productivity, purity and concentrations. It turned out that the illustrated process worked at its best when it was carried out at 50 % recirculation. The temperature played a fundamental role. The membrane unit was thermoregulated at 50 °C to eliminate precipitation of the product. Product **27** crystallized in the collection vessel at room temperature, which allowed the final purity to reach 98 %, with an enantiomeric excess of 99 %. The robustness and

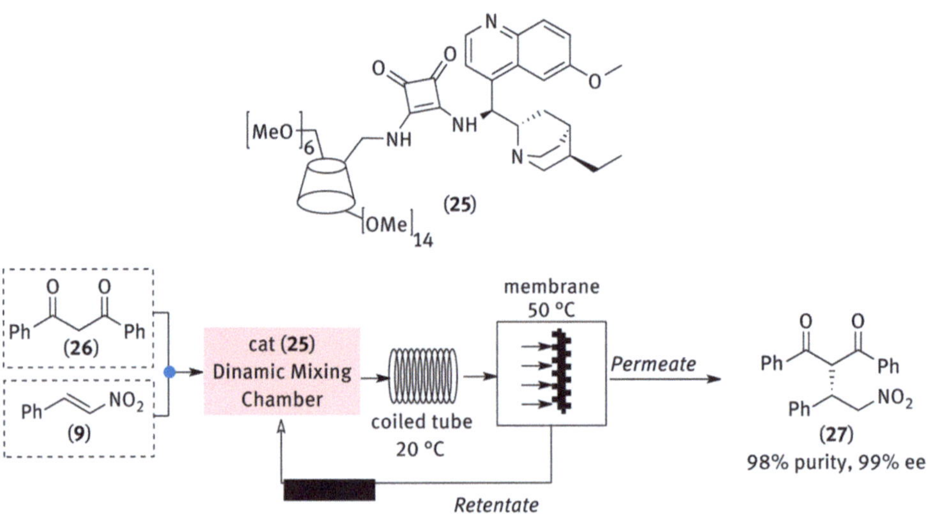

Figure 10.7: Stereoselective Michael addition with membrane separation.

reusability of catalyst **25** was demonstrated up to 100 °C in the flow reactor, and over 18 days of operation.

10.3.1 Continuous-flow organophotoredox transformations

The combination of photocatalysis and flow chemistry has become a well-established procedure for chemical transformations. Performing photochemical transformations under continuous-flow conditions is particularly advantageous since the flow set-up ensures uniform irradiation to the entire reaction solution. The small depth of a microreactor allows maximum penetration of light, and thus irradiation even of relatively concentrated solutions can straightforwardly be achieved. Importantly, the production rate of a photochemical process can easily be adjusted in a microphotoreactor. To change the irradiation time of the photochemical processes it is sufficient to vary the flow rate of the system. As microstructured reactors possess high heat transfer coefficients, cooling that may be required during a photochemical process is achieved efficiently and without greater efforts. Further miniaturization is possible with the use of light-emitting diodes (LEDs) instead of conventional light sources. Moreover, the concept of numbering up by using several microreactors in parallel is regarded to be ideally suited to achieve the industrial production of large amounts of photochemical products. For all these reasons, it is not surprising that, currently, speaking about photocatalytic processes means speaking of flow chemistry.

During the last decade, the concept of dual catalysis of photoredox and organocatalysis emerged. MacMillan et al. [36] first showed that the combination of those two catalytic cycles result in having enantioenriched products with structures that would otherwise be challenging to make. After the initial discoveries of MacMillan, a number of dual organophotoredox catalytic reactions were published, often with high yields and stereoselectivities. Many reviews were published in the last few years, summarizing the findings [37–41].

While there are many reports of non-stereoselective continuous-flow processes involving the combination of microreactors with light-mediated photocatalysis, the use of chiral organocatalysts is still underdeveloped. To our knowledge, up to this day, only two examples exist where research groups developed a successful continuous-flow setup for this dual catalysis methodology, and they will be commented here.

Neumann and Zeitler [42] studied the in-flow reaction between bromo-malonate **29** and octanal **30** in the presence of the second generation MacMillan imidazolidinone (triflate salt) **31**, Eosin Y **28** as photocatalyst and 2,6-lutidine to afford α-alkylated aldehyde **32**. Two different reactor setups were designed, one employing glass micro-reactor technology together with LEDs at 530 nm and the other employing FEP (Fluorinated Ethylene Propylene) HPLC tubing wrapped in coils around the 23 W CFL household bulb which was immersed in a cooling bath, as depicted in Figure 10.8(a) and Figure 10.8(b), respectively. FEP is a copolymer of hexafluoropropylene and tetrafluoroethylene, that

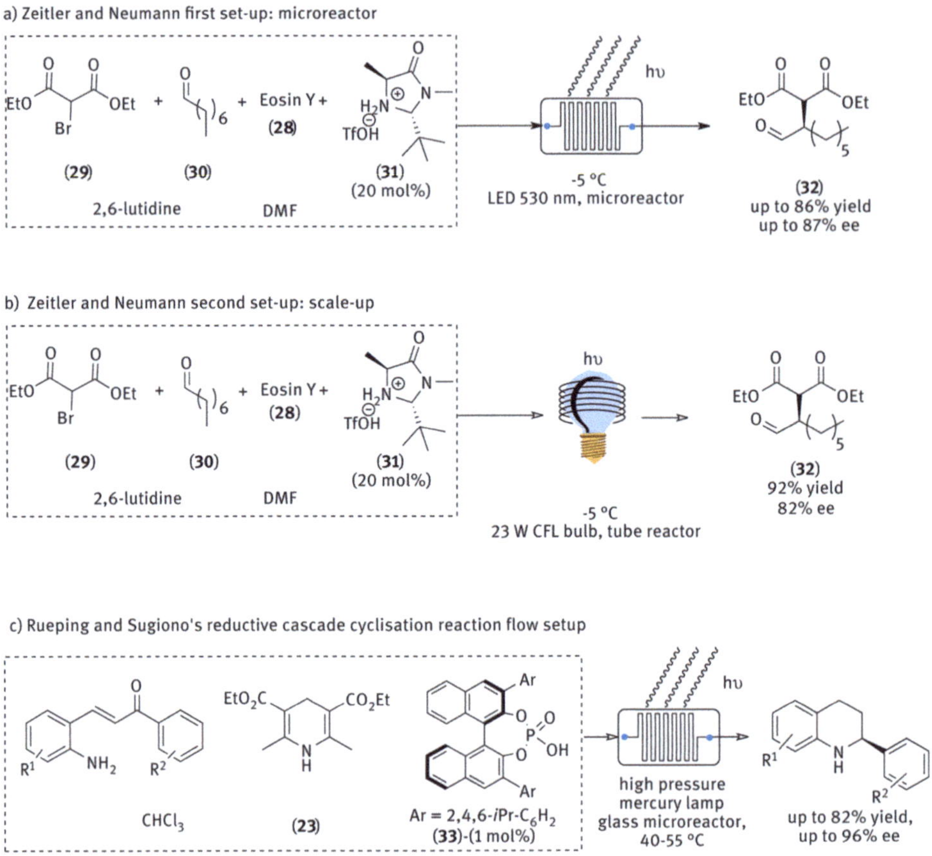

Figure 10.8: Continuous-flow organophotoredox transformations.

is highly transparent, solvent-resistant, and flexible. The reaction under flow condition still exhibits high enantioselectivity and high yields, leading to product **32** in up to 86 % yield and up to 87 % ee in the microreactor with a residence time of 45 min. The corresponding reaction in batch afforded the products with comparable results but required 18 hours. In the second setup (Figure 10.8(b)), the authors used a reactor with large reactor volume and length (10.5 mL and 21 m of length) wrapped in two layers around the lamp and they were able to easily scale-up the reaction. It was possible to generate 1.92 mmolPRODUCT/h in 92 % yield and 82 % ee, with a 107-fold increase in productivity with respect to the batch conditions, thanks to the more efficient irradiation.

Rueping and Sugiono reported the organocatalytic photocyclization–transfer hydrogenation cascade reaction starting from 2-amino-chalcones in the presence of the chiral phosphoric acid **33** and Hantzsch ester **23** [43]. For the continuous-flow setup, they used a glass microreactor immersed in a thermostated water bath. Directly next to it, a high-pressure mercury lamp irradiated the reactor from the side. This methodology allowed to prepare differently substituted isoquinolines in very high yield and ee, starting from readily available 2-aminochalcones (Figure 10.8(c)). Also in this case, the flow setup showed a significant increase in productivity. Moreover, the continuous removal of the product from the irradiation source avoided over-irradiation that can lead to undesired background reactions.

Those continuous-flow experiments showed a drastic increase of productivity, up to 100-fold, with respect to the batch systems. This is due to the high molar extinction coefficients of organic dyes and metal complexes which prevent most of the internal reactor volume to receive efficient irradiation: under batch conditions, most of the light is absorbed in the first few millimetres of the solution. In continuous flow, the increased surface-to-volume-ratio of the microreactors is exploited to achieve much higher levels of irradiation. The two examples by Zeitler and Rueping well highlight the benefits of performing stereoselective photoredox catalysis under continuous-flow conditions: the more efficient irradiation in the reaction vessel that leads to an easy scalability and the continuous removal of the product from the light source to avoid photodegradation and/or undesired side-reactions.

10.4 Solid-supported stereoselective organic catalysts in flow

The preparation of solid-supported catalysts requires a greater synthetic effort if compared to the use of the same catalysts under homogenous conditions. However, in our opinion, the advantages largely overcome the drawbacks. This is particularly evident in the case of continuous-flow processes, in which the catalyst permanently resides in the catalytic reactors [44, 45]. First, as mentioned in the Introduction section, the reaction product is not contaminated by the catalyst at the reactor output, so the separation step is avoided. The product-catalyst separation step is not only time-consuming but also not favorable in terms of solvent using. Second, by choosing the suitable concentrations, during the catalytic process the substrate is continuously exposed to a (super)stoichiometric amount of the catalyst inside the reactor: this may lead to a faster reaction time and a high turnover number (TON). Moreover, side reactions can be suppressed, thus achieving an overall cleaner process. Finally, the confinement of the supported catalyst in the reactor is the most obvious and general way to achieve its continuous recycle. It is not surprising, then, that already back in 1996 Itsuno, one of the pioneers in the use of polymer-supported ligands, reported the preparation of an enantiopure oxazaborolidine supported on an insoluble resin **34**; this system was used to pack a column, in which a solution of cyclopentadiene and methacrolein was slowly percolated [46]. A solution of the chiral Diels-Alder adduct was continuously eluted from the column and collected in a flask. Using 5.7 mmol of the polymeric catalyst **34**, 138 mmol of (*R*)-adduct **35** with 71 % ee were obtained, in what can be considered, in a visionary approach, the precursor of a continuous-flow process (Figure 10.9).

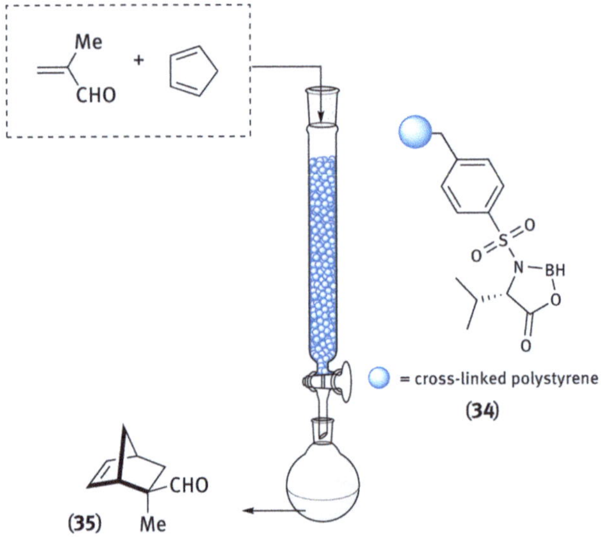

Figure 10.9: Solid-supported chiral catalyst 34 to promote stereoselective Diels-Alder transformation under fluidic conditions.

The choice of the support is obviously crucial for determining the overall performances of the catalyst and the continuous process. In principle, one can envision to use all the kind of different supports that have been developed and reported in the literature to build the catalytic reactor. In practice, if one looks up in the literature, the choice of the support for performing stereoselective organocatalyis in flow is limited to the classical silica and polymeric organic materials (polystyrenes and polyacrylates), according to the research group's expertise. On the contrary, the preparation and the modification of these materials are well-documented in the literature, so they allow a fast anchoring of the desired catalyst onto the insoluble support.

One exception is represented by textile fibers such as nylon and polyacrilonitrile (see Section 4.1).

10.4.1 Packed-bed reactors

Packed-bed reactors are the most popular catalytic reactors for flow since they are easily prepared starting from the heterogenized catalyst. The catalyst is immobilized onto an insoluble support (usually silica or a polymeric matrix) and then loaded ("packed") into the reactor. The method for preparing a packed-bed reactor is quite straightforward and requires the use of classical catalyst immobilization techniques. For this reason, packed-bed reactors were developed by those research groups active in the field of catalyst recovery and recycle: in the packed-bed reactors, the immobilized catalysts found an immediate application in flow. The main drawback consists in the inhomogeneous packing of the reactor: this leads to the formation of stagnation zones and hot spots, resulting in a broad residence time distribution and, in general, in an uncontrolled fluid dynamics. In the presence of an organic polymer as a support, the problem can be enhanced by resin swelling properties, especially with gel-type polymers, that are lightly cross-linked resins. On the contrary, the easy functionalization of inorganic supports like silica gel, zeolites, alumina and carbon, made them popular materials for developing (organo) catalytic continuous processes. However, the low catalyst charge on the inorganic support can be a limiting factor, especially when a high loading is required. Moreover, the high polar nature of these materials can affect the course of the catalyzed reaction, either unfavorably or unexpectedly [47] so a careful study of the influence of the support should be performed.

Already back in 2000, while the term "organocatalysis" was in the process of being coined, Lectka et al. [48] developed what is commonly considered the first example of stereoselective, continuous flow, organocatalytic process. The system consisted in three columns assembled in series: the first containing a polymer-supported base (BEMP), the second a supported organocatalyst (a cinchona derivative), the third a scavenger (a supported aniline), to retain unreacted material. By simply pouring a THF solution of acyl chloride and imine along the system, enantioenriched β-lactams could be prepared in a gravity-driven process. After this pioneering example (and few other examples from the same group), it took some time to assist to the growth of continuous-flow organocatalytic processes that make use of pumps to force the fluid through the system, thus leading to a more accurate control over the flow rate and, consequently, the residence time. Since many specific reviews [44, 45] on this topic [49–51] and a personal account [52] appeared in the last few years, this Section will discuss only selected significant examples.

Fiber materials deserve a special mention in this section. In recent years, the application of fiber materials as supports for immobilization of homogeneous catalysts has become an area of interest in the design and synthesis of heterogenized catalysts [53], and, later on, in the development of continuous processes [54]. List [55] proposed a very uncommon flow setup, consisting of 20 pieces of nylon 6,6 as a solid support for the anchoring of a sulfonamide derivative of cinchona alkaloid. The sulfonamide derivative was UV-irradiated in the presence of penta-erithritol triacrylate (PETA) as a cross-linker to obtain the so-called "organotextile catalyst" **36**, that was tested in the desymmetrization of cyclic anhydrides. As a first study, the authors proved the robustness of the heterogenized catalyst by performing more than 300 batch recycling experiments. The catalyst was then applied to the preparation of a valuable precursor of statin derivatives according to the experimental setup depicted in Figure 10.10. 20 sheets of nylon-supported sulfonamide **36** were packed into a polypropylene cartridge of a BÜCHI-Flash System and flushed with MTBE before use. A solution of anhydride **37** and MeOH in MTBE was continuously recirculated for 48 h into the reactor, until

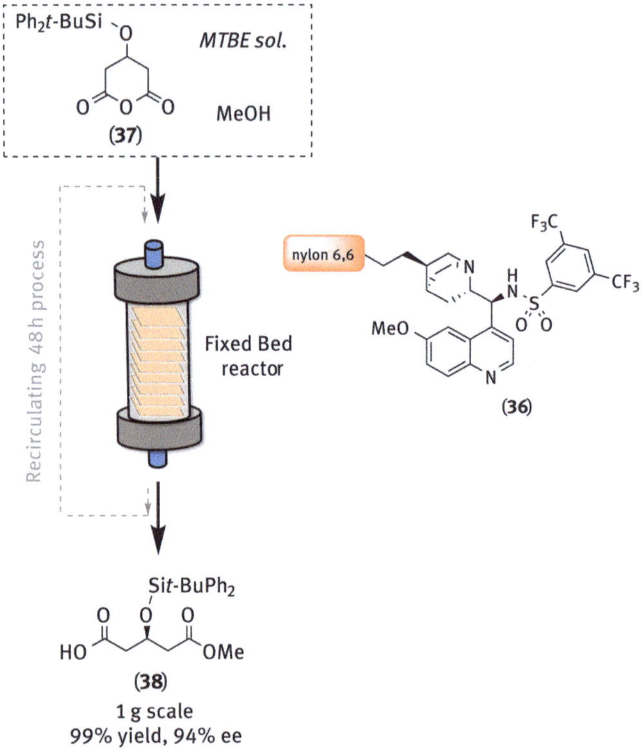

Figure 10.10: Packed-bed reactor with organocatalyst supported on Nylon fibers.

full conversion of the starting material **37** was achieved. After reaction time, it was possible to isolate 1 g of the desired product **38** in 99 % yield and 94 % ee. The column was then washed with MTBE and reused for 10 times with no loss of activity. This system resulted in a multi-gram scale synthesis of product **38** and proved the long-term stability of organocatalyst **36.**

Polyacrylonitrile fiber (PANF), which is well known as "artificial wool", as a synthetic fiber material, offers an ideal starting material for preparing various functionalized catalysts since it contains abundant modifiable cyano groups. Different from the nylon-supported catalyst which was modified by photochemical immobilization with low catalyst loadings, PANF can be modified by chemical immobilization with various functional groups with high catalyst loading in green solvents. Wang and Zhang group reported the design and preparation of chiral pyrrolidine functionalized polyacrylonitrile fiber catalysts, evaluated for their catalytic performance in asymmetric Michael addition of ketones to nitrostyrenes in water. Then the fiber catalysts were further applied to a packed-bed reactor for continuous-flow Michael addition [56].

The family of Cinchona alkaloids gave to organocatalysis some of the most active and stereoselective catalysts of the recent years [57]. 9-amino-9-deoxy-9-*epi*-cinchona alkaloids gave a boost to the field of asymmetric aminocatalysis since they are easily derived from natural sources and allowed the stereoselective functionalization of a variety of sterically hindered carbonyl compounds, which cannot be functionalized using secondary amines [58]. It is not surprising, then, to learn that there are many examples of supported versions, applied both in batch and in flow processes.

Benaglia research group prepared 9-amino-9-deoxy-9-*epi*-quinine supported on highly cross-linked polystyrene **39a** [59]. Catalyst **39a** was prepared by synthesizing a modified 9-amino-*epi*-quinine bearing a triazole as a linker and a styrene moiety, by an *ad hoc* procedure starting from commercially available quinine. The immobilization step was achieved by copolymerization of the chiral monomer at 70 °C in the presence of divinylbenzene (DVB) as a co-monomer, AIBN as a radical initiator, 1-dodecanol and toluene as the porogenic solvents, according to Frechèt procedure [60]. The solid catalyst **39a** was employed in a packed-bed catalytic reactor to perform, for the first time, the enantioselective Michael addition of an aldehyde to *trans*–β–nitrostyrene under continuous-flow conditions. The same catalytic packed-bed reactor was used for the preparation of (*S*)-Warfarin **41** under continuous-flow conditions. A solution of 4-OH-coumarin **42**, benzalacetone **43** and trifluoroacetic acid as a cocatalyst in dioxane was flowed into the reactor containing the polystyrene-supported 9-amino-epi-quinine **39a**. With a residence time of 5 h at 50 °C, the authors were able to isolate the product in up to 90 % yield and up to 87 % ee. (Figure 10.11) In the same year, the Pericàs group reported the sequential preparation of a small library of enantioenriched Michael adducts in flow mode by using a packed-bed reactor with catalyst **39b** [61]. They used polystyrene with a low degree of cross-linking as the support. The Cinchona-based primary amines require the presence of a Bronsted acid co-catalyst that helps the substrate activation. Ciogli et al. [62] reported the preparation of a silica-based catalyst containing both supported 9-amino-9-deoxy-9-*epi*-quinine and a benzoic acid derivative **40** through radical thiol-ene reaction. The bifunctional material **40** was first tested in batch, showing stereoselectivity comparable to the homogeneous system, and then in flow, in the addition of cyclohexanone to *trans*–β–nitrostyrene. The packed-bed reactor was then used for the preparation of warfarin, affording the desired product in 95 % isolated yield and 78 % ee after 16 h at room temperature (Figure 10.11).

The revolutionary work of MacMillan in 2000, on the use of chiral imidazolidinones as "organic catalysts" is a milestone in the field of (organo)catalysis, where he reported, for the first time, the reaction of α,β-unsaturated aldehydes through the formation of a transient, chiral iminium ion [63]. In this paper MacMillan used the term "organocatalysis" for the first time. Chiral imidazolidinones proved to be very efficient and versatile catalysts, therefore several authors have become interested in immobilizing these species onto a solid support with the aim of recovery and recycle the precious material [64]. In 2013 Cozzi and Benaglia reported the continuous-flow stereoselective organocatalyzed Diels Alder reactions in a chiral catalytic "homemade" HPLC column [65]; the silica-supported MacMillan imidazolidinone **44** was packed into an empty

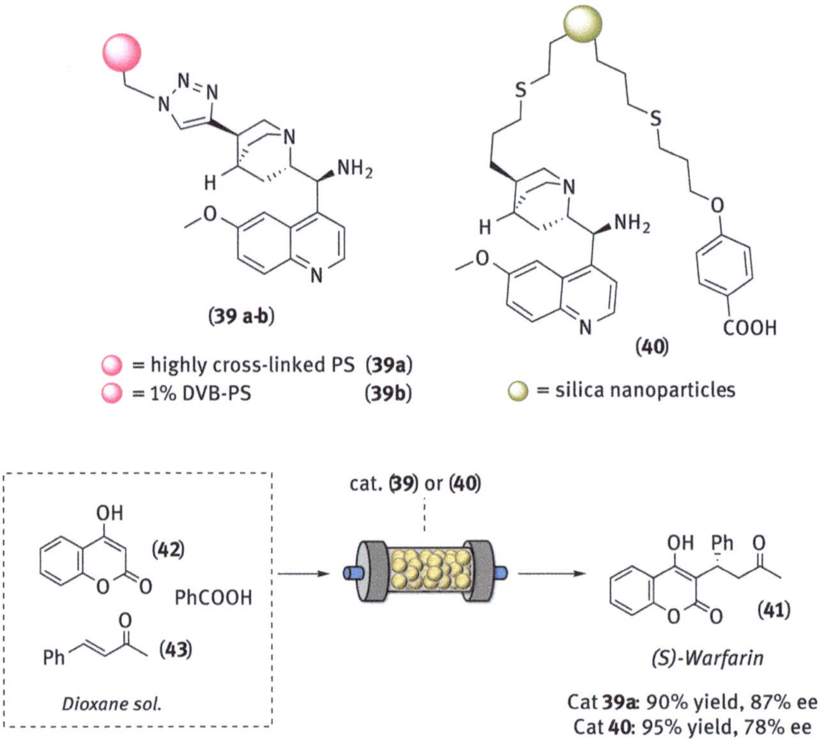

Figure 10.11: Continuous-flow synthesis of (*S*)-Warfarin by supported quinines.

stainless steel HPLC column and used to efficiently promote stereoselective Diels-Alder reactions on different substrates for more than 150 h of continuous operation [66]. Finelli et al. [67] reported the *in-flow* preparation of the first-generation MacMillan's organocatalyst immobilized onto silica through a carbamate linkage (Figure 10.12). Their synthesis started from the esterification of *L*-phenylalanine **45** that was carried out at 50 °C in a 3 mL coil (1 hour residence time) in 90% yield, followed by the formation of the corresponding ethanolamide **46** at 60 °C, (3 mL coil, 1 hour residence time, 90% yield). Amide **46** was then reacted with acetone in the presence of *p*-toluensulfonic acid (*p*TsOH) in a 2.4 mL column containing 4 Å molecular sieves (70 °C, 1 hour residence time, 90 % yield) to afford the desired 4-imidazolidinone **47**. The anchoring step was conducted by activating the 4-imidazolidinone **47** with carbodiimide **48** in a 3 mL coil (60 °C, 1.5 hours residence time, 75% yield) followed by pumping intermediate **49** in a 1.2 mL packed-bed reactor containing 3-aminopropyl-functionalized silica together with Et_3N (room temperature, 1.1 hours residence time). The in-flow synthesis of the organocatalyst **50** was accomplished in 4.5 hours and an overall yield of 54 % yield. As a proof of concept, they also evaluated the performance of the organocatalyst **50** in the benchmark reaction: the Diels-Alder between *E*-cinnamaldehyde **51** and

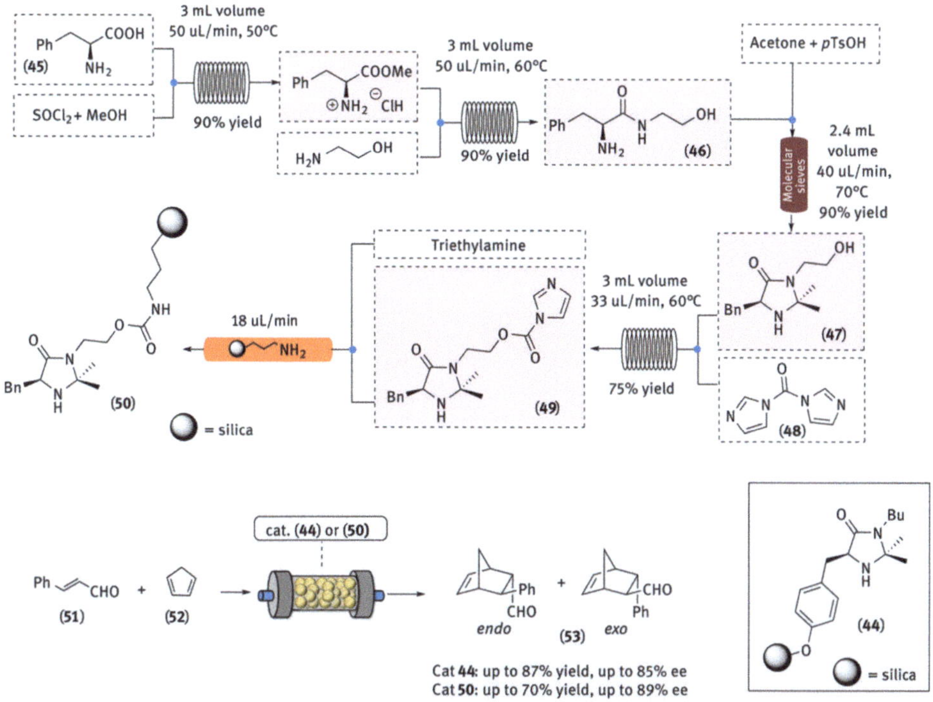

Figure 10.12: In-flow preparation of MacMillan's organocatalyst 50 immobilized onto silica and its use in the Diels-Alder reaction between *E*-cinnamaldehyde and cyclopentadiene.

cyclopentadiene **52**. The in-flow prepared catalyst **50** needed a long conditioning time (10 hours). After this time, adduct **53** was produced with yield from 60 % to 70 % and enantioselectivity from 84 % to 89 % for around 10 hours. However, a drastic drop in yields and enantioselectivity was observed after this time, with no possibility to regenerate the catalyst.

Hayashi–Jørgensen catalyst is a chiral catalyst belonging to the class of diarylprolinol silyl ethers. This organocatalyst was independently reported in 2005 by the groups of Jørgensen and Hayashi [68–70], and its popularity increased during the time since it is able to promote reactions both via the enamine and iminium ion activation modes, thus opening the possibility to use a variety of different substrates. It also proved to be excellent promoter of tandem processes. The Pericàs group was very active in the support and application of this catalyst and its modifications [71, 72]. In 2016 they reported a more challenging transformation, the asymmetric cyclopropanation promoted by catalyst **54** [73]. The authors prepared different types of supported catalysts, with different solid supports, having different swelling properties (microporous *vs* macroporous polystyrene) and the presence or absence of a triazole moiety as a linker/spacer. After extensive studies, catalyst **54** was selected for the continuous-flow cyclopropanation reaction between α,β-unsaturated aldehydes and bromo methyl malonate in the presence of *N*-methylimidazole as a base, as illustrated in Figure 10.13. To improve the process, at the end of the column the authors quenched with acid and introduced a liquid-liquid separator, in order to perform an in-line work-up. After evaporation, the desired products were purified by chromatography.

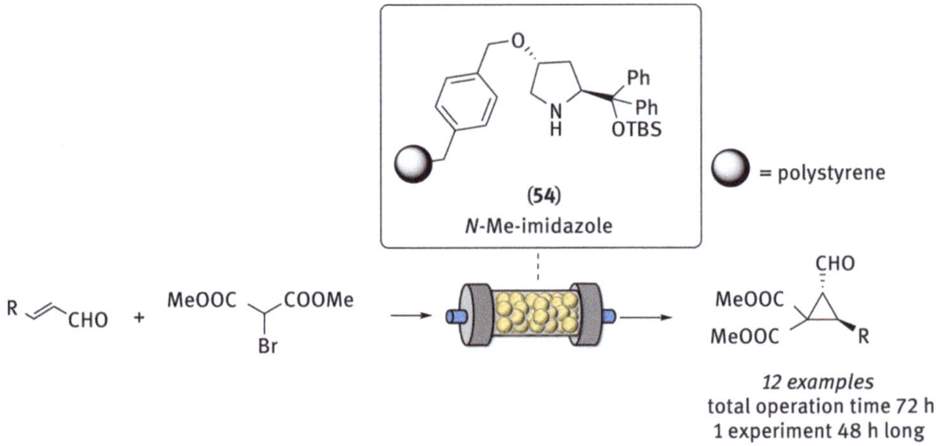

Figure 10.13: Stereoselective cyclopropanation with supported catalyst 54.

The Szczesniak group reported the preparation of a Wang resin-supported Hayashi–Jørgensen catalyst as optimal catalyst for the Micheal addition of aldehydes to *trans*–β–nitrostyrene [75]. Synthetic elaboration of the produced adducts lead to functionalized products, such as cyclic nitrones and pyrrolidines. The authors envisaged that also the *trans*-3,4-disubstituted piperidine ring systems of (+)-Paroxetine and (+)-Femoxetine could be built using the same methodology [74]. (–)-Paroxetine and (–)-Femoxetine are two selective serotonin reuptake inhibitors, widely used for the treatment of depression, anxiety, and panic disorders. For their investigations, they studied the Wang resin-supported Hayashi–Jørgensen catalyst **55**, in a simple homemade flow setup, consisting of feeding stream connected to a syringe pump and column packed with immobilized catalyst. (Figure 10.14(a)) Aldehyde **56** reacted with **57** to afford (+)-Paroxetine precursor **58** in 100 % conversion after 4 hours with 81:19 *syn:anti* ratio and with 93 % ee for the *syn* isomer. The synthetic modifications of **58** to (+)-Paroxetine were then conducted in batch mode. Almost simultaneously, Ötvös, Pericàs and Kappe reported a similar strategy for the preparation of a key intermediate of (-)-Paroxetine: the enantioselective asymmetric conjugate addition between 4-fluorocinnamaldehyde **59** and dimethyl malonate **60** to yield chiral aldehyde **61** [74]. As the chiral catalyst, they selected a polystyrene-supported *cis*-4-hydroxydiphenylprolinol TBS ether **62** developed as a modified version of the classical *trans* analogue. The authors performed extensive reaction parameters screening in flow, using a packed-bed Omnifit glass column as a flow reactor.

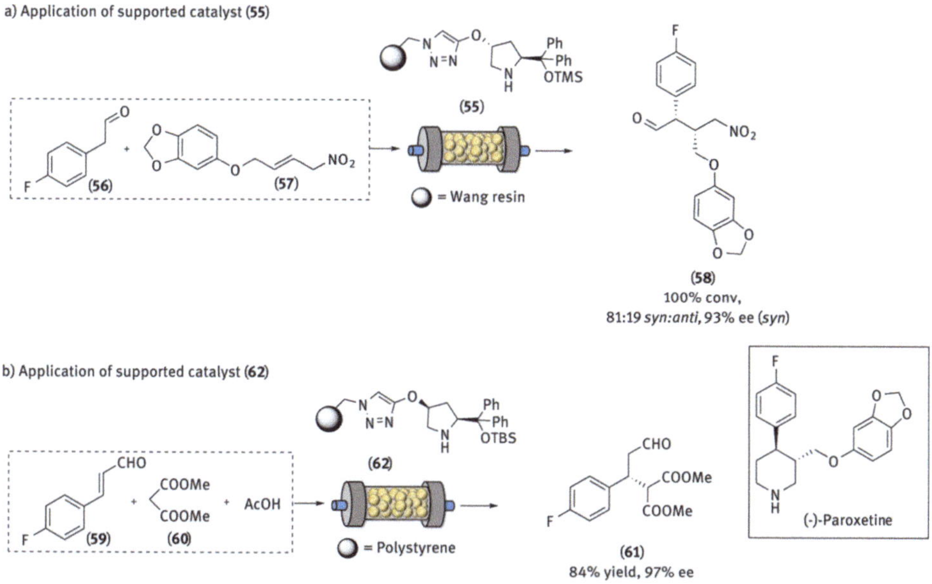

Figure 10.14: Continuous-flow synthesis of (+)- and (-)-Paroxetine precursors 58 and 61.

They found that heating at 60 °C was beneficial for enhancing the reaction rate without erosion of the enantioselectivity; moreover, running the reaction under solvent-free conditions allowed them to reduce the waste and to speed-up the product isolation. They also assessed the robustness of the catalyst and the preparative capability of their system by running a scale-out experiment. A mixture of aldehyde **59**, dimethyl malonate **60** (eq. (2)) and acetic acid as an additive was pumped into the flow reactor containing 1 g of the supported catalyst **62** maintained at 60 °C, with a flow rate of 70 μL/min in order to have 20 min residence time. (Figure 10.14(b)) After 7 h long continuous-flow experiment, they isolated more than 17 g of the chiral aldehyde **61** by simply removing unreacted components by evaporation (84 % isolated yield, 97 % ee). The large-scale synthesis offered a productivity of 2.47 g/h of the chiral aldehyde. Notably, adduct **61** was processed further via a telescoped reductive amination-lactamization-amide/ester reduction sequence to afford the chiral key intermediate of (-)-Paroxetine. These two last examples highlight the great potentiality of flow chemistry in the preparation of APIs or APIs precursors [13, 19, 21].

The Dixon group recently developed a new class of bifunctional catalysts [76], the chiral iminophosphorane (BIMP) organosuperbase catalysts, carrying a thiourea and a powerful Bronsted base, as an alternative to bifunctional tertiary amine/H-bond donor organocatalysts in Michael addition reactions. The authors developed a polystyrene-supported version of this new class of catalysts [77] that proved very active and that were, later on, applied to the continuous flow, stereoselective Michael additions of low-reactive pro-nucleophiles [78]. Polystyrene-supported catalyst **63** was tested in the reaction between **64** and *trans*–β–nitrostyrene **9** under flow chemistry conditions (Figure 10.15). The authors demonstrated the continuous-flow production of 8.6 g of **65** over the course of 13 h, in 79:21 dr and 78 % ee, with a productivity of 7.14 $mmol_{product}\ h^{-1}\ mmol_{catalyst}^{-1}$ (or 7.20 $mmol_{product}\ h^{-1}\ g_{catalyst}^{-1}$) with an effective catalyst loading of 0.8 mol %.

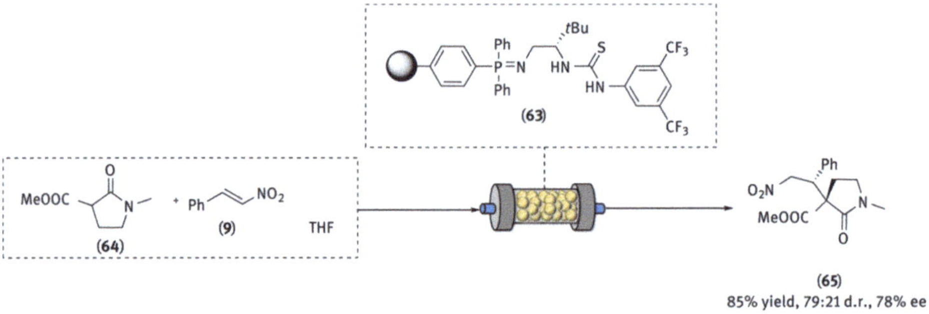

Figure 10.15: Polystyrene-supported iminophosphorane bifunctional catalyst 63 in stereoselective Michael additions under flow conditions.

As discussed this far, it should be clear that, to prepare a packed-bed reactors, a certain amount of the solid material, containing the catalytic supported species, is inserted in a reactor, typically an empty stainless steel or glass column, and used in a continuous process. Therefore, most of the packed-bed devices reported so far belong to the class of meso-reactors. Only recently, the "lab-on-a-chip" approach was also introduced to study heterogenized stereoselective catalysts at the microscale by the group of Massi and Belder [79]. The authors developed a lab-on-a-chip platform integrating a nanoliter-sized packed bed reactor which was seamlessly interfaced with a HPLC functionality using a chiral stationary phase for in-situ separation of enantiomeric products in the reactor effluent. The chip device, that was engineered based on the authors' expertise in designing glass chips incorporating HPLC columns, included two packed microcolumns, one for catalysis in flow and the other for direct downstream HPLC enantioseparation. The amount of supported catalyst packed in the microcolumn was in the ng scale, a massive reduced amount with respect to the classic packed bed reactor. For their studies, they used two different organocatalysts supported onto silica, a proline tetrazole derivative and the Hayashi-Jørgensen diaryl prolinol silylether. The novel approach, integrating immobilized heterogeneous organocatalysis with HPLC-MS on a single microfluidic device, enabled the rapid analysis of minute amounts of reactor output, by using little quantities of precious catalysts. Notably, it was possible to reuse a single chip over a time period of a couple months and the determined diastereo- and enantioselectivities were in good agreement with batch or up-scaled flow experiments. Given the possibility to easily change both stationary phases, namely the chiral catalyst and selector, we can expect that, in the future, this sort of device may serve as a useful screening tool.

10.4.2 Monolithic reactors

In a monolithic reactor, the catalyst occupies the space of the reactor in the form of a "monolith", a structured material possessing a regular or irregular network of channels, that is generally synthesized inside the reactor. The material has two types of pores: large flow-through pores, that allow the flow of the solution through the reactor, and smaller pores (meso- or micropores) that provide large surface area (typically $> 100\,m^2$) and allow the diffusion of reactants to the catalytic sites. Different types of monolithic materials were developed for applications in chromatography: because of their porous nature, they guarantee a great tolerance to high flow rates and an efficient mass transfer through their pores. The use of monolithic reactors for catalytic purposes offers several advantages over conventional packed beds: the greater tolerance to high flow rates allow higher back pressure and higher productivity.

Typically, a monolithic reactor is built by the copolymerization of different monomers, one containing the catalyst, and a large excess of cross-linking agent, in the presence of porogens inside the reactor. After the removal of the porogens, the resulting materials possess high void volume and large surface area that prevent pressure drops along the reactor, making them suitable for flow reactions. Moreover, due to the high cross-linking degree and the rigidity of the macromolecular network, these materials swell to a limited extent and are more resistant to mechanical stress. Of course, the synthesis of a properly modified catalyst appropriate for the polymerization step is required. Drawbacks of the monolithic reactors include pore clogging and non-uniformity of radial permeability. Moreover, the reduced accessibility of the catalytic sites buried deeply inside the micropores of the monolith lead to a "loss" of potentially active stereoselective sites. There are several reliable methods in the literature for the preparation of monoliths by radical addition polymerization of vinyl monomers and most of them are reported by Fréchet et al. [60].

These methods are often slightly modified to allow the convenient direct incorporation of the chiral catalyst into the polymer. However, when a different combination of monomers is attempted, some efforts can be required to obtain the desired polymeric structure. For this reason, also post-functionalization of the monolith is possible, where the monolith is prepared according to well-established standard procedures, based on polymerization of chloromethylstyrene/divinylbenzene [80], methacrylates [81] or norbornene-type monomers [82].

While there are different examples of metal-based monolithic catalytic reactors [44, 83] the application of monolithic reactors for stereoselective organocatalysis is still limited.

In the first example, Benaglia and Mandoli reported an *ad hoc* designed MacMillan type catalyst bearing a triazole spacer and a styrene moiety (compound **66**, Figure 10.16), easily synthesized starting from (*S*)-tyrosine methyl ester [84]. Compound **66** was used as a monomer in a copolymerization reaction inside an empty HPLC column at 70 °C in the presence of divinylbenzene (DVB) as a co-monomer, AIBN as a radical initiator, 1-dodecanol and toluene as the porogenic solvents (3: 1 v/v, approx. 60 vol % of the feed mixture). All the chiral monomer **66** was incorporated into the monolith, so it was possible to determine the absolute amount of the MacMillan derivative immobilized inside the flow device and the loading onto the polymeric support directly from the feed composition. The prepared flow devices were used in the stereoselective Diels-Alder cycloaddition between *trans*-cinnamaldehyde **51** and cyclopentadiene **52** as a model reaction. The first flow device was activated by trifluoroacetic acid and tested under different flow rates. After a conditioning time of 4–6 hours, a steady-state regime was reached, that allowed to produce in continuo the desired cycloadducts in 54–61 % yield. In agreement with the results obtained with the non-supported catalyst, product **53** was obtained as a rough 1:1 mixture of *endo*/*exo* isomers, both with enantioselectivities higher than 90 % ee. In order to improve the chemical yield, the flow rate was reduced so as to increase the residence time: indeed with a flow rate of 2 μL/min the product was isolated in 77 % yield that was further incremented to 91 %, by operating

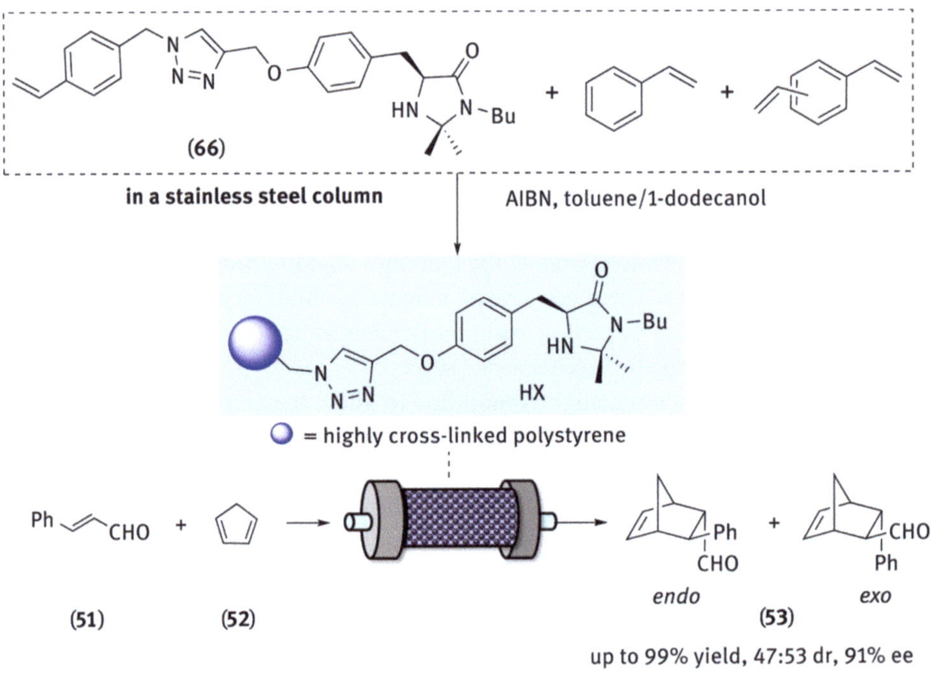

Figure 10.16: Preparation of a chiral monolithic flow reactor and its use in stereoselective Diels-Alder cycloaddition of *trans*-cinnamaldehyde with cyclopentadiene.

at 1 μL/min. This reactor was used to catalyze two additional organocatalytic reactions: the 1,3-dipolar cycloaddition between *N*-benzyl-*C*-phenyl nitrone and crotonic aldehyde, and the Friedel–Crafts alkylation of *N*-methyl pyrrole with cynnamaldehyde. A second flow device was activated by tetrafluoroboric acid and used for performing the reaction *in continuo* between cyclopentadiene and three different aldehydes. After 3 days of continuous operation the reactor was washed and used for carrying out the reaction between cyclopentadiene and crotonic aldehyde, affording the cycloadduct in yields higher than 94 % and enantioselectivity up to 85 % ee. Finally, in order to verify the activity of the system after a prolonged TOS the reactor was washed once more and used to promote again the initial reaction between cyclopentadiene and cinnamic aldehyde. Indeed, after 150 working hours of the catalytic reactor, the product was isolated in yield and stereoselectivity totally comparable to those of the first 24 hours of activity. Despite the low flow rates necessary to reach satisfactory conversions, the authors demonstrated that the tetrafluoborate chiral monolithic reactor was tolerant to flow rate increase, and a remarkable level of productivity of 338 h^{-1} was reached by using a flow rate of 18.8 μL/min. The monolithic reactor outperformed the corresponding packed-bed reactor with a 3-fold increase in productivity.

In a similar approach, Massi et al. [85] prepared a polystyrene monolithic pyrrolidinyl-tetrazole reactor to promote the aldol reactions of various cyclic and acyclic ketones with differently substituted aromatic aldehydes. Interestingly, integration of an analytical platform for monitoring the reaction progress in the flow regime was an important objective of this study. The polymerization mixture containing monomer **67** (Figure 10.17) was transferred into the stainless steel column, which was heated at 70 °C for24 h in a standard convection oven. After cooling, the resulting monolithic microreactor was connected to a HPLC instrument and then washed with THF to remove the porogens and residual non-polymeric material. The *N*-Boc deprotection step was next performed by sequentially flowing TFA/THF and Et_3N/THF solutions through the column. The authors observed nearly no swelling in water–ethanol (1: 1) solvent. The outflow of the microreactor was redirected to a 6-port 2-position switching valve, connected to a second pump for dilution and a HPLC. This set-up allowed to easily monitor the reaction course and to speed up the optimization conditions. The scope of the reaction was extended to 16 aldol products that were easily isolated in ee comparable to the ee reported in the literature for the non-supported catalyst in batch. The authors also assessed the remarkable long-term stability of the catalytic bed (ca. five days on stream).

In Section 4.1 it was already reported the work by Pericàs and Rodrìguez-Escrich, who developed different reactors with supported Jørgensen-Hayashi diarylprolinol for the continuous preparation of enantioenriched cyclopropanes [73]. In this same work, they also prepared the monolithic version of catalyst **54** (see Figure 10.13). Catalyst preparation was achieved under the classic Frechèt-type polymerization conditions of the proper monomer in the presence of styrene and divinylbenzene as the co-monomers. The authors prepared three different monomers, exploiting different anchoring sites

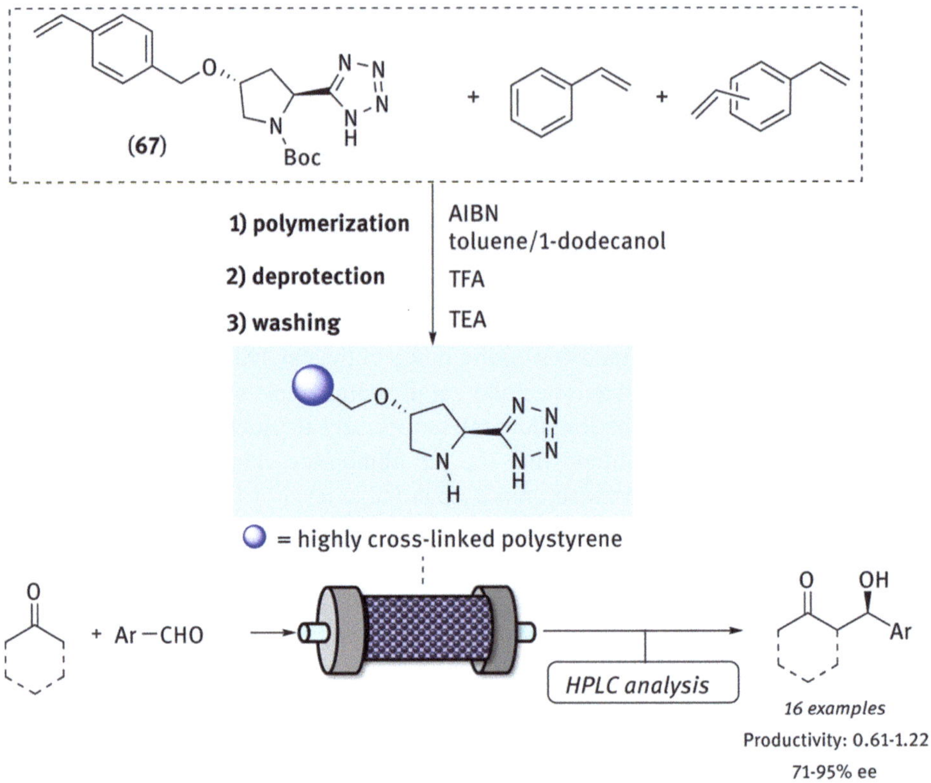

Figure 10.17: Preparation of a chiral polystyrene monolithic pyrrolidinyl-tetrazole flow reactor and itsuse in stereoselective Aldol reactions.

and the presence of a spacer (a triazole moiety). As mentioned, they also synthesized the analogue resins for comparison purpose and tested the 6 supported catalysts in the cyclopropanation reaction between α,β-unsaturated aldehydes and bromomalonate with *N*-methylimidazole as a base. They found out that the resin (microporous polymer) was more performing that the monolith (macroporous polymer), that showed a sharp drop in catalytic activity after 24 h operation, probably due to mechanical collapse of the system under the operating pressure.

10.4.3 Inner wall-functionalized reactors

One of the major drawbacks of the flow devices of the smallest section is their strong propensity to clog, especially if the narrow channels are packed with a powdered catalyst or filled for a significant volume fraction (e. g 30–40 %) by a monolithic porous material. From the fluid-dynamic point of view, a better solution would be the catalyst deposited onto the channels' inner walls to leave a free central bore for unhindered flow. On the contrary, given the confinement of the catalyst in a seemingly small fraction of the overall channels volume, this choice might look less than optimal in view of the reduced amount of active species which can be immobilized onto the continuous-flow reactor walls and the limited contact between the reactants and the catalyst sites.

Inner wall-functionalized, also called "wall-coated" reactors, represent a fascinating, still almost unexplored opportunity in which the catalyst is covalently attached onto the reactor inner walls. The fabrication of this type of catalytic reactor can simply involve a deposition of the catalyst onto the bare reactor walls. The word "deposition" can be immediately associated with metals: and indeed, most of the reported examples involve the use of metal or metal oxides catalysts confined inside borosilicate glass or metal capillaries [83].

Regarding organocatalysis, the only reported example belong to Verboom group, who developed a method to covalently immobilize poly(glycidyl methacrylate) (PGMA) polymer brushes on the interior of a microreactor to get a brush film [86]. The oxirane groups of the PGMA brushes were used for the anchoring of an achiral catalytic base, namely, 1,5,7-triazabicyclo[4.4.0]dec-5-ene (TBD) **68**. The authors observed that the catalytic brush film thickness linearly increased on increasing the polymerization time; this lead to an enhanced catalytic loading on the microchannel walls. The preparation procedure is summarized in Figure 10.18: several microreactors (100 µm width and depth, 103 cm length) were filled with a solution of glycidyl methacrylate (GMA) monomer **69** in MeOH:H_2O 4:1 in the presence of CuBr and 2,2′-dipyridyl. The solution was left inside the channels for 20–120 min. Afterward, a 0.1 M solution of TBD **68** in EtOH was flowed through the channel at 65 °C for 17 h, leaving the inner walls coated with PGMA polymer brushes. By varying the polymerization time, it was possible to tune the thickness of the polymer

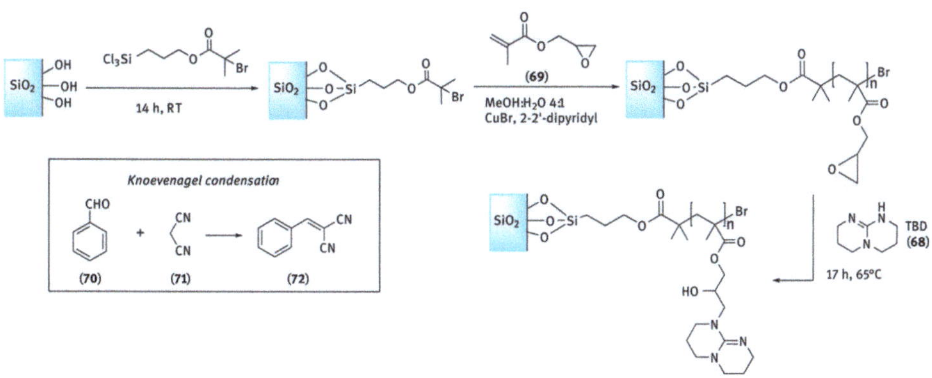

Figure 10.18: Preparation of inner-wall functionalized microreactor and application in Knoevenagel condensation.

brushes and the amount of catalyst. The amount of catalyst in the brushes was measured by X-ray photoelectron spectroscopy (XPS). An acid-base titration procedure allowed to estimate the number of TBD units in the polymer and to correlate this number to the brushes thicknesses. The Knoevenagel reaction between benzaldehyde **70** and malononitrile **71** to give 2-benzylidene malononitrile **72** was selected as a model reaction to study the performance of these catalytic microreactors and was carried out in acetonitrile at 65 °C, in continuous flow (Figure 10.18). The formation of the condensation product was monitored in real time by in-line ultraviolet–visible (UV–vis) detection and the reaction times were varied by changing the flow rates from 20 to 0.2 µL/min. The polymeric coating turned out to be highly effective in the catalysis and all the experiments carried out with the microreactors bearing coatings with different thicknesses, measured by high-resolution scanning electron microscopy (HR-SEM), showed a linear dependence with the catalytic activity. According to the authors, the whole nanostructure was involved in the catalysis and the reaction did not occur only at the interface, but the reagents diffused throughout the coating to reach all catalytic units driven by the complete swelling of the PGMA polymer brushes in acetonitrile. After each experiment, the catalytic system was regenerated by flushing a 0.1 M solution of triethylamine through the microchannel. The PGMA-TBD coated devices showed no decreasing in the catalytic activity or leaching after being used for 25 times. The catalytic device, stored under nitrogen, was able to reproduce the same results of the experiments also after 30 days.

Later, the same group reported the hydrolysis of 4-nitrophenyl acetate catalyzed by a lipase supported on poly(methacrylic acid) polymer brushes [87].

Applications of the polymer-brushes methodology in the development of catalytic mini or microreactors to be used in flow processes for promoting stereoselective transformations can be easily envisaged; however, the use is still unreported at the moment.

10.5 Outlook and perspectives

In the last 10 years, we assisted to a blossoming of organocatalytic processes in flow. In one sense, the use of flow chemistry for organocatalytic reactions helped to overcome the main problem associated with organocatalysis, the high catalyst loading. Although some of the studies discussed in this chapter reported that grams of relatively complex organic products could be obtained at a reasonable productivity rate with mini- and-micro-continuous-flow reactors developed by academic research groups, there is still the need of improving the productivity of these systems. Some issues have to be overcome if we want to develop a reliable organocatalytic flow process, for example activation, efficiency, lifetime, degradation of the catalysts and possible reactivation of supported chiral catalysts. Improvements can be achieved by either (i) improving the catalyst efficiency or

(ii) improving the support stability in order to obtain long-lasting continuous process in case of supported catalysts.

Regarding the organocatalysts, by going through the data one may have the impression that the relevance of much of the work published so far is essentially limited to the proof-of-concept stage. This is true if we consider that most of the new devices are tested for benchmark reactions; the transformation of compounds of practical interest often turns out to be more challenging than with the model substrates. However, we have demonstrated, with selected examples, that useful intermediates or precursors of interesting molecules can be easily prepared in useful amounts by exploiting chiral organocatalysts in flow. The potential for discovery of these organocatalytic systems is still not fully exploited.

The use of supported (chiral) organocatalysts is undoubtedly more appealing: at the cost of an *ad hoc* designed and prepared catalyst, the organic chemist has the opportunity to develop a more efficient and convenient reaction, in which the product is recovered almost pure. Regarding the possible supports, the choices appear almost unlimited; however, it is basically the expertise of the users, together with the final application (reaction, solvent, etc.) that will drive the decision towards one material or another. We showed that, at the moment, only few materials are usually employed.

Another big issue to consider is the robustness of the catalysts. Although some systems proved to be more robust than others, the problem of catalyst's degradation and/or deactivation is a huge limit for the application of organocatalysts in industry. We expect that, in the near future, some of the most promising catalytic systems will undergo studies for their deactivation pathways, in order to bring deeper knowledge and thus leading to improved protocols for extending the lifespan of the catalyst.

We assisted to a great improvement of organocatalytic processes, that now can bear the competitions with metal-based systems. Chiral catalyst immobilization emerged as a step toward the solution of the longstanding problem of homogeneous catalysis: recovery and reuse. The subsequent application of organocatalytic systems in continuous-flow reactions contributed to opening the way towards industrial applications. We expect that the combination of organocatalysis with flow chemistry and other enabling technologies will greatly implement the arsenal of the synthetic organic chemists towards the development of sustainable and efficient processes.

References

[1] Wegner J, Ceylan S, Kirschning A. Flow chemistry – a key enabling technology for (multistep) organic synthesis. Adv Synth Catal. 2012;354:17–57.

[2] Jas G, Kirschning A. Continuous flow techniques in organic synthesis. Chem Eur J. 2003;9:5708–23.

[3] Fitzpatrick DE, Battilocchio C, Ley SV. Enabling technologies for the future of chemical synthesis. ACS Cent Sci. 2016;2:131–8.

[4] Kirschning A, Solodenko W, Mennecke K. Combining enabling techniques in organic synthesis: continuousflow processes with heterogenized catalysts. Chem Eur J. 2006;12:5972–90.
[5] Wegner J, Ceylan S, Kirschning A. Ten key issues in modern flow chemistry. Chem Commun. 2011;47:4583–92.
[6] Hartman RL, McMullen JP, Jensen KF. Deciding whether to go with the flow: evaluating the merits of flow reactors for synthesis. Angew Chem Int Ed. 2011;50:7502–19.
[7] Wiles C, Watts P. Recent advances in micro reaction technology. Chem Commun. 2011;47:6512–35.
[8] Darvas F, Dorman G, Hessel V. Flow chemistry organic: volume 1: fundamentals. Walter de Gruyter GmbH, 2014. (De Gruyter Textbook).
[9] Wirth T. Microreactors in organic synthesis and catalysis. 2nd ed. Weinheim: Wiley-VCH, 2013.
[10] Plutschack MB, Pieber B, Gilmore K, Seeberger PH. The Hitchhiker's guide to flow chemistry. Chem Rev. 2017;117:11796–893.
[11] Lee SL, O'Connor TF, Yang X, Cruz CN, Chatterjee S, Madurawe RD, et al. Modernizing pharmaceutical manufacturing: from batch to continuous production. J Pharm Innov. 2015;10:191–200.
[12] https://www.in-pharmatechnologist.com/Article/2015/05/01/FDA-calls-on-manufacturers-to-begin-switch-from-batch-to-continuous-production.
[13] Malet-Sanz L, Susanne F. Continuous flow synthesis. a pharma perspective. J Med Chem. 2012;55:4062–98.
[14] Rodrigues T, Schneider P, Schneider G. Accessing new chemical entities through microfluidic systems, angew. Chem Int Ed. 2014;53:5750–8.
[15] Welch CJ, Hawkins JM, Tom J. Precompetitive collaboration on enabling technologies for the pharmaceutical industry. Org Process Res Dev. 2014;18:481–7.
[16] Baumann M, Baxendale IR. The synthesis of active pharmaceutical ingredients (APIs) using continuous flow chemistry, Beilstein. J Org Chem. 2015;11:1194–219.
[17] Gutmann B, Cantillo D, Kappe CO. Continuous-flow technology-a tool for the safe manufacturing of active pharmaceutical ingredients. Angew Chem Int Ed. 2015;54:6688–728.
[18] Baxendale IR, Braatz RD, Hodnett BK, Jensen KF, Johnson MD, Sharratt P, et al. Achieving continuous manufacturing: technologies and approaches for synthesis, workup, and isolation of drug substance. J Pharm Sci. 2015;104:781–91.
[19] Porta R, Benaglia M, Puglisi A. Flow chemistry: recent developments in the synthesis of pharmaceutical products. Org Process Res Dev. 2016;20:2–25.
[20] Badman C, Cooney CL, Florence A, Konstantinov K, Krumme M, Mascia S, et al. Why we need continuous pharmaceutical manufacturing and how to make it happen. J Pharm Sci. 2019;108:3521–3.
[21] Baumann M, Moody TS, Smyth M, Wharry S. A perspective on continuous flow chemistry in the pharmaceutical industry. Org Process Res Dev. 2020. DOI:10.1021/acs.oprd.9b00524.
[22] Anderson NG. Using continuous processes to increase production. Org Process Res Dev. 2012;16:852–69.
[23] Gérardy R, Emmanuel N, Toupy T, Kassin V-E, Tshibalonza NN, Schmitz M, et al. Continuous flow organic chemistry: successes and pitfalls at the interface with current societal challenges. Eur J Org Chem. 2018;2018:2301–51.
[24] Tsukanov SV, Johnson MD, May SA, Rosemeyer M, Watkins MA, Kolis SP, et al. Development of an Intermittent-flow enantioselective Aza-Henry reaction using an arylnitromethane and homogeneous brønsted acid–base catalyst with recycle. Org Process Res Dev. 2016;20:215–26.
[25] Rossi S, Puglisi A, Benaglia A. Additive manufacturing technologies: 3D printing in organic synthesis. ChemCatChem. 2018;10:1512–25.

[26] Odedra A, Seeberger PH. 5-(Pyrrolidin-2-yl)tetrazole-Catalyzed Aldol and Mannich reactions: acceleration and lower catalyst loading in a continuous-flow reactor. Angew Chem Int Ed. 2009;48:2699–702.
[27] Carroccia L, Musio B, Degennaro L, Romanazzi G, Luisi R. Microreactor-mediated organocatalysis: towards the developmentof sustainable domino reactions. J Flow Chem. 2013;3:29–33.
[28] Nakashima E, Yamamoto H. Process catalyst mass efficiency by using proline tetrazole column-flow system. Chem Eur J. 2018;24:1076–9.
[29] Pirola M, Puglisi A, Raimondi L, Forni A, Benaglia M. Evaluation of in-batch and in-flow synthetic strategies towards the stereoselective synthesis of a fluorinated analogue of retro-thiorphan. Molecules. 2019;24:2260–9.
[30] Schobera L, Ratnama S, Yamashitaa Y, Adebara N, Piepera M, Berkessel A, et al. An asymmetric organocatalytic aldol reaction of a hydrophobic aldehyde in aqueous medium running in flow mode. Synthesis. 2019;51:1178–84.
[31] See, for example: Hall AM, Broomfield-Tagg R, Camilleri M, Carbery DR, Codina A, Whittaker DT, et al. Online monitoring of a photocatalytic reaction by real-time high resolution flowNMR spectroscopy. Chem Commun. 2017; 54: 30–3.
[32] Rueping M, Bootwicha T, Sugiono E. Continuous-flow catalytic asymmetric hydrogenations: reaction optimization using FTIR inline analysis. Beilstein J Org Chem. 2012;8:300–7.
[33] Ormerod D, Lefevre N, Dorbec M, Eyskens I, Vloemans P, Duyssens K, et al. On the potential of homogeneous Pd catalyst separation by ceramic membranes. Application to down-stream and continuous flow processes. Org Process Res Dev. 2016;20:911–20.
[34] Kragl U, Dreisbach C. Continuous asymmetric synthesis in a membrane reactor. Angew Chem Int Ed. 1996; 35: 642. For the first organocatalytic asymmetric synthesis with membrane separation, see:.
[35] Kisszekelyi P, Alammar A, Kupai J, Huszthy P, Barabas J, Holtzl T, et al. Asymmetric synthesis with cinchona-decorated cyclodextrin in a continuous-flow membrane reactor. J Catal. 2019;371:255–61.
[36] Nicewicz DA, MacMillan DW. Merging photoredox catalysis with organocatalysis: the direct asymmetric alkylation of aldehydes. Science. 2008;322:77–80.
[37] Brimioulle R, Lenhart D, Maturi MM, Bach T. Enantioselective catalysis of photochemical reactions. Angew Chem Int Ed Engl. 2015;54:3872–90.
[38] Meggers E. Asymmetric catalysis activated by visible light. Chem Commun. 2015;51:3290–301.
[39] Yoon TP. Photochemical stereocontrol using tandem photoredox-chiral lewis acid catalysis. Acc Chem Res. 2016;49:2307–15.
[40] Skubi KL, Blum TR, Yoon TP. Dual catalysis strategies in photochemical synthesis. Chem Rev. 2016;116:10035–74.
[41] Zou YQ, Hormann FM, Bach T. Iminium and enamine catalysis in enantioselective photochemical reactions. Chem Soc Rev. 2018;47:278–90.
[42] Neumann M, Zeitler K. Application of microflow conditions to visible light photoredox catalysis. Org Lett. 2012;14:2658–61.
[43] Sugiono E, Rueping M. A combined continuous microflow photochemistry and asymmetric organocatalysis approach for the enantioselective synthesis of tetrahydroquinolines. Beilstein J Org Chem. 2013;9:2457–62.
[44] Puglisi A, Benaglia M, Chiroli V. Stereoselective organic reactions promoted by immobilized chiral catalysts in continuous flow systems. Green Chem. 2013;15:1790–813.
[45] Rodríguez-Escrich C, Pericàs MA. Organocatalysis on tap: enantioselective continuous flow processes mediated by solid-supported chiral organocatalysts. Eur J Org Chem. 2015;2015:1173–88.

[46] Kamahori K, Ito K, Itsuno S. Asymmetric diels-alder reaction of Methacrolein with cyclopentadiene using polymer-supported catalysts: design of highly enantioselective polymeric catalysts. J Org Chem. 1996;61:8321–4.
[47] Fraile JM, Garcia JI, Herrerias CI, Mayoral JA, Pires E. Enantioselective catalysis with chiral complexes immobilized on nanostructured supports. Chem Soc Rev. 2009;38:695–706.
[48] Hafez AM, Taggi AE, Wack H, Drury III WJ, Lectka T. Column asymmetric catalysis for β-Lactam synthesis. Org Lett. 2000;2:3963–5.
[49] Puglisi A, Benaglia M, Chiroli V. Stereoselective organic reactions promoted by immobilized chiral catalysts in continuous flow systems. Green Chem. 2013;15:1790–813.
[50] Puglisi A, Benaglia M, Porta R, Coccia F. Organocatalysis in flow. Curr Organocatalysis. 2015;2:79–101.
[51] Atodiresei I, Vila C, Rueping M. Asymmetric organocatalysis in continuous flow: opportunities for impacting industrial catalysis. ACS Catal. 2015;5:1972–85.
[52] Rodríguez-Escrich C, Pericàs MA. Catalytic enantioselective flow processes with solid-supported chiral catalysts. Chem Rec. 2018;18:1–20.
[53] Zhang L, Han L, Zhao G, Chai R, Zhang Q, Liu Y, et al. Structured Pd–Au/Cu-fiber catalyst for gas-phase hydrogenolysis of dimethyl oxalate to ethylene glycol. Chem Commun. 2015;51:10547–50.
[54] Mason BP, Price KE, Steinbacher JL, Bogdan AR, McQuade DT. Greener approaches to organic synthesis using microreactor technology. Chem Rev. 2007;107:2300–18.
[55] Lee J-W, Mayer-Gall T, Opwis K, Song CE, Gutmann JS, List B. Organotextile catalysis. Science. 2013;341:1225.
[56] Du J, Shuai B, Tao M, Wang G, Zhang W. Pyrrolidine modified PANF catalyst for asymmetric Michael addition of ketones to nitrostyrenes in aqueous phase. Green Chem. 2016;18:2625–31.
[57] Hiemstra H, Marcelli T. Cinchona alkaloids in asymmetric organocatalysis. Synthesis. 2010;8:1229–79.
[58] Melchiorre P. Cinchona-based primary amine catalysis in the asymmetric functionalization of carbonyl compounds. Angew Chem Int Ed. 2012;51:9748–70.
[59] Porta R, Benaglia M, Coccia F, Cozzi F. Puglisi A solid supported 9-Amino-9-deoxy-epi-quinine as efficient organocatalyst for stereoselective reactions in batch and under continuous flow conditions. Adv Synth Catal. 2015;357:377–83.
[60] Viklund C, Svec F, Fréchet JM, Irgum K. Monolithic, "Molded", porous materials with high flow characteristics for separations, catalysis, or solid-phase chemistry: control of porous properties during polymerization. Chem Mater. 1996;8:744–50.
[61] Izquierdo J, Ayats C, Henseler AH, Pericas MA. A polystyrene-supported 9-amino (9-deoxy) epi quinine derivative for continuous flow asymmetric Michael reactions. Org Biomol Chem. 2015;13:4204–9.
[62] Ciogli A, Capitani D, Di Iorio N, Crotti S, Bencivenni G, Donzello MP, et al. A silica-supported catalyst containing 9-Amino-9-deoxy-9-epi-quinine and a benzoic acid derivative for stereoselective batch and flow heterogeneous reactions. Eur J Org Chem. 2019;2019:2020–8.
[63] Ahrendt KA, Borths CJ, MacMillan DW. New strategies for organic catalysis: the first highly enantioselective organocatalytic Diels– Alder reaction. J Am Chem Soc. 2000;122:4243–4.
[64] Benaglia M, Puglisi A. Editors Catalyst immobilization: methods and applications. Wiley-VCH Verlag GmbH & Co., 2020.
[65] Chiroli V, Benaglia M, Cozzi F, Puglisi A, Annunziata R, Celentano G. Continuous-flow stereoselective organocatalyzed Diels–Alder reactions in a chiral catalytic "Homemade" HPLC column. Org Lett. 2013;15:3590–3.

[66] Porta R, Benaglia M, Chiroli V, Coccia F, Puglisi A. Stereoselective Diels-Alder reactions promoted under continuous-flow conditions by silica-supported chiral organocatalysts. Isr J Chem. 2014;54:381–94.

[67] Oliveira PH, Santos BM, Leão RA, Miranda LS, San Gil RA, Souza RO, et al. From immobilization to catalyst use: a complete continuous-flow approach towards the use of immobilized organocatalysts. ChemCatChem. 2019;11:5553–61.

[68] Franzen J, Marigo M, Fielenbach D, Wabnitz TC, Kjaersgaard A, Jorgensen KA. A general organocatalyst for direct α-Functionalization of Aldehydes: stereoselective C–C, C–N, C–F, C–Br, and C–S bond-forming reactions. Scope and mechanistic insights. J Am Chem Soc. 2005;127:18296–304.

[69] Marigo M, Wabnitz TC, Fielenbach D, Jorgensen KA. Enantioselective organocatalyzed α-Sulfenylation of Aldehydes. Angew Chem Int Ed. 2005;44:794–7.

[70] Hayashi Y, Gotoh H, Hayashi T, Shoji M. Diphenylprolinol silyl ethers as efficient organocatalysts for the asymmetric Michael reaction of aldehydes and nitroalkenes. Angew Chem Int Ed. 2005;44:4212–5.

[71] Fan X, Rodriguez-Escrich C, Sayalero S, Pericas MA. Paraldehyde as an acetaldehyde precursor in asymmetric Michael reactions promoted by site-isolated incompatible catalysts. Chem Eur J. 2013;19:10814–17.

[72] Fan X, Rodriguez-Escrich C, Wang S, Sayalero S, Pericas MA. Highly enantioselective cross-Aldol reactions of Acetaldehyde mediated by a dual catalytic system operating under site isolation. Chem Eur J. 2014;20:13089–93.

[73] Llanes P, Rodriguez-Escrich C, Sayalero S, Pericas MA. Organocatalytic enantioselective continuous-flow cyclopropanation. Org Lett. 2016;18:6292–5.

[74] a) Szcześniak P, Staszewska-Krajewska O, Furman B, Mlynarski J. Solid supported Hayashi–Jørgensen catalyst as an efficient and recyclable organocatalyst for asymmetric Michael addition reactions. Tetrahedron Asymmetry. 2017;28:1765–73. b) Szcześniak P, Buda S, Lefevre L, Staszewska-Krajewska O, Mlynarski J. Total asymmetric synthesis of (+)-Paroxetine and (+)-Femoxetine. Eur J Org Chem. 2019;2019:6973–82.

[75] Otvos S. B., Pericàś M. A., Kappe C. O. Multigram-scale flow synthesis of the chiral key intermediate of (-)-paroxetine enabled by solvent-free heterogeneous organocatalysis. Chem. Sci. 2019;10:11141–6.

[76] Yang J, Farley AJ, Dixon DJ. Enantioselective bifunctional iminophosphorane catalyzed sulfa-Michael addition of alkyl thiols to unactivated β-substituted-α, β-unsaturated esters. Chem Sci. 2017;8:606–10.

[77] Goldys AM, Núñez MG, Dixon DJ. Creation through immobilization: A new family of high performance heterogeneous bifunctional iminophosphorane (BIMP) superbase organocatalysts. Org Lett. 2014;16:6294–7.

[78] Farley AJ, Jakubec P, Goldys AM, Dixon DJ. Bifunctional iminophosphorane superbases: potent organocatalysts for enantio- and diastereoselective Michael addition reactions. Tetrahedron. 2018;74:5206–12.

[79] Warias R, Zaghi A, Heiland JJ, Piendl SK, Gilmore K, Seeberger PH, et al. An integrated lab-on-a-chip approach to study heterogeneous enantioselective catalysts at the microscale. ChemCatChem. 2018;10:5382–5.

[80] See, for example: Nikbin N, Ladlow M, Ley SV. Continuous flow ligand-free Heck reactions using monolithic Pd [0] nanoparticles. Org Process Res Dev. 2007; 11: 458–62.

[81] See, for example: Poupart R, Nour El Houda D, Chellapermal D, Guerrouache M, Carbonnier B, Le Droumaguet B. Novel in-capillary polymeric monoliths arising from glycerol carbonate methacrylate for flow-through catalytic and chromatographic applications. RSC Adv. 2016; 6: 13614–17.

[82] Anderson EB, Buchmeiser MR. Catalysts immobilized on organic polymeric monolithic supports: from molecular heterogeneous catalysis to biocatalysis. ChemCatChem. 2012;4:30–44.
[83] Munirathinam R, Huskens J, Verboom W. Supported catalysis in continuous-flow microreactors. Adv Synth Catal. 2015;357:1093–123.
[84] Chiroli V, Benaglia M, Puglisi A, Porta R, Jumde RP, Mandoli A. A chiral organocatalytic polymer-based monolithicreactor. Green Chem. 2014;16:2798–806.
[85] Greco R, Caciolli L, Zaghi A, Pandoli O, Bortolini O, Cavazzini A, et al. A monolithic 5-(pyrrolidin-2-yl)tetrazole flow microreactor for the asymmetric aldol reaction in water-ethanol solvent. React Chem Eng. 2016;1:183–93.
[86] Costantini F, Bula WP, Salvio R, Huskens J, Gardeniers HJGE, Reinhoudt DN, et al. Nanostructure based on polymer brushes for efficient heterogeneous catalysis in microreactors. J Am Chem Soc. 2009;131:1650–1.
[87] Costantini F, Benetti EM, Reinhoudt DN, Huskens J, Vancso GJ, Verboom W. Enzyme-functionalized polymer brush films on the inner wall of silicon-glass microreactors with tunable biocatalytic activity. Lab Chip. 2010;10:3407–12.

Armando Carlone, Luca Bernardi

11 Enantioselective organocatalytic approaches to active pharmaceutical ingredients – selected industrial examples

Abstract: Catalysis is, often, the preferred approach to access chiral molecules in enantioenriched form both in academia and in industry; nowadays, organocatalysis is recognised as the third pillar in asymmetric catalysis, along with bio- and metal-catalysis. Despite enormous advancements in academic research, there is a common belief that organocatalysis is not developed enough to be applicable in industry. In this review, we describe a selection of industrial routes and their R&D process for the manufacture of active pharmaceutical ingredients, highlighting how asymmetric organocatalysis brings added value to an industrial process. The thorough study of the steps, driven by economic stimuli, developed and improved chemistry that was, otherwise, believed to not be applicable in an industrial setting. The knowledge discussed in the reviewed papers will be an invaluable resource for the whole research community.

Keywords: organocatalysis, API synthesis, industrial processes, active pharmaceutical ingredient, asymmetric synthesis, industrial chemistry, sustainable chemistry

11.1 Introduction

Chiral molecules are instrumental to a range of chemical industries, including pharmaceuticals, flavours and fragrances, agrochemicals and fine chemical ones. The ability to access them efficiently is a fundamental problem that can be tackled in a number of ways and catalysis is, for many reasons, often the preferred one. Its sustained development is at the forefront of research both in academia and industry; bio- and metal-catalysis have played a major role in asymmetric catalysis for decades [1].

Nowadays, organocatalysis [2–4] is recognised as the third pillar in asymmetric catalysis, along with bio- and metal-catalysis. The complementarity of the three pillars and their synergy has opened new horizons to asymmetric catalysis and enriched the toolbox of the chemists. Triggered by the generalization of the concept of aminocatalysis [5] – that is the activation of carbonyl compounds by enamine and iminium ion intermediates – in 2000 [6, 7], the field of asymmetric organocatalysis

This article has previously been published in the journal Physical Sciences Reviews. Please cite as: Carlone, A., Bernardi, L. Enantioselective organocatalytic approaches to active pharmaceutical ingredients – selected industrial examples *Physical Sciences Reviews* [Online] 2019, 8. DOI: 10.1515/psr-2018-0097

https://doi.org/10.1515/9783110590050-011

has gained a prominent role in academic research over the last two decades [8]. The enthusiasm and expertise demonstrated by a large number of research groups have advanced its development at a very fast pace. Asymmetric organocatalysis has become an established and powerful synthetic tool for the chemo- and enantioselective functionalisation of organic compounds, and has already had significant impact in the synthesis of natural products, intermediates for pharmaceuticals and other structurally complex biologically active compounds [9–13]. The great benefits of organocatalysis are widely appreciated; they encompass, among many others, the use of generally non-toxic catalysts, the introduction of new, powerful, generic modes of activation and the fact that the reactions are most of the times robust, scalable and easy to perform. All these aspects render organocatalysis a very appealing and powerful technology platform for industry [14].

Interestingly, some of the landmarks in organocatalysis were developed in industrial settings – not in academia – much earlier than the field and the term «organocatalysis» were put forward in 2000 [6]. Two versions of the intramolecular asymmetric proline-catalysed aldol cyclisation of tri-ketones (Hajos–Parrish–Eder–Sauer–Wiechert reaction) were invented already in the early 1970s at Hoffmann-La Roche [15] and Schering AG [16]. The first highly enantioselective phase-transfer catalysed (PTC) reaction (alkylation of an indanone catalysed by quaternary ammonium salts derived from *Cinchona* alkaloids) was developed at Merck Sharp & Dohme in the middle of the 1980s [17], spurring the disclosure of a number of multi-Kg, operationally simple, PTC processes [18, 19].

However, there is a tendency to believe that the full industrial application of organocatalysis is hampered by the generally low activation that organocatalysts exert on reaction substrates, resulting in restricted scope, low turnover numbers (TON), and turnover frequencies (TOF). On the other hand, in the field of synthetic organic chemistry, only a small number of literature procedures has the potential to be transferred at industrially relevant scale [20]. In fact, many literature methods do not satisfy one or more of the SELECT criteria [21] (Safety considerations, Environmental concerns, Legal aspects, Economical constraints, Control requirements, Throughput) and are, therefore, unsuitable for scale-up.

Based on recently published applied research, we describe, in this chapter, a selection of industrial routes and their R&D process to show how organocatalysis had an impact on the prospective manufacture of commercial active pharmaceutical ingredients (APIs). We limited our choice to recent examples, wherein the processes were either demonstrated on a multi-Kg scale or stated to be the manufacturing process. We redirect the reader to previous monographies and reports for other equally important and meaningful cases [19, 22–31], which altogether highlight the power of this relatively new asymmetric synthesis technology platform for the production of APIs and building blocks for the pharma and fine chemical industry.

11.2 Asymmetric organocatalysis in the industrial synthesis of APIs

11.2.1 Letermovir

Letermovir (**6**) (AIC-246 and MK-8228, Figure 11.1) [32] is the active principle of the antiviral drug Prevymis. Disclosed by Baeyer/AiCuris, letermovir was given both fast track and orphan drug status. Merck & Co. have recently advanced it to a recent approval for the treatment of human cytomegalovirus infections. Letermovir is constituted by an 8-fluorodihydroquinazoline core, bearing an acetyl stereogenic centre at C-4, an aryl group at N-3 and a piperazine at C-2. Early synthetic approaches by the developers applied a late stage classical resolution of the corresponding ester **5** with a succinic acid [33]. A synthetic sequence affording ester **5** in racemic form is sketched in Figure 11.1. Urea formation from two aniline derived building blocks **1** and **2** followed by Heck reaction delivers a compound (**3**) prone to suffer a base promoted intramolecular aza-Michael reaction, affording **4**. Condensation with the piperazine delivers the dihydroquinazoline ester **5**.

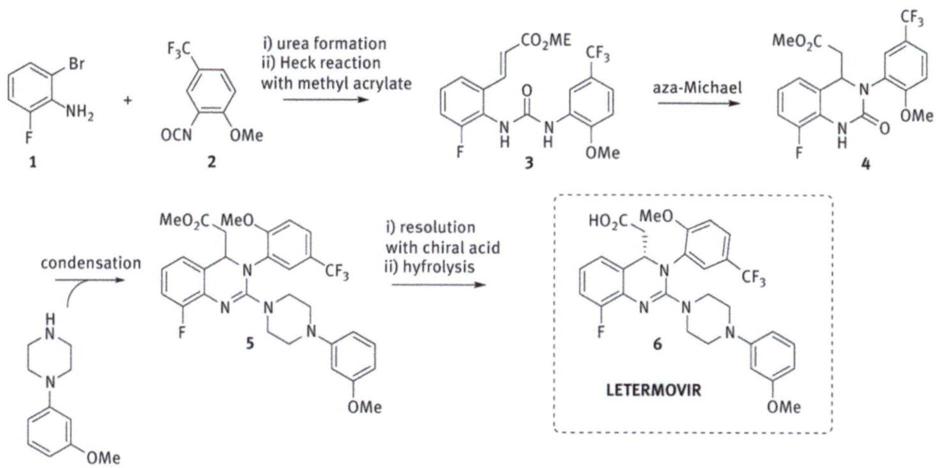

Figure 11.1: Early synthetic approach to letermovir **6**.

The requirement of a late stage resolution prompted the investigation of alternative strategies at Merck. Attempts to render the aza-Michael reaction **3** → **4** enantioselective using chiral phase-transfer catalysts were at least in part successful (56 % ee for product **4**, which could be upgraded to >98 % ee via crystallisation of the racemate). However, significant racemization in the PCl_5-mediated conversion of **4** to **5** was observed, unless a specific solvent (2,2,2-trifluoroethanol) was used. Therefore, approaches aiming at the cyclisation of guanidine substrates (i.e. anticipating the racemizing condensation with the piperazine before the enantioselective step) were considered and investigated (Figure 11.2) [34]. Metal-mediated reactions (aminocarbonylation, reductive Heck, sequential Heck-hydrogenation) did not seem promising. The catalyst activity in the aminocarbonylation reaction of **7** was not sufficient; simple debromination was found to severely compete with the reductive Heck reaction of **8**, while hydrogenation was at least in part precluded by the poor stability of alkene **9**. In contrast, the organocatalytic aza-Michael reaction of guanidine **10** turned out to be highly successful. Two distinct strategies proved to be feasible, as summarized in the following paragraphs.

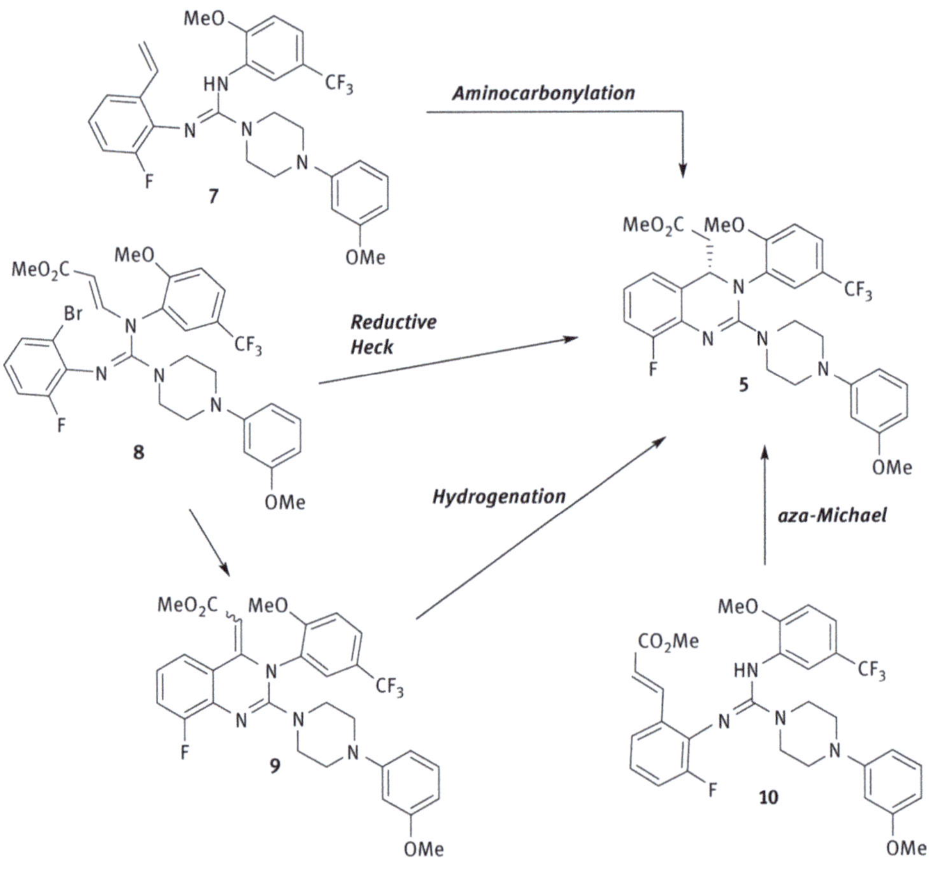

Figure 11.2: Catalytic enantioselective cyclisations of guanidine derivatives **7–10**.

A process chemistry to substrate **10** is reported in Figure 11.3 [34, 35]. Previously described urea **3** was used as a key intermediate; however, its preparation was improved compared to the first approach reported in Figure 11.1. In more detail, Heck reaction of aniline **1**, formation of the corresponding phenyl carbamate **12** and subsequent addition of the second aniline **13** provided crystalline urea **3** in 88 % yield without isolation of intermediates **11** and **12**. Then, dehydration of the urea was achieved with PCl_5, providing di-imide **14** which smoothly underwent addition by the piperazine. Use of toluene as solvent was found to be essential to avoid undesired early racemic cyclisation of guanidine **10**, which could be isolated in good yield as a salicylate salt.

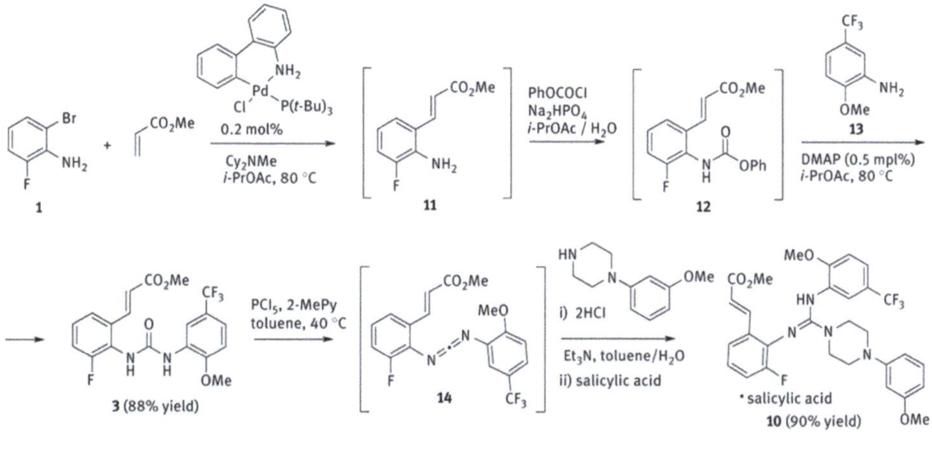

Figure 11.3: Process chemistry to aza-Michael substrate **10**.

Then, taking the results of the previously attempted enantioselective aza-Michael reaction of urea **3** as a starting point, the PTC aza-Michael reaction of guanidine **10** targeting dihydroquinolizidine **5** in enantioenriched form was investigated. A screening of >200 chiral phase-transfer catalysts and reaction conditions was undertaken, giving the following hints:

- an apolar aprotic solvent such as toluene was required, to avoid background cyclisation promoted by protic solvents via Brønsted acid activation of the substrate (*vide infra*);
- amongst catalysts tested, standard *Cinchona* alkaloids mono-quaternised at the quinuclidine nitrogen gave moderate selectivities (10–40 % ee). Bis-quaternised *Cinchona* alkaloids, recently introduced by the same process chemistry research department [36], gave distinct improvements in terms of both reaction rate and selectivity. The electronics of the aromatic substituents showed a strong influence on the reaction outcome; electron-deficient derivatives were the most effective. Although enantioselectivities >90 % ee could not be reached, catalyst **15** derived from cinchonidine (Figure 11.4) gave a moderate 72–76 % ee even at 3–5 mol% loadings and was considered optimal. Lower loadings were not applicable, perhaps due to a slow degradation of the catalyst via Hofmann elimination during the reaction;

- with aqueous potassium phosphate as the base, a strong effect of the catalyst counterion on the reaction outcomes was observed; using a counterion other than bromide lowered the enantioselectivity;
- an efficient stirring/agitation rate was essential to achieve optimal results.

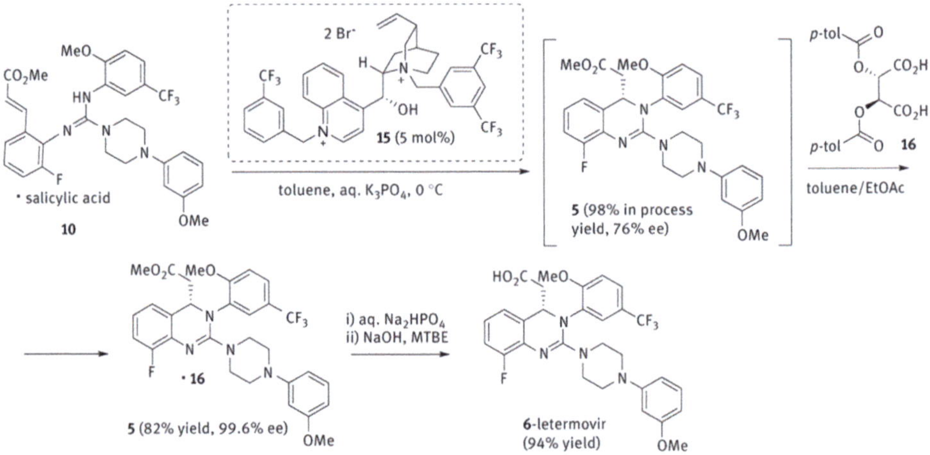

Figure 11.4: Process chemistry to letermovir **6** based on the asymmetric phase-transfer catalysed aza-Michael reaction.

Conditions claimed to be suitable for a manufacturing process were thus established, enabling the obtainment of the cyclized ester **5** in 98 % in-process yield and 76 % ee (Figure 11.4). Crystallisation with the tartaric acid derivative **16** upgraded the enantioenrichment of **5** to 99.6 % ee. This salt was converted to letermovir **6** via the formation of the free base **5**, followed by ester hydrolysis.

Despite the moderate enantioselectivity of the catalytic asymmetric reaction, this approach is still advantageous compared to the resolution, as evidenced by the overall 82 % yield of the aza-Michael reaction – crystallisation process, giving **5** with sufficient enantiopurity to be carried on to letermovir **6**. Indeed, the overall process (Figure 11.3 and Figure 11.4), which entails >60 % yield over seven steps, was used to prepare over 1 tonne of letermovir **6** to support clinical trials.

Nevertheless, the PTC reaction showcases suboptimal features, such as the high catalyst loading (likely related to poor catalyst stability under the reaction conditions), moderate enantioselectivity as well as the requirement of an efficient agitation of the biphasic mixture. Therefore, additional efforts were undertaken to identify a more effective catalytic asymmetric approach for the aza-Michael reaction **10** → **5**. Whereas chiral Lewis acids and bases were not competent/selective catalysts for this transformation, the significant background reactivity observed in protic solvents (2,2,2-trifluoroethanol: 100% conversion in 15 minutes, MeOH: 100% conversion in 1 hour), prompted an investigation with chiral hydrogen bond donor catalysts [37]. In fact, several of these catalysts were found to be able to promote the reaction in an enantioselective fashion (Figure 11.5). Relatively strong hydrogen bond donors such as phosphoric acid **17** were less active/selective than weaker donors **18–20**, with C_2-symmetric bis-triflamide **20** providing the most promising results.

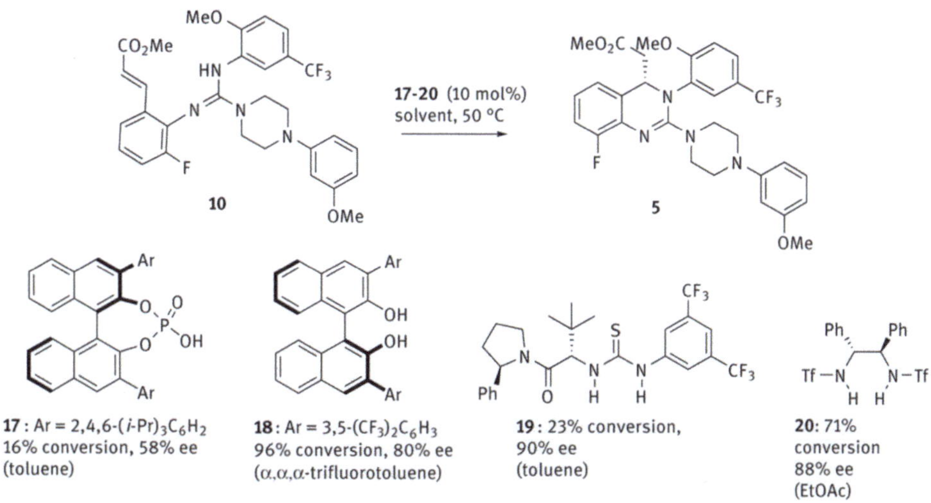

Figure 11.5: Screening of chiral hydrogen bond donors in the aza-Michael reaction: selected results with catalysts **17–20**.

The structure of the bis-triflamide catalyst was thoroughly refined, and reaction conditions optimized, leading to the protocol reported in Figure 11.6 that, on a small scale, afforded the target ester **5** in 95 % yield and 94 % ee under the action of 5 mol% of catalyst **21**. After having demonstrated the applicability of this protocol to broad range of guanidines related to **10**, a thorough mechanistic investigation was carried out. The structural requirements for a catalyst to be efficient in this reaction were first identified: both N-H of the bis-triflamide are essential, and a flexible (non-necessarily C_2-symmetric) 1,2-diarylethane chiral backbone is required. Then, DFT calculations combined with kinetic isotope effect and control experiments suggested coordination of the ester moiety to the catalyst via double hydrogen bond, giving transition state **22** as depicted in Figure 11.6. This type of activation was considered to be favoured over a reaction pathway involving tautomerisation and formation of an aza-*ortho*-quinone methide, prone to undergo an electrocyclisation according to transition state **23**.

It is remarkable that two distinct organocatalytic approaches, one involving basic and the other one acidic conditions, were both found to be effective for the asymmetric aza-Michael reaction **10** → **5**, the key stereoestablishing step in letermovir's preparation. These developments were recognised by a 2017 Presidential green

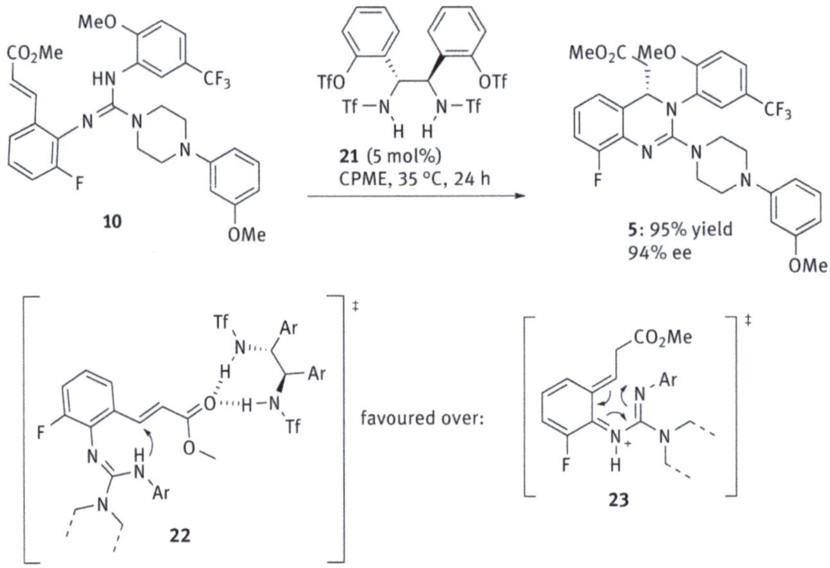

Figure 11.6: Optimised hydrogen bond donor catalysed aza-Michael reaction and plausible transition state.

chemistry challenge award from the US Environmental Protection Agency, given to the overall chemical process to letermovir encompassing the hydrogen bond donor catalysis.

11.2.2 Censavudine

Censavudine (**24**) (Figure 11.7, BMS-986,001, formerly known as festinavir) is a nucleoside reverse transcriptase inhibitor disclosed [38, 39] at Yale University (US) and further developed at Bristol Myers-Squibb (BMS). This antiviral nucleoside analogue is under clinical investigation for the treatment of HIV-infected patients showing resistance to current treatments [40]. Besides, censavudine seems to feature less toxicity compared to other nucleoside reverse transcriptase inhibitors [41].

A recent article [42] highlights the strategy behind the development of a scalable process to censavudine at BMS: extensive patent and literature data describing preparations of related nucleoside analogue drugs existed; these extensive precedents mandated thorough considerations on intellectual property and freedom to operate. Ultimately, a *de novo* synthetic plan was considered to be required for an efficient manufacturing process to compound **24**. Two retrosynthetic proposals were devised (Figure 11.7): one leading to 5-methyluridine as starting material and based on more or less reliable transformations (alkyne formation, Claisen rearrangement, regioselective acetal opening followed by dehydration via iodide), and one more

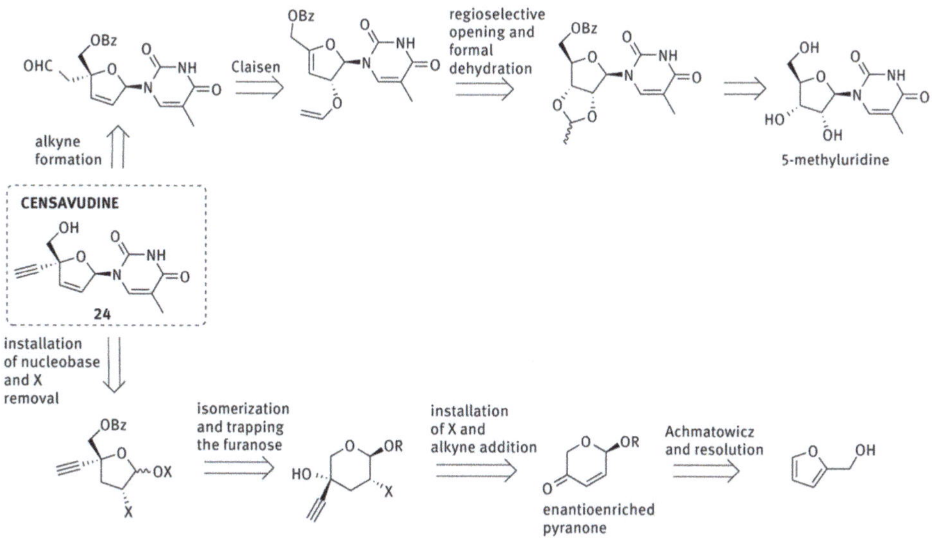

Figure 11.7: Retrosynthetic proposals to censavudine **24**.

direct which ingeniously engineered a poorly efficient previous approach by Oncolys Biopharma Inc [43]. The latter route involved more challenging chemistry (use of a stereorelay group X, furanose trapping of an equilibration furanose – pyranose mixture, etc.). Besides, an enantioenriched pyranone starting material was required, obtainable from furfural in racemic form but for which no appealing enantioselective access existed.

Both routes were investigated and both showed viability. The first approach [44, 45] starting from 5-methyluridine appeared more readily to be developed but less appealing than the second one via pyranone, which involved fewer steps and purifications. However, this route evidently required a considerable investment and was much more risky; its commericial applicability rested, in fact, on the invention of methodologies enabling feasible access to the pyranone, furanose trapping and X-group removal. Nevertheless, it was taken to further development ultimately leading to a successful process [46, 47] as described in the following paragraphs.

Aiming at pyranone, an efficient Achmatowicz rearragement of furfurol **25** giving lactol pyranone **26** was first required. This goal was realized by changing a standard unappealing stoichiometric *m*-CPBA oxidation to a catalytic protocol based on vanadium, that employs *tert*-butyl hydroperoxide as terminal oxidant (Figure 11.8) [48]. The preparation of pyranone acetal **27** in enantioenriched form was then tackled, at first with enzymatic approaches. A kinetic resolution of the benzoylated derivative rac-**27a**, wherein undesired (*R*)-**27a** is selectively deacylated under the action of a lipase, could be developed (Figure 11.8) [49]. However, the two-step process was characterized by a low overall yield of **27a**

(26 %) due to the intrinsic limitations of a kinetic resolution. Reasoning that racemization of **26** via ring opening can occur even under acylative conditions, a more straightforward and efficient dynamic resolution process of lactol **26** was investigated, aiming to trap the desired enantiomer of **26** via acylation with a stereoselective catalyst. A screening of more than 100 enzymes did not reveal any useful results, delivering either the acylated products **27** with poor enantioselectivities, or the wrong (*R*)-**27** enantiomer. Conversely, a screening of organic catalysts for the same reaction indicated somehow surprisingly that an exceedingly simple and cheap isothiourea Lewis base, levamisole **28**, a commercial anthelmintic pharmaceutical, could be used as efficient catalyst to perform this transformation. Aliphatic anhydrides performed much better than benzoyl, with iso-butyric anhydride giving best selectivity (**27b**-88 % ee). Ultimately, 2-phenylacetyl anhydride was preferred due to crystallinity of the corresponding product **27c**, enabling enantioenrichment by crystallisation and providing a satisfactory access to an acylated pyranone **27** in nearly enantiopure form.

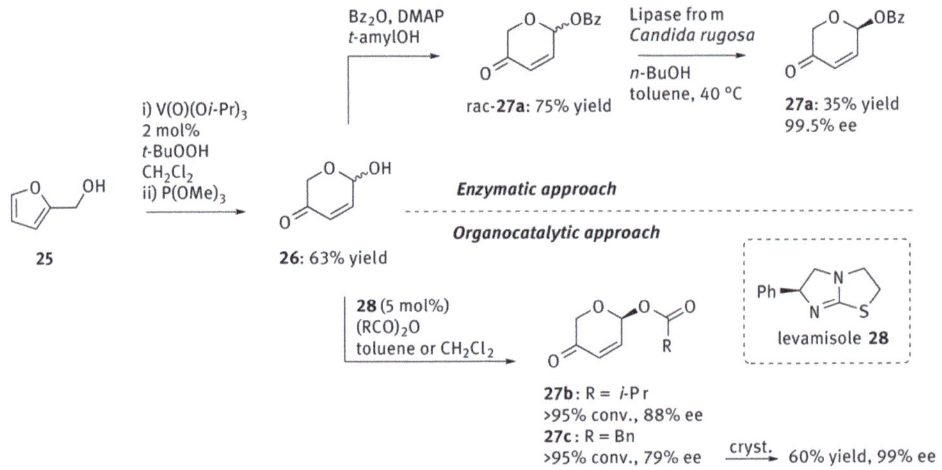

Figure 11.8: Enantioselective access to acylated pyranones **27** based on Achmatowicz rearrangement followed by enzymatic and organic catalysis.

The stereodirecting group, a thiophenol, was installed by Michael addition to pyranone **27c**, delivering **29** which was added to a pre-formed lithium acetylide resulting in compound **30**, isolated in good yield (Figure 11.9). The excellent diastereoisomeric ratio highlights the utility of the thiocresol as stereodirecting group. Compound **30** could be deacylated by aqueous HCl treatment, giving an equilibrating mixture of pyranoses **31** and furanoses **32** (both present as diastereoisomeric mixture). The ratio between **31** and **32** was found to be dependent on the polarity of the medium, with the pyranose form **31** highly favoured in a polar medium such as acetonitrile, and the desired furanose form **32** slightly prevailing in apolar solvent. These observations are in accordance with the hypothesis that an intramolecular hydrogen bond could stabilise the furanose form **32**, as also evidenced by gas-phase calculations. For these reasons, a solvent switch to toluene was performed before the key furanose trapping event. Interestingly, levamisole **28** was again proposed as the optimal catalyst to effect this reaction, resulting in the double benzoylated product **33** highly prevalent (98:2) over its pyranose counterpart, and as a single α-isomer. Interestingly, the related patent [47] discloses that good results can be obtained using DMAP as promoter for this transformations. Compound **33** was directly treated with a base removing the TMS group and affording **34** which could be isolated as a single diastereoisomer in an impressive 91 % yield.

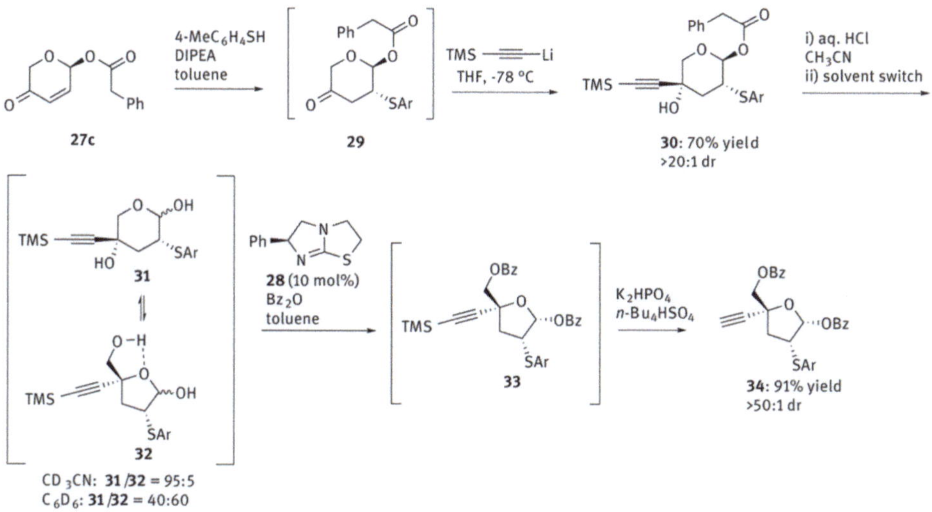

Figure 11.9: Installation of the stereodirecting group, addition of the acetylide, trapping of furanose and TMS removal.

The stereodirecting thiocresol group exterted a key role also in the following Vorbrüggen reaction, by anchimerically directing the addition of the silylated nucleobase to the β-face of the nucleoside (see **35**, Figure 11.10) [50]. The corresponding product **36**, obtained in very good yield and diastereoselectivity, was then subjected to the removal of the stereodirecting group via sulfilimine **37**, generated with chloramine-T, followed by mild thermolysis, delivering compound **38** which could be converted to censavudine **24** by a DBU catalysed transesterification.

The overall synthesis, claimed to be applicable on a 250 Kg scale, encompasses 5 steps from key starting material pyranone **27c** and a remarkable 44 % overall yield, without any chromatographic purification. The significance of these advances goes beyond the straight chemistry; this story can be taken as an example of

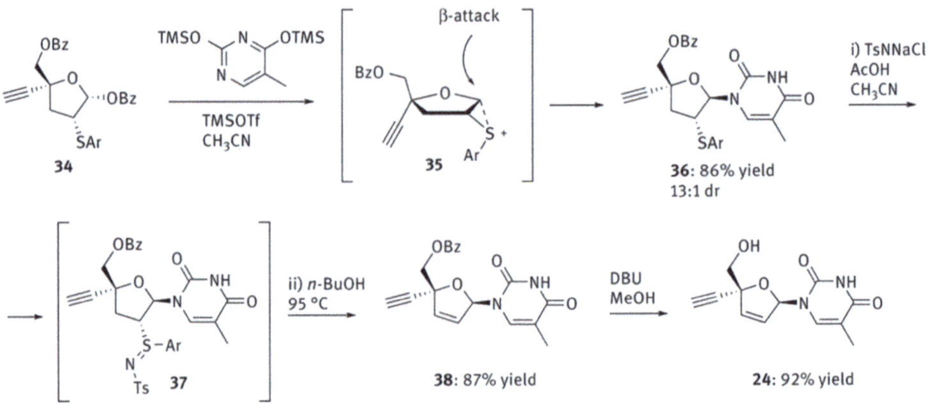

Figure 11.10: Vorbrüggen reaction, removal of the stereodirecting group and completion of the synthesis.

successful unconventional strategic decisioning in process chemistry development, as reported here with the words by one of the research coordinators [42]:

> The multiple innovations of this route highlight the need for vision in selecting a commercial process: the focus on the pyranone strategy came at a time when the furan oxidation catalyzed by vanadium, the levamisole chemistries for enantioselection and the chloramine sulfur elimination were yet to be invented.

Remarkably, organocatalysis was the indispensable tool that enabled some of these inventions.

11.2.3 Uprifosbuvir

Another antiviral nucleoside analogue for which enantioselective Lewis base catalysis has played a key role in the development of an efficient synthesis is uprifosbuvir **39** (MK-3682, Figure 11.11). This compound was being developed by Merck & Co. for the treatment for chronic hepatitis C virus (HCV) infections [51, 52], advanced until Phase II clinical trials but was then discontinued based on the analysis of the commercial/medicinal scenario and the clinical results [53]. Structurally speaking, compound **39** is consitituted by a 2′-chloro nucleoside core, linked at 5′ with a stereodefined phosphoramidate moiety (Figure 11.11). The antiviral active form of this and related antiviral nucleoside analogues are their triphosphorylated derivatives. However, these compounds are rather unstable and difficult to transfer across cell membranes, and cannot thus be administered. These issues can be solved by using phosphoramidate substituents, rendering pro-drugs (ProTide – PronucleoTide) [54] with increased cell permeability and stability. These groups are enzymatically cleaved inside the cells, giving monophosphorylated nucleotides prone to suffer further phosphorylation ultimately rendering the active drugs.

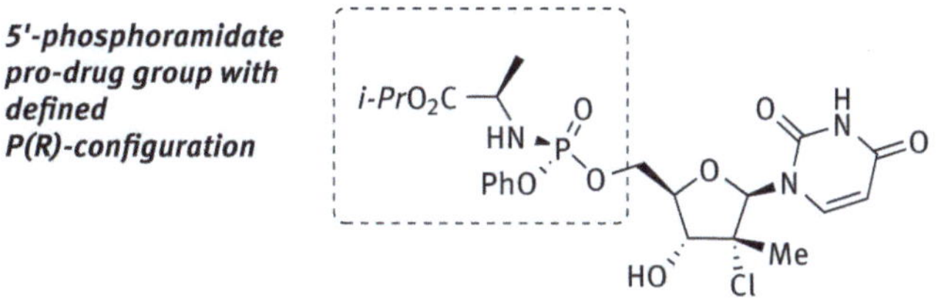

39: uprifosbuvir

Figure 11.11: The structure of the former anti-HCV candidate uprifosbuvir **39**, highlighting the phosphoramidate prodrug group.

The overall activity of these prodrugs is influenced by the P-stereochemistry of the phosphoramidate group, which must thus be controlled in the synthesis. A standard approach to effect the installation of the phosphoramidate group in this and related drugs involves a stereospecific S_N2 reaction at P(V) with a stereomerically pure phosphoramidating agent, ususally resolved from its diastereomeric mixture by crystallisation [55]. As shown in Figure 11.12, a proposed synthesis of uprifosbuvir **39** involves treatment of the 2′-chlorouridine derivative **42** with a strong magnesiated base and the isomerically pure phosphoramidate **44**, resulting in a regioselective (at 5′-O) and

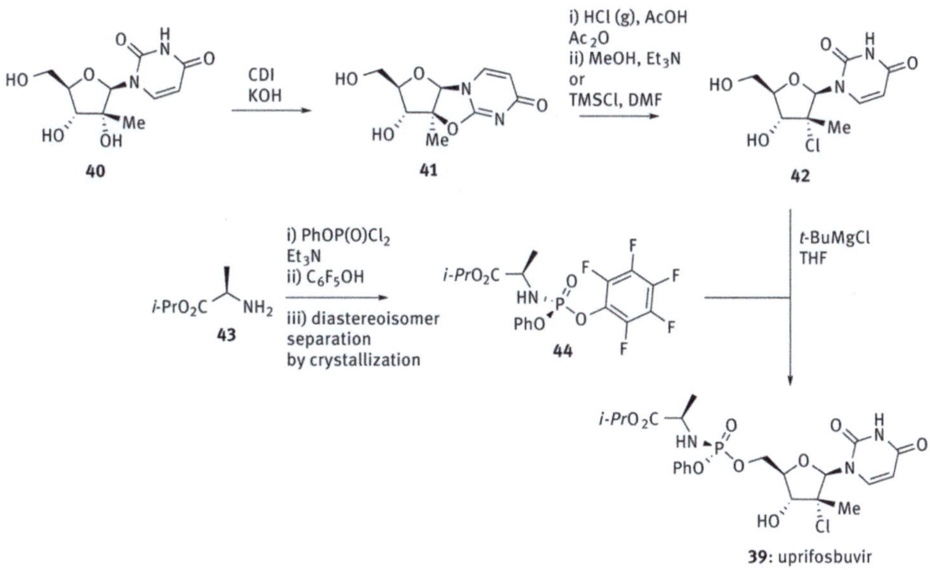

Figure 11.12: Reported synthesis of uprifosbuvir **39** using a phosphoramidate reagent stereochemically pure at phosphorous.

stereocontrolled phosphoramidation reaction. The key 2′-chlorouridine **42** can be obtained by dehydration of 2′-methyl uridine **40** with 1,1′-carbonyldiimidazole (CDI), followed by chloride ring opening of the dehydrouridine **41** [56].

To avoid the requirement of a non-readily obtainable enantiopure phosphoramidating reagent **44**, scientists at Merck have recently tackled a more direct approach by devising a catalytic stereoselective reaction between nucleoside **42** and a racemic (at P) chlorophosphoramidate **45** (Figure 11.13) [57]. Despite lack of solid literature precedents, they envisioned that a catalyst enabling control of the stereochemistry at P in the reaction could also ensure continuous racemization of **45** during the reaction, resulting in a dynamic stereoselective process (Dynamic Kinetic Asymmetric Transformation, DYKAT) converting all **45** to the desired stereoisomer of **39**. Besides, the same catalyst could be able to control the regioselectivity, resulting in a 5′-regioselective functionalisation without resorting to protecting group strategies.

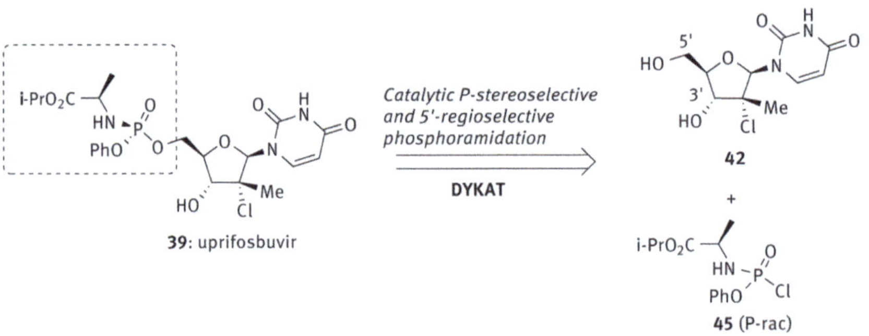

Figure 11.13: Stereoselective introduction of the phosphoramidate group using racemic **45**.

Initial experiments showed the minimal substrate bias on the stereoselectivity of this reaction. Using typical non-stereoselective conditions for the formation of phosphoramidates (*N*-methylimidazole catalyst **46**, 2,6-lutidine as stoichiometric base), the product **39** was obtained in a 52:48 diastereomeric ratio at P (Figure 11.14). This result was favourably interpreted, leaving space for catalyst-exerted stereocontrol. A catalyst screening identified some specific imidazole structures as suitable promoters. Catalyst **47**, featuring a silyl protected hydroxy group, gave a moderate selectivity. The non-symmetric results obtained with its enantiomer *ent*-**47** indicated a likely recognition of the nucleoside by the catalyst. Ultimately, more promising results were obtained with its carbamate derivative **48**. In all cases, regioselectivity favouring the 5′-isomer was excellent (>94:6).

In order to improve the performance of the catalyst, a thorough mechanistic study was then undertaken, whose results can be summarized as follows:

- kinetics: 0th order (saturation) in chlorophosphoramidate **45** and 2,6-lutidine; 1st order in nucleoside **42**; 2nd order in catalyst **48**. In fact, higher stereosectivity and yields were obtained increasing the catalyst **48** loading (5 mol%: 12 %

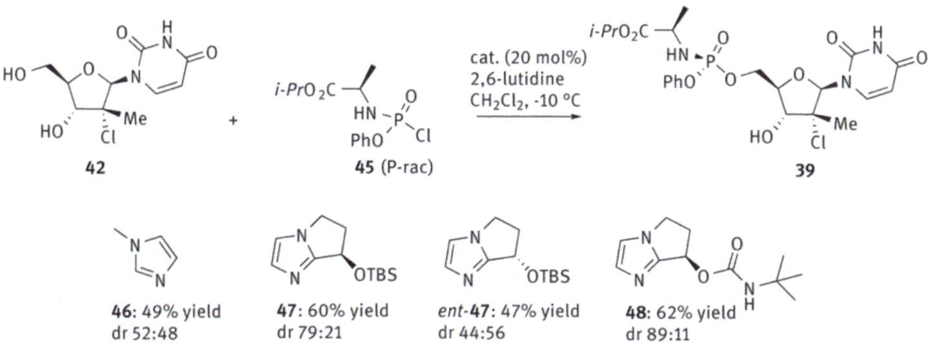

Figure 11.14: Stereoselective phosphoramidate formation: selected results from first catalyst screening.

yield, 71:29 dr; 10 mol%: 23 % yield, 79:21 dr; 20 mol%: 62 % yield, 89:11 dr; 40 mol%: 82 % yield, 94:6 dr), confirming the 2nd order in catalyst;
- the epimeric ratio at P was found to be constant over the course of the reaction, for both chlorophosphoramidate **45** and its catalyst adduct (*vide infra*). Furthermore, rapid interconversion between the two catalyst-bound epimers was detected;
- using a chiral pyridine Brønsted base co-catalyst influenced the stereoselectivity of the reaction (matched/mismatched effects).

These data pointed to the reaction pathway shown in Figure 11.15, wherein chlorophosphoramidate **45** combines with the Lewis base catalyst **48**, giving two epimeric adducts **49** and **50**. These compounds are rapidly interconverting by the action of another molecule of catalyst **48** (addition – elimination resulting in equilibration between **49** and **50**). Adduct **50**, with the right stereochemistry at P, reacts with nucleoside **42** in the turnover limiting P-O forming step according to transition state **51**, wherein a second catalyst molecule acts as a Brønsted base on the proton of the attacking 5′ hydroxy group. DFT calculations showed additional interactions between the strongly Lewis basic P=O and some aromatic C–H of the catalysts and of the phenolic substituent, assisting the reaction by oxyanion stabilisation. On the other hand, the carbamate N-H does not have any obvious role; in fact, a corresponding N-methylated catalyst gave very similar results. In this reaction model, stereoselectivity is not due to the equilibrating ratio of **49** and **50**, but to the lower energy of transition state **51** compared to its counterpart giving opposite configuration at phosphorous.

Taking into consideration that two catalyst molecules are involved in the rate and stereo-determining step of the reaction, dimeric structures, that is catalysts bearing two imidazole moieties linked via the carbamate functions, were prepared and tested in the reaction. This study identified catalyst **52**, wherein the two imidazoles are linked via a semiflexible 1,3-disubstituted phenyl chain, as optimal (Figure 11.16).

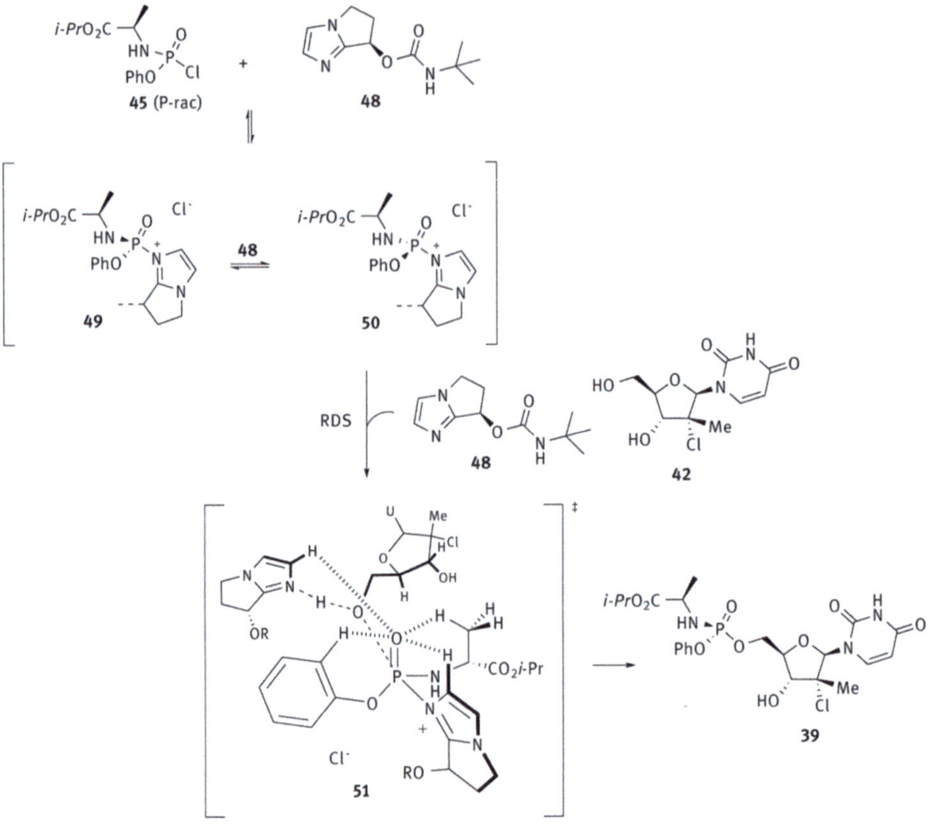

Figure 11.15: Reaction pathway and transition state model.

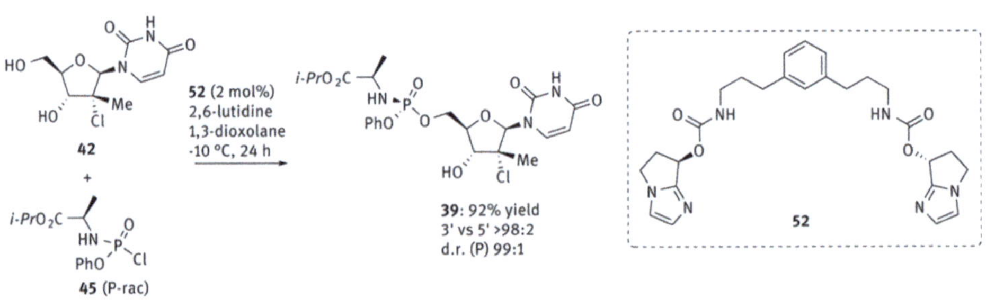

Figure 11.16: Optimised reaction with dimeric catalyst **52**.

Catalyst **52** was not only more active than monomeric **48** (k_{rel} ca. 10), but also more stereoselective (99:1 d.r.). The higher activity allowed to lower the catalyst loading as low as 2 mol%, still providing a reasonably fast and highly regio- and stereoselective reaction.

A kinetic study showed that the reaction catalysed by dimeric catalyst **52** is first order in catalyst, as expected. However, somewhat surprisingly, the reaction featured saturation kinetics (0th order) in nucleoside **42**, and first order in chlorophosphoramidate **45**. Therefore, chlorophosphoramidate activation is the rate determining step of this reaction, wherein the P–O bond formation is faster. Despite slow phosphoramidate activation, it seemed that equilibration between the two epimeric activated phosphoramidate adducts is still fast, thus presumably excluding a role of the equilibration as a stereodeterming factor in this reaction.

The protocol described in Figure 11.16 could be applied to several other nucleoside analogues, demonstrating the robustness of the methodology and its generality. Therefore, despite the interruption of the clinical trials of urpifosbuvir **39**, the development of this highly innovative stereoselective phosphoramidate forming reaction can have a great impact in drug manufacturing. Indeed, other commercial nucleoside analogues (e.g. sofosbuvir) feature the same prodrug group which could be introduced in principle with this elegant and efficient Lewis base catalysed reaction.

11.2.4 Funapide

Funapide **53** (former developmental code names TV-45,070 and XEN-402) was in-licensed by Teva from Xenon and in 2017 was under development in Phase-II clinical trials for postherpetic neuralgia (PHN), as antagonist of the Nav1.7 sodium ion channel protein. Its development was then discontinued by Teva and, in March 2018, Xenon and Teva entered into an agreement terminating by mutual agreement the collaborative development of Funapide [58]. Funapide **53** is a small-molecule lactam containing a chiral spiro-ether (Figure 11.17); Teva describes the development of its synthesis from the early-phase one, through the first asymmmetric route, to finally present the optimised plant synthesis [59].

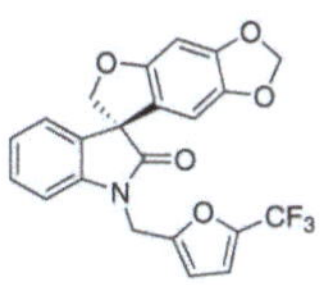

53: FUNAPIDE

Figure 11.17: Structure of funapide **53**.

Understandably, the early-phase route (Figure 11.18) [60] suffers from few drawbacks, but it was critical to supply material for the Phase I clinical studies. While all the steps were successfully carried out on multi-kilo scale, there were few points to address:

- the removal of the hydroxy group in **54** was carried out in neat TFA;
- the base-promoted aldol reaction between **55** and formaldehyde provided racemic **56** with no chemoselectivity, giving rise to the hemiaminal formation from the addition of a second equivalent of formaldehyde;
- the phosphine by-products from the Mitsunobu cyclisation needed large-scale chromatography to be removed;
- SMB chromatography was crucial for the resolution of the desired enantiomer;
- alkylation of **57** with the bromomethyl furan derivative was necessary to enchance solubility for the SMB purification, meaning that the chromatography had to be run on a larger scale; significant yield losses late in the synthesis;
- furthermore, use of the genotoxic bromide alkylating agent at the end of the synthesis meant that potential trace contamination posed a further issue to the process.

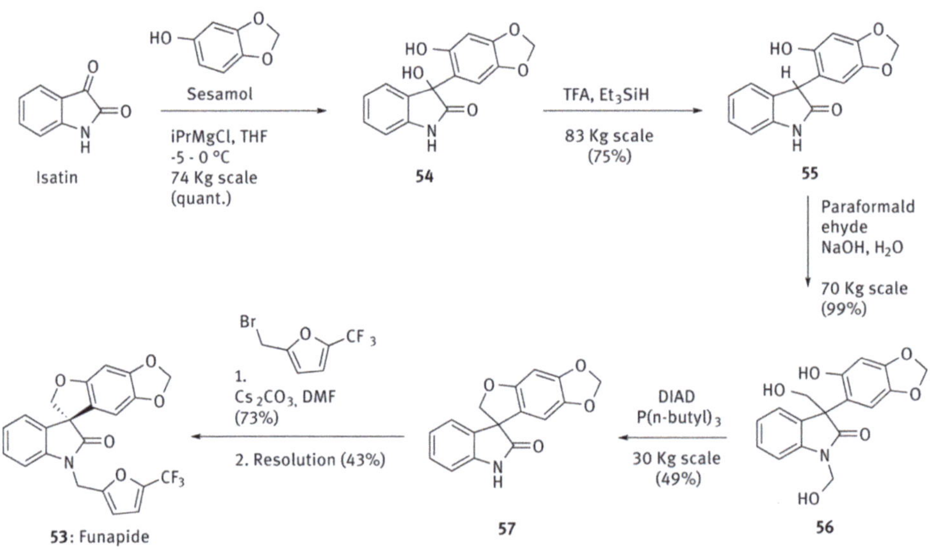

Figure 11.18: Early-phase route to funapide **53**.

A first-generation asymmetric route [61] was developed with the primary aim of avoiding the need for SMB chromatography in separating the desired enantiomer of **53**. The strategy relied on the installation of the hydroxymethyl moiety via PTC asymmetric alkylation of **55** (Figure 11.19).

However, it was necessary to introduce protecting groups in the synthesis due to the presence of two additional sites that could be alkylated in the asymmetric step; hence, both the nitrogen and the hydroxy group were protected. The hydroxy group was derivatized with a benzyl group so that it could be deprotected simultaneously with the installed alkyl group. On the other hand, the route relied on the

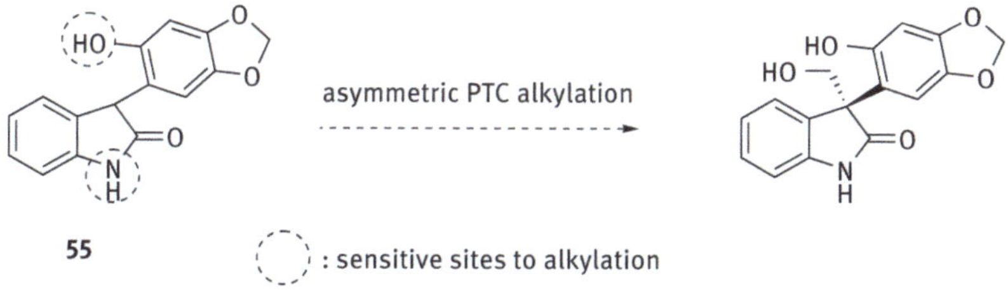

Figure 11.19: Strategy for the asymmetric step.

deprotection of the nitrogen after the Mitsunobu reaction. Therefore, compound **58** was prepared and subjected to PTC conditions using Lygo phase-transfer catalyst [62] that forged the stereogenic centre with moderately high ee (Figure 11.20).

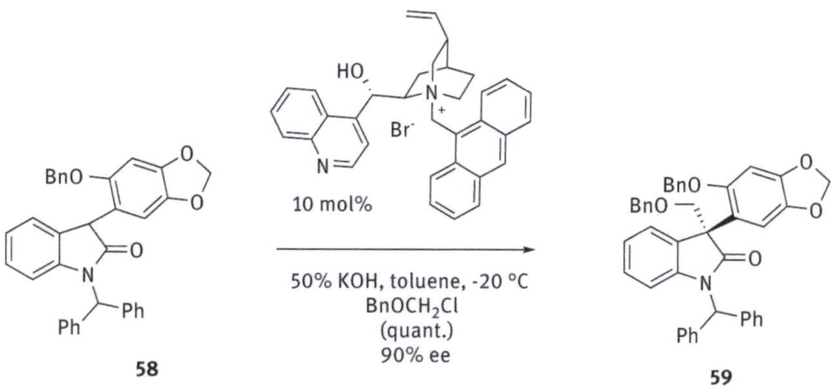

Figure 11.20: PTC asymmetric alkylation step.

The development of the Mitsunobu step introduced the innovative use of a phosphine which could be removed by treatment with aqueous acid (Figure 11.21); this improvement addressed the issue of the chromatographic purification that was previously needed for the removal of impurities in this step. It is worth noting that the probably higher cost and complexity of the phosphine introduced in this step, as opposed to the tributylphospine, is compensated by the advantages that it brings in the process in terms of purification costs and time, and impurities carried through.

The overall first-generation asymmetric route is depicted in Figure 11.22. Isatin and sesamol are again the key starting materials; however, isatin was first protected to deliver intermediate **62** that was, then, coupled with sesamol. Protection with benzyl bromide and removal of the hydroxy group produced the key intermediate **58** needed for the asymmetric step. It can already be noted, from the slightly different conditions reported, the first attempts of the process chemists to solve the issue of using neat TFA for the reduction step of **64**.

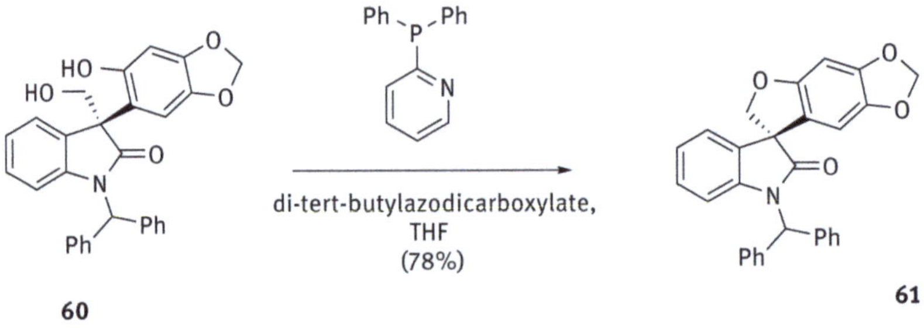

Figure 11.21: Mitsunobu reaction exploiting a phosphine which can be removed by aqueous acid wash.

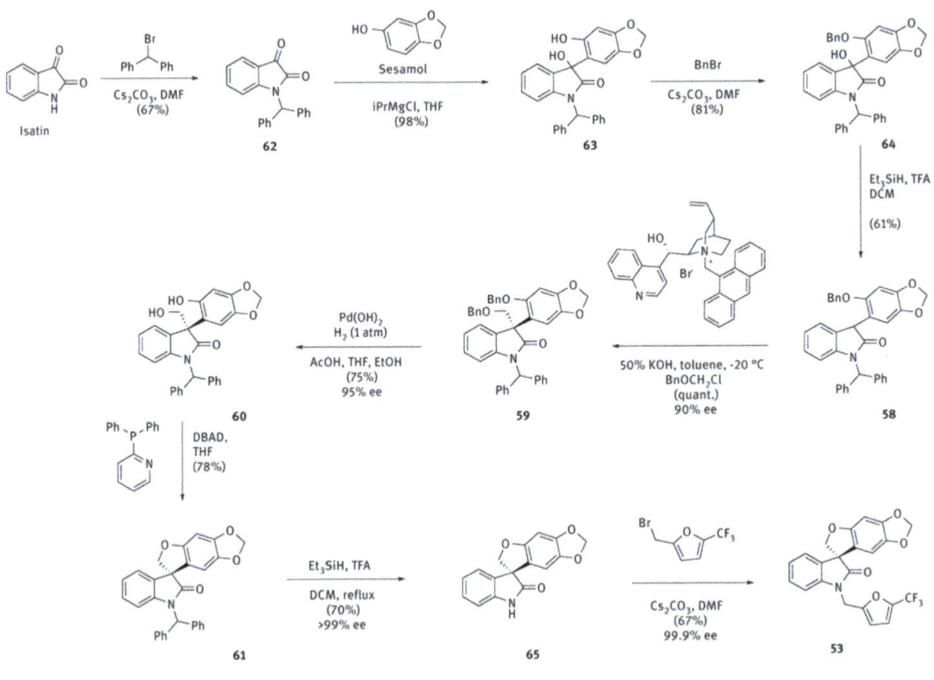

Figure 11.22: First-generation asymmetric route.

The PTC asymmetric alkylation step previously described, followed by debenzylation and Mitsunobu reaction, gave advanced intermediate **61.** Subsequent deprotection and alkylation with the bromomethyl furan derivative yielded desired funapide **53.**

This first-generation asymmetric route provided important learnings; the most important one, that provided the argument to develop a new generation process, is that enrichment from moderately high ee to the required ee for the API is possible over several steps through crystallisations (i.e. from 90 % ee in **59** to 99.9 % ee in **53**), thus avoiding the need for SMB. There were also some drawbacks in this synthesis:

the genotoxic bromo-derived alkylating agent is still used at the end of the synthesis and the route has become more convoluted due to the installation and removal of the protecting groups to accommodate the asymmetric step.

During the development of the two routes described above, it was further observed that, despite the number of protic sites in the starting materials, only a slight excess of one equivalent of Grignard reagent was necessary; this suggested that it was merely acting as a base and this learning prompted the improvement of this step. In fact, it was found that a slight excess of potassium carbonate in THF was sufficient to carry out the desired reaction (Figure 11.23); this finding allowed to avoid the use of an expensive base and to carry out the step in less moisture-sensitive conditions, along with an improvement on the safety avoiding the emission of propane from the process. The authors note that potassium carbonate with small particle size was needed for best results, highlighting the impact of some factors that sometimes may be neglected in small-scale runs. All efforts to carry out the following step in a different solvent, reducing the amount of TFA needed, were unsuccessful due to poor reactivity or the formation of the dimeric impurity **66** that was difficult to remove by crystallisation.

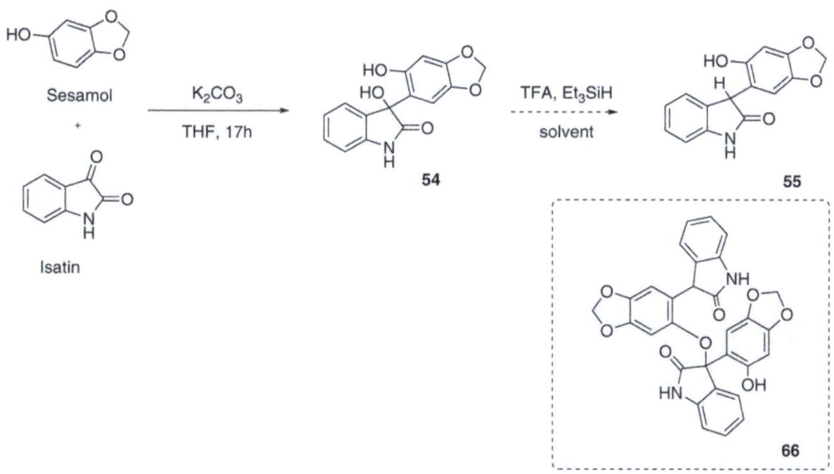

Figure 11.23: Potassium carbonate-promoted coupling of sesamol and isatin and subsequent reduction.

The researchers, therefore, focused on the following asymmetric aldol reaction step (Figure 11.24), postponing the development of the reduction step.

They tested Sharpless ligands (*Cinchona*-derived $(DHQ)_2PHAL$ and $(DHQ)_2Pyr$) on the aldol reaction with **55** but, despite encouraging results, significant level of impurities were produced along with hemiaminal **56a**. Bifunctional catalyst **68** showed promising improvements eliminating the formation of impurities, albeit

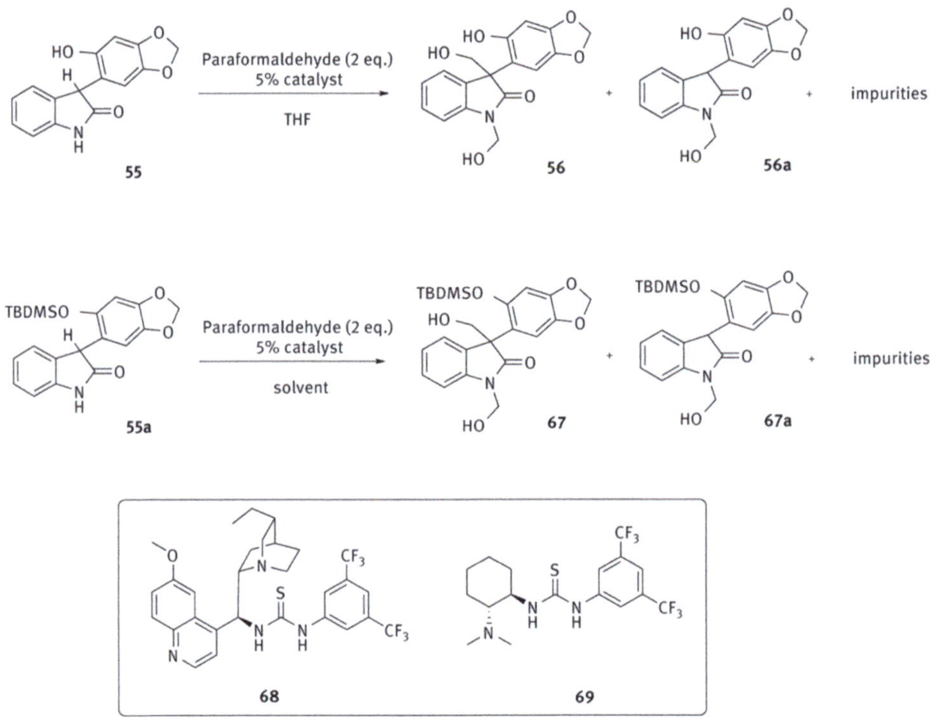

Figure 11.24: First improvements on asymmetric aldol reaction.

product **56** was formed in racemic form (Figure 11.24). It was rationalised that protection of the phenol could be beneficial to the reaction avoiding the competition of hydrogen bonding; in fact, TDBMS-protected substrate **55a** showed improved reactivity with Sharpless ligands, and catalysts **68** and **69**, giving rise to a cleaner impurity profile, albeit lower ee with Sharpless ligands.

Hemiaminal formation could impact on the steroselectivity by hydrogen bonding to catalysts **68** and **69**. Therefore, functionalisation of the oxindole was performed earlier in the synthetic route in order to avoid hemiaminal formation; this strategy would also allow to use the genotoxic furan alkylating agent earlier in the synthesis and to avoid potential trace contamination in the final product. Furthermore, this would also remove the need for a nitrogen protecting group that should have been removed later, avoiding two further steps in the process. As speculated, oxindole **70** showed a significant improvement in ee. The reaction was best carried out as a slurry in heptane, with **68** as catalyst, yielding **71** in 73 % ee (Figure 11.25); higher dilution of heptane were detrimental to the enantioselectivity suggesting that the competitive hydrogen bonding of the installed hydroxymethyl group was reduced by limiting the amount of product in solution. Given the moderate ee obtained, a preferential crystallisation was investigated; fortunately, after dissolution in methanol and cooling, a scalemic mixture (5 % ee) of product precipitated, leaving in solution the desired *S*-isomer in 98–99 % ee. Subsequent addition of water allowed the isolation of the product by precipitation and filtration.

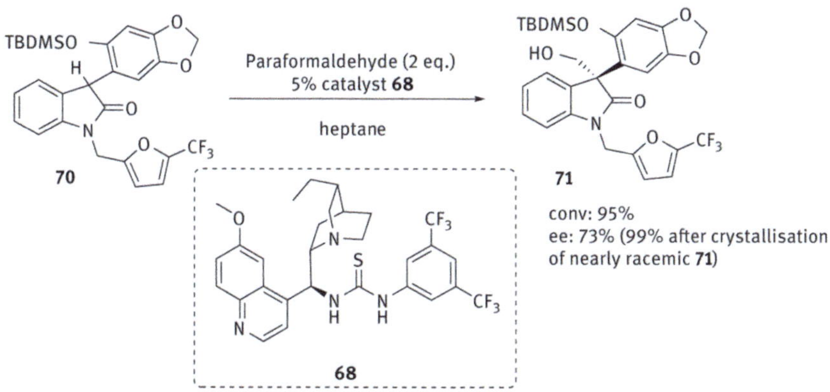

Figure 11.25: Asymmetric aldol reaction with catalyst **68**.

With optimal conditions for the asymmetric aldol reaction in hand, attention was then turned to improving and tuning all the steps for the plant process (Figure 11.26). The optimised plant synthesis features a concise process which benefits from telescoping of several steps.

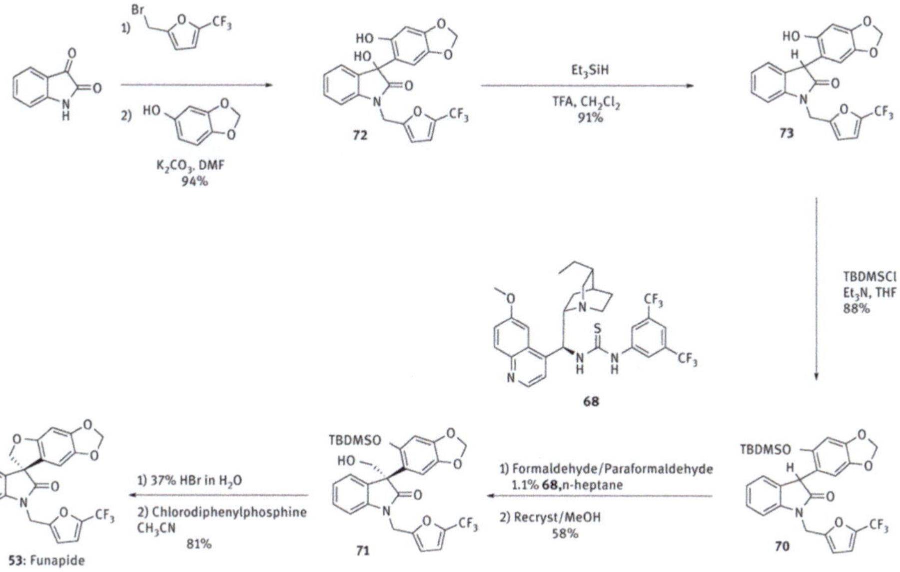

Figure 11.26: Optimised plant synthesis.

In the first step, isatin is alkylated installing the furyl appendage, and then reacted with sesamol in the same pot, in DMF, in presence of K_2CO_3. Once consumption of isatin is confirmed by an in-process check, sesamol is added as a solution in DMF; this avoided the use of Cs_2CO_3 that was previously reported. **72** was isolated directly from the reaction mixture after pH adjustment and addition of *i*-PrOH.

After a careful screening of cosolvents to perform the subsequent reduction step, it was necessary to revert to the use of dichloromethane that, although an undesirable class 2 solvent, had the advantage of poorly solubilizing both the starting material and the product, thus minimising the formation of dimeric **66** (see Figure 11.23).

The key asymmetric aldol step showed best performance when a mix of solid paraformaldehyde and aqueous formaldehyde was used, although the researchers do not give an explanation on potential reasons. Under these conditions, the reaction could be carried out with only 1.1 % mol of **68** in *n*-heptane at 25 °C. The crude product showed 70.5 % ee that could be improved to 99 % ee in 58 % yield, by preferential crystallisation of the racemic mixture. It is noteworthy that discarded racemic **71** could be converted back to **70** via a retro-aldol reaction, thus recovering precious material, albeit the conditions are not described.

The silyl-group deprotection was optimised to avoid a retro-aldol reaction; deprotection in 37 % aqueous HBr was found to be optimal. The amount of water in the reaction is critical to the success as below 0.5 equivalents resulted in a stalled deprotection; by charging the proper amount of H_2O, the final content of water was ca. 0.1 % (as measured by Karl Fischer titration) and this allowed the following step to be carried out successfully.

To perform the final ring closure, a thorough study was carried out to avoid the use of Mitsunobu reagents which complicated isolation through the generation of coloured by-products. After careful screening and investigation, it was observed that a phosphorus-containing leaving group on the installed hydroxymethyl could allow the ring closure to occur. The best reagent was found to be chlorodiphenylphosphine that gave clean **53** at 37 °C. After solvent exchange in MeOH and seeding, the desired **53** was isolated in 81 % yield, 99.6 % purity by HPLC area, and >99.9 % ee.

In conclusion, the reported process for the synthesis of funapide **53** is the shortest synthesis reported to date, despite the use of a protecting group. All the products were directly isolated, avoiding any purification by chromatography. The catalyst loading is interestingly low for an organocatalysed reaction that performs well at room temperature; all the studies carried out to improve the efficiency of the reaction are very insightful and show how an interesting process could be finely tuned in industry when significant benefits can be expected. Unfortunately the authors do not comment on any attempt to recycle the catalyst that, albeit used at 1.1 % mol, might have a significant impact both on cost and on mass utilisation of the process. It is also interesting to note, to understand how the selection process works in industry, that the catalytic step was selected despite the moderate enantioselectivity;

in fact, the ee could be upgraded via preferential crystallisation. Additionally, the racemic product **71** was reported to be converted back to **70** via a retro-aldol reaction, thus allowing the recovering of precious material.

As an interesting addition to this process, a process to synthesise catalyst **68** (and its related quinine derivative **68a**) on scale was reported [63]. The authors worked to develop the original synthesis reported by Soós (Figure 11.27) [64] so that it could be amenable for manufacture. One of the main drawbacks of the lab-scale synthesis was the need for at least one chromatographic purification; the synthesis of the amine **75** avoiding a chromatographic purification has been previously reported on gram-scale by Connon [65] and Melchiorre [66].

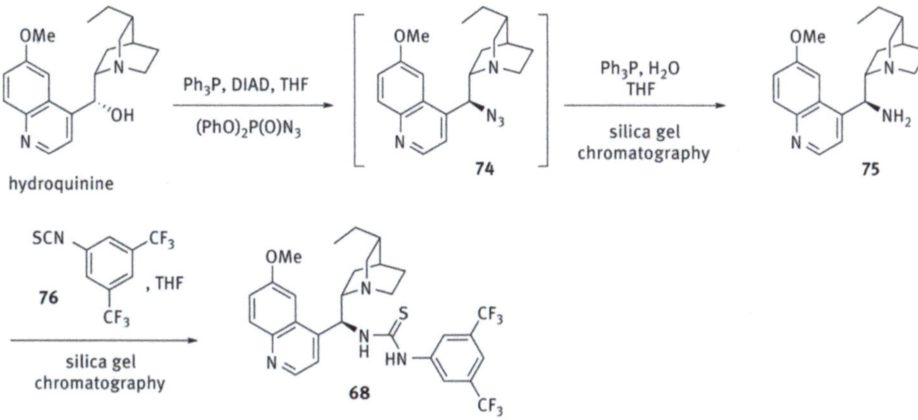

Figure 11.27: Original lab-scale synthesis reported by Soós.

The development work was first focused on replacing the water-miscible solvent THF in the first step to simplify the workup. A solvent screening identified CH_2Cl_2 as an optimal, albeit non ideal because of its toxicity, solvent for this step; it provided the best impurity profile and the solvent volume could be reduced by 50 %. The workup was carried out by charging 3N HCl to the reaction mixture; this allowed the desired amine **75** to remain, as a hydrochloride salt, in the aqueous upper layer while, by carefully controlling the pH, the unreacted triphenylphosphine and the sideproduct triphenylphosphine oxide remained in the bottom organic wash. Treatment of the aqueous phase with aqueous ammonia liberated **75**, and back-extraction with ethyl acetate furnished a solution that, after removal of water, was used directly in the next step. The workup of the last step was carried out via an acid wash to remove the undesired impurities in the aqueous phase while the **68•HCl** remained in the organic layer and **68** could be liberated via an ammonia wash and recrystallised with CH_3CN to reach the target purity.

The process, applicable to both dihydroquinine derivative **68** and its more available quinine counterpart **68a** (Figure 11.28), was successfully demonstrated in the pilot plant, affording over 14 kg of **68a** over four stages in a single reactor. The new process, consisting of three chemical reactions and one recrystallisation, was performed in a single reactor, significantly improving the practicability and productivity of the synthesis. An interesting comment from the authors is that the pilot plant process provides catalyst **68a** at a cost of $2750/Kg; this provides useful insight into the cost contribution that this brings catalyst to an industrial process.

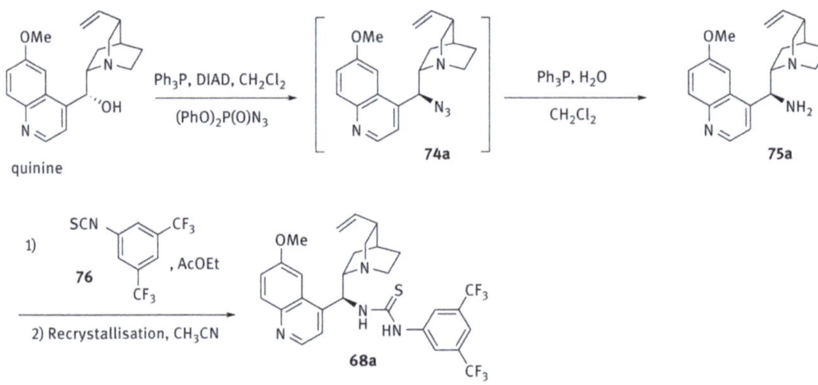

Figure 11.28: Optimised plant process for catalyst **68a**.

11.3 General considerations

Asymmetric organocatalysis, starting with the early work in the early 1970s and with more impetus in the last twenty years contributes to develop cost-effective and efficient processes in industry. Industry shows a genuine interest in organocatalysis by routinely screening it during the decision process for route selection; many companies contribute to its funding for basic research and they have an active participation in collaborative funding applications.

Despite the great interest in industry, most of the R&D work is, understandably, not disclosed; however, we can infer from published research that organocatalysis is a fundamental tool and has already demonstrated a tremendous impact. Industry is carrying out cutting-edge applied research to bring innovative solutions and great benefits in commercial processes both at the economical and at the environmental level.

The works highlighted in this chapter show indeed that organocatalysis is a useful tool in asymmetric synthesis with considerable applicative potential. It is remarkable that these examples exploited different organocatalytic species and activation modes: phase-transfer catalysis, hydrogen bond activation, Lewis base catalysis, bifunctional catalysis. Together with other examples not highlighted here, this strongly indicates the broad utility of organocatalysis in asymmetric synthesis on scale. It is also remarkable that, in these examples, relatively high catalyst loadings were applied, somehow contrasting the general (academic) belief of the requirement of very high TONs for a process to be competitive on scale. It can be inferred that the relative simplicity of the catalyst structures, as well as the complexity of the target molecules, compensates such low catalyst proficiency.

It is worth comparing the examples of this chapter with a couple of famous and highly successful academia-driven stories. The Strecker-type synthesis developed by Jacobsen and co-workers in the late 1990s [67] was taken to an efficient and scaleable process after a decade [68, 69] through a series of improvements regarding: i) simplification of catalyst structure and efficiency through extensive mechanistic studies; ii) milder reaction conditions; iii) cheaper cyanide source; iv) easier downstream chemistry to target N-Boc amino-acid (Figure 11.29).

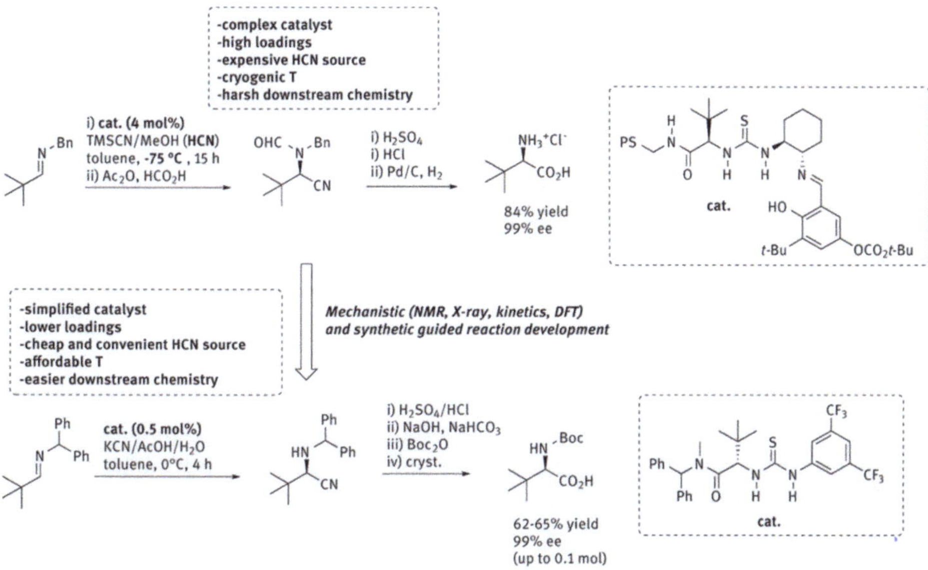

Figure 11.29: Scaleable Strecker synthesis under thiourea catalysis.

Similarly, the phase-transfer alkylation of benzophenone imines catalysed by binaphthylic quaternary ammonium salts disclosed by Maruoka and co-workers in 1999 [70] was implemented after several years to a commercially attractive process [71], in the frame of an academia-industry partnership, by a series of dramatic developments on catalyst synthesis and structure, as well as substrate simplification (Figure 11.30).

In these examples, many years of very intense research activity turned two conceptually innovative, but with limited applicability, asymmetric transformations into appealing chemical processes. These methodologies target some relatively

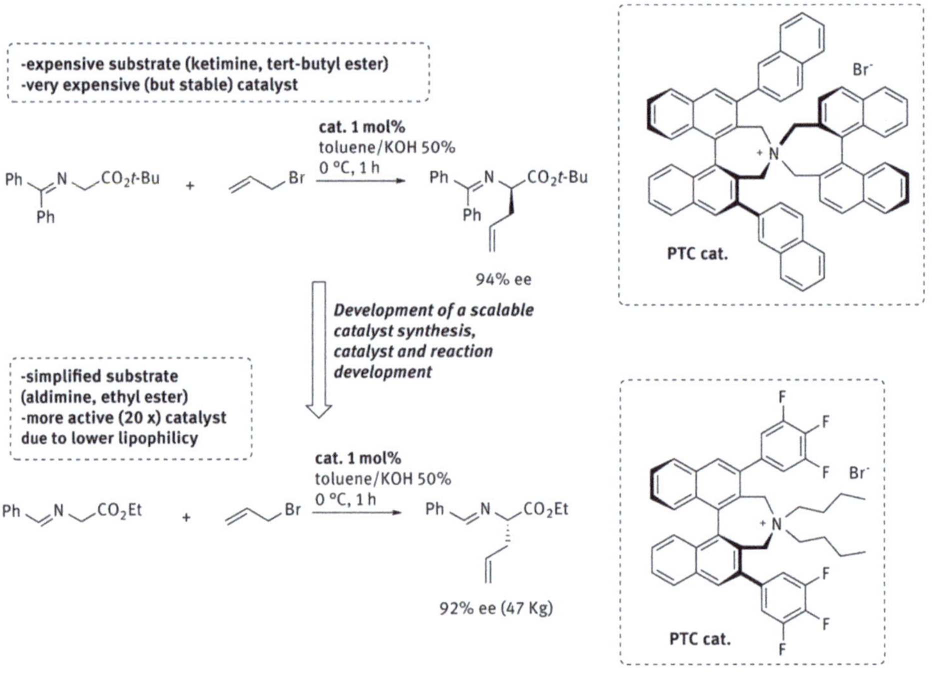

Figure 11.30: Scaleable alkylation of glycine imines.

simple chiral building blocks of broad utility. In contrast, the examples highlighted in the previous sections were directed towards specific targets of considerable complexity, and probably suffered extreme time constraints.

Process chemistry and industrial R&D drives the discovery of new chemistries with the goal of accessing synthetic targets for commercial interest. Moreover, the systematic and thorough study of impurities in each synthetic step, along with the crucial understanding of each parameter, leads to important improvements and discoveries. Unfortunately, not all of the knowledge acquired is shared, for obvious reasons, but what is published is an invaluable resource for further understanding of the research developed in academia. The commercial impact is also only mentioned in reports from industry; the choices made, however, are a good indicator of what makes a technology successful. The observation of how industry develops their processes, through incremental steps, where success is met with additional problems to tackle, to fine tune a process in every detail, provides us with a greater understanding and rare information.

11.4 Outlook and perspectives

Asymmetric organocatalysis has already had a significant impact both in basic and applied research as testified by numerous reports. It is noteworthy that many different activation modes found application; PTC, enamine/iminium ion, hydrogen bond, and Lewis base catalytic reactions have been reported in the manufacture of commercial compounds. The relatively young age of organocatalysis, may suggest that only few experienced researchers in organocatalysis are now working in industry. A rapid and direct mechanism that makes a technology platform adopted and applied in industrial process is the involvement in R&D of people expert and educated in the field that act as evangelists. In fact, more reports on the use of organocatalysis in industry are being published and this is also due, obviously, to the slower publishing rate of industry.

Despite the relatively high loadings of organocatalysis, recent developments show that low-loading organocatalysis is achievable. This, coupled with the fact that the catalysts are rather simple to prepare and new activation modes are being discovered, suggests that more reports and interesting developments are expected in the near future. A recent survey of the patent literature dealing with asymmetric organocatalysis in drug development [14] confirms the utility of this technology and allows to foresee growing industrial developments. Furthermore, going back to the cases of Strecker-type reaction and glycine imine alkylation, it can be envisioned that research efforts can improve dramatically organocatalytic transformations which at the state-of-the-art do not appear feasible and make them convenient. To simplify, we feel that there is a tendency to report organocatalytic reactions in the literature at high (5–20 mol%) loadings, even if higher TONs could readily be achieved [72, 73].

Altogether, we believe that asymmetric organocatalysis will have an increasing role in the production of APIs, meeting sustainability and cost requirements; this technology will likely flank other more established methodologies providing exciting opportunities. Society can only have a sustainable development if we exploit modern technologies and asymmetric organocatalysis promises to contribute considerably to it.

References

[1] In: Blaser H-U, Schmidt E, editor(s). Asymmetric catalysis on industrial scale: challenges, approaches and solutions. Weinheim, Germany: Wiley-VCH, 2004

[2] In: Dalko P, editor(s). Comprehensive Enantioselective Organocatalysis (3 Volumes). Weinheim, Germany: Wiley-VCH, 2013

[3] In: List B, editor(s). Asymmetric organocatalysis, in topics in current chemistry, 2010. Berlin, Germany and New York, USA: Springer-Verlag, 2009:191

[4] In: Rios Torres R, editor(s). Stereoselective Organocatalysis. Hoboken, New Jersey, USA: Wiley, 2013

[5] Melchiorre P, Marigo M, Carlone A, Bartoli B. Asymmetric aminocatalysis—gold rush in organic chemistry. Angew Chem Int Ed. 2008;47:6138–71.

[6] Ahrendt KA, Borths CJ, MacMillan DWC. New strategies for organic catalysis: the first highly enantioselective organocatalytic Diels–Alder reaction. J Am Chem Soc. 2000;122:4243–4.

[7] List B, Lerner RA, Barbas III CF. Proline-catalyzed direct asymmetric aldol reactions. J Am Chem Soc. 2000;122:2395–6.

[8] MacMillan DWC. The advent and development of organocatalysis. Nature. 2008;455:304–8.

[9] Marqués-López E, Herrera RP, Christmann M. Asymmetric organocatalysis in total synthesis – a trial by fire. Nat Prod Rep. 2010;27:1138–67.

[10] Alemán J, Cabrera S. Applications of asymmetric organocatalysis in medicinal chemistry. Chem Soc Rev. 2013;42:774–93.

[11] Abbasov ME, Romo D. The ever-expanding role of asymmetric covalent organocatalysis in scalable, natural product synthesis. Nat Prod Rep. 2014;31:1318–27.

[12] Sun BF. Total synthesis of natural and pharmaceutical products powered by organocatalytic reactions. Tetrahedron Lett. 2015;56:2133–40.

[13] Merad J, Lalli C, Bernadat G, Maury J, Masson G. Enantioselective Brønsted acid catalysis as a tool for the synthesis of natural products and pharmaceuticals. Chem Eur J. 2018;24:3925–43.

[14] Hughes DL. Asymmetric organocatalysis in drug development – highlights of recent patent literature. Org Process Res Dev. 2018;22:574–84.

[15] Hajos ZG, Parrish DR. Asymmetric synthesis of bicyclic intermediates of natural product chemistry. J Org Chem. 1974;39:1615–21.

[16] Eder U, Sauer G, Wiechert R. New type of asymmetric cyclization to optically active steroid CD partial structures. Angew Chem Int Ed. 1971;10:496–7.

[17] Dolling U-H, Davis P, Grabowski EJJ. Efficient catalytic asymmetric alkylations. 1. Enantioselective synthesis of (+)-indacrinone via chiral phase-transfer catalysis. J Am Chem Soc. 1984;106:446–7.

[18] Maruoka K. Practical aspects of recent asymmetric phase-transfer catalysis. Org Process Res Dev. 2008;12:679–97.

[19] Tan J, Yasuda N. Contemporary asymmetric phase transfer catalysis: large-scale industrial applications. Org Process Res Dev. 2015;19:1731–46.

[20] For a selection of examples, see the special feature section. Org Process Res Dev. 2011;15:1088–211.

[21] Butters M, Catterick D, Craig A, Curzons A, Dale D, Gillmore A, et al. Critical assessment of pharmaceutical processess - a rationale for changing the synthetic route. Chem Rev. 2006;106:3002–27.

[22] In: Blaser H-U, Federsel H-J, editor(s). Asymmetric catalysis on industrial scale: challenges, approaches and solutions, Second ed. Weinheim, Germany: Wiley-VCH . see Chapters 9, 26, 27 2011

[23] Xu F. Organocatalysis for asymmetric synthesis – from lab to factory. In: Dunn PJ, Hi KK (M), Krische MJ, Williams MT, editor(s). Sustainable catalysis - challenges and practices for the pharmaceutical and fine chemical industries. Hoboken, New Jersey, USA: Wiley. Chapter 14 2013:317–38.

[24] Scott JP, Ashwood MS, Brands KM, Brewer SE, Cowden CJ, Dolling UH, et al. Development of a phase transfer catalyzed asymmetric synthesis for an estrogen receptor beta selective agonist. Org Process Res Dev. 2008;12:723–30.

[25] Xu F, Zacuto M, Yoshikawa N, Desmond R, Hoerrner S, Itoh T, et al. Asymmetric synthesis of telcagepant, a CGRP receptor antagonist for the treatment of migraine. J Org Chem. 2010;75:7829–41.

[26] Xu F, Corley E, Zacuto M, Conlon DA, Pipik B, Humphrey G, et al. Asymmetric synthesis of a potent, aminopiperidine-fused imidazopyridine dipeptidyl peptidase IV inhibitor. J Org Chem. 2010;75:1343–53.
[27] Yue T-Y, McLeod DD, Albertson KB, Beck S R, Deerberg J, Fortunak JM, et al. Stereoselective process for a CCR3 antagonist. Org Process Res Dev. 2006;10:262–71.
[28] Jiang X, Gong B, Prasad K, Repič O. A practical synthesis of a chiral analogue of FTY720. Org Process Res Dev. 2008;12:1164–9.
[29] Ishii Y, Fujimoto R, Mikami M, Murakami S, Miki Y, Furukawa Y. Practical syntheses of chiral α-amino acids and chiral half-esters by kinetic resolution of urethane-protected α-amino acid N-carboxyanhydrides and desymmetrization of cyclic meso-anhydrides with new modified Cinchona alkaloid catalysts. Org Process Res Dev. 2007;11:609–15.
[30] Ager DJ, Anderson K, Oblinger E, Shi Y, VanderRoest J. An epoxidation approach to a chiral lactone: application of the Shi epoxidation. Org Process Res Dev. 2007;11:44–51.
[31] Mittendorf J, Benet-Buchholz J, Fey P, Mohrs KH. Efficient asymmetric synthesis of β-amino acid BAY 10-8888/PLD-118, a novel antifungal for the treatment of yeast infections. Synthesis. 2003;1:136–40.
[32] Verghese PS, Schleiss MR. Letermovir. Treatment of human cytomegalovirus infection, Antiinfective agent. Drugs Future. 2013;38:291–8.
[33] Goossen K, Kuhn O, Berwe M, Krüger J, Militzer H-C. Process for the preparation of dihydroquinazolines. US20090221822.
[34] Humphrey GR, Dalby S M, Andreani T, Xiang B, Luzung MR, Song ZJ, et al. Asymmetric synthesis of letermovir using a novel phase-transfer-catalyzed aza-Michael reaction. Org Process Res Dev. 2016;20:1097–103.
[35] Luzung M, Humphrey GR, Xiang B, Belyk K M, Dalby SM, Schwab W, et al. Process for the preparation of substituted quinazoline compounds. WO2015088931 A1.
[36] Xiang B, Belyk K M, Reamer RA, Yasuda N. Discovery and application of doubly quaternized cinchona-alkaloid-based phase-transfer catalysts. Angew Chem Int Ed. 2014;53:8375–8.
[37] Chung CK, Liu Z, Lexa KW, Andreani T, Xu Y, Ji Y, et al. Asymmetric hydrogen bonding catalysis for the synthesis of dihydroquinazoline-containing antiviral, letermovir. J Am Chem Soc. 2017;139:10637–40.
[38] Dutschman GE, Grill SP, Gullen EA, Haraguchi K, Takeda S, Tanaka H, et al. Novel 4-substituted stavudine analog with improved anti-human immunodeficiency virus activity and decreased cytotoxicity. Antimicrob Agents Chemother. 2004;48:1640–6.
[39] Cheng Y-C, Tanaka H, Baba M Antiviral nucleosides analogs and methods for treating viral infections, especially HIV. US7,589,078.
[40] Cotte L, Dellamonica P, Raffi F, Yazdanpanah Y, Michel Molina J-M M, Boué F, et al. Randomized placebo-controlled study of the safety, tolerability, antiviral activity, and pharmacokinetics of 10-day monotherapy with BMS-986001, a novel HIV-NRTI, in treatment-experienced HIV-1–infected subjects. Immune Defic Syndr. 2013;63:346–54.
[41] Wang F, Flint OP. BMS-986001, an HIV nucleoside reverse transcriptase inhibitor, does not degrade mitochondrial DNA in long-term primary cultures of cells isolated from human kidney, muscle, and adipose tissue. Antimicrob Agents Chemother. 2013;57:6205.
[42] Eastgate MD, Schmidt MA, Fandrick KR. On the design of complex drug candidate syntheses in the pharmaceutical industry. Nature Rev. 2017;1:1–16.
[43] Nagai K, Kiguchi S, Koyama H, Hume WE, Tsujimoto S. Method for producing 4′ethynyl d4T. US20100280235.
[44] Gallagher WP, Deshpande PP, Li J, Katipally K, Sausker J. A Claisen approach to 4′Ed4T. Org Lett. 2015;17:14–7.

[45] Gallagher WP, Deshpande PP, Li J, Katipally K Method for producing festinavir using 5-methyluridine as starting material. WO2014/172264.
[46] Ortiz A, Benkovics T, Beutner GL, Shi Z, Bultman M, Nye J, et al. Scalable synthesis of the potent HIV inhibitor BMS-986001 by non-enzymatic dynamic kinetic asymmetric transformation (DYKAT). Angew Chem Int Ed. 2015;54:7185–8.
[47] Ortiz A, Benkovics T, Shi Z, Deshpande PP, Guo Z, Kronenthal DR, et al. Sulfilimine and sulfoxide methods for the preparation of festinavir. WO2013177243A1.
[48] Ji Y, Benkovics T, Beutner GL, Sfouggatakis C, Eastgate MD, Blackmond DG. Mechanistic insights into the vanadium-catalyzed Achmatowicz rearrangement of furfurol. J Org Chem. 2015;80:1696–702.
[49] Benkovics T, Ortiz A, Guo Z, Goswami A, Deshpande P. Enantioselective preparation of (*S*)-5-oxo-5,6-dihydro-2 H-pyran-2-yl benzoate. Org Synth. 2014;91:293–306.
[50] Wilson LJ, Liotta D. A general method for controlling glycosylation stereochemistry in the synthesis of 2′-deoxyribose nucleosides. Tetrahedron Lett. 1990;31:1815–8.
[51] Mayes BA, Stewart AJ, Moussa AM D-Amino acid compounds for liver disease. WO2013177219A1.
[52] Gosselin G, Parsy CC, Alexandre F-R, Rahali H, Griffon J-F, Milhau J, et al. 2′-Chloro nucleoside analogs for HCV infection. WO2014058801A1.
[53] Press release Sep 2017: http://investors.merck.com/news/press-release-details/2017/Merck-Discontinues-MK-3682B-and-MK-3682C-Development-Programs/default.aspx.
[54] Cahard D, McGuigan C, Balzarini J. Aryloxy phosphoramidate triesters as Pro-Tides. Mini Rev Med Chem. 2004;4:371–81.
[55] Ross BS, Reddy PG, Zhang H-R, Rachakonda S, Sofia MJ. Synthesis of diastereomerically pure nucleotide phosphoramidates. J Org Chem. 2011;76:8311–19.
[56] Simmons BL, Campos KR, Klapars A, Stewart AJ, Mayes BA, Maligres PE, et al. Process for making nucleoside phosphoramidate compounds. WO2016064797A1.
[57] DiRocco DA, Ji Y, Sherer EC, Klapars A, Reibarkh M, Dropinski J, et al. A multifunctional catalyst that stereoselectively assembles prodrugs. Science. 2017;356:426–30.
[58] Press release March 2018: http://investor.xenon-pharma.com/phoenix.zhtml?c=253202&p=irol-newsArticle&ID=2336851.
[59] Sclafani JA, Chen J, Levy DV, Reese H, Dimitri M, Mudipalli P, et al. The first asymmetric pilot-scale synthesis of TV-45070. Org Process Res Dev. 2017;21:1616–24.
[60] Cadieux J-J, Chafeev M, Chowdhury S, Fu J, Jia Q, Abel S, et al. Synthetic methods for spiro-oxindole compounds. U.S. Patent 8,445,696, May 21, 2013.
[61] Sun S, Fu J, Chowdhury S, Hemeon IW, Grimwood ME, Mansour TS Asymmetric syntheses of spiro-oxindole compounds useful as therapeutic agents. U.S. Patent 9,487,535, Nov 08, 2016.
[62] Lygo B, Wainwright PG. A new class of asymmetric phase-transfer catalysts derived from Cinchona alkaloids — application in the enantioselective synthesis of α-amino acids. Tetrahedron Lett. 1997;38:8595–8.
[63] Wang Y, Milkiewicz KL, Kaufman ML, He L, Landmesser NG, Levy DV, et al. Plant process for the preparation of cinchona alkaloid-based thiourea catalysts. Org Process Res Dev. 2017;21:408–13.
[64] Vakulya B, Varga S, Csámpai A, Soós T. Highly enantioselective conjugate addition of nitromethane to chalcones using bifunctional cinchona organocatalysts. Org Lett. 2005;7:1967–9.
[65] McCooey SH, Connon SJ, Accessible R. 9-epi-amino cinchona alkaloid derivatives promote efficient, highly enantioselective additions of aldehydes and ketones to nitroolefins. Org Lett. 2007;9:599–602.

[66] Cassani C, Martín-Rapún R, Arceo E, Bravo F, Melchiorre P. Synthesis of 9-amino(9-deoxy)epi Cinchona alkaloids, general chiral organocatalysts for the stereoselective functionalization of carbonyl compounds. Nat Protoc. 2013;8:325–44.
[67] Sigman MS, Vachal P, Jacobsen EN. A general catalyst for the asymmetric Strecker reaction. Angew Chem Int Ed. 2000;39:1279–81.
[68] Zuend SJ, Jacobsen EN. Mechanism of amido-thiourea catalyzed enantioselective imine hydrocyanation: transition state stabilization via multiple non-covalent interactions. J Am Chem Soc. 2009;131:15358–74.
[69] Zuend SJ, Coughlin MP, Lalonde MP, Jacobsen EN. Scaleable catalytic asymmetric Strecker syntheses of unnatural α-amino acids. Nature. 2009;461:968–71.
[70] Ooi T, Kameda M, Maruoka K. Molecular design of a C_2-symmetric chiral phase-transfer catalyst for practical asymmetric synthesis of α-amino acids. J Am Chem Soc. 1999;121:6519–20.
[71] Ikunaka M, Maruoka K. Asymmetric phase-transfer catalysis for the production of non-proteinogenic α-amino acids, chapter 9. In: Blaser H-U, Federsel H-J, editor(s). Asymmetric catalysis on industrial scale: challenges, approaches and solutions, 2nd ed. Weiheim, Germany.: Wiley-VCH, 2011:151–69.
[72] Bernardi L, Fochi M, Carbone R, Martinelli A, Fox ME, Cobley CJ, et al. Organocatalytic asymmetric conjugate additions to cyclopent-1-enecarbaldehyde: a critical assessment of organocatalytic approaches towards the telaprevir bicyclic core. Chem Eur J. 2015;21:19208–22.
[73] Stevanović D, Bertuzzi G, Mazzanti A, Fochi M, Bernardi L. Catalytic enantioselective povarov reactions of ferrocenecarbaldehyde-derived imines – brønsted acid catalysis at parts-per-million level loading. Adv Synth Catal. 2018;360:893–900.

Index

https://doi.org/10.1515/9783110590050-012

Bei Fragen zur Produktsicherheit wenden Sie sich bitte an:
If you have any questions regarding product safety,
please contact:

Walter de Gruyter GmbH
Genthiner Straße 13
10785 Berlin
productsafety@degruyterbrill.com